Coral Reef Remote Sensing

James A. Goodman · Samuel J. Purkis
Stuart R. Phinn
Editors

Coral Reef Remote Sensing

A Guide for Mapping, Monitoring and Management

Springer

Editors
James A. Goodman
Electrical and Computer Engineering
University of Puerto Rico at Mayagüez
Mayagüez, PR
USA

and

HySpeed Computing
Miami, FL
USA

Samuel J. Purkis
Oceanographic Center, National Coral Reef Institute
Nova Southeastern University
Dania Beach, FL
USA

Stuart R. Phinn
School of Geography, Planning and Environmental Management
University of Queensland
Brisbane, QLD
Australia

ISBN 978-90-481-9291-5 ISBN 978-90-481-9292-2 (eBook)
DOI 10.1007/978-90-481-9292-2
Springer Dordrecht Heidelberg New York London

Library of Congress Control Number: 2012954527

Printed on acid-free paper

Springer is part of Springer Science+Business Media (www.springer.com)

Foreword and Onward

While sailing around the world in 1834, Charles Darwin was so impressed with the pristine coral reefs that he observed in the Atlantic, Pacific, and Indian oceans that he devoted his first book, "*The Structure and Distribution of Coral Reefs*", published in 1842, to an analysis of how atolls, fringing and coast-hugging barrier reefs were formed. In *Voyage of the Beagle*, he notes, ". . .*travellers tell us of the vast dimensions of the Pyramids and other great ruins, but how utterly insignificant are the greatest of these, when compared to these mountains of stone accumulated by the agency of various minute and tender animals*!"

If Darwin could magically retrace his voyage in the twenty-first century, he would likely be shocked at how much has been lost of that lacey garland of reef plants and animals that once spanned the shallow blue waters of the globe between 30° North and 30° South. But he would likely also be heartened by the technologies that now exist that for the first time make it possible to accurately assess the nature of the changes and to take actions to reverse the alarming trends.

This remarkable book,*Coral Reef Remote Sensing*: *A Guide for Mapping*, *Monitoring* and *Management* for the first time documents the full range of remote sensing systems, methodologies and measurement capabilities essential to understanding more fully the status and changes over time of coral reefs globally. Such information is essential and provides the foundation for policy development and for implementing management strategies to protect these critically endangered ecosystems.

In Darwin's time, the ocean seemed so vast, so resilient, that it seemed there was little humans could do to alter its nature. Even more than a century later, Rachel Carson in her 1951 Classic, "The Sea Around Us", imagined the ocean was somehow too big to fail. "*Eventually man . . .found his way back to the sea*", she wrote. "*And yet he has returned to his mother sea only on her terms. He cannot control or change the ocean as, in his brief tenancy of earth, he has subdued and plundered the continents*".

Now we know: unprecedented human activity is driving global warming, climate change, sea level rise, pollution, acidification of the ocean, and loss of species and entire ecosystems, including the focus of this volume—coral reefs.

While occupying less than 1 % of the area of the ocean, shallow coral reef systems host an extraordinary diversity of fish (about 25 % of the known marine species) and twice as many phyla of animals as the notably rich and diverse rainforests of the world. If coral reefs are in trouble, the ocean is in trouble. That is bad news for the planet, and therefore for us.

A recent Joint Statement by the heads of the Intergovernmental Oceanographic Commission, the United Nations Environment Program, and the International Union for the Conservation of Nature states:

> *Globally, best estimates suggest that about 10 % of coral reefs are already degraded, many beyond recovery, and another 20 % are likely to decline further within the next 20 years. At least two thirds of the world' s coral reefs may collapse ecologically within the lifetime of our grandchildren, unless we implement effective management of these ecosystems as an urgent priority.*

This assessment may be optimistic. The 2011 World Resources Institute report, "*Reefs at Risk*" notes that three-quarters of all of the world's coral reefs are at risk from overfishing, pollution, and the effects of climate change. In the Philippines, 70 % of the coral reefs are gone and only 5 % are estimated to be in good condition. In the Caribbean and Gulf of Mexico, the decline may be as much as 80 % in 30 years. More than 100 countries have coral reefs within their jurisdiction, and some island nations, with small land mass have large ocean assets centered on coral reefs. Effective management of these valuable ecosystems requires the ability to document their present status and monitor changes over time.

During the past two decades, there have been remarkable advances in the ability to remotely measure and monitor the health and status of coral reef systems across the planet as well as the conditions of the water and surrounding environment. The techniques documented in this volume complement on-site observations and together, provide critically important insights necessary for effective management.

Included here is an overview of technologies for reef mapping, technical information useful for scientists and other research and policy development experts, ideas for application of remote sensing to resolve questions, and thoughts about future remote sensing technologies and their applications.

I wholeheartedly recommend this book to scientists, students, managers, remote sensing specialists, and anyone who would like to be inspired by the ingenious new ways that have been developed and are being applied to solve one of the world's greatest challenges: how to take care of the ocean that takes care of us.

If it had been available in 1834, Charles Darwin would surely have had a copy on his shelf.

Sylvia A. Earle
National Geographic Explorer in Residence
Founder, Mission Blue

Preface

Overview. Remote sensing stands as the defining technology in our ability to monitor coral reefs, as well as their biophysical properties and associated processes, at regional to global scales. With overwhelming evidence that much of Earth's reefs are in decline, our need for large scale, repeatable assessments of reefs has never been so great. Fortunately, the last two decades have seen a rapid expansion in the ability for remote sensing to map and monitor coral reef ecosystems, as well as the overlying water column and surrounding environment.

Remote sensing is now a fundamental tool for the mapping, monitoring and management of coral reef ecosystems. Remote sensing offers repeatable, quantitative assessments of habitat and environmental characteristics over spatially extensive areas. As the multi-disciplinary field of coral reef remote sensing continues to mature, results demonstrate that the techniques and capabilities continue to improve. New developments allow reef assessments and mapping to be performed with higher accuracy, across greater spatial areas, and with greater temporal frequency. The increased level of information that remote sensing now makes available also allows more complex scientific questions to be addressed.

As defined for this book, remote sensing includes the vast array of geospatial data collected from land, water, ship, airborne, and satellite platforms. The book is organized by technology, including: visible and infrared sensing using photographic, multispectral and hyperspectral instruments; active sensing using light detection and ranging (LiDAR); acoustic sensing using ship, autonomous underwater vehicle (AUV), and in-water platforms; and thermal and radar instruments.

Emphasis and Audience. This book serves multiple roles. It offers an overview of the current state-of-the-art technologies for reef mapping, provides detailed technical information for coral reef remote sensing specialists, imparts insight on the scientific questions that can be tackled using this technology, and also includes a foundation for those new to reef remote sensing. The individual sections of the book include introductory overviews of four main types of remotely sensed data used to study coral reefs, followed by specific examples demonstrating practical applications of the different technologies being discussed. Guidelines for selecting the most appropriate sensor for particular applications are provided, including an overview of how to utilize remote data as an effective tool in science and management. The text is richly illustrated with examples of each sensing technology applied to a range of scientific, monitoring and management questions in reefs around the world. As such, the book is broadly accessible to a general audience, as well as students, managers, remote sensing specialists, and anyone else working with coral reef ecosystems.

Outline and Roadmap. The book is divided into five sections, with the first four highlighting different suites of remote sensing technologies and the fifth section discussing the best use of remote sensing in effective science and management. Each of the four technology sections begins with an introductory chapter followed by a series of application chapters that discuss each technology in more detail and define the applications for which they are best suited.

- *Section I. Visible and Infrared Remote Sensing.*
  Chapters 1–4 introduce passive optical remote sensing (i.e., sensing technologies that rely on visible and infrared spectra of reflected sunlight as the basis for image measurements). These technologies include aerial and astronaut *photography*, and *multispectral* and *hyperspectral* aerial and satellite imaging. Such techniques are best suited for assessing habitat characteristics (e.g., habitat type, composition, and distribution) in clear, shallow (<20 m) water. The typical spatial resolution, or pixel size, employed in visible and infrared remote sensing ranges from fine (<0.5–5 m) to moderate (10–30 m) scale. Visible and infrared remote sensing is also suitable for assessing associated surrounding environmental conditions at varying degrees of detail (e.g., water properties, bathymetry, and coastal/island land characteristics).
- *Section II. LiDAR Remote Sensing.*
  Chapters 5–7 introduce active optical remote sensing (i.e., technologies that measure the return signal of an actively emitted energy source). This technology is centered on LiDAR, but also includes an evolving set of techniques that merge LiDAR and hyperspectral imagery. LiDAR data are commonly acquired from aircraft platforms and are best suited for measurements of water depth, seafloor topography, geomorphology, and recently general habitat categories (e.g., broad cover types, such as reef versus seagrass versus sand) in clear, moderately deep (<40 m) water. Spatial resolution varies as a function of depth, but is typically at relatively fine (1–5 m) scale. Emerging techniques in LiDAR sensors and analysis techniques are also extending the level of detail that can be achieved for reef characteristics and surrounding water properties.
- *Section III. Acoustic Remote Sensing.*
  Chapters 8–10 present the field of acoustic remote sensing (i.e., sensors that measure sound, either sound emitted directly from objects/organisms or return signals from actively emitted pulses of sound). *Acoustic* remote sensing is primarily a ship-based technology, but is also deployed on AUVs or using in-water platforms. The breadth of platforms available for acoustic remote sensing allows

collection of acoustic data across a range of water depths, from shallow (5–20 m) to deep (>100 m). Spatial resolution for this technology typically ranges from fine (1–10 m) to moderate (20–50 m) scale. Acoustic techniques are best suited for assessing water depth, seafloor topography, geomorphic zones (with variable roughness), general habitat categories (e.g., distinguishing biological cover types, as well as hard- versus soft-bottom habitats), water velocity, and fish presence/distribution. As with other technologies, advances in this field are currently extending the level of detail and types of information that can be retrieved from acoustic remote sensing.

- *Section IV. Thermal and Radar Remote Sensing.*
  Chapters 11–13 introduce two technologies that commonly center on measuring the environment surrounding coral reef ecosystems. *Thermal* remote sensing is an example of passive optical remote sensing, which focuses on measuring emitted heat, and hence temperature of the water surface. Thermal remote sensing is predominantly a satellite-based technology that typically operates in coarse (>1 km) scale spatial resolutions. Radio detection and ranging (*Radar*) remote sensing is an active sensing technology, which uses radio waves to measure the range, altitude, direction, and speed of sea surface characteristics (i.e., waves and currents). Radar remote sensing includes both ground-based systems and satellite platforms, ranging from moderate (25–50 m) to coarse (>1 km) scale spatial resolution. Both thermal and radar technologies provide valuable information on processes affecting coral reefs.
- *Section V. Effective Use of Remote Sensing in Science and Management.*
  Chapters 14–15 explain and demonstrate the concepts of validation and accuracy assessment of image-based map products, as well as how to measure these parameters and effectively utilize them in science and management applications. The need to understand the accuracy and reliability of remote sensing products is highlighted as a fundamental component of effective decision-making using this data. Discussion in these chapters focuses on recognizing the different strengths and weaknesses of the various remote sensing techniques and illustrating which techniques are best suited for different specific goals. Additional discussion in this section includes the importance of developing an understanding between the needs and expectations of those using remote sensing image products and those producing products.

The following tables represent the book outline in a condensed format, providing a quick-look roadmap to the typical capabilities that characterize each technology. Note that this is a summary only, and thus represents a simplified view of the different aspects of coral reef remote sensing.

Data type	Assessment level	Appropriate technology
Habitat	Detailed species assessment	Field Observation and Measurement
	Habitat type and composition Shallow water (<20 m)	Chapter 3. Multispectral Chapter 4. Hyperspectral Chapter 7. LiDAR/Hyperspectral
	General habitat categories Shallow water (<20 m)	Chapter 2. Photography Chapter 6. LiDAR
	General habitat categories Moderate depth (20–50 m)	Chapter 6. LiDAR Chapter 9. Acoustic
	General habitat categories Deep water (>50 m)	Chapter 9. Acoustic Chapter 10. Deep Acoustic
Geomorphology	Reef and landscape level Shallow water (<20 m)	Chapter 2. Photography Chapter 3. Multispectral Chapter 4. Hyperspectral Chapter 6. LiDAR Chapter 7. LiDAR/Hyperspectral
	Reef and landscape level Moderate depth (20–50 m)	Chapter 6. LiDAR Chapter 9. Acoustic
	Reef and landscape level Deep water (>50 m)	Chapter 9. Acoustic Chapter 10. Deep Acoustic
Water properties	General water composition Shallow water (<20 m)	Chapter 4. Hyperspectral Chapter 7. LiDAR/Hyperspectral
	General water composition Moderate depth (20–50 m)	Chapter 6. LiDAR
Bathymetry	High accuracy depth Shallow water (<20 m)	Chapter 6. LiDAR Chapter 7. LiDAR/Hyperspectral
	Moderate accuracy depth Shallow water (<20 m)	Chapter 3. Multispectral Chapter 4. Hyperspectral
	High accuracy depth Moderate depth (20–50 m)	Chapter 6. LiDAR Chapter 9. Acoustic
	High accuracy depth Deep water (>50 m)	Chapter 9. Acoustic
Water temperature	Water surface temperature	Chapter 12. Thermal
Currents/waves	Water surface characteristics	Chapter 13. Radar

	Application strength(s)	Deployment platform(s)	Spatial resolution
Section I. Visible-Infrared			
Chapter 1. Introduction			
Chapter 2. Photography	Habitat characteristics	Aircraft	0.05–30 m
Chapter 3. Multispectral	Habitat type	Satellite	
Chapter 4. Hyperspectral	Habitat composition	ISS/Shuttle	
	Habitat distribution		
	Water properties		
Section II. LiDAR			
Chapter 5. Introduction			
Chapter 6. LiDAR	Water depth	Aircraft	1–5 m
Chapter 7. LiDAR/ Hyperspectral	Geomorphology		
	General habitat categories		
	General reef types		
	Habitat distribution		
	General water properties		
Section III. Acoustic			
Chapter 8. Introduction			
Chapter 9. Acoustic	Water depth	Ship	1–40 m
Chapter 10. Deep Acoustic	Geomorphology	AUV	
	General habitat categories	In-Water	
	General reef types		
	Habitat distribution		
	Water velocity		
	Fish presence/distribution		
Section IV. Thermal and Radar			
Chapter 11. Introduction			
Chapter 12. Thermal	Water temperature	Satellite	>1 km
	Water salinity		
Chapter 13. Radar	Surface wind and currents	Satellite	25 m – >1 km
	Particle tracking	Ground station	
	Waves		
Section V. Effective Use			
Chapter 14. Validation			
Chapter 15. Science and Management			

In all cases, emerging analysis techniques and improved sensor systems are expanding the scope of what is capable for each technology. There is also a strong movement toward integrating multiple technologies for assessing a given area, leveraging the strengths of each technology to provide a more comprehensive view of the reef ecosystem (e.g., merging LiDAR, hyperspectral and acoustic technologies to provide a broad view of the reef, from the shallow emergent reef crest down to the deep offshore reefs). As a result, a full grasp of what can be achieved using remote sensing is possible only through delving into the details presented in each chapter.

James A. Goodman
University of Puerto Rico at Mayagüez
Mayagüez, PR, USA

HySpeed Computing
Miami, FL, USA

Samuel J. Purkis
Nova Southeastern University
Dania Beach, FL, USA

Stuart R. Phinn
University of Queensland
Brisbane, QLD, Australia

Acknowledgments

This book project has been a remarkable journey. Since its inception following the International Coral Reef Symposium in 2008, the book quickly gained momentum and grew in scope and dimension, taking almost 4 years in the making. All of the individual authors and editors contributing to the book deserve thanks for scraping time out of their busy schedules to create a unique resource for the coral reef community. Their efforts persevered despite the intrusion of earthquakes, tsunamis, floods, sicknesses, deaths, and births. A heartfelt debt of gratitude also goes out to reviewers Dr. Deepak Mishra, Department of Geosciences at Mississippi State University, and Dr. Pete Mumby, School of Biological Sciences at University of Queensland, who expertly took on the task of reviewing the book as a whole. The result is an achievement of which we are all proud. Thank you.

J. Goodman is grateful for support from: the University of Puerto Rico at Mayagüez, particularly colleagues and staff, Miguel Vélez-Reyes, Samuel Rosario-Torres, Maribel Feliciano-Ruiz, Richard Appeldoorn, Francisco Pagán, Milton Carlo and Godoberto López-Padilla; the Bernard M. Gordon Center for Subsurface Sensing and Imaging Systems, under the Engineering Research Centers Program of the National Science Foundation #EEC-9986821; colleagues and staff at Northeastern University, Michael Silevitch, Anne Magrath, John Beaty, Phil Cheney, and Mariah Nobrega; the National Oceanic and Atmospheric Administration, Center for Sponsored Coastal Ocean Research, under award #NA05NOS4261159 to the University of Puerto Rico for the Caribbean Coral Reef Institute; the Puerto Rico, National Aeronautics and Space Administration, EPSCoR Program award #NNX09AV03A; HySpeed Computing for time and resources to complete this book project; the staff at Springer, Petra van Steenbergen, Cynthia de Jonge, and Hermine Vloemans; the Department of Marine Geology and Geophysics at the Rosenstiel School of Marine and Atmospheric Science, University of Miami, particularly Pamela Reid, for providing an academic home in Miami; Susan Ustin at the University of California at Davis who was an instrumental influence to my remote sensing career; editing assistants Bubba, Demi, Ras and Alley; and a vast array of coral reef and remote sensing colleagues who have contributed their knowledge throughout the years. And especially to my family for indulging my impulse to pursue this book while simultaneously starting a new company, without you it would not have been possible. My warmest thanks go out to my parents, Jay and Sarajane, my in-laws Nancy Maynard and Robert Corell, and especially to my wonderful wife, Jennifer, for your love and unwavering support.

S. Purkis wishes to thank the National Coral Reef Institute and Nova Southeastern University's Oceanographic Center for supporting his scientific endeavors. He is especially indebted to the members of his remote sensing lab for surrounding him with stimulating science and wishes to thank Gwilym Rowlands, Alexandra Dempsey, and Jeremy Kerr, in particular. Bernhard Riegl continues to provide inspiration and, no matter what the topic, is acknowledged as a source of advice. It has been a pleasure to work with so many talented authors and ultimate thanks must go to them for making this book a reality.

S. Phinn is particularly grateful to the staff and students from the Biophysical Remote Sensing Group, specifically Dr. Chris Roelfsema, who have focused on reefs in their studies and work projects—our group has learned a lot from collaborating and sharing knowledge in the field and lab all over the world. With Chris's assistance, Ian Leiper, Robert Canto, Julie Scopelitis and Karen Joyce have significantly advanced our capabilities and understanding of how to map and monitor coral reefs. The support of Drs. Peter Mumby, Serge Andréfouet, and Ove Hoegh-Guldberg was critical in enabling our group to establish coral reef remote sensing expertise and links to ecology and management. Finally, it is very pleasing to see multiple authors of this book who were Ph.D. students 5–10 years ago, and have progressed to established research scientists, working to understand and manage coral reefs.

Contents

Section II LiDAR

Contributors

Tim Battista National Oceanic and Atmospheric Administration, 1305 East West Highway, Silver Spring, MD 20910, USA, e-mail: tim.battista@noaa.gov

Ray Berkelmans Australian Institute of Marine Science, PMB 3, Townville, QLD 4810, Australia, e-mail: r.berkelmans@aims.gov.au

John C. Brock U.S. Geological Survey, Coastal and Marine Geology Program, USGS National Center, Mail Stop 915-B, 12201 Sunrise Valley Drive, Reston, VA 20192, USA, e-mail: jbrock@usgs.gov

Susan A. Cochran USGS Pacific Coastal and Marine Science Center, 400 Natural Bridges Dr., Santa Cruz, CA 95060, USA, e-mail: scochran@usgs.gov

Thiago B. S. Correa Rosenstiel School of Marine and Atmospheric Science, University of Miami, 4600 Rickenbacker Cswy, Miami, FL 33149, USA; ConocoPhillips, 600 North Dairy Ashford, Houston, TX 77029, USA, e-mail: tcorrea@rsmas.miami.edu

Bryan Costa NOAA/NOS/NCCOS/CCMA Biogeography Branch, 1305 East West Highway, Silver Spring, MD 20910, USA, e-mail: bryan.costa@noaa.gov

Gregor P. Eberli Rosenstiel School of Marine and Atmospheric Science, University of Miami, 4600 Rickenbacker Cswy, Miami, FL 33149, USA, e-mail: geberli@rsmas.miami.edu

Greg Foster Oceanographic Center, National Coral Reef Institute, Nova Southeastern University, 8000 N. Ocean Drive, Dania, FL 33004, USA, e-mail: fjohn@nova.edu

Arthur Gleason Department of Physics, University of Miami, 1320 Campo Sano Ave, Coral Gables 33146, USA, e-mail: art.gleason@miami.edu

James A. Goodman Electrical and Computer Engineering, University of Puerto Rico at Mayagüez, PO Box 3535, Mayagüez, PR 00681, USA; HySpeed Computing, PO Box 431824, Miami, FL 33243, USA, e-mail: james.goodman1@upr.edu

Mark Grasmueck Rosenstiel School of Marine and Atmospheric Science, University of Miami, 4600 Rickenbacker Cswy, Miami, FL 33149, USA, e-mail: mgrasmueck@rsmas.miami.edu

Humberto Guarin Bert Instruments Inc., 2646 Sherman St., Hollywood, FL 33020, USA, e-mail: hguarin@bertinst.com

John D. Hedley ARGANS Ltd. Tamar Science Park, Derriford, Plymouth, Devon PL6 8BT, UK, e-mail: jhedley@argans.co.uk

Scott F. Heron Coral Reef Watch, National Oceanic and Atmospheric Administration, 675 Ross River Rd, Townsville, QLD 4817, Australia; Marine Geophysical Laboratory, Physics Department, School of Engineering and Physical Sciences, James Cook University, Townsville, QLD 4811, Australia, e-mail: scott.heron@noaa.gov

Malcolm L. Heron Marine Geophysical Laboratory, School of Environmental and Earth Sciences, James Cook University, Townsville, QLD 4811, Australia; Australian Institute of Marine Science, Townsville, QLD 4810, Australia, e-mail: mal.heron@ieee.org

Eric Hochberg Bermuda Institute of Ocean Sciences, 17 Biological Station, St. George's GE 01, Bermuda, e-mail: eric.hochberg@bios.edu

Stacy Jupiter Wildlife Conservation Society, Fiji Country Program, 11 Ma'afu St., Suva, Fiji, e-mail: sjupiter@wcs.org

Joong Yong Park Optech, Inc., 7225 Stennis Airport Road, Suite 300, Kiln, MS 39556, USA, e-mail: joongyongpark@optech.com

Stuart R. Phinn School of Geography, Planning and Environmental Management, Centre for Spatial Environmental Research, The University of Queensland, Brisbane, QLD 4072, Australia, e-mail: s.phinn@uq.edu.au

William G. Pichel Center for Satellite Applications and Research, National Oceanic and Atmospheric Administration, Room 102 World Weather Building, 5200 Auth Road, Camp Springs, MD 20746, USA, e-mail: william.g.pichel@noaa.gov

Simon J. Pittman NOAA/NOS/NCCOS/CCMA Biogeography Branch, 1305 East West Highway, Silver Spring, Maryland, MD 20910, USA; Marine Science Center, University of the Virgin Islands, 2 John Brewers Bay, St. Thomas, VI 00802, U.S. Virgin Islands, e-mail: simon.pittman@noaa.gov

Samuel J. Purkis Oceanographic Center, National Coral Reef Institute, Nova Southeastern University, 8000 N. Ocean Drive, Dania, FL 33004, USA, e-mail: purkis@nova.edu

Bernhard Riegl Oceanographic Center, National Coral Reef Institute, Nova Southeastern University, 8000 N. Ocean Drive, Dania, FL 33004, USA, e-mail: rieglb@nova.edu

Chris M. Roelfsema School of Geography, Planning and Environmental Management, Centre for Spatial Environmental Research, The University of Queensland, Brisbane, QLD 4072, Australia, e-mail: c.roelfsema@uq.edu.au

Chris Taylor National Oceanic and Atmospheric Administration, Center for Coastal Fisheries and Habitat Research, 101 Pivers Island Road, Beaufort, NC 28516, USA, e-mail: chris.taylor@noaa.gov

Klaas Verwer Rosenstiel School of Marine and Atmospheric Science, University of Miami, 4600 Rickenbacker Cswy, Miami, FL 33149, USA; Statoil, Sandsliveien 90, 5254 Bergen, Norway, e-mail: klver@statoil.com

Lisa M. Wedding NOAA/NOS/NCCOS/CCMA Biogeography Branch, 1305 East West Highway, Silver Spring, Maryland, MD 20910, USAInstitute of Marine Science, University of California at Santa Cruz, 100 Shaffer Rd., Santa Cruz, CA 95060, USA; NOAA/SWFSC Fisheries Ecology Division, 110 Shaffer Rd., Santa Cruz, CA 95060, USA, e-mail: lisa.wedding@noaa.gov

Scarla J. Weeks Biophysical Oceanography Group, School of Geography, Planning and Environmental Management, University of Queensland, Brisbane, QLD 4072, Australia, e-mail: s.weeks@uq.edu.au

Jennifer M. Wozencraft Coastal and Hydraulics Laboratory, U.S. Army Corps of Engineers, Joint Airborne lidar Bathymetry Technical Center of Expertise, Engineer Research and Development Center, 7225 Stennis Airport Road, Suite 100, Kiln, MS 39556, USA, e-mail: jennifer.m.wozencraft@usace.army.mil

Hiroya Yamano Center for Environmental Biology and Ecosystem Studies, National Institute for Environmental Studies, 16-2 Onogawa, Tsukuba, Ibaraki 305-8506, Japan, e-mail: hyamano@nies.go.jp

Section I
Visible and Infrared

Chapter 1
Visible and Infrared Overview

Stuart R. Phinn, Eric M. Hochberg and Chris M. Roelfsema

Abstract This chapter introduces visible and infrared remote sensing, specifically photographic, multispectral and hyperspectral imaging systems (Chaps. 2–4), and the situations in which they do and don't work for mapping and monitoring coral reefs. Spectral dimensions of imaging sensors are explained, along with their fundamental control on the amount and type of information able to be mapped on coral reefs from airborne and satellite sensors. A specific set of coral reef biophysical environmental variables capable of being mapped by visible and infrared imaging systems is also defined. Examples are provided of image processing approaches that deliver science and management relevant data for monitoring coral reefs.

1.1 Introduction

This chapter provides an introduction to photographic, multispectral and hyperspectral image data and how they can be used to map and monitor coral reefs and their surrounding environment. The intent is to provide both a conceptual overview

S. R. Phinn (✉) · C. M. Roelfsema
Centre for Spatial Environmental Research, School of Geography, Planning and Environmental Management, The University of Queensland, Brisbane, QLD 4072, Australia
e-mail: s.phinn@uq.edu.au

C. M. Roelfsema
e-mail: c.roelfsema@uq.edu.au

E. M. Hochberg
Bermuda Institute of Ocean Sciences, 17 Biological Station, St. George's GE01, Bermuda
e-mail: eric.hochberg@bios.edu

J. A. Goodman et al. (eds.), *Coral Reef Remote Sensing*,
DOI: 10.1007/978-90-481-9292-2_1,

and the technical underpinning for the reader to understand applications presented in Chaps. 2–4. This chapter also presents a fundamental basis for Chaps. 14 and 15.

1.1.1 Visible and Infrared Imaging Systems

In remote sensing, a sensor records the intensity of light reflected from a distant object in several spectral bands. The resulting spectral response "signature" is used to provide the identity and other information about the object. This is possible because in principle every object exhibits a unique characteristic spectral response pattern. This pattern is a function of the object's structure and component materials, as well as the electromagnetic energy falling on the object. The spectral response pattern can be so characteristic of the object's physical and chemical properties that it provides a spectral signature with which to identify the object.

Most people are familiar with the concept of spectral response in the form of color. The human eye has specialized cells (cones) that are generally sensitive to three colors: blue, green, and red. If an object, such as a plant, absorbs blue and red light and reflects green light, then only green light is available to be seen, only the green cones are stimulated, and the object thus appears green to human perception.

By design, color film photography replicates the sensitivity of the human eye. In this case, the light entering a camera induces chemical change on the photographic film, with blue, green, and red light each inducing different specific changes. Through chemical processing and developing, the film is converted to a "true color" representation of the scene originally imaged in the camera's field of view.

Digital photographs mimic the color of those derived from analog film. In a digital camera, light captured by a photosensitive element induces an electrical charge with an intensity that is proportional to the incident light intensity. In modern digital cameras, millions of photosensitive elements are arranged in a two-dimensional array; the individual elements are referred to as picture elements, or pixels. The charge induced in each pixel is converted to a numerical value that is recorded digitally. These components are often referred to as charge-coupled-devices (CCD). Together, this array of digital values represents the image captured by the camera. The actual photosensitive elements are typically made of silicon, which is sensitive to light across the visible and near-infrared (NIR) portions of the spectrum (400–700 and 700–1,000 nm, respectively; Fig. 1.1). Optical filters are used to limit and separate the wavelengths reaching the detector array into blue ($\sim$400–500 nm), green ($\sim$500–600 nm), and red ($\sim$600–700 nm). The result is a set of three images that are composited as red: green: blue (RGB) to produce a true color scene. It is useful to note that different optical filters could be employed so that the camera would image a different set of wavelengths, for example the infrared.

Spectral imaging follows the same principle and often utilizes the same technology as digital photography. The main conceptual differences are that generally

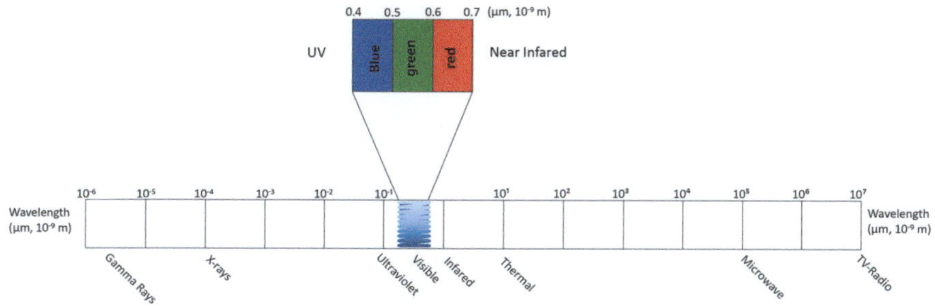

Fig. 1.1 The electromagnetic spectrum as shown by wavelength units and corresponding portions of the spectrum measured by remote sensing instruments (modified from Lillesand et al. 2008)

more than three wavebands are simultaneously imaged and that the wavebands are generally chosen specifically for their utility to discern the identity or biophysical status of the objects being imaged. A technological difference is that, whereas a digital camera instantaneously acquires a two-dimensional image, spectral imagers typically scan a scene to build an image pixel-by-pixel or line-by-line. A "whiskbroom" imager uses a mirror to scan side-to-side along the sensor's path, reflecting light into a one-dimensional array of photosensitive elements representing the image's spectral dimension, thus recording the digital data one pixel at a time. A "push-broom" imager uses a two-dimensional array of photosensitive elements; the side-to-side elements correspond to the image's spatial dimension, while the top-to-bottom elements correspond to the image's spectral dimension. The push-broom sensor thereby scans a scene one line at a time.

The terms multispectral and hyperspectral describe the spectral characteristics of the imaging system. Multispectral sensors typically have few (3–10) wavebands that are each relatively broad (~20–100 nm). The wavebands are not necessarily contiguous, but are placed in regions of the spectrum that are deemed important for a particular science measurement. In contrast, hyperspectral sensors image relatively narrow (~10 nm or less) wavebands across a continuous spectral range, typically including the visible, near-infrared, and often shortwave-infrared (1,000–2,500 nm). The key difference is that multispectral sensors measure in discrete wavebands for each pixel, while hyperspectral sensors measure a continuous spectrum for each pixel.

Photography, multispectral and hyperspectral imaging are passive remote sensing techniques, in that they rely on the reflection of ambient sunlight to illuminate objects for measurement (Fig. 1.2). Passive sensors are therefore only useful if there is a clear, well-lit view of the object of interest. Passive sensors cannot be used through cloud cover or at night. Table 1.1 outlines the attributes of several commonly used remote sensing photographic, multispectral and hyperspectral imaging systems.

All of these technologies have been used successfully for remote sensing of coral reef ecosystems. The properties and processes of coral reefs that have been

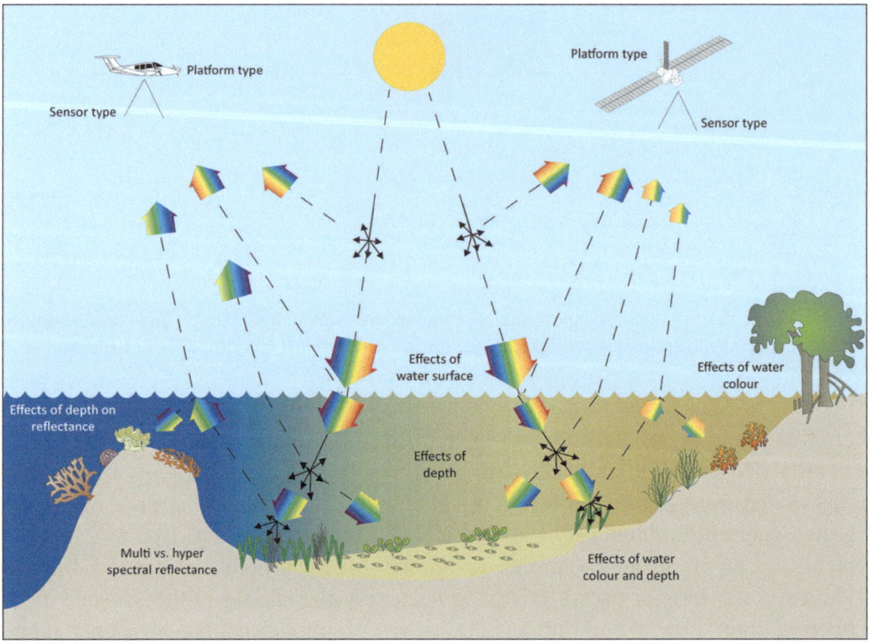

Fig. 1.2 Environmental features and processes in coral reefs affecting the radiative transfer processes recorded by passive optical remote sensing instruments, including photographic, multispectral and hyperspectral imaging systems. This diagram identifies features able to be measured, along with factors that reduce the ability to use images of coral reefs (Remote Sensing Toolkit www.gpem.uq.edu.au/cser-rstoolkit)

mapped using remote sensing data include their extent, composition (e.g., benthic cover, habitat characteristics), biophysical attributes (e.g., bathymetry, water quality, sea surface temperature), biogeochemistry (e.g., primary production, calcification), and geology (e.g., morphology, sedimentary diversity). Remote sensing products are also becoming increasingly recognized for their usefulness to monitor changes in reef composition over time. Table 1.2, several review papers (Kuchler et al. 1988; Green et al. 2000; Mumby et al. 2004b; Andréfouët et al. 2005a; Eakin et al. 2010; Hochberg 2011), and the Remote Sensing Toolkit (www.gpem.uq.edu.au/cser-rstoolkit) all provide a good history and critical assessment of coral reef remote sensing research and application.

Research on remote sensing for coral reefs has followed two fundamental paths. The first has been development of techniques to compensate for water column and atmosphere effects on the remotely sensed signal (Lyzenga 1978, 1985; Gordon and Clark 1980; Bierwirth et al. 1993; Gordon 1997; Lee et al. 1999; Louchard et al. 2003; Gao et al. 2009; Dekker et al. 2011). For coral reefs, an important implication from this research is that for passive sensors to be useful the seafloor must be visibly observable in the imagery. Optically deep areas, or areas with high turbidity, cannot be mapped using passive techniques alone, and active systems

Table 1.1 Summary table listing sensor types and associated spatial, spectral, radiometric and temporal resolution

Example sensors	Spatial scale	Spectral resolution	Radiometric resolution	Temporal resolution
Aerial photography				
Pan Colour stereo CIR stereo	**Extremely fine to fine**: (local) 1:5,000–1:25,000 **Extent**: 1.3–33 km^2 per photo **GRE**: 0.05–20 m	**>100 nm** **Low -broad band**: - Visible - Colour - Green, Red, NIR	**High**: > 10 bit (1,024 levels)	**User controlled**: (subject to weather and aircraft availability)
Airborne multispectral				
SpecTerra DMSV Daedalus-1268 ADAR	**Extremely fine to fine**: (local) **Extent**: 100 km^2 **GRE**: 0.5–10 m	**>100 nm** **Medium range**: 350–2,500 nm **Total bands**: 3–20	**Medium**: >8 bit (256 levels)	**User controlled**: (subject to weather and aircraft availability)
Airborne hyperspectral				
CASI HyMap AVIRIS AISA	**Extremely fine to fine**: (local) **Extent**: 100 km^2 **GRE**: 0.5–10 m	**5–50 nm** **High range**: 350–2500 nm **Total bands**: >20	**High**: >12 bit (4,096 levels)	**User controlled**: (subject to weather and aircraft availability)
High spatial resolution multispectral				
QuickBird 2 Ikonos Rapid Eye GeoEye-1 Worldview-1, -2	**Extremely fine**: (local) **Extent**: >25 km^2 **GRE**: 0.5–1 m (pan) 1.5–5 m (multi)	**>100 nm** **Medium range**: 400–1,000 nm **Total bands**: 1–8	**High**: 11–12 bit (2,048–4,096 levels)	**Programmable**: 1–3 day repeat (subject to weather)

(continued)

Table 1.1 (continued)

Example sensors	Spatial scale	Spectral resolution	Radiometric resolution	Temporal resolution
Moderate spatial resolution multispectral				
Landsat 7 ETM+	**High to medium**:	**>100 nm**	**Medium to high**:	**Programmable**:
Landsat TM	(local, province, region)	**Medium to high range**:	8–12 bit	1–46 day repeat
SPOT	**Extent**:	450 nm–12.5 um	(256–4,096 levels)	Sensor dependent
Resourcesat-1	>100 km^2	**Total bands**:		(subject to weather)
ALOS	**GRE**:	3–14		
ASTER	2.5–15 m (pan)			
	10–30 m (multi)			
	90 m (thermal)			
Low spatial resolution multispectral				
SPOT VMI	**Coarse**:	**>50 nm**	**High**:	**Programmable**:
NOAA AVHRR	(region)	**Medium to high range**:	10 bit	1–3 day repeat
SeaWifs	**Extent**:	400 nm–12.5 um	(1,024 levels)	Sensor dependent
OrbView-2	>1,000 km^2	**Total bands**:		(subject to weather)
Seastar	**GRE**:	4–15		
MERIS	300 m–1 km			
Moderate-low spatial resolution hyperspectral				
Hyperion	**Medium to coarse**:	**10–100 nm**	**High**:	**Programmable**:
MODIS	(province, region)	**Medium to high range**:	12 bit	1–3 day repeat
	Extent:	400 nm–14.4 um	(4,096 levels)	Sensor dependent
	>1,000 km^2	**Total Bands**:		(subject to weather)
	GRE:	36–220		
	30 m–1 km			

GRE ground resolution element or pixel size

must instead be considered (see Sects. 1.2 and 1.3). The second research path has been development of higher-level products that provide insight to reef status or function (Atkinson and Grigg 1984; Bour et al. 1986; Loubersac et al. 1988; Mumby et al. 1997; Hochberg and Atkinson 2000, 2008; Roelfsema et al. 2002; Isoun et al. 2003; Andréfouët et al. 2004a; Lesser and Mobley 2007; Palandro et al. 2008; Purkis et al. 2008). A number of basic products produced from this research are currently in routine use, while others remain in development.

Chapters 2– 4 cover the most commonly available, and frequently used, sources of passive visible and infrared remote sensing data, including film based aerial and space photography, digital cameras, and multispectral and hyperspectral imaging systems. The progression through these technologies also represents a progression in detail of coral reef information that can be retrieved from the respective data sources. Generally, the detail in information content is controlled by the spatial and spectral capabilities of the sensor. A main focus of this chapter is describing the control that spatial resolution and spectral band number, band width and band position have on the information able to be mapped.

Aerial photography is the simplest and most historically relevant data set, typically covering local to regional scales (several km^2 to 100s of km^2), sometimes with records stretching back to the 1930s (Hernandez-Cruz et al. 2006). Astronaut photography is also available, and although collection is opportunistic rather than systematic, valuable information can be extracted from this imagery. Recent advances in aerial photography have also seen large format digital cameras being adopted and used extensively by survey companies and governments. These systems provide larger area coverage, less processing and more consistent spectral data than previous generations of cameras. Multispectral systems generally cover the same tasks as aerial photography, but over larger areas (10^4–10^6 km^2), with significant repeat capacity. Hyperspectral sensors provide added spectral bands, with much narrower bandwidths and greater ability to identify specific targets.

1.1.2 Chapter Outline

This chapter provides the technical basis for understanding Chaps 2–4, and an overview of the situations in which passive visible and infrared remote sensing does and doesn't work for mapping and monitoring coral reefs. The chapter starts by explaining the spectral dimensions of remote sensing instruments in detail, along with how these dimensions control the amount and type of information on coral reefs that can be effectively mapped. A specific set of biophysical environmental variables, relevant to coral reef science and management, and able to be mapped by multispectral and hyperspectral systems, are then defined. Examples are provided of image based map products and processing approaches required to deliver science and management data for monitoring coral reefs. The chapter finishes with an overview of future directions.

Table 1.2 Table linking remote sensing instruments and coral reef biophysical properties, listing the feasibility and processing approaches used to derive the products from the different input data types

	Film photography	Digital photography	Multispectral imaging	Hyperspectral imaging
Reef/non-reef	Operational Manual interpretation	Operational Manual interpretation Per-pixel classification Object based mapping	Operational Per-pixel classification Object based mapping	Operational Per-pixel classification Object based mapping
Reef type	Operational Manual interpretation	Operational Manual interpretation Per-pixel classification Object based mapping	Operational Per-pixel classification Object based mapping	Operational Per-pixel classification Object based mapping
Reef composition (e.g., geomorphic zones, benthic communities)	Operational Manual interpretation	Operational Manual interpretation Per-pixel classification Object based mapping	Operational Per-pixel classification Object based mapping	Operational Per-pixel classification Object based mapping Sub-pixel analysis
Patterns of reef composition	Operational Manual interpretation	Operational Manual interpretation Per-pixel classification Object based mapping	Operational Per-pixel classification Object based mapping	Operational Per-pixel classification Object based mapping
Bathymetry and derived variables	Not operational	Operational Empirical Semi-analytic Analytic	Operational Empirical Semi-analytic Analytic	Operational Empirical Semi-analytic Analytic

(continued)

Table 1.2 (continued)

	Film photography	Digital photography	Multispectral imaging	Hyperspectral imaging
Biophysical reef properties	Not operational	Operational Empirical Semi-analytic Analytic	Operational Empirical Semi-analytic Analytic	Operational Empirical Semi-analytic Analytic
Biophysical reef processes	Not operational	Research	Operational Empirical Semi-analytic Analytic	Operational Empirical Semi-analytic Analytic
Surrounding water properties	Not operational	Research	Operational Empirical Semi-analytic Analytic	Operational Empirical Semi-analytic Analytic
Surrounding land properties	Not operational	Operational Empirical Semi-analytic Analytic	Operational Empirical Semi-analytic Analytic	Operational Empirical Semi-analytic Analytic

1.2 Physical and Technical Principles

1.2.1 Imaging Sensor Dimensions

As discussed in the introductory section of this chapter, remote sensing data can be differentiated by the dimensions of the imaging sensor used to capture the image data (Table 1.1). These dimensions are outlined below and are critical for understanding the relationship to the environmental feature being mapped, as the dimensions control the type and level of detail of information able to be extracted from images.

- *Spectral:* the location, width and number of spectral bands used to record light.
- *Spatial:* pixel size and image extent.
- *Radiometric:* levels of brightness detected.
- *Temporal:* the time and repetition frequency at which image data are acquired.

The spectral dimension of remotely sensed data is the primary control of the type(s) of information able to be measured and mapped. You will notice that the chapters in this book correspond to remote sensing instruments differentiated by their spectral dimensions. In this chapter we introduce two primary forms of passive or optical data: multispectral and hyperspectral. Note that aerial photography in its film-based and more recent digital format is considered to be a multispectral system. All of these sensors can be mounted on boats, underwater ROVs and AUVs, people (e.g., divers, snorkelers), aircraft and satellites. The primary differences between multispectral and hyperspectral image data are shown in Fig. 1.3, where a comparison of reflectance signatures clearly shows the improved ability of the hyperspectral band-set to discriminate different reef features, such as bleached versus un-bleached corals.

The other fundamental control on the mapping and monitoring of coral reefs using remote sensing is spatial dimension. This includes pixel size and image extent (Fig. 1.4), as well as the size of the target features. Generally speaking, image pixel size must be smaller than the length or breadth of the target feature you wish to map. For example, to detect small coral patches, pixels <1 m are required, while geomorphic zones can be mapped with image pixels of 10–30 m (Fig. 1.4). Spatial and spectral dimensions also interact to define the features able to be discriminated on reefs, where given the same spectral resolution more information can be derived using higher spectral resolution.

Radiometric dimensions relate to the level of precision used to record light reaching a sensor (e.g., recording 256 vs. 1,024 levels of brightness). A higher radiometric resolution (e.g., 1,024 brightness levels) is required for detecting subtle changes in reflection or absorption of sunlight by coral reef features. Temporal dimension refers to the frequency with which an imaging sensor can revisit or re-image the same location. For more dynamic reef features you may need daily acquisitions, while yearly images may be sufficient for longer term changes.

Another factor controlling the type of information able to be extracted from remotely sensed images of coral reefs is the image processing algorithm used to transform the images into maps of benthic cover types, water depth, percent macro-algal cover, or other relevant parameters. This is the process of transforming an image from a qualitative picture into a quantitative digital map that can be used for science and management (Table 1.2). The image processing algorithm is an equation, or series of equations, applied to every pixel in an image to identify habitat characteristics and/or estimate environmental parameters.

1.2.2 Spectral Characteristics

The spectral dimension of a remotely sensed image determines if it can be used to map particular coral reef biophysical variables. As introduced above, spectral dimension refers to the quantities of light or electromagnetic energy measured in each image pixel. More specifically, spectral dimension refers to the location, width and number of spectral bands measured by the sensor. Remote sensing instruments use detectors, including light-sensitive film and light-sensitive detector materials (e.g., silicon) to measure the strength of electromagnetic energy, or number of photons per unit time, in selected portions of the electromagnetic spectrum. These film and solid detector materials are sensitized to specific regions (i.e., bands) of the electromagnetic spectrum for measurement purposes. Traditionally, remote sensing science uses a wavelength notation (as opposed to frequency) to denote the different portions of the electromagnetic spectrum.

Significant amounts of work have been completed on radiative transfer processes in gases, liquids, solids and plants; hence there is a high level of understanding about how specific structural and chemical attributes of these features control absorption and scattering at specific wavelengths. Radiative transfer refers to the processes of transmission, absorption and scattering of electromagnetic energy. Based on this knowledge, remote sensing detectors, especially multispectral and hyperspectral systems, are designed to measure electromagnetic energy in pre-defined portions of the spectrum known to be sensitive to specific structural and chemical attributes of features or associated processes in the environment.

The individual spectral bands used for any particular sensor cover a set range of wavelengths. For example, the multispectral system shown in Fig. 1.3 covers the blue, green, red and near-infrared portions of the electromagnetic spectrum using 100 nm wide spectral bands. In contrast, the hyperspectral system in Fig. 1.3 covers the same range of wavelengths using hundreds of 10 nm wide spectral bands. Multispectral systems provide broadly applicable spectral reflectance signatures suitable for mapping coral reef benthic features at a coarse level (e.g., geomorphic zones; Table 1.1 and Chaps. 2 and 3). Hyperspectral systems provide highly detailed spectral reflectance signatures enabling better discrimination of coral reef benthic features, and improved quantitative estimation of biophysical,

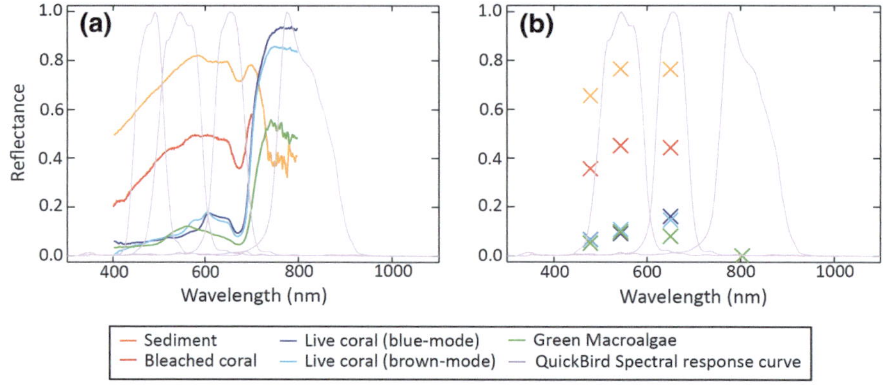

Fig. 1.3 The spectral dimensions of visible and infrared remote sensing data: **a** Shows the reflectance signature using a full-spectral resolution field spectrometer; **b** Shows the same reflectance signatures using a multispectral band set (provided by Ian Leiper)

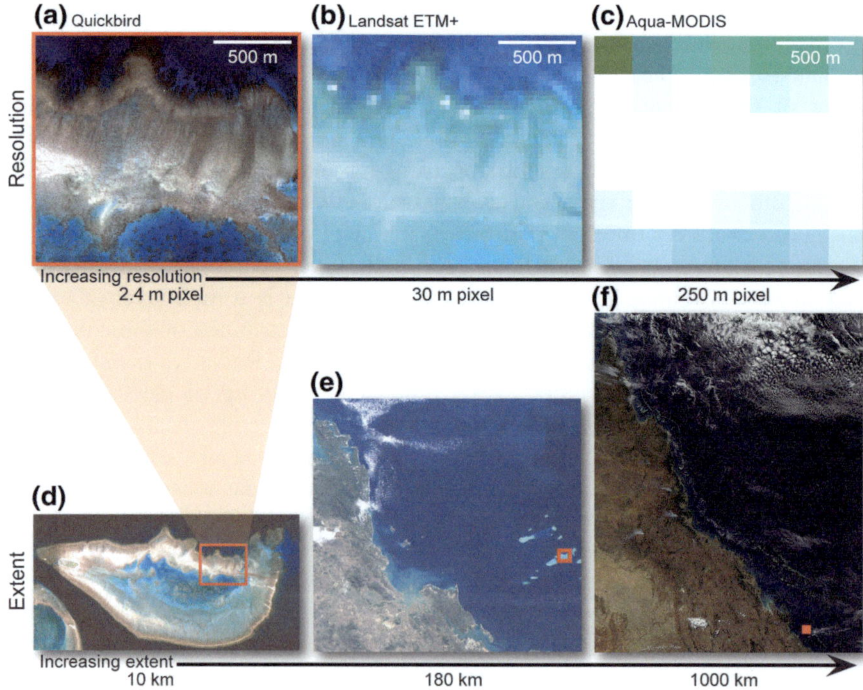

Fig. 1.4 The different spatial dimensions of remote sensing data for an image of Heron Reef, Australia. Images (**a**–**c**) show the effects of progressively larger pixel sizes for a 1.5 km long section of Heron Reef. Images (**d**–**e**) show different image extents, starting at Heron Reef (**d**) and moving to the entire Great Barrier Reef (**f**). The *red box* indicates the same area as shown in images (**a**–**c**) (provided by Ian Leiper)

structural, chemical and process attributes (Hochberg and Atkinson 2000; Hedley and Mumby 2002; Hochberg et al. 2003, 2004; Mumby et al. 2004b).

Mapping coral reef features, either by discriminating benthic features or estimating biophysical properties, such as depth and pigment concentration, requires remotely sensed data with the appropriate spectral and spatial dimensions. Once these are identified, a suitable image processing algorithm can be selected. For mapping coral reef features, a significant amount of work has been completed globally to show that as you increase the number of spectral bands and decrease the pixel size the greater number of benthic and substrate cover types you will be able to map (Andréfouët et al. 2003). This corresponds to a progression from mapping reef/non-reef, to mapping geomorphic zones and reef biotope zones, to mapping benthic communities. A similar pattern is observed when mapping coral reef biophysical properties, whether in the water column, benthos or substrate; increasing the number of spectral bands enables more detailed and precise estimation of biophysical properties. Large numbers of spectral bands with narrow band widths also permits specific absorption features or inflection points, produced by photosynthetic or non-photosynthetic pigments, to be resolved (Hochberg and Atkinson 2000, 2008; Hedley and Mumby 2002; Hochberg et al. 2003, 2004; Mumby et al. 2004b; Hochberg and Atkinson 2008). Research on hydro-optics in water bodies and photosystems in corals has established which wavelength regions are absorbed by specific chemicals and processes; hence reflectance signatures resolving these features can be used in algorithms to estimate or map them for each pixel. Figure 1.5 further illustrates the relative differences in spectral content for multispectral versus hyperspectral images using example data from Heron Reef, Australia.

1.2.3 Photography (Film and Digital)

Aerial and space photography in its film-based form cannot display a spectral-reflectance signature; however their simple format and long term collection worldwide make them a unique resource for coral reef mapping and monitoring over time. Film based products are typically transformed to maps of benthic cover through systematic interpretation keys for specific features based on subjective, context specific, visual interpretation cues. To accomplish this, photographs, or negatives, are often scanned into digital format and processed into maps using image processing or geographic information system (GIS) software. If historic, thematically simple maps (e.g., geomorphic zones, sand, coral, etc.) are required, aerial photographs are highly suitable for this application. Any detailed mapping of coral reef features using photography requires extensive site-specific context and field knowledge, along with high spatial resolution (<1:5,000 scale) aerial photographs in either color or black and white formats.

Aside from field survey data, photography is often the only systematically collected, long term archive of spatial information available for coral reefs in many areas. It should be noted, however, that standard format photographs contain

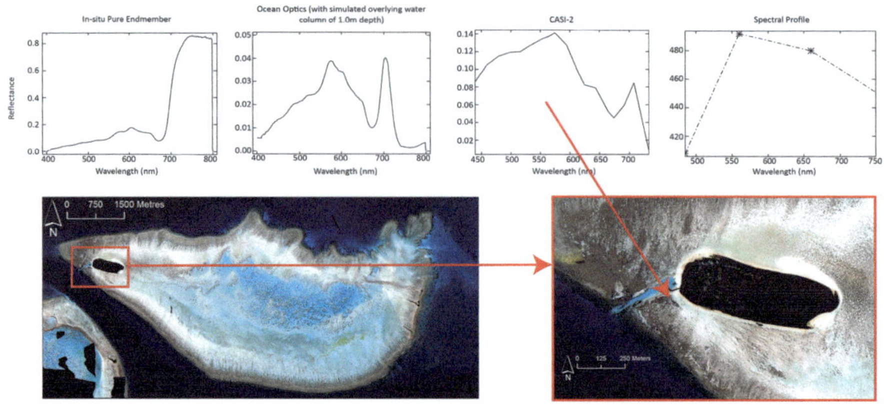

Fig. 1.5 Example spectral signatures from the same patch of live coral. The progression of reflectance signature graphs, from left to right is: in-situ reflectance from field spectrometry, modelled at-surface reflectance with 1.0 m of water, at-surface reflectance from airborne hyperspectral (CASI 2); and at-surface reflectance (x 10,000) from satellite multispectral (QuickBird 2) image (provided by Ian Leiper)

significant spatial distortions due to the geometry of the photograph acquisition process. As a result, the scale, or relation of ground distance to the same distance in a photograph, may vary. Uncorrected photographs can therefore not be used to produce spatially accurate maps for comparison over time or for integration with other spatial data until they are first ortho-corrected. The ortho-correction process transforms photographs to a digital format with consistent spatial scale, allowing them to be more effectively used for comparative mapping purposes.

1.2.4 Multispectral Imaging Systems

Multispectral systems on airborne and satellite platforms, including the current generation of large-format digital mapping cameras, typically have 3–10 spectral bands per pixel, resulting in a simplified spectral reflectance signature (Figs. 1.3 and 1.5). For thematic mapping of coral reef features, image pixel size and spectral band placement will control the type and amount of information able to be discriminated. Several published papers, including the images in Fig. 1.4, show that multispectral data with moderate pixel sizes (20–30 m) can be used to map 5–6 coral reef benthic classes at accuracy levels of 80 %, while multispectral data with smaller pixel sizes (<5.0 m) can map 10–12 classes of coral reef benthic cover features at comparable accuracy (Andréfouët et al. 2003; Roelfsema and Phinn 2010).

Due to the broad spectral bands used in multispectral systems, their utility for mapping quantitative biophysical properties (e.g., pigment concentration) is limited since the narrow width of absorption features associated with photosynthetic and non-photosynthetic pigments cannot be resolved. Multispectral data do contain suitable bands for use in empirical and semi-analytic methods for estimating the depth of the water column in each pixel (Stumpf et al. 2003; Dekker et al. 2011); however, the limitations associated with these approaches should be clearly noted in terms of depth restrictions and errors introduced by heterogeneous benthic features.

1.2.5 Hyperspectral Imaging Systems

Hyperspectral airborne and satellite systems typically have 10–1,000 spectral bands per pixel, resulting in detailed spectral reflectance signatures (Figs. 1.3 and 1.5). This increased level of precision in the spectral dimension allows small deviations in reflectance signatures to be detected and the magnitude of different absorption and reflectance features to be quantified. Like multispectral systems, the pixel size of an image will also control the types and number of features that can be mapped. Hyperspectral data allow more detailed mapping of benthic cover types since the differences in the structure or chemistry of coral reef features can be better detected. Hence, mapping of benthic communities to the level of live coral, different coral structural forms, dead coral, and macro- and micro-algae is possible (Mumby et al. 1998; Hochberg and Atkinson 2000, 2003; Goodman and Ustin 2003; Hochberg et al. 2003; Andréfouët et al. 2004b; Mumby et al. 2004a). Although a significant amount of work has been completed on field spectrometry to more explicitly relate hyperspectral signatures of coral reefs to pigment content and other functional properties, little work has been published scaling that up to image-based mapping (Brock et al. 2006; Hochberg and Atkinson 2008).

1.3 Image Processing

When using remotely sensed data on coral reefs it is essential to understand how the image or image-based map was produced. Chapters 2– 4 outline the types of processing applied to multispectral and hyperspectral images to produce thematic or quantitative maps of coral reef properties. Understanding the suitability and quality of these map products requires knowledge of both the forms of remote sensing data and the processing steps used to generate the maps. As an example of this process, Fig. 1.6 illustrates an overview of the different steps used for generating a benthic cover map from multispectral QuickBird imagery.

1.3.1 Image Preprocessing

Image data sets directly output from airborne or satellite imaging systems are first subjected to a series of image preprocessing operations where algorithms are

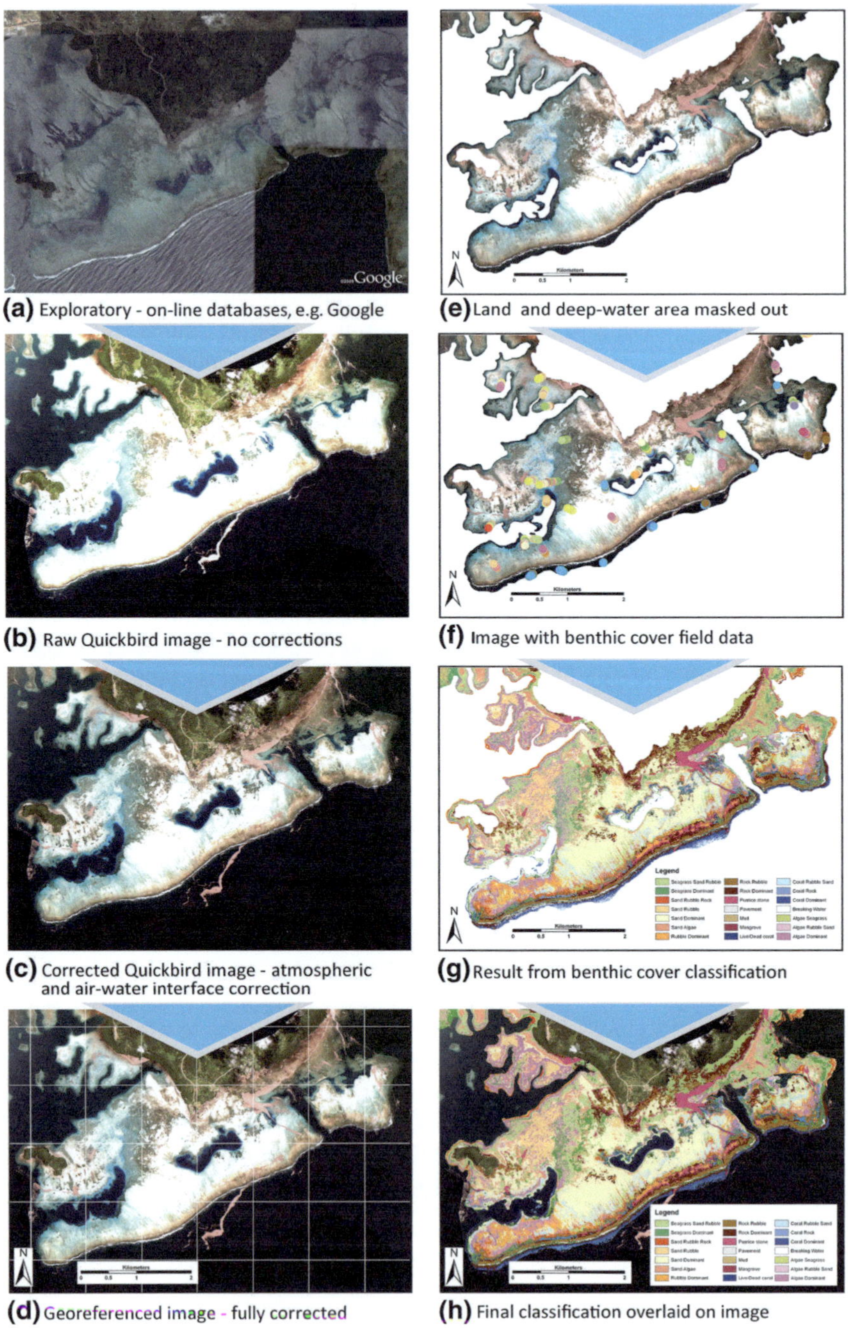
(a) Exploratory - on-line databases, e.g. Google
(b) Raw Quickbird image - no corrections
(c) Corrected Quickbird image - atmospheric and air-water interface correction
(d) Georeferenced image - fully corrected
(e) Land and deep-water area masked out
(f) Image with benthic cover field data
(g) Result from benthic cover classification
(h) Final classification overlaid on image
Google
N

◀ **Fig. 1.6** Complete remote sensing image processing flow from image collection to map production. (*Source* Phinn et al. 2010) Steps in the processing sequence: **a** Browse image from Google Earth (Landsat TM/QuickBird combination); **b** Raw QuickBird image with no corrections; **c** Corrected QuickBird image after atmospheric and air–water interface corrections: **d** Georeferenced QuickBird image after atmospheric and air–water interface corrections; **e** Fully corrected image **d**, with non-reef areas masked out; **f** Shallow water and exposed reef image with calibration and validation field data; **g** Benthic cover map produced by image classification of **f**; **h** Benthic cover map overlaid on the original image

applied to each image pixel to correct several types of distortions. Some image processing operations, such as geometric correction (translating pixel coordinates into a known geographic coordinate system, projection and datum), are essential if you are planning to link field data or other spatial data with your remotely sensed images. A good starting point to explain what these are and why they are essential can be found at: www.ga.gov.au/earth-monitoring/geodesy/geodetic-datums.html. Additional processing operations (e.g., atmospheric correction) are required if the image is going to be used to estimate biophysical properties of the water column or corals (e.g., depth and pigment concentrations).

Raw image data: This is the first output from an imaging sensor, which typically has no coordinate system, projection or datum, and cannot be used or displayed with other spatial data such as field survey GPS points. The image pixel values also represent relative measures of reflected light, and cannot be related to light interactions on the water surface or reef. Nonetheless, this data can still be used for basic visual assessments.

Corrected data (geometric, radiometric and atmospheric): These are the first stages in image processing, referred to as image preprocessing steps. Geometric correction involves aligning the image to an established coordinate system, projection and datum, which allows the image to be used or overlaid with other spatial data and field data. An accuracy or error level should be provided as part of this correction. Radiometric correction translates the relative pixel values to absolute measures of radiance per unit wavelength of light. Atmospheric correction removes atmospheric effects and thereby transforms the radiometric values into surface radiance or reflectance. This allows field based measurements or biophysical parameters to be estimated. In some cases, additional corrections may be required to remove sunglint or attenuation due to the water column.

1.3.2 Processing Types

Transforming photographic, multispectral and hyperspectral images from preprocessed, or corrected, images to maps showing specific coral reef biophysical properties requires the application of manual and/or software driven image processing operations. Details of these operations and their output map products for reef science and management are provided in Chaps. 2– 4. This section introduces the types of operations, their output products and associated validation needs as a basis for understanding the application and management chapters. Image processing operations are applied once the geometric, radiometric and atmospheric correction operations are complete.

Two general types of processing operations can be applied, with the distinction based on the type of output map required for science or management purposes. In this context, all of the processing and output data are in digital format and can be referred to as digital maps or spatial information.

Processing to thematic maps: In this processing option, a variety of techniques, ranging from manual to automated, are used to group pixels representing the same feature on a coral reef into pre-defined sets of thematic classes. The output is an imaged-based map of the different classes as defined for a given level of detail, such as geomorphic zones or benthic communities (e.g., Ahmad and Neil 1994; Andréfouët et al. 2003; Andréfouët et al. 2005a). These maps are often referred to as categorical or thematic, and show discrete boundaries.

Processing to biophysical property maps: In this processing option, either empirical relationships or established models are applied to each image pixel to produce an estimate of a biophysical property. Examples include bathymetry or chlorophyll-a concentration in the water column (e.g., Purkis et al. 2002; Mumby et al. 2004a; Kutser and Jupp 2006; Kutser et al. 2006). These are often referred to as continuous maps since each pixel has a unique value.

In each approach there is also capacity to include other forms of remote sensing imagery and spatial data (e.g., boat–based sonar, airborne LiDAR depth sounding, or pre-existing maps) to improve map accuracy or expand the types of features or processes able to be mapped (Brock and Purkis 2009; Bejarano et al. 2010). Each of the output map products can also be used to produce maps for the same area over time and then used in the detection and measurement of changes or trends in coral reef properties and associated processes over time (Palandro et al. 2003, 2008; Scopelitis et al. 2007, 2009; Chap. 15).

1.3.3 Thematic Mapping

Thematic maps can be produced using two general approaches: manual digitizing of boundaries of an image or photo displayed on the screen using a pre-set list of classes and interpretation cues; or utilizing mapping algorithms provided in image-processing software. The choice of which method to use depends on: the output coral reef map classes required; the type of photograph or image data being used; amount of background knowledge and experience of the person(s) doing the mapping; and availability of field data for the area to be mapped. A more detailed outline of this process and its options is provided in the Remote Sensing Toolkit (www.gpem.uq.edu.au/cser-rstoolkit).

Manual digitizing can be applied to all forms of photography, multispectral and hyperspectral images, but has most frequently been used in higher spatial resolution aerial photography and satellite image data. These applications have focused on mapping benthic communities and coral reef benthic cover types, such as live and dead coral, where high levels of detail and local context are available to identify specific reef features (Cuevas-Jimenez and Ardisson 2002; Knudby et al. 2007; Scopélitis et al. 2009). In some cases, regionally and globally applicable mapping programs using broad levels of detail have used manual digitizing to produce reef maps, such as the Millennium Coral Reef Mapping Project, which utilizes the global archive of Landsat Thematic Mapper and Landsat Enhanced Thematic Mapper data with 30 × 30 m pixels (Andréfouët et al. 2005b; Andréfouët 2008).

More recent developments have seen image processing systems provide semi-automatic processes that replicate manual interpretation, in the form of geographic object-based image analysis (GEOBIA). These approaches enable hierarchical segmentation of images into pre-set features or objects at specific spatial scales (e.g., reef/non-reef, geomorphic zones and benthic community zones and patches) (Benfield et al. 2007). After segmentation the image objects or features are then labeled manually or automatically.

Image classification is the most common algorithmic approach to producing thematic maps from multispectral and hyperspectral data sets. Image classification is used to assign a pre-defined thematic class label to each pixel in an image. The classification algorithms are based on two assumptions: (1) each image pixel contains only one type of coral reef benthic feature (i.e., that a pixel is smaller than the feature to be mapped); and (2) all image pixels containing that type of coral reef feature have a similar spectral reflectance signature. Since hyperspectral images produce spectral signatures with higher degree of detail and precision than multispectral and photographic images (e.g., Fig. 1.5), classification algorithms using hyperspectral data can discriminate more coral reef benthic cover types. Increased thematic detail can also be achieved by adding contextual information into the process, including measures such as image texture or roughness and other forms of image and spatial information. Image classification routines can further include post-classification manual editing to increase the level of thematic detail and accuracy of coral reef maps.

The final stage in the mapping process should always be some form of validation, where the output coral reef map is compared to a suitable form of reference data, either from field survey or other spatial data, so that the overall and individual class mapping accuracies are known (Andrefouet 2008; Mumby et al. 1998; Roelfsema and Phinn 2010).

1.3.4 Biophysical or Continuous Variable Mapping

Production of maps quantifying biophysical properties or processes on coral reefs and their surrounding environments can only be done from fully corrected airborne or satellite images. This type of processing applies one or more equations to each image pixel to transform the pixel value from a measurement of reflectance to a measurement of a biophysical property of the coral reef or surrounding water column, atmosphere or land (Phinn et al. 2010). These approaches are based on the assumption that the measured spectral reflectance in certain bands has a direct relationship to the biophysical property being estimated. For example, absorption of light at specific wavelengths have known relationships to: water column depth; concentrations of absorbing and scattering organic and inorganic materials; concentrations of photosynthetic and non-photosynthetic pigments in coral, seagrass and algae; and processes such as photosynthesis (Mobley 1994; Hedley and Mumby 2002).

Several approaches are commonly used to deliver maps of coral reef biophysical properties. In the first case, the relative area of each pixel occupied by a set of coral reef benthic cover types (e.g., coral, sand, algae) is estimated using "unmixing" techniques. These techniques assume the image pixel is larger than the features to be mapped and are applied to images which have had the influence of the water column removed (Hedley and Mumby 2003; Hedley et al. 2004; Goodman and Ustin 2007; Lesser and Mobley 2007). The mathematical solutions required for these techniques become more accurate as the number of un-correlated input variables (spectral bands in this case) increases; hence hyperspectral image data are used predominantly in this approach. The remaining approaches, commonly referred to as "inversion" techniques use empirical or analytic mathematical solutions to extract biophysical information from image pixels, including water depth, concentrations of organic and inorganic material in the water column, and benthic/substrate reflectance signatures. Empirical approaches are mainly used for estimating depth or bathymetric surfaces, require calibration against field measured depths, and typically only function accurately over homogeneous substrates to depths of 5–10 m. These techniques can be applied to both multispectral and hyperspectral data. Analytic and semi-analytic approaches function more effectively on hyperspectral image data sets, and often require locally specific field data on optical properties of the water column and benthic spectral reflectance signatures to produce accurate results. These results, however, are more robust than empirical approaches and produce accurate maps to depths of 20–25 m in areas with heterogeneous benthic and substrate features (Kutser et al. 2006; Dekker et al. 2011).

1.4 Future Directions

Advances in science and technology will affect the sensors, data types, data accessibility, processing techniques and, collectively, our ability to transform remotely sensed images into maps of coral reef biophysical properties. Scientific advances pertain to the ongoing development and testing of image processing algorithms to more accurately map and monitor biophysical properties of coral reefs. Advances in technology relate to changes to the spatial, spectral and radiometric dimensions of imaging sensors on airborne or satellite platforms, and the capabilities of the platforms themselves. The Committee on Earth Observation Satellites (CEOS), a global collective of scientists building and using satellites to map and monitor earth's ecosystems, maintains an online database of all current and planned sensors, along with their dimensions and links to data download sites (known as the Mission, Instruments and Measurements database, which can be found at http://database.eohandbook.com/).

1.4.1 Technological Advances

Improvements in available spatial dimensions of multispectral and hyperspectral imaging sensors will continue to fill the scale-gaps evident in Fig. 1.4, providing the potential for global-scale coverage of images with varying pixel sizes: 0.05–0.5 m (digital aerial photography), 0.5–10 m (high spatial resolution satellites), 10–100 m (moderate spatial resolution satellites), and 100–1,000 m (low spatial resolution satellites).

Improvements in spectral dimensions will remain predominantly in the multispectral domain, with satellite imaging sensors continuing to move beyond the traditional four band set (blue-green–red and NIR) towards 10–20 spectral band sets designed to address specific environmental applications and to maximize sensor sensitivity. Hyperspectral sensors will continue to be used mainly from airborne platforms, while several long awaited satellite systems will be launched in 2012–2015, providing moderate spatial resolution global hyperspectral coverage (EnMAP, HyspIRI). In all cases, sensor radiometric resolution and radiometric calibration consistency will also be improved, allowing increased detection of reflectance/absorption differences and more accurate detection of changes in image time-series.

The temporal dimensions, or repeat frequency, of satellite imaging systems will continue to be expanded; most single sensor/platform high spatial resolution systems already provide almost daily repeat acquisition. This is made possible by use of pointable imaging sensors and more agile satellite platforms (e.g., GeoEye-1, Worldview 2), as well as systems with constellations of multiple satellite platforms of the same sensor. Daily repeat coverage of an area maximizes the user's ability to collect cloud-free, low-wind, low-wave and low-sunglint coral reef images.

Associated advances in image storage, search/archive capacity across networks, and more frequent use of open access software and image archives is providing users with greater ability to locate, check and download archive satellite image data from coral reefs around the world. Acquisition of new images, especially airborne or high spatial resolution imagery, is currently still confined primarily to research or commercial service providers. Advances in GPS and digital photography, especially in terms of low-cost, accurate, waterproof systems has allowed field survey data of biophysical reef properties to be easily collected, georeferenced and placed in a format able to be integrated with coral reef airborne or satellite images. Continued improvements in the integration of field data with image data are essential for the calibration and validation of thematic mapping and biophysical applications on coral reefs.

1.4.2 Scientific Advances

At a scientific level there are two driving forces: (1) advances in image processing algorithms; and (2) development of applications/algorithms/models for mapping specific biophysical properties of coral reefs.

In the first case, image processing algorithms continue to be developed within and external to the remote sensing field. Digital image processing spans mathematics, physics, computer vision, signal processing, astronomy and medical imaging, to name a few; hence development of image correction, enhancement, thematic mapping and modeling is widespread. The most recent advances finding their way into coral reef applications are object-based image analysis, multivariate data fusion and new forms of spatially explicit regression analysis and unmixing. Once these new approaches have been identified, the next stage is testing their applicability for mapping, monitoring or modeling relevant coral reef biophysical properties. Thematic mapping of coral reef zones from multispectral and hyperspectral images will continue as the main application area in reef remote sensing, but with increased integration of other image data sets (e.g., LiDAR; Chap. 7) into object-based image analysis algorithms (e.g., segmentation then classification) and classification models allowing multiple forms of data (e.g., support vector machines, random forest). The application of analytic and semi-analytic modeling approaches to estimate per-pixel water depth, water properties and bottom reflectance is moving to operational status and the output data present a new set of variables to be fully tested with thematic mapping approaches (Chap. 4).

The area of multispectral and hyperspectral coral reef remote sensing with the most potential is the further development of techniques for mapping reef properties such as: the amount of live coral, algae and sediment cover; structural forms of coral cover; benthic micro-algae biomass; and coral and algae light absorption efficiency. These properties provide key links for studies assessing coral productivity, coral reef biochemistry, carbon-fluxes and nutrient dynamics on reefs. Advancements in these areas will require close collaboration between coral reef ecosystem scientists and the biophysical remote sensing community.

Acknowledgments Ian Leiper for provision of selected figures and graphics for the chapter.

Suggested Reading

Remote Sensing Toolkit website: www.gpem.uq.edu.au/cser-rstoolkit
CEOS Sensor List website: database.eohandbook.com/measurements/overview.aspx
Green EP, Mumby PJ, Edwards AJ, Clark CD (2000a) Remote sensing handbook for tropical coastal management. UNESCO, Paris
Mumby PJ, Skirving W, Strong AE, Hardy JT, LeDrew E, Hochberg EJ, Stumpf RP, David LT (2004a) Remote sensing of coral reefs and their physical environment. Mar Pollut Bull 48:219–228
Phinn SR, Roelfsema CM, Stumpf RP (2010) Remote sensing: discerning the promise from the reality. In: Longstaff BJ, Carruthers TJB, Dennison WC, Lookingbill TR, Hawkey JM, Thomas JE, Wicks EC, Woerner J (eds) Integrating and applying science: a handbook for effective coastal ecosystem assessment. IAN Press, Cambridge, pp 201–222

References

Ahmad W, Neil DT (1994) An evaluation of Landsat Thematic Mapper (TM) digital data for discriminating coral reef zonation: Heron Reef (GBR). Int J Remote Sens 15:2583–2597
Andrefouet S (2008) Coral reef habitat mapping using remote sensing: a user vs. producer perspective. Implications for research, management and capacity building. J Spatial Sci 53:113–129
Andréfouët S, Kramer P, Torres-Pulliza D, Joyce KE, Hochberg EJ, Garza-Perez R, Mumby PJ, Riegl B, Yamano H, White WH, Zubia M, Brock J, Phinn SR, Naseer A, Hatcher BG, Muller-Karger FE (2003) Multi-sites evaluation of IKONOS data for classification of tropical coral reef environments. Remote Sens Environ 88:128–143
Andréfouët S, Zubia M, Payri C (2004a) Mapping and biomass estimation of the invasive brown algae *Turbinaria ornata* (Turner) J. Agardh and *Sargassum mangarevense* (Grunow) setchell on heterogeneous Tahitian coral reefs using 4-meter resolution IKONOS satellite data. Coral Reefs 23:26–38
Andréfouët S, Payri C, Hochberg EJ, Hu C, Atkinson MJ, Muller-Karger FE (2004b) Use of in situ and airborne reflectance for scaling up spectral discrimination of coral reef macroalgae from species to communities. Mar Ecol Prog Ser 283:161–177
Andréfouët S, Hochberg EJ, Chevillon C, Muller-Karger FE, Brock JC, Hu C (2005a) Multi-scale remote sensing of coral reefs. In: Miller RL, Castillo CED, McKee BA (eds) Remote sensing of coastal aquatic environments: technologies, techniques and applications. Springer, The Netherlands, pp 299–317
Andréfouët S, Muller-Karger FE, Robinson JA, Kranenburg CJ, Torres-Pulliza D, Spraggins S, Murch B (2005b) Global assessment of modern coral reef extent and diversity for regional science and management applications: a view from space. 10th international coral reef symposium, pp 1732–1745
Atkinson MJ, Grigg RW (1984) Model of coral reef ecosystem. II. Gross and net benthic primary production at French Frigate Shoals, Hawaii. Coral Reefs 3:13–22
Bejarano S, Mumby P, Hedley J, Sotheran IS (2010) Combining optical and acoustic data to enhance the detection of Caribbean forereef habitats. Remote Sens Environ 114:2768–2778
Benfield SL, Guzman HM, Mair JM, Young JAT (2007) Mapping the distribution of coral reefs and associated sublittoral habitats in Pacific Panama: a comparison of optical satellite sensors and classification methodologies. Int J Remote Sens 28:5047–5070
Bierwirth PN, Lee TJ, Burne RV (1993) Shallow sea-floor reflectance and water depth derived by unmixing multispectral imagery. Photogram Eng Remote Sens 59:331–338

Bour W, Loubersac L, Rual P (1986) Thematic mapping of reefs by processing of simulated SPOT satellite data: application to the *Trochus niloticus* biotope on Tetembia Reef (New Caledonia). Mar Ecol Prog Ser 34:243–249

Brock J, Purkis S (2009) The emerging role of lidar remote sensing in coastal research and resource management. J Coastal Res: Special issue 53—Coast Appl Airborne Lidar 53:1–5

Brock J, Yates K, Halley R, Kuffner I, Wright C, Hatcher B (2006) Northern Florida reef tract benthic metabolism scaled by remote sensing. Mar Ecol Prog Ser 312:123–139

Cuevas-Jimenez A, Ardisson PL (2002) Mapping shallow coral reefs by colour aerial photography. Int J Remote Sens 23:3697–3712

Dekker A, Phinn SR, Anstee J, Bissett P, Brando VE, Casey B, Fearns P, Hedley J, Klonowski W, Lee ZP, Lynch M, Lyons M, Mobley C (2011) Inter-comparison of shallow water bathymetry, hydro-optics, and benthos mapping techniques in Australian and Caribbean coastal environments. Limnol Oceanogr Methods 9:396–425

Eakin CM, Nim CJ, Brainard RE, Aubrecht C, Elvidge CD, Gledhill DK, Muller-Karger F, Mumby PJ, Skirving WJ, Strong AE, Wang MH, Weeks S, Wentz F, Ziskin D (2010) Monitoring Coral Reefs from Space. Oceanography 23:118–133

Gao BC, Montes MJ, Davis CO, Goetz AFH (2009) Atmospheric correction algorithms for hyperspectral remote sensing data of land and ocean. Remote Sens Environ 113:S17–S24

Goodman J, Ustin S (2003) Airborne hyperspectral analysis of coral reef ecosystems in the Hawaiian Islands. International symposium on remote sensing of environment

Goodman J, Ustin SL (2007) Classification of benthic composition in a coral reef environment using spectral unmixing. J Appl Remote Sens 1:17

Gordon HR (1997) Atmospheric correction of ocean color imagery in the Earth observing system era. J Geophys Res Atmos 102:17081–17106

Gordon HR, Clark DK (1980) Atmospheric effects in the remote sensing of phytoplankton pigments. Bound-Layer Meteorol 18:299–313

Green EP, Mumby PJ, Edwards AJ, Clark CD (2000b) Remote sensing handbook for tropical coastal management. UNESCO, Paris

Hedley JD, Mumby PJ (2002) Biological and remote sensing perspectives of pigmentation in coral reef organisms. Adv Mar Biol 43:277–317

Hedley JD, Mumby PJ (2003) A remote sensing method for resolving depth and subpixel composition of aquatic benthos. Limnol Oceanogr 48:480–488

Hedley J, Mumby P, Joyce K, Phinn S (2004) Determining the cover of coral reef benthos through spectral unmixing. Coral Reefs 23:21–25

Hernández-Cruz LR, Purkis SJ, Riegl BM (2006) Documenting decadal spatial changes in seagrass and *Acropora palmata* cover by aerial photography analysis in Vieques, Puerto Rico: 1937–2000. Bull Mar Sci 79(2):401–404

Hochberg EJ (2011) Remote sensing of coral reef processes. In: Dubinsky Z, Stambler N (eds) Coral Reefs: an ecosystem in transition. Springer, Dordrecht, pp 25–35

Hochberg EJ, Atkinson MJ (2000) Spectral discrimination of coral reef benthic communities. Coral Reefs 19:164–171

Hochberg EJ, Atkinson MJ (2003) Capabilities of remote sensors to classify coral, algae, and sand as pure and mixed spectra. Remote Sens Environ 85:174–189

Hochberg E, Atkinson M (2008) Coral reef benthic productivity based on optical absorptance and light-use efficiency. Coral Reefs 27:49–59

Hochberg EJ, Atkinson MJ, Andréfouët S (2003) Spectral reflectance of coral reef bottom-types worldwide and implications for coral reef remote sensing. Remote Sens Environ 85:159–173

Hochberg EJ, Atkinson MJ, Apprill A, Andréfouët S (2004) Spectral reflectance of coral. Coral Reefs 23:84–95

Isoun E, Fletcher C, Frazer N, Gradie J (2003) Multi-spectral mapping of reef bathymetry and coral cover; Kailua Bay, Hawaii. Coral Reefs 22:68–82

Knudby A, LeDrew E, Newman C (2007) Progress in the use of remote sensing for coral reef biodiversity studies. Prog Phys Geogr 31:421

Kuchler DA, Biña RT, Claasen DvR (1988) Status of high-technology remote sensing for mapping and monitoring coral reef environments. In: Proceedings of 6th international coral reef symposium, vol 1, pp 97–101

Kutser T, Jupp DLB (2006) On the possibility of mapping living corals to the species level based on their optical signatures. Estuar Coast Shelf Sci 69:607–614

Kutser T, Miller I, Jupp DLB (2006) Mapping coral reef benthic substrates using hyperspectral space-borne images and spectral libraries. Estuar Coast Shelf Sci 70:449–460

Lee ZP, Carder KL, Mobley CD, Steward RG, Patch JS (1999) Hyperspectral remote sensing for shallow waters: 2. Deriving bottom depths and water properties by optimization. Appl Optics 38:3831–3843

Lesser MP, Mobley CD (2007) Bathymetry, water optical properties, and benthic classification of coral reefs using hyperspectral remote sensing imagery. Coral Reefs 26:819–829

Lillesand TM, Kiefer RW, Chipman JW (2008) Remote sensing and image interpretation. 6th edn. Wiley

Loubersac L, Dahl AL, Collotte P, Lemaire O, D'Ozouville L, Grotte A (1988) Impact assessment of Cyclone Sally on the almost atoll of Aitutaki (Cook Islands) by remote sensing. In: Proceedings of 6th international coral reef symposium, vol 2, pp 455–462

Louchard EM, Reid RP, Stephens FC, Davis CO, Leathers RA, Downes TV (2003) Optical remote sensing of benthic habitats and bathymetry in coastal environments at Lee Stocking Island, Bahamas: a comparative spectral classification approach. Limnol Oceanogr 48:511–521

Lyzenga DR (1978) Passive remote sensing techniques for mapping water depth and bottom features. Appl Optics 17:379–383

Lyzenga DR (1985) Shallow-water bathymetry using combined lidar and passive multispectral scanner data. Int J Remote Sens 6:115–125

Mobley C (1994) Light and water: radiative transfer in natural waters. Academic Press, San Diego

Mumby PJ, Green EP, Edwards AJ, Clark CD (1997) Coral reef habitat-mapping: how much detail can remote sensing provide? Mar Biol 130:193–202

Mumby PJ, Green EP, Clark CD, Edwards AJ (1998) Digital analysis of multispectral airborne imagery of coral reefs. Coral Reefs 17(1):59–69

Mumby PJ, Hedley J, Chisholm JRM, Clark CD, Ripley HT, Jaubert J (2004b) The cover of living and dead corals from airborne remote sensing. Coral Reefs 23:171–183

Mumby PJ, Skirving W, Strong AE, Hardy JT, LeDrew E, Hochberg EJ, Stumpf RP, David LT (2004c) Remote sensing of coral reefs and their physical environment. Mar Pollut Bull 48:219–228

Palandro D, Andréfouët S, Muller-Karger F, Dustan P, Hu C, Hallock P (2003) Detection of changes in coral reef communities using Landsat 5/TM and Landsat 7/ETM + Data. Can J Remote Sens 29:207–209

Palandro DA, Andréfouët S, Hu C, Hallock P, Muller-Karger FE, Dustan P, Callahan MK, Kranenburg C, Beaver CR (2008) Quantification of two decades of shallow-water coral reef habitat decline in the Florida Keys National Marine Sanctuary using Landsat data (1984–2002). Remote Sens Environ 112:3388–3399

Phinn SR, Roelfsema CM, Stumpf RP (2010) Remote sensing: discerning the promise from the reality. In: Longstaff BJ, Carruthers TJB, Dennison WC, Lookingbill TR, Hawkey JM, Thomas JE, Wicks EC, Woerner J (eds) Integrating and applying science: a handbook for effective coastal ecosystem assessment. IAN Press, Cambridge, pp 201–222

Purkis S, Kenter JAM, Oikonomou EK, Robinson IS (2002) High-resolution ground verification, cluster analysis and optical model of reef substrate coverage on Landsat TM imagery (Red Sea, Egypt). Int J Remote Sens 23:1677–1698

Purkis SJ, Graham NAJ, Riegl BM (2008) Predictability of reef fish diversity and abundance using remote sensing data in Diego Garcia (Chagos Archipelago). Coral Reefs 27:167–178

Roelfsema CM, Phinn SR (2010) Integrating field data with high spatial resolution multi spectral satellite imagery for calibration and validation of coral reef benthic community maps. J Appl Remote Sens 4(1):043527 doi:10.1117/1.3430107

Roelfsema CM, Phinn SR, Dennison WC (2002) Spatial distribution of benthic microalgae on coral reefs determined by remote sensing. Coral Reefs 21:264–274

Scopelitis J, Andrefouet S, Largouet C (2007) Modelling coral reef habitat trajectories: evaluation of an integrated timed automata and remote sensing approach. Ecol Model 205:59–80

Scopélitis J, Andréfouët S, Phinn S, Chabanet P, Naim O, Tourrand C, Done T (2009) Changes of coral communities over 35 years: integrating in situ and remote-sensing data on Saint-Leu Reef (la Réunion, Indian Ocean). Estuar Coast Shelf Sci 84:342–352

Stumpf R, Holderied K, Sinclair M (2003) Determination of water depth with high resolution satellite image over variable bottom types. Limnol Oceanogr 48:547–556

Chapter 2
Photography Applications

Susan A. Cochran

Abstract Photographic imaging is the oldest form of remote sensing used in coral reef studies. This chapter briefly explores the history of photography from the 1850s to the present, and delves into its application for coral reef research. The investigation focuses on both photographs collected from low-altitude fixed-wing and rotary aircraft, and those collected from space by astronauts. Different types of classification and analysis techniques are discussed, and several case studies are presented as examples of the broad use of photographs as a tool in coral reef research.

2.1 Introduction

Aerial photography is the oldest form of remote sensing and can be dated back to 1858, when a French photographer, printmaker, writer, caricaturist, and balloonist by the name of Gaspard-Felix Tournachon (also known as Nadar; Fig. 2.1) used a tethered hot-air balloon to capture an image of the Bievre Valley in France (PAPA International 2010; Daunier 1862). Since that time, photographers have placed cameras on objects ranging from pigeons to kites to rockets and satellites, in order to capture the "bird's eye view" from above.

The first time an airplane was used as a platform for aerial photography was in 1909 when Wilbur Wright, who was in Italy for business, took up a passenger collecting motion pictures of a military airfield near Rome. Aerial photography

S. A. Cochran (✉)
USGS Pacific Coastal and Marine Science Center,
400 Natural Bridges Dr., Santa Cruz, California 95060, USA
e-mail: scochran@usgs.gov

J. A. Goodman et al. (eds.), *Coral Reef Remote Sensing*,
DOI: 10.1007/978-90-481-9292-2_2,

Fig. 2.1 "Elevating photography to the condition of art", 1862 caricature of Nadar by Honoré Daunier (Daunier 1862)

began to be used extensively by the military in WWI. Cameras were developed specifically for use in airplanes, and aerial photography soon replaced traditional maps drawn by aerial observers. After the end of WWI in 1918, aerial cameras began to be used for commercial purposes and surveying.

Aerial photographs of the Great Barrier Reef were collected as early as 1928 (Stephenson et al. 1931; Fairbridge and Teichert 1948), yet it was the onset of WWII in 1939 that precipitated the photography of many Pacific islands and reefs, including low-angle oblique photographs, for military purposes. Initially used to chart beach landings, these military photographs soon became the basis of scientific studies, from tracking morphological changes and the influence of wave patterns, to investigating the structure and orientation of platform reefs, and many others (Steers 1945). Since that time, aerial photographs have become an important source of information for scientists studying coral reefs. This chapter includes discussion on the use of aerial photography acquired through the use of either fixed-wing or rotary aircraft, as well as space-based photography, to capture images of coral reefs for mapping baseline information, change detection, and other scientific and management applications. Case studies are presented to illustrate several uses of aerial photography in coral reef environments.

2.2 Photography of Coral Reefs

Fixed-wing or rotary aircraft are the most common platforms from which to collect low-altitude photographs. However, other platforms have also been useful in coral reef studies. Rützler (1978) attached a camera to an aluminum frame that was suspended from a tethered helium balloon to successfully collect images at an altitude of 50 m above Carrie Bow Cay, Belize. Scoffin (1982) suspended a radio-operated camera from the cord of a kite to collect vertical and oblique photographs at altitudes from 50 to 200 m over reef flats in the Cook Islands. Such low-cost methods are shown to be especially useful in locations where winds of 7–25 knots might occur, such as those commonly found around tropical islands exposed to the trade winds.

High-altitude photographs of Earth taken by astronauts from the windows of spacecraft, such as the International Space Station and the Space Shuttle, have also played a role in studying coral reef ecosystems. The photographs are in the public domain from NASA's Office of Earth Sciences at the Johnson Space Center (http://eol.jsc.nasa.gov). Robinson et al. (2000) estimate that nearly 30,000 of over 375,000 photographs in the database at that time had potential value for coral reef studies (e.g., not over- or under-exposed, not too oblique, and with limited sun-glint). As of October 2011, the database contains over one million photographs of Earth. If the percentage (8 %) estimated by Robinson et al. (2000) holds true, the number of photographs with potential value for coral reef studies may now be as high as 80,000. Cloud cover is often a problem when acquiring images in tropical regions using commercial satellites with pre-determined orbits. However, because astronauts can be visually selective when collecting images, photographs acquired in this manner over coral reef areas tend to have low or no cloud cover (Robinson et al. 2000). Astronaut-acquired photographs provide a low-cost alternative to commercial imagery, which can be important for developing countries, and can be used stand-alone or for supplementing analyses using other data sources (Robinson et al. 2002).

One of the most common uses of aerial photography in coral reef environments is for mapping the location and extent of a variety of marine and coastal habitats. Chauvaud et al. (1998) used aerial photographs for thematic mapping of marine communities in the Bay of Robert on the island of Martinique in the French West Indies. They state that high-resolution aerial photographs are well suited to map the intricate complexities of tropical coastal habitats. Ekebom and Erkkilä (2003) summarize the advantages and disadvantages of numerous remote-sensing techniques used for habitat mapping. Their test study, comparing habitat delineations from multiple user-interpreters, concluded that the high spatial resolution of aerial photographs provided the best data source for reliable habitat identification.

Aerial photograph collections are often acquired with an overlap ranging from 60 to 65 %, making them ideal for viewing and mapping the three-dimensional perspective of reef morphology using stereo-pairs (Sheppard et al. 1995). Stereo-pairs and photogrammetry methods have also been used to calculate depths to underwater objects (Tewinkel 1963) and estimate underwater topography (Murase et al. 2008), which could be potentially helpful in mapping the geomorphology of reef systems. Andréfouët et al. (2002) show that using high-resolution spatial information from aerial photographs, rather than optical data from other airborne or satellite sensors, results in more accurate detection and mapping of the spatial patterns found during coral bleaching events. Fletcher et al. (2003) merged aerial orthophoto mosaics with NOAA topographic surveys (T-sheets) and hydrographic surveys (H-sheets) to map historical shoreline position and calculate coastal erosion rates on the Hawaiian island of Maui.

Since aerial photography has the longest history in coral reef remote sensing, it offers unparalleled usefulness to those studies documenting change through time. Armstrong (1981) used vertical panchromatic aerial photographs to calculate over 40 years of change, including the effects from two hurricanes, on a coral reef off Puerto Rico. To compute the exact scale of each photograph, which is necessary for accurate calculations of habitat areas and measuring change, known man-made structures were measured in the field and compared to their image on the photographs. In addition to the vertical photographs, oblique photographs were used to assist in defining different reef features. Yamano et al. (2000) used aerial photography to track a 21 year change in coral zonation. Lewis (2002) used aerial photographs to track the loss of geomorphologic reef structure over a 40 year period, while Hernandez-Cruz et al. (2006) used aerial photographs to document seagrass extents over a 63 year period.

Most recently, commercial and government aerial photography has transitioned away from film-based systems to digital cameras. This transition has significantly improved the extent of coverage, consistency of photo quality, ability to conduct automated processes, and ease of integration of aerial photographs with other spatial information. For example, Palandro et al. (2003) combined aerial photographs with IKONOS imagery from an orbital satellite to track changes on a coral reef over a 19 year period. In order to merge film-based prints with modern digital data, it is first necessary to scan the prints. Often, different sets of aerial photographs are acquired at different altitudes using different cameras and lenses, resulting in different spatial resolutions and scales. Scanning the photographs at a high resolution (typically 300 dpi or higher) and then employing resampling and geographic correction provides direct comparison on a pixel-to-pixel basis between different sets of imagery. Advanced computer techniques simplify the merging process of different datasets.

2.3 Photography Analysis and Classification Techniques

Low-altitude aerial photography is an excellent resource for studies of coral reef habitats due to its high resolution. As with other remote sensing technologies, aerial photography also enables a synoptic view of large areas of study, something that is not typically possible when conducting on-the-ground or in-the-water observations (usually due to the limitations of manpower). Photography is relatively inexpensive, and because of its on-demand nature, allows planning for optimum conditions (e.g., sun angle, cloud cover, sea-surface state, tide level) during image collection. Distortions present in photographs are primarily a function of camera properties (e.g., radial lens distortions), which can be corrected through use of camera models and geometric algorithms accounting for variations in aircraft positioning (e.g., altitude, roll, pitch and yaw). Photographs are commonly collected from a vertical viewpoint (i.e., directly overhead), but in some instances are collected from an oblique angle, which introduces additional distortions to be considered during analysis but provides a different viewpoint that can be useful in some applications.

Computer analysis of film-based photographs first requires that they are scanned, or digitized, at a specified resolution (i.e., pixel resolution) to assign a digital number (DN) representing the relative intensity of reflected light for each pixel in the image. Alternately, for digital-based photographs, DN values already represent the relative light intensity for specific spectral bands. The spatial resolution of each pixel is dependent on the resolution of the camera, the scanner, and the flying altitude upon collection, but is typically high enough (ranging from 0.1 to 1 m per pixel) to be able to resolve many complex reef features.

Aerial photographs are usually collected on-demand, and scheduled when both sky and water conditions are at their best. Nonetheless, since film photography lacks detailed radiometric information, it is infeasible to apply preprocessing corrections to the water column to account for attenuation of wavelengths with depth due to absorption and scattering. In some instances, such corrections can be applied when using digital aerial photography, but are best suited for multispectral and hyperspectral sensors, which are spectrally better characterized than photography. As an example, Lyzenga (1978, 1981) describes the use of the image itself to produce a depth-invariant bottom index to compensate for differences in variable depths when discriminating between bottom types.

Registration, or georeferencing, of aerial photographs collected for coral reef studies may be difficult without supplemental data. Aerial photographs of on-land features are georegistered using ground control points, previous orthophotographs, and/or digital terrain models (DTMs). However, due to the offshore location of many coral reef areas, images of these study sites may not contain any referencing land or shoreline that can be matched to image locations, or the images may only contain a limited linear extent of coastline, which, if paired with on-land ground control points or a DTM and stretched to match, may spatially distort the seaward edge of an image. In these instances, it may be necessary to use georeferenced bathymetric data, such as LiDAR (Light Detection and Ranging; see Chaps. 5–7) or other underwater terrain models, to correctly georeference the aerial images. Additionally, because astronaut-acquired photographs are acquired using varying camera lenses from different altitudes, resulting in different scales, and are also taken from different look angles through the windows of spacecraft, which leads to different spatial resolutions between the near and far edges of a photograph, additional steps must be taken to resample and geometrically correct these images to a known map projection (Robinson et al. 2000). A complete background on the use of astronaut-acquired photography and subsequent digital data for remote sensing purposes is explored fully in Robinson et al. (2002).

After preprocessing, georeferenced digital images may be analyzed individually, or mosaicked together to cover an entire coral reef region (Chavez et al. 2000). Occasionally, the differences in cross-scene illumination due to bidirectional sun angles, common on low-altitude aerial photography as an aircraft flies back and forth across a study area, are apparent when creating a mosaic of individual aerial images. In such cases, additional radiometric corrections will be required to normalize the pixels across the individual images prior to mosaicing (Lillesand and Kieffer 1994; Beisl and Woodhouse 2004; Beisl et al. 2006). Finally, if desired, a land mask can be applied in order to focus subsequent computer analysis on the in-water features only.

Analysis of aerial photography for coral reef studies can range from a basic visual interpretation to more complex computer-aided classifications. At its simplest and most effective form, aerial photography can be used for mapping coral reefs by hand-drawing polygons on the hard photographic copies themselves, with or without transparent overlays (e.g., Manoa Mapworks 1984), or digitized on a visual display in either image processing or geographic information software, a technique known as 'heads up' digitizing (e.g., Coyne et al. 2003; Scopélitis et al. 2009).

Supervised or unsupervised computer classification may be performed using the spectral information in digitized aerial photographs. Color film records light reflected in three wavelengths—red, green, and blue (RGB). Color-infrared (CIR) film, also known as false-color photography, records light reflected in the near-infrared, red, and green wavelengths. CIR imaging is often used in land-mapping applications, as healthy vegetation reflects more radiation in the near-infrared (NIR) wavelengths than green wavelengths. However, NIR wavelengths are absorbed within the first few cm of water, and the use of false-color photography for coral reef applications must take this into consideration. Hopley and Catt (1988) describe the use of NIR photography to monitor the ecological response of sea-level rise on several sites in the Great Barrier Reef, noting that the acquisitions were timed to coincide with negative daytime tides. So while CIR images are useful for mapping mangroves or other emergent vegetation, as well as some shallow (less than ~3 m) reef flat habitats (Fig. 2.2), CIR photography is typically not used in coral reef management applications where it is necessary to image reefs in greater than ~3 m water depth (Hopley 1978).

Because lack of light penetration in deeper water is a limitation when using aerial photographs to study coral reefs (as well as multispectral and hyperspectral remote sensing), they are limited in their use as the sole source of information for inferring coral abundance, species types, and other benthic studies across the entire reef. For these types of investigations, it is helpful to combine aerial photography with other remotely sensed or field-collected data. For example, Fig. 2.3 shows (a) aerial photography of a coral reef ecosystem off the west coast of the island of Hawai'i, and (b) the same imagery merged with high-resolution LiDAR

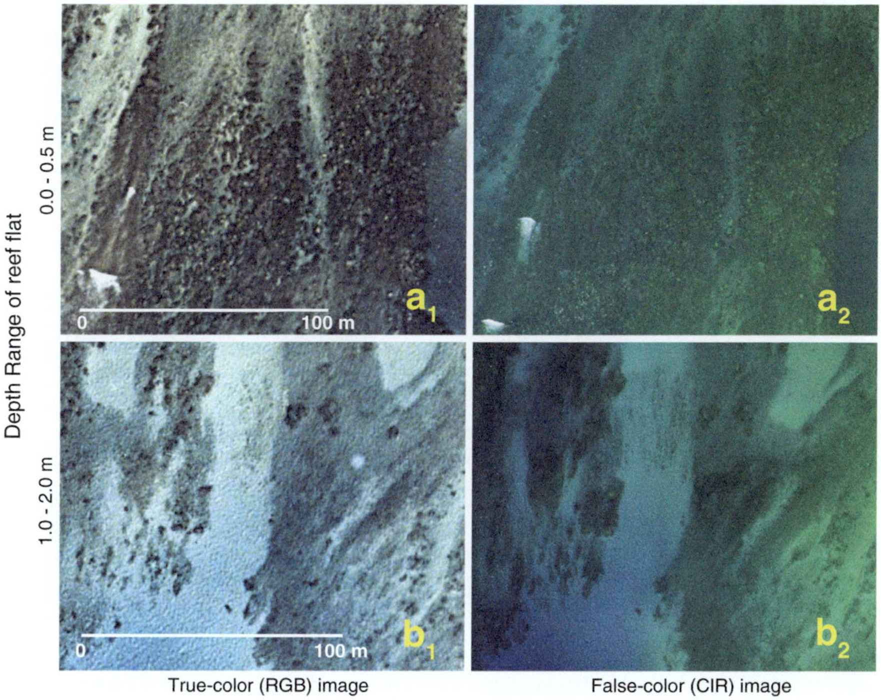

Fig. 2.2 Comparison of true-color (RGB) and false-color (CIR) images at different depths on a reef flat off the south coast of the island of Moloka'i. Water depth of reef flat shown in the top two images $\mathbf{a}_1$ and $\mathbf{a}_2$ ranges from 0.0 to 0.5 m; water depth of reef flat shown in the bottom two images $\mathbf{b}_1$ and $\mathbf{b}_2$ ranges from 1.0 to 2.0 m. At extremely shallow depths, it is possible to discern features on the reef flat in the 0.0 to 0.5 m CIR image $\mathbf{a}_2$, but in water depths as little as 1 m $\mathbf{b}_2$ features begin to blur due to attenuation of the near-infrared wavelengths

bathymetric data. At approximately 15–20 m depth, it becomes difficult to discriminate coral features in the aerial photographs. The LiDAR bathymetry, however, extends to nearly 40 m depth. The combination of spectral information from the aerial photography and shaded relief created from the bathymetric data extends the depth at which underwater features can be "seen" and assists with interpretation and analysis (Cochran et al. 2007). The three-dimensional detail provided by the LiDAR data allows easy discrimination of topographic features, such as smaller patch reefs, pits, and channels. It also allows for calculations of rugosity, or topographic variability, an important factor for habitat complexity (Brock et al. 2004). Additional applications of the use of LiDAR data to map and quantify coral habitats are presented in Chaps. 5–7.

Recent advances have been made in computer software to classify images using multi-scale object-oriented (pattern or texture) recognition rather than using spectral data alone. Object-based image analysis (OBIA) uses a two-step process where the objects in an image are first defined, or segmented, using vectors rather than raster pixels. The objects are then used as training blocks in a classification

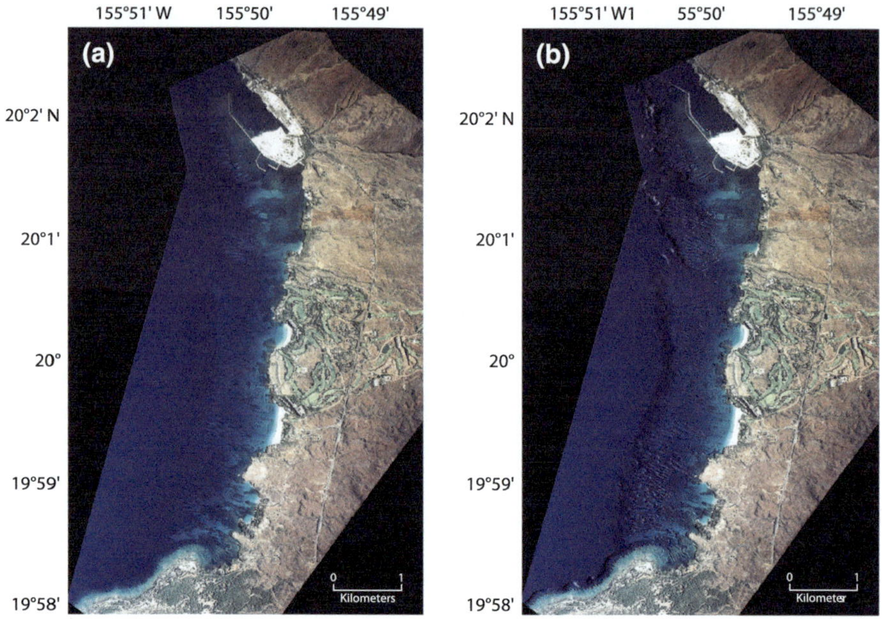

Fig. 2.3 Example comparing **a** aerial photography of a nearshore coral reef system along the northwest coast of the island of Hawai'i, and **b** the same imagery merged with bathymetric data. The combination of spectral information from the aerial photography and shaded relief from the bathymetric data extends the depth at which underwater features can be "seen" and assists with interpretation and analysis (modified from Cochran et al. 2007)

system based on fuzzy logic (Wang et al. 2004; Benfield et al. 2007). In this type of analysis, the size, shape, and neighborhood context of a training block is taken into account, as well as the spectral information, leading to increased classification accuracy. This approach can be especially helpful when analyzing aerial photographs where spectral information is limited (relative to other multispectral or hyperspectral imagery). In addition, because low-altitude aerial photographs have relatively high spatial resolution, resulting in pixels being typically smaller than the features they represent, they are well suited to OBIA. These advances suggest that continued research in object-oriented analysis has promising potential as an addition to the photographic toolbox for reef applications.

2.4 Example Applications of Photography

The case studies presented in this chapter were chosen to illustrate the broad range of uses of aerial photography in coral reef management, but are by no means inclusive of the many potential uses of this important data source. The first study uses aerial photographs to create maps of coral reef environments in order to provide baseline spatial information about benthic habitat types and extent. The second set of studies use several years of aerial photographs to track changes over time. The third study incorporates astronaut-acquired photographs as a secondary, complementary data source when mapping tropical atolls with satellite imagery. The final case study uses the spectral information from digital aerial images, combined with bathymetric and field data, to quantitatively measure suspended sediment in the water column of a reef-flat environment.

2.4.1 Baseline Spatial Mapping

In order to document change in any ecosystem, one must first have a baseline inventory of resources. Benthic habitat maps, which allow spatial documentation of the location of coral, other biologic and geologic zones, percentage of live coral cover, and potentially the overall health of a system, are important tools used by coral reef managers to assess change. In 1998, while recognizing the lack of such baseline information for many United States coral reefs, the President of the U.S. issued Executive Order 13089 establishing the U.S. Coral Reef Task Force, whose primary duty is to preserve and protect coral reefs in the U.S. and U.S. Trust Territories. In response to this mandate, the National Oceanic and Atmospheric Administration (NOAA) National Ocean Service (NOS) implemented a program to provide digital maps of United States coral reefs for use in a geographic information system (GIS). The first benthic habitat maps completed for this program were of Puerto Rico and the U.S. Virgin Islands (Kendall et al. 2001), followed by the Main Hawaiian Islands (Coyne et al. 2003), the Northwestern Hawaiian Islands (National Oceanic and Atmospheric Administration 2003), American Samoa, Guam, and the Northern Mariana Islands (NOAA National Center for Coastal Ocean Science 2005), and the Republic of Palau (Battista et al. 2007). Mapping of other U.S. coral reefs is ongoing. The earliest of these mapping projects were based only on aerial photography, while more recent benthic habitat mapping projects utilized satellite imagery in addition to aerial photography. The benthic habitat mapping of the Main Hawaiian Islands (Coyne et al. 2003) is presented here.

Approximately 1,500 aerial photographs of the Main Eight Hawaiian Islands were collected by NOAA in 2000 at a scale of 1:24,000. Adjacent flight lines were flown with a 30 % overlap, and consecutive photos on each flight line were taken with a 60 % overlap to minimize distortion and spectral variations from sunglint. Diapositives (color transparencies) created from the original negatives were scanned at 500 dpi to create images with a spatial resolution of 1 m per pixel. Orthomosaics were created for each island by first applying lens parameter corrections to each image, and using airborne GPS, along with aerial triangulation software, to provide preliminary georegistration of the images.

To provide continuity to the coral reef scientific community, a hierarchical classification scheme was developed through a series of workshops incorporating input from coral reef scientists, resource managers, local experts, and others.

The classification scheme describes four broad classes, eleven specific categories, and a number of detailed sub-categories, for a total of twenty-seven distinct habitats (Table 2.1). The hierarchical scheme allows users to expand or collapse the level of thematic detail as needed. The classification scheme also describes eleven zones, which refer to a habitat's location within the coral reef ecosystem. These zones correspond to typical reef geomorphology terms found in current scientific literature (e.g., reef flat, fore reef, back reef, reef crest), but are not indicators of the substrate or biological cover type.

To incorporate the hierarchical classification scheme into a GIS, the authors utilized the Habitat Digitizer Extension, a tool created for ArcView/ArcGIS by NOAA in order to facilitate heads-up (on-screen) digitizing of benthic habitat polygons (public domain software; search the ArcScripts database at www.esri.com to download for free). The extension allows users to assign attributes to the benthic habitat polygons based on a predetermined and customizable classification scheme using a point-and-click window dialog. The extension also allows the user to set a minimum mapping unit (MMU) restriction, if desired, so that polygons smaller than the MMU will not be digitized.

Habitat boundaries (areas with similar specific colors or patterns) are hand-digitized and assigned attributes based on the classification scheme. A 1-acre minimum mapping unit (MMU) was chosen for the NOAA benthic habitat mapping projects based on the scale of the aerial photography and the objectives of the mapping program (Kendall et al. 2001). Occasionally the brightness, contrast, and color of the digital image were adjusted in order to enhance subtle features and assist with interpretation. In addition, the original 1:24,000 scale prints and diapositives were available for viewing as needed, as was external information such as nautical charts and supplemental anecdotal evidence.

After the first draft of a habitat map was completed, scientists and local experts visited sites in order to field check and validate information contained in the map. The primary purposes of these visits included visually documenting those areas in the aerial photographs that were difficult to interpret for one reason or another, as well as verifying assigned polygon attributes. After navigating by GPS to a particular field site, scientists made observations either by snorkel from the surface, free diving, or directly from the boat using a viewing box when conditions (depth and water clarity) permitted. Field check observations were then used to revise and correct the draft maps as needed. Local scientific and management experts reviewed the final drafts, and their recommendations, especially for those polygons labeled as "unknown", were incorporated into the final maps (Fig. 2.4).

The validity, or usefulness, of any interpretation or classification map may be determined with an accuracy assessment that compares the map with what is actually found in the field. In addition to determining the overall accuracy of a map, it is helpful to determine the accuracy of a map from both the producer and user points of view. Producer accuracy denotes which points on a map are classified correctly during a field assessment; user accuracy signifies the probability that a point in a given class is actually represented by that class in the field (Lillesand and Kiefer 1994).

Table 2.1 List of broad classes and detailed category levels in the hierarchical classification scheme for the Main Eight Hawaiian Islands (adapted from Coyne et al. 2003)

Broad class	Specific category	Category details
Unconsolidated Sediment (0 to <10 % submerged vegetation)		
	Mud	
	Sand	
Submerged vegetation		
	Seagrass	
		Continuous seagrass (90–100 % cover)
		Patchy (discontinuous) seagrass (50 to <90 % cover)
		Patchy (discontinuous) seagrass (10 to <50 % cover)
	Macroalgae (fleshy and turf)	
		Continuous macroalgae (90–100 % cover)
		Patchy (discontinuous) macroalgae (50 to <90 % cover)
		Patchy (discontinuous) macroalgae (10 to <50 % cover)
		Coral reef and hardbottom
	Coral reef and colonized hardbottom	
		Linear reef
		Aggregated coral
		Spur and groove
		Individual patch reef
		Aggregated patch reef
		Scattered coral/rock in unconsolidated Sediment
		Colonized pavement
		Colonized volcanic rock/boulder
		Colonized pavement with sand channels
	Uncolonized hardbottom	
		Reef rubble
		Uncolonized pavement
		Uncolonized volcanic rock/boulder
		Uncolonized pavement with sand channels
	Encrusting/coralline algae	
		Continuous encrusting/coralline algae (90–100 % cover)
		Patchy (discontinuous) encrusting/coralline algae (50 to >90 % cover)
		Patchy (discontinuous) encrusting/coralline algae (10 to >50 % cover)
Other delineations		
	Land	
	Emergent vegetation	
	Artificial	
	Unknown	

For the accuracy assessment in this case study, an independent third party surveyed a total of 1,225 randomly selected sample points using the same methods as those used during field check validations. Overall accuracy for the Main Eight Hawaiian Islands project was estimated to be 90 % at the major habitat level of the classification scheme hierarchy and 80 % at the most detailed habitat level.

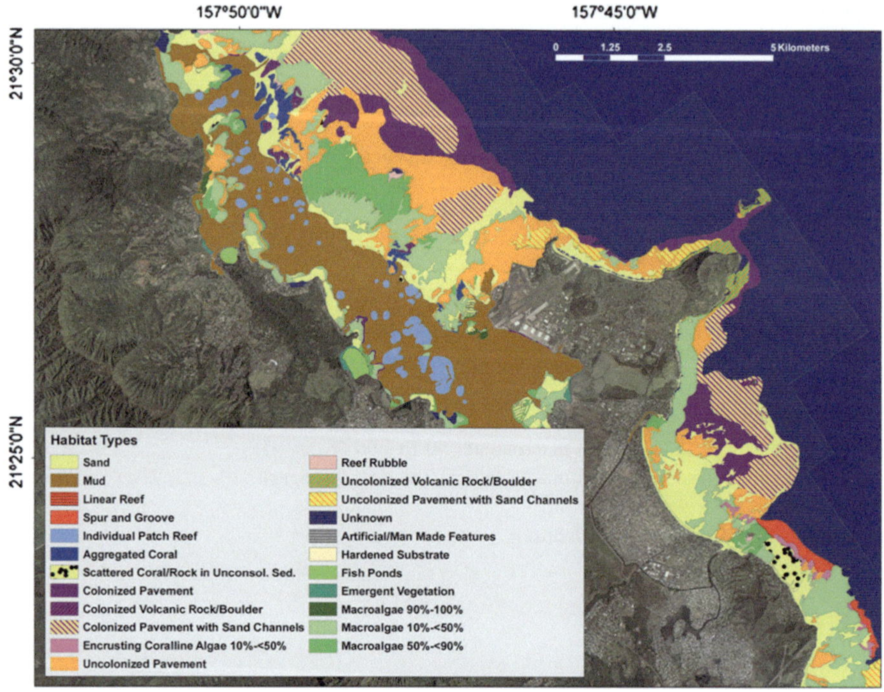

Fig. 2.4 Benthic habitat map of the Kane'ohe area of the island of O'ahu; created using data from Coyne et al. (2003); overlain on an IKONOS satellite image

The maps created in this study (Coyne et al. 2003) were among the first to quantitatively document the baseline status of the shallow coral reefs along the coasts of the Main Eight Hawaiian Islands. They revealed a total of 171 km^2 of submerged vegetation, 204 km^2 of unconsolidated sediment, and 415 km^2 of coral reef and colonized hardbottom. The maps may be viewed digitally in a GIS system or printed and used as stand-alone products to provide useful information to scientists, managers and other decision makers, and the general public.

2.4.2 Time Series Analysis

Today's coral reefs are faced with challenges from pressures such as sedimentation, sea-level rise, increased storm frequency, ocean acidification, and temperature changes, many of which are long-term (>10 year) disturbances directly resulting from global climate change (Wilkinson 2008). It is essential for managers to be able to monitor changes from these pressures at the same temporal scales in which they occur in order to fully understand the trends. Because of its extended history of data acquisitions, aerial photography is ideally suited to detect change through time on coral reefs.

Scopélitis et al. (2009) incorporated aerial photographs covering a span of 30 years (1973, 1978, 1989, 1997, and 2003) with two QuickBird satellite images (2002, 2006) to scale-up quantitative in situ data to document changes on Saint-Leu, a high-energy, fringing coral reef off La Réunion in the South West Indian Ocean. During the 35 year time span covered by the imagery, Saint Leu was regularly impacted by seasonal storms; however, two large tropical cyclones caused significant additional disturbances. Sedimentation from storm run-off, in addition to the mechanical wave damage caused by category 5 tropical cyclone Firinga in 1989, caused severe mortality to the reefs (Naim et al. 1997). A significant bleaching event occurred 2 months after category 4 tropical cyclone Dina (2002), but physical damage to the coral reef during that storm was not as severe as the damage from Firinga. Table 2.2 summarizes the overlap of these storms' time periods with the data used in this study.

A pan-sharpened 2006 QuickBird image was used as a reference to geometrically correct the historical imagery. A minimum of 10 control points were selected on each image, choosing easily identifiable objects such as buildings, roads, and sand patches when possible, and a first-order polynomial algorithm was applied ($RMSE < 0.5$). Because the images were individually interpreted in GIS using visual analysis, no attempt at radiometric normalization was made between imagery.

Scopélitis et al. (2007) defined different coral habitats using a combination of coral growth form, life status, taxonomy, and substratum. Because of the challenges in identifying some of these parameters consistently throughout the historical images, a broader classification scheme was created on a community basis for this study, defined only by coral growth form and substratum. This reduced level of detail resulted in 15 thematic classes that could be tracked throughout most of the historical imagery.

Beginning with the 2006 QuickBird image and working backward through time, coral community maps were created from all the images. Prior to visual interpretation, a mask was applied to the images to exclude any land, exposed beaches and breaking waves. An identical, comprehensive mask was used for all the images in order to eliminate variability between these features. Field-survey data from 2007 were considered to be representative of coral communities seen in the 2006 QuickBird image since no major disturbance occurred between the date of image acquisition and the in situ data observations; these data were used to guide the 2006 map creation. To create the 2003 thematic map, the 2006 polygons

Table 2.2 Summary of available data and major events on Saint-Leu reef (modified from Scopélitis et al. 2009)

	1973	1978	1987	1989	1993	1997	2000	2002	2003	2006	2007
Aerial photos	X	X		X		X			X		
Satellite images								X		X	
In situ data			X		X	X	X	X			X
Tropical cyclone				Firinga				Dina			

were overlaid on the 2003 image. The polygon boundaries were then edited to match the distribution of features in the 2003 image and corroborated using field-survey data. Any changes in polygon boundaries were considered to be temporal changes in the spatial extent and/or position of a community. Subsequently, the edited 2003 polygons were used, along with field-survey data, to interpret the 2002 imagery and the 2002 polygons were used, along with field-survey data, to interpret the 1997 imagery.

The 1989 image, acquired soon after tropical cyclone Firinga, was so different from the 1997 image that the polygon overlay method could not be used. Because of the lack of contemporaneous field-survey data, the 1987 map from Naim (1989) was used as a reference to guide the visual interpretations prior to 1989. The lack of historical field-survey data, a common limitation when using archived historical imagery in studies, also prevented accuracy assessments from being performed for these images.

GIS methods were used to identify community polygon changes between each successive map pair. Community polygon differences were considered noise when the area/perimeter ratio was near zero (± 1 %). The resulting difference maps allow users to track changes in reef community composition through time.

Scopélitis et al. (2007) showed that, despite the challenges of identifying features found consistently throughout the imagery and a lack of historical in situ data, aerial photographs can be a valuable resource for a time series analysis of the spatial extent of coral reef communities. By pushing back the baseline, a better picture of a coral reef's resilience (or lack of) can be observed and documented, allowing managers the opportunity to better understand the long-term trends with respect to current-day stressors.

2.4.3 Astronaut Photography as a Secondary Data Source

Coral pinnacles rising up from the lagoon floor and dotting the sea surface or near-surface are an important feature, both biologically and navigationally, of many atolls. Mapping of these elements using remotely sensed imagery can be a challenging task due to their spectral similarity with frequently present small clouds. Comparing multi-temporal satellite datasets can help solve this problem; however, the cost of obtaining multiple images can be prohibitive. In order to address this problem, Andréfouët and Robinson (2003) undertook a study combining freely available astronaut-acquired space photography and video from the NASA Space Shuttle with satellite imagery to distinguish clouds from coral pinnacles.

For this study, both astronaut-acquired hard-copy photographs and deinterlaced digital still images extracted from high-definition television (HDTV) video were merged with previously acquired SPOT HRV and/or Landsat ETM+ imagery for 84 atolls throughout the South Pacific. The hard photographs were digitized at 2,400 ppi (10.6 μm/pixel). Geometric rectification of the Space Shuttle imagery (both digitized hard-copy photographs and deinterlaced stills pulled from the

HDTV video) was accomplished using the satellite image as a reference by means of easily identifiable control points found on the atoll rims.

The near infrared (NIR) bands from the satellite imagery and the red bands from the Space Shuttle imagery were stretched and thresholded to create binary images with the lagoon water as one endmember (i.e., 0 = black) and with clouds and sub-surface pinnacles as the other (i.e., 255 = white). The binary images were compared and any pixels that were not the same across images were considered to be clouds. A tolerance factor was included to compensate for any errors in geometric rectification, and newly confirmed pinnacles were used as additional control points to refine the geometric rectification of the Space Shuttle imagery.

Andréfouët and Robinson (2003) noted that this method of evaluation required the images to be relatively cloud-free, as clouds that appear on the same pixel across multiple images could be misinterpreted as pinnacles. Additional temporally different imagery can help further refine the interpretation. The authors also found that deeper sub-surface pinnacles were sometimes difficult to distinguish using the processed red bands due to their weaker spectral signatures. Stretching and thresholding the green bands in these instances helped with the interpretation.

Astronaut-acquired space photography provides a low-cost alternative to commercial satellite imagery, which can be important for developing countries and for other budget-restricted studies requiring multi-temporal sets of imagery. Although limited in its usefulness as a primary data source for coral reef mapping studies due to challenges with spatial rectification, astronaut-acquired photography is well suited as a secondary data source to complement other forms of remotely sensed data.

2.4.4 Suspended Sediment Studies

Historical changes to tropical island watersheds often result in increased land-based pollution, including sediment, nutrients, and other pollutants that threaten many coral reefs worldwide. Terrestrial sediment is introduced to the nearshore environment through erosion and storm runoff, and while a natural background amount is expected in a healthy ecosystem, excessive amounts can lead to degradation of reefs (Field et al. 2008). Sediment that settles on coral can inhibit feeding, reduce recruitment, and decrease calcification rates. Fine sediment suspended in the water column can block light and thereby decrease sunlight needed for photosynthesis (Fabricius 2005).

Historical land-use changes on the Hawaiian island of Moloka'i have resulted in large amounts of sediment delivered from the uplands to the coast and deposited on the inner reef flat. This sediment is then resuspended in a daily cycle of tides, wind, and waves. In 2005, the U.S. Geological Survey used digital aerial photography, along with LiDAR bathymetry data and in situ water sampling, to quantitatively map suspended sediment concentrations in the top 0.5 m of the water column on the south shore of Moloka'i (Cochran et al. 2008).

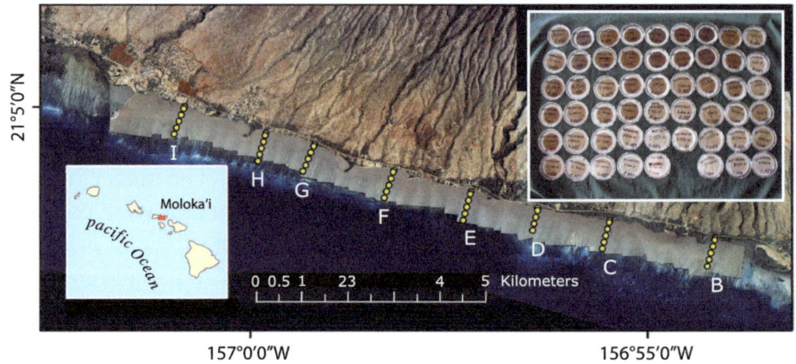

Fig. 2.5 Map showing natural-color mosaic created from aerial video stills and location of water samples collected 7 April 2005; overlain on 1999 aerial photography of the south coast of the island of Moloka'i. Water samples from transect A were not used in this analysis as they fall outside the boundaries of the mosaic. *Inset upper right*: photo of sediment-laden filters displayed as transects I (*left*) to A (*right*), from nearshore (*top*) to offshore (*bottom*) (Cochran et al. 2008)

A digital camera system was used to acquire high-resolution digital images at a resolution of approximately 15 cm per pixel. The camera system was set to collect data in three bands: 455–565 nm (blue/green), 560–640 nm (red), and 760–900 nm (near IR). Mounted on a helicopter, the images were collected from an altitude of 1,500–2,000 ft (457–609 m), minimizing the need for post-processing due to atmospheric conditions. A high-resolution digital video camera was mounted next to the multispectral camera to simultaneously record natural-color video. Multiple in-water teams collected in situ water samples from the upper 0.5 m of the water column during the overflight, and the sample locations were marked using hand-held GPS units. Water samples were filtered in the laboratory to determine the suspended sediment concentrations and total sediment mass per sample.

Although no post-processing for atmospheric corrections was necessary because of the low-altitude collection, a geometric correction was applied in order to spatially match and mosaic the multispectral photos to each other. The process was repeated on overlapping still images pulled from the natural-color video to create a second mosaic to assist with image interpretation and visualization. The resulting mosaics were then further corrected spatially using existing high-resolution LiDAR bathymetry, choosing image-to-image control points and "slaving" the mosaics onto the LiDAR master (Fig. 2.5).

In order to calibrate the image mosaic, suspended sediment concentration (SSC) values from the in situ water samples were plotted against the digital numbers (DN values) of various image bands, band combinations, and band ratios. Several types of regressions were applied to the various plots, with results indicating that a exponential regression applied to the ratio of Band 1 (blue/green) to Band 2 (red) resulted in the best fit ($r^2 = 0.75$, $p < 0.001$) (Fig. 2.6). This exponential regression equation was applied to the entire mosaic image resulting in a digital map representing the SSC in the upper 0.5 m of the water column (Fig. 2.7a).

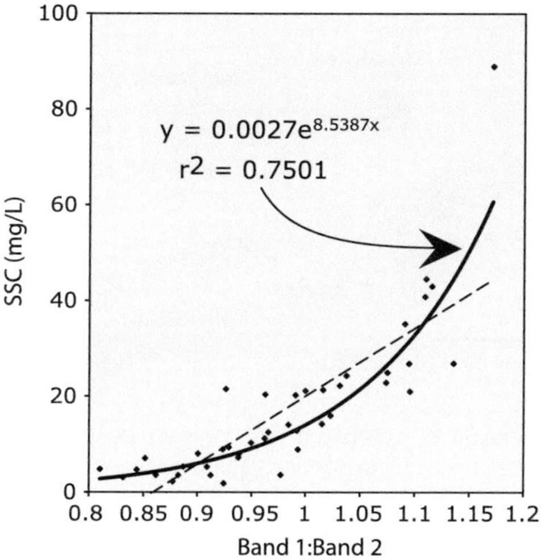

Band Combination	Exponential regression r2 value	Linear regression r2 value
B1 (BlueGreen)	0.1576	0.4007
B2 (Red)	0.7433	0.4754
B3 (NIR)	0.7495	0.4664
Grayscale	0.6711	0.0518
B1:B2	0.7501	0.6587
B1:B3	0.7457	0.6587
B2:B1	0.7341	0.5998
B2:B3	0.0706	0.0225
B3:B1	0.7118	0.5415
B3:B2	0.0776	0.0266
B1+B2+B3	0.6386	0.3747

Fig. 2.6 (*left*) Graph of suspended sediment concentrations plotted against a ratio of Band 1 (*blue/green*) to Band 2 (*red*), resulting in the best-fit exponential regression. This equation was then applied to the entire image mosaic for further analyses. (*right*) Results of exponential and linear regression r^2 values for SSC to various image band combinations (Cochran et al. 2008)

LiDAR bathymetric data were used to calculate the volume of water in the upper 0.5 m of the water column in the study area. The data were first resampled to the same resolution as the image mosaic (Fig. 2.7b). For depths greater than 0.5 m, the data were truncated to create a false bottom, thus representing the top 0.5 m of the water column. Standard GIS raster calculations were used to create a layer representing the volume of water per pixel (Fig. 2.7c). The SSC layer was divided pixel-by-pixel by the Water Volume layer to determine the mass of sediment in the upper 0.5 m water column per pixel (Fig. 2.7d). By summing all the pixels in the sediment mass layer, calculations showed a total of nearly 120 kg of suspended sediment in the upper 0.5 m of the water column in the study area (covering nearly 10 km^2), with an average concentration of 13.4 mg/L. This has important implications, as concentrations greater than 10 mg/L have been shown to be deleterious to coral reef systems (Rogers 1990).

The combination of aerial imagery and in situ sampling provides a method to calculate the concentration and total mass of suspended sediment in the upper water column and has particular application for creating "snap-shots" of turbidity levels on remote or large reef tracts. Note that this approach allows interpolation of SSC from a limited number of in situ measurements, and that different image band combinations may work better in different study areas or seasons and should be tested accordingly. Comparison of levels and distribution of SSC from two different time periods also provides a potential technique for monitoring long-term changes.

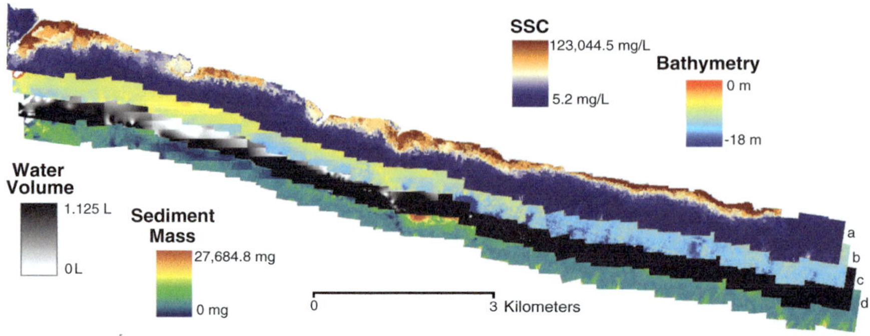

Fig. 2.7 **a** Map layer showing suspended sediment concentration (SSC) per pixel, **b** LiDAR bathymetry layer used to determine bottom for water column volume calculations, **c** map layer showing volume of water in upper 0.5 m of water column per pixel, and **d** map layer showing total mass of sediment for the upper 0.5 m of the water column in each pixel (Cochran et al. 2008)

2.5 Conclusions and Future Directions

Aerial photography provides a unique tool for coral reef management applications. Its on-demand nature allows managers to plan data collection during optimum solar conditions, or during or following specific events (e.g., storms, bleaching, etc.), and the low relative cost for high spatial resolution images is within range of many project budgets. While case-specific, other high-resolution data acquisitions may cost anywhere from two to three times as much depending on the sensor or logistical issues (Mumby et al. 1999, 2000). The case studies presented in this chapter represent the broad range of uses of aerial photography in coral reef management, from spatial mapping to establish baseline conditions or monitor change and track changes over time, to using spectral information inherent in digital images to measure water conditions or other environmental parameters.

As technology advances in the collection and computer analysis of aerial photography, more and more studies are acquiring images digitally, thus eliminating the work of developing and scanning hard-copy photographs. Many studies that historically used film-based aerial photography as a research analysis tool are now using advanced digital techniques, as well as transitioning to multispectral and hyperspectral images as these types of remotely sensed images become more readily available and within budget (for more details, see Chaps. 3 and 4). Using GIS software and combining digital aerial images with other remotely sensed or field-collected data layers gives scientists and coral reef managers one more tool to assist with decision-making.

Acknowledgments This contribution was supported by the U.S. Geological Survey's Geologic Processes on Pacific Coral Reefs Project. Pat Chavez and Rian Bogle (USGS Flagstaff, AZ) collected the digital photographs and natural-color video for the Moloka'i sediment study.

Ann Gibbs, Amy Foxgrover, and Curt Storlazzi (USGS Santa Cruz, CA) contributed excellent suggestions and a timely review of this work, as did the book editors.

Suggested Reading

Aber JS, Marzolff I, Ries J (2010) Small-format photography: principles, techniques and geoscience applications. Elsevier, Amsterdam

Berlin GLL, Avery TE (2003) Fundamentals of remote sensing and airphoto interpretation, 6th edn. Prentice Hall, Upper-Saddle River

Paine DP (1981) Aerial photography and image interpretation for resource management. Wiley, New York

Paine DP, Kiser JD (2003) Aerial photography and image interpretation, 2nd edn. Wiley, New York

References

Andréfouët S, Robinson JA (2003) The use of Space Shuttle images to improve cloud detection in mapping of tropical coral reef environments. Int J Remote Sens 24:143–149

Andréfouët S, Berkelmans R, Odriozola L, Done T, Oliver J, Müller-Karger F (2002) Choosing the appropriate spatial resolution for monitoring coral bleaching events using remote sensing. Coral Reefs 21:147–154

Armstrong RA (1981) Changes in a Puerto Rican coral reef from 1936 to 1979 using aerial photoanalysis. In: Proceedings of the 4th international coral reef symposium 1, pp 309–316

Battista TA, Costa BM, Anderson SM (2007) Shallow-water benthic habitats of the Republic of Palau. NOAA, Silver Spring

Beisl U, Woodhouse N (2004) Correction of atmospheric and bidirectional effects in multispectral ADS40 images for mapping purposes. In: Proceedings of the 20th congress ISPRS, Istanbul

Beisl U, Woodhouse N, Lu S (2006) Radiometric processing scheme for multispectral ADS40 data for mapping purposes. In: Annual conference on ASPRS, Reno

Benfield SL, Guzman HM, Mair JM, Young JAT (2007) Mapping the distribution of coral reefs and associated sublittoral habitats in Pacific Panama; a comparison of optical satellite sensors and classification methodologies. Int J Remote Sens 28:5047–5070

Brock JC, Clayton TD, Nayegandhi A, Wright CW (2004) LIDAR optical rugosity of coral reefs in Biscayne National Park, Florida. Coral Reefs 23:48–59

Chauvaud S, Bouchon C, Maniére R (1998) Remote sensing techniques adapted to high resolution mapping of tropical coastal marine ecosystems (coral reefs, sea grass beds and mangrove). Int J Remote Sens 19:3625–3639

Chavez PS, Isbrecht J, Velasco MG, Sides SC, Field ME (2000) Generation of digital image maps in clear coastal waters using aerial photography and laser bathymetry data, Moloka'i, Hawai'i. In: Saxena NK (ed) Recent advances in marine science and technology, 2000. PACON, Honolulu

Cochran SA, Gibbs AE, Logan JB (2007) Geologic resource evaluation of Pu'ukoholā Heiau National Historic Site, Hawai'i Part II Benthic habitat mapping. USGS, California

Cochran SA, Chavez PS, Isbrecht J, Bogle RC (2008) Mapping sediment concentration on a fringing coral reef using airborne multispectral remote sensing and in situ sampling. In: Proceedings, Moloka'i, Hawai'i. Ocean Sci, Orlando

Coyne MS, Battista TA, Anderson M, Waddell J, Smith W, Jokiel P, Kendall MS, Monoco ME (2003) Benthic habitats of the Main Hawaiian Islands. NOAA, Silver Spring

Daunier H (1862) Le Boulevard, http://www.brooklynmuseum.org, last accessed 24 Jan 2012
Ekebom J, Erkkilä A (2003) Using aerial photography for identification of marine and coastal habitats under the EU's habitats directive. Aquatic Conserv Mar Freshw Ecosyst 13:287–304
Fabricius KE (2005) Effects of terrestrial runoff on the ecology of corals and coral reefs; review and synthesis. Mar Poll Bull 50:125–146
Fairbridge RW, Teichert C (1948) The low Isles of the Great Barrier Reef; a new analysis. Geogr J 111:67–88
Field ME, Calhoun RS, Storlazzi CD, Logan JB, Cochran SA (2008) Sediment on the Moloka'i reef. In: Field ME, Cochran SA, Logan JB et al (eds) The coral reef of south Moloka'i, Hawai'i—portrait of a sediment-threatened fringing reef. USGS, California
Fletcher C, Rooney J, Barbee M, Lim SC, Richmond B (2003) Mapping shoreline change using digital orthophotogrammetry on Maui, Hawaii. J Coast Res SI38:106–124
Hernandez-Cruz LR, Purkis SJ, Reigl BM (2006) Documenting decadal spatial changes in sea grass and *Acropora palmata* cover by aerial photography analysis in Vieques, Puerto Rico: 1937–2000. Bull Mar Sci 79:401–414
Hopley D (1978) Aerial photography and other remote sensing techniques. In: Stoddart DR, Johannes RE (eds) Coral reefs: research methods. UNESCO, Paris
Hopley D, Catt PC (1988) Use of near infra-red aerial photography for monitoring ecological changes to coral reef flats on the Great Barrier Reef. In: Proceedings of the 6th international coral reef symposium 3, pp 503–508
Kendall MS, Monaco ME, Buja KR, Christensen JD, Druer CR, Finkbeiner M, Warner RA (2001) Methods used to map the benthic habitats of Puerto Rico and the US Virgin Islands. NOAA, Silver Spring
Lewis JB (2002) Evidence from aerial photography of structural loss of coral reefs at Barbados, West Indies. Coral Reefs 21:49–56
Lillesand TM, Kieffer RW (1994) Remote sensing and image interpretation, 3rd edn. Wiley, New York
Lyzenga DR (1978) Passive remote sensing techniques for mapping water depth and bottom features. Appl Optics 17:379–383
Lyzenga DR (1981) Remote sensing of bottom reflectance and water attenuation parameters in shallow water using aircraft and Landsat data. Int J Remote Sens 2:71–82
Mapworks Manoa (1984) Molokai coastal resource atlas. USACE, Honolulu
Mumby PJ, Green EP, Edwards AJ, Clark CD (1999) Cost-effectiveness of remote sensing for coastal management. J Environ Manag 55:157–166
Mumby PJ, Green EP, Edwards AJ, Clark CD (2000) Cost-effectiveness of remote sensing for coastal management. In: Green EP, Mumby PJ, Edwards AJ et al (eds) Remote sensing handbook for tropical coastal management. UNESCO, Paris
Murase T, Tanaka M, Tani T, Miyashita Y, Ohkawa N, Ishiguro S, Suzuki Y, Kayanne H, Yamano H (2008) A photogrammetric correction procedure for light refraction effects at a two-medium boundary. Photogramm Eng Remote Sens 74:1129–1136
Naim O (1989) Les platiers recifaux de la Reunion: geomorphologie, contexte hydrodynamique et peuplements benthiques. Université de la Réunion, Laboratoire d'Écologie Marine, Agence d'Urbanisme de la Réunion
Naim O, Cuet P, Letourneur Y (1997) Experimental shift in benthic community structure. In: Proceedings of the 8th international coral reef symposium 2:1873–1878
National Oceanic and Atmospheric Administration (2003) Atlas of the shallow-water benthic habitats of the Northwestern Hawaiian Islands (Draft). NOAA, Silver Spring
NOAA National Centers for Coastal Ocean Science (2005) Shallow-water benthic habitats of American Samoa, Guam, and the Commonwealth of the Northern Mariana Islands. NOAA, Silver Spring
Palandro D, Andréfouët S, Dustan P, Müller-Karger FE (2003) Change detection in coral reef communities using Ikonos satellite sensor imagery and historic aerial photographs. Int J Remote Sens 24:873–878

PAPA International (2010) History of aerial photography. Professional aerial photographers association. http://www.papainternational.org. Accessed 24 Jan 2012
Robinson JA, Feldman GC, Kuring N, Franz B, Green E, Noordeloos M, Stumpf RP (2000) Data fusion in coral reef mapping: working at multiple scales with SeaWiFS and astronaut photography. In: Proceedings of the 6th international conference remote sensing for marine and coastal environment 2:473–483
Robinson JA, Amsbury DL, Liddle DA, Evans CA (2002) Astronaut-acquired orbital photographs as digital data for remote sensing: spatial resolution. Int J Remote Sens 23:4403–4438
Rogers CS (1990) Responses of coral reefs and reef organisms to sedimentation. Mar Ecol Prog Ser 62:185–202
Rützler K (1978) Photogrammetry of reef environments by helium balloon. In: Stoddart DR, Johannes RE (eds) Coral reefs: research methods. UNESCO, Paris
Scoffin TP (1982) Reef aerial photography from a kite. Coral Reefs 1:67–69
Scopélitis J, Andréfouët S, Largouët C (2007) Modelling coral reef habitat trajectories: evaluation of an integrated timed automata and remote sensing approach. Ecol Model 205:59–80
Scopélitis J, Andréfouët S, Phinn S, Chabanet P, Naim O, Tourrand C, Done T (2009) Changes of coral communities over 35 years: integrating in situ and remote-sensing data on Saint-Leu Reef (la Réunion, Indian Ocean). Est Coast Shelf Sci 84:342–352
Sheppard CRC, Matheson K, Bythell JC, Murphy P, Myers CB, Blake B (1995) Habitat mapping in the Caribbean for management and conservation: use and assessment of aerial photography. Aquatic Conserv Mar Freshw Ecosyst 5:277–298
Steers JA (1945) Coral reefs and air photography. Geogr J 106:223–238
Stephenson TA, Tandy G, Spender MA (1931) The structure and ecology of Low Islands and other reefs: scientific reports of the Great Barrier Reef expedition 1928–1929. Brit Mus Nat Hist 3:17–112
Tewinkel GC (1963) Water depths from aerial photographs. Photogramm Eng 29:1037–1042
Wang L, Sousa WP, Gong P (2004) Integration of object-based and pixel-based classification for mapping mangroves with IKONOS imagery. Int J Remote Sens 25:5655–5668
Wilkinson C (2008) Status of coral reefs of the world. GCRMN, Townsville
Yamano H, Kayanne H, Yonekura N, Kudo K (2000) 21 year changes of back sreef coral distribution: causes and significance. J Coast Res 16:99–110

Chapter 3
Multispectral Applications

Hiroya Yamano

Abstract Multispectral satellite sensors have been used widely for coral reef applications. There is a long history of multispectral image acquisition, second to photography, with data available at global coverage for relatively low cost. This chapter summarizes the history, observation targets, and image processing methods of multispectral remote sensing of coral reefs. A variety of applications are presented in a progression from image classification and mapping to monitoring and modeling. Insight into the future directions of multispectral coral reef imaging is then provided, with a focus on higher spatial resolution, higher spectral resolution, greater data acquisition capacity, and integration with other data sources.

3.1 Introduction

Large-scale, synoptic mapping and monitoring are a basic requirement for the management and conservation of coral reefs (Green et al. 2000; Newman et al. 2006). One of the most practical solutions for meeting this need is multispectral remote sensing. Multispectral sensors are typically broadband, with three or four 60–100 nm wide wavebands in the visible to near infrared regions of the electromagnetic spectrum. Although the broadband limits detailed discrimination of specific benthic features (e.g., coral, seagrass and macroalgae) that have similar reflectance characteristics (Hochberg et al. 2003), multispectral sensors have been used widely for mapping, monitoring and management of coral reefs, mostly

H. Yamano (✉)
Center for Environmental Biology and Ecosystem Studies,
National Institute for Environmental Studies, 16-2 Onogawa, Tsukuba,
Ibaraki 305-8506, Japan
e-mail: hyamano@nies.go.jp

J. A. Goodman et al. (eds.), *Coral Reef Remote Sensing*,
DOI: 10.1007/978-90-481-9292-2_3,

because of the long history of data acquisition, global coverage of coral reef areas, and relatively low cost. Multispectral satellite sensors are still evolving. Most notably, spatial resolution of the sensors is getting higher (i.e., smaller pixel sizes), contributing to higher habitat classification accuracy and more detailed mapping of spatial features on coral reefs (Mumby and Edwards 2002; Andréfouët et al. 2003).

Multispectral satellite remote sensing of coral reefs has approximately 40 years of history, dating back to the launch of the first Landsat satellite in 1972. The first application using multispectral imagery for coral reef mapping was achieved in the Great Barrier Reef of Australia using Landsat MSS data with spatial resolution of 80 m (Smith et al. 1975; Jupp et al. 1985), where analysis focused on products for approximating bathymetry and reef geomorphologic zones. Deployment of SPOT HRV and Landsat TM in the 1980s gave the scientific community access to data with improved spatial resolution of 20–30 m. Though the data contributed significantly to improved regional land cover mapping, mapping habitats on coral reefs was limited to several basic classes in the 1980s (Vercelli et al. 1988). During this period the potential for characterization and quantification of coastal tropical environments for management and planning was suggested (Loubersac and Populus 1986), but nonetheless Kuchler et al. (1988) concluded "that current technologies are unable to fulfill this goal because of their lack of refinement for coral reef environments." In the early 1990s, some studies demonstrated mapping and change detection using Landsat TM (e.g., Zainal et al. 1993; Ahmad and Neil 1994), but they did not include quantitative classification accuracy and the use of remote sensing approaches to management of coral reefs remained scarce.

Significant advances occurred in the late 1990s from both application and technology perspectives. Green et al. (1996) reviewed applications of remote sensing for the assessment of tropical coastal resources and pointed out the importance of comparing the capabilities of different sensors in order to aid managers' decisions in selecting a remote sensing technique. Quantitative evaluation of cost (inclusive of both image and processing costs), benefit and classification accuracy of several remote sensors was assessed by Mumby et al. (1997). A systematic classification scheme of benthic features for habitat mapping of coral reefs, which is widely used today, was also developed (Mumby and Harborne 1999). A series of studies serving as guidelines were summarized and included in a remote sensing handbook for tropical coastal management (Green et al. 2000).

Significant advances in technology were made by the Landsat 7 Enhanced Thematic Mapper (ETM+) and IKONOS sensors, which were both launched in 1999. Though the spatial resolution and spectral arrangement of Landsat ETM+ was the same as Landsat TM, the acquisition schedule specifically targeted coral reefs throughout the world (Arvidson et al. 2001). This meant that mapping all the coral reefs was now possible by satellite data (Andréfouët et al. 2006). IKONOS provided the first publicly-available commercial multispectral data with high spatial resolution (4 m), which delivered high classification accuracy and a scale suitable for mapping details within geomorphic zones (Mumby and Edwards 2002; Andréfouët et al. 2003).

More recently, availability of high spatial resolution satellites has been enhanced, and at the present there are ~10 readily available types of multispectral satellite data, whose costs and benefits have been rigorously evaluated in numerous peer-reviewed publications (Table 3.1). Enhanced data availability and guidelines for their ability are contributing to improved application to coral reefs.

Other types of multispectral satellite sensors relevant to coral reefs are the low spatial resolution sensors (0.2–1 km pixels) (Table 3.1). Satellites carrying these sensors were also launched as early as the 1970s, and these and more recent sensors have been successfully used for observing biological oceanographic variables such as chlorophyll concentration. Recent studies showed their effectiveness for understanding cause-and-effect for coral reef processes, because they provide information on the environment surrounding the reefs (Abram et al. 2003; Hu et al. 2003; Maina et al. 2008).

Table 3.1 Attributes and characteristics of multispectral satellite sensors

Sensor	Deployment years	Spatial resolution (m)	No. bands[a]	Revisit time (d)[b]	Price ($US) 50 km^2	Price ($US) 5,000 km^2	Accuracy ~6 classes[c]
Medium resolution							
Landsat MSS	1972–2012	80	3	16	0	Free	30[1]
Landsat TM	1982–2012	30	4	16	0	Free	60–75[1,2,3]
Landsat ETM+	1999–2003	30	4	16	0	Free	60–75[1,4]
Landsat OLI	2013–	30	5	16	0	Free	N/A
IRS LISS-III	1995–	24	3	24	600	600	N/A
SPOT HRV	1986–1996	20	3	26	1,680	3,360	50–55[1]
SPOT HRVIR	1998–	20	3	26	2,660	5,320	50–55[1]
SPOT HRG	2002–	10	3	26	1,428	7,560	N/A
Terra ASTER	1999–	15	3	16	120	240	60–65[5]
ALOS AVNIR2	2007–2011	10	4	46	500	1,000	70–75[6]
FORMOSAT-2	2004–	8	4	1	3,500	7,000	N/A
High resolution							
IKONOS	1999–	4	4	3	1,000	100,000	75–90[2,4,7]
KOMPSAT-2	2006–	4	4	28	375	37,500	N/A
QuickBird	2001–	2.5	4	1–3.5	700	70,000	80[8]
GeoEye-1	2008–	2	4	3	625	62,500	N/A
WorldView-2	2009–	2	7	1.1–3.7	1,450	145,000	N/A
Low resolution							
Nimbus-7 CZCS	1978–1986	825	5	1	0	Free	N/A
ADEOS OCTS	1996–1997	700	8	3	0	Free	N/A
OrbView-2 SeaWIFS	1997–2011	1130	8	1	0	Free	N/A
Terra MODIS	1999–	250–1000	2–9	1–2	0	Free	N/A

[a] Bands located in the visible to NIR wavelengths

[b] High resolution sensors are pointable, allowing flexibility in revisit time

[c] Habitat classification accuracy, where available, for approximately 6 classes: [1] Mumby et al. (1997); [2] Mumby and Edwards (2002); [3] Call et al. (2003); [4] Andréfouët et al. (2003); [5] Capoisini et al. (2003); [6] Ministry of the Environment (2008); [7] Maeder et al. (2002); [8] Mishra et al. (2006)

3.2 Multispectral Analysis and Classification

3.2.1 Types of Analysis

Remote sensing of coral reefs can be divided into two categories: direct and indirect (Andréfouët and Riegl 2004; Andréfouët et al. 2005; Table 3.2). Direct remote sensing is where the coral reef itself is the target of remote sensing, while indirect remote sensing refers to studies that focus on the oceanic and atmospheric environment surrounding the reef. Targets that can be measured by multispectral remote sensing are described in Mumby et al. (2004).

Direct remote sensing. The objectives of direct remote sensing are the biotic and morphologic features (Table 3.2), where benthic features such as habitats may be of primary interest to management. Multispectral satellite sensors with blue bands can differentiate three to six habitat classes with reasonable overall accuracy (60–75 %, Table 3.1). Factors affecting the classification accuracy are the spectral band arrangement of the sensor, spatial resolution of the sensor, and characteristics of the benthic features to be classified. The broad width of these sensor's spectral bands limits accurate discrimination of more specific benthic features, since many biotic features (coral, seagrass and macroalgae) have similar reflectance characteristics, and it is suggested that hyperspectral sensors can better differentiate such features (Hochberg et al. 2003). Instruments lacking blue bands, which penetrate water to see the bottom features, results in significant decrease in classification accuracy (Table 3.1). In addition, because of the high heterogeneity and spatial scale of reef features, spectral signatures in one pixel are often mixed, meaning that studies using medium spatial resolution sensors typically produce lower classification accuracy (Hochberg and Atkinson 2003; Fig. 3.1). Increasing the number of habitat categories also results in decreased classification accuracy (Mumby et al. 1997; Andréfouët et al. 2003). A compilation of results suggests a predictive relationship between overall accuracy versus number of classes ($r^2 = 0.63$) for Landsat ETM+ and for IKONOS ($r^2 = 0.82$) (Andréfouët et al. 2003).

Bleached corals show significantly higher reflectance than healthy corals in all wavelength regions (Holden and LeDrew 1998; Hochberg et al. 2003; Yamano and Tamura 2004), and thus occurrence of bleaching is expected to be detectable using multispectral remote sensing. In order to avoid spectral mixing in one pixel, Andréfouët et al. (2002a) suggested the use of high ($\sim$2 m) spatial resolution sensors to detect bleaching in one image. No studies using multispectral satellite remote sensing have succeeded in detecting coral mass spawning yet, though observation by airplanes and by Synthetic Aperture Radar (SAR) produced potentially promising results (Willis and Oliver 1990; Jones et al. 2006).

Morphologic features provide the basis for examining habitat distributions within and between coral reefs. The location of shallow reef areas is easily detected by multispectral satellite sensors, and many coral reefs exhibit a common,

Table 3.2 Observation targets that can be measured by multispectral satellite remote sensing and their applications

Target	Spatial Res.			Applications		
	High	Med.	Low	Mapping	Monitoring	Modeling
Direct remote sensing						
Coral species				not feasible [1]		
Coral/algal cover	x			see Chap. 4		
Habitat classes (<5)	x	x		Habitat mapping[2–9]	Habitat change[22–26]	Population dynamics[28–29]
Habitat classes (>5)	x*	x*		Resource inventory[10–11]	Reef productivity[27]	
				Reef productivity[12–15]		
				Habitat diversity[16–17]		
				MPA evaluation[18]		
				Biodiversity[19–21]		
Bleaching	x	x**			Bleaching detection[30–32]	
Structural complexity	x	x**			Habitat change[33]	
Geomorphology	x	x		Typology[34–35]		
				MPA planning[36]		
Location of shallow reefs	x	x	x**	MPA planning[37]		
				Risk assessment[38]		
Bathymetry	x	x		Mapping[39]		
Shoreline	x	x		Reef topography[40]	Shoreline change[41]	Sediment transport[42]
Indirect remote sensing						
Land use/change	x	x	x			
Sea surface temperature			x	see Chap. 12		
UV radiation			x		Bleaching potential[43]	Susceptibility[44]
PAR			x		Bleaching potential[43]	Susceptibility[44]
Light attenuation	x	x	x	Productivity[14]		
Cloud/dust cover	x***	x***	x		Coral decline[45]	
Sea level			x			
Salinity			x			

(continued)

Table 3.2 (continued)

Target	Spatial Res.			Applications		
	High	Med.	Low	Mapping	Monitoring	Modeling
Chl-a concentration	x***	x***	x	Connectivity[46]	Coral decline[47–48]	Susceptibility[44]
Algal blooms	x***	x***	x		Coral decline[47–48]	
Turbidity/suspended sediment	x***	x***	x	Connectivity[46]		Sediment transport[49]
Wind speed			x			Susceptibility[44]
Ocean circulation			x	Connectivity[46]		
Coastal circulation			x	Connectivity[46]		
Precipitation			x			

Notes on application capabilities: x = existing applications; x* = contextual editing is needed to achieve reasonable accuracy; x** = limited application because of lower spatial resolution; x*** = limited acquisition interval, which is insufficient for oceanographic and climatologic monitoring

References for mapping, monitoring and modeling applications: [1] Kutser and Jupp (2006); [1] Mumby et al. (1997); [2] Mumby and Edwards (2002); [3] Call et al. (2003); [4] Andréfouët et al. (2003); [5] Capoisini et al. (2003); [6] Ministry of the Environment (2008); [7] Maeder et al. (2002); [8] Mishra et al. (2006); [10] Andréfouët et al. (2004); [11] Andréfouët et al. (2009a); [12] Andréfouët and Payri (2000); [13] Brock et al. (2006); [14] Hochberg and Atkinson (2008); [15] Moses et al. (2009); [16] Mumby (2001); [17] Harborne et al. (2008); [18] Rioja-Nieto and Sheppard (2008); [19] Mumby et al. (2008); [20] Mellin et al. (2009); [21] Dalleau et al. (2010); [22] Dustan et al. (2001); [23] Palandro et al. (2003); [24] Palandro et al. (2008); [25] Schuyler et al. (2006); [26] Sharma et al. (2008); [27] Moses et al. (2008); [28] Riegl and Purkis (2005); [29] Scopelitis et al. (2007); [30] Elvidge et al. (2004); [31] Yamano and Tamura (2004); [32] Rowlands et al. (2008); [33] LeDrew et al. (2004); [34] Andréfouët et al. (2001a); [35] Yamano et al. (2006b); [36] Beger et al. (2006); [37] Mora et al. (2006); [38] Burke et al. (2011); [39] Stumpf et al. (2003); [40] Yamano (2007); [41] Webb and Kench (2010); [42] Yokoki et al. (2006); [43] Masiri et al. (2008); [44] Maina et al. (2008); [45] Shinn et al. (2000); [46] Andréfouët et al. (2002b); [47] Abram et al. (2003); [48] Hu et al. (2003); [49] Ouillon et al. (2004)

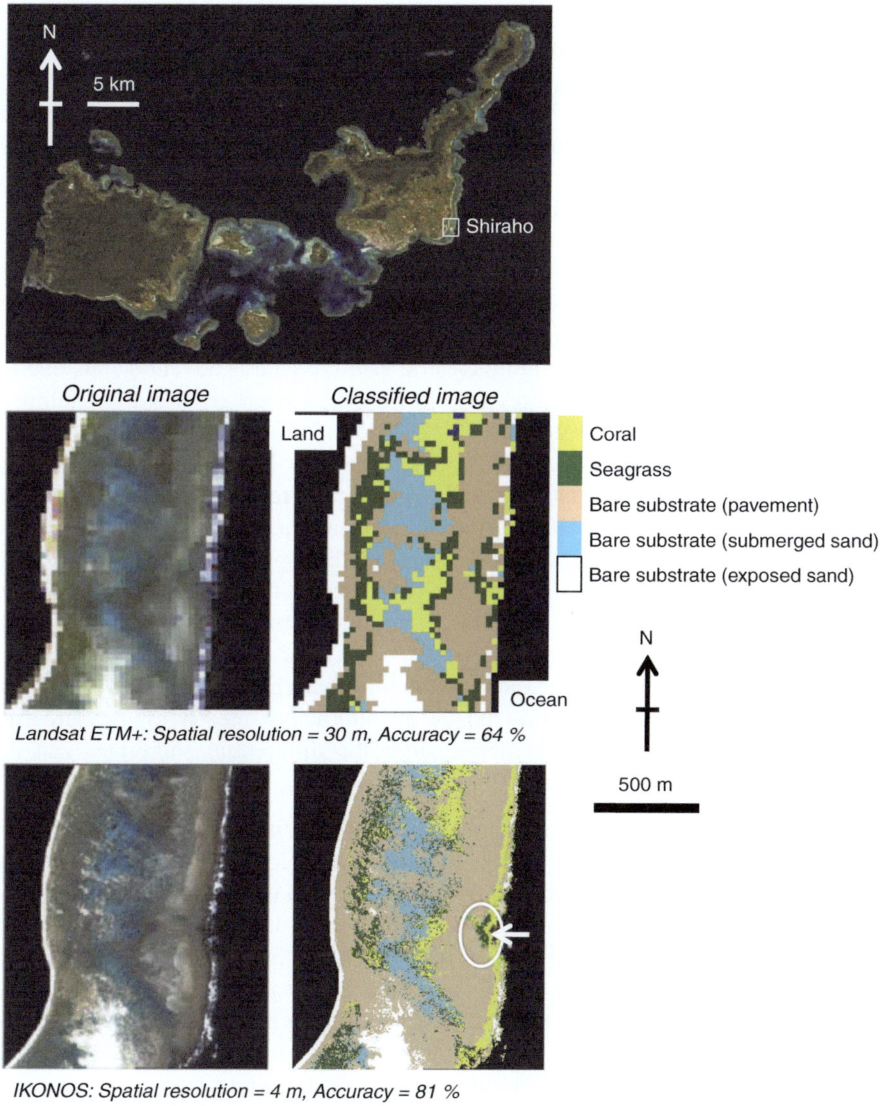

Fig. 3.1 Example of simple habitat classification using Landsat ETM+ and IKONOS images at Shiraho Reef, Ishigaki Island, Japan (24°22′N, 124°15′E). Higher spatial resolution delivers higher classification accuracy. To improve accuracy further, the habitat classified as "seagrass" in the ellipsoid indicated by the arrow on the IKONOS classified image can be corrected to "coral" using contextual editing (Andréfouët et al. 2003)

distinctive pattern of geomorphic zonation (i.e., forereef, reef crest, back reef, and lagoon with pinnacles), which may be easily visible and occur at spatial scales of tens to hundreds of meters. Investigating reef geomorphologic zonation (geomorphology) is one of the more successful applications of multispectral remote sensing to reef environments during its history from the 1970s to the present (Smith et al. 1975; Andréfouët et al. 2006).

Wavelength-dependency of light penetration offers the ability to estimate bathymetry of reefs, as it is possible to estimate water depths using multispectral data by tuning some parameters using ground truth data (e.g., light attenuation coefficients, water depth, and bottom albedo) (Philpot 1989). Recent applications of this method demonstrate that it is possible to discern water depths with root-mean-square-error (RMS) <30 % of true depths in <25 m of water (Stumpf et al. 2003). Although other methods to measure water depths are effective, and in some cases significantly more accurate (i.e., Section III Acoustic and Section II LiDAR), estimation of bathymetry using multispectral satellite data is advantageous due to its availability in otherwise inaccessible locations by boats or aircrafts equipped with acoustic or LiDAR sensors.

Extracting shorelines (i.e., the low and high tide limits) are important for examining dynamics of coasts. Shorelines in coral reef environments can be extracted successfully using NIR sensor bands, where lack of foam and suspended sediments (which can affect the NIR wavelength region) and the presence of remnant water on reef flats during low tide (which can affect the SWIR wavelength region) contribute to improved shoreline detection. Yamano et al. (2006a) found a strong linear relationship ($r^2 = 0.81$) between the error of shoreline position and spatial resolution of the NIR bands. Extracting waterlines under various water level conditions can also be utilized as an alternative means to measure the topography of shallow intertidal areas, since the differential shorelines can be regarded as contours (Yamano 2007).

Indirect remote sensing. The objective of indirect remote sensing is the environment surrounding the reefs (Table 3.2). For example, coastal land use affects coral reefs through watershed-based pollution. As its name suggests, Landsat has been designed for mapping land areas, and Landsat-derived global land cover estimates have been prepared (Robinson et al. 2006), which can be used to provide estimates of terrestrial inputs to reef ecosystems. Oceanic and atmospheric environments (e.g., photosynthetically active radiation, light attenuation coefficient, cloud/dust cover, chlorophyll concentration, algal bloom, turbidity/suspended sediment concentration, ocean circulation and coastal circulation) are generally large-scale and variable through space and time. As a result, coarse spatial resolution satellite sensors (e.g., MODIS, MERIS, SeaWiFS) with short observation intervals (~ 1 day) may be more feasible for observing these large-scale oceanic and environmental features, which change on a daily basis. In contrast, most high- and moderate-spatial resolution multispectral sensors have relatively long (~ 10 days) observation intervals, which limit the ability to examine change at a fine temporal scale.

3.2.2 Image Processing

The general procedures for image preprocessing (i.e., geometric correction, radiometric correction, atmospheric correction, and sunglint removal) were introduced in Chap. 1 and are summarized in Green et al. (2000). This section is focused on selected examples for the processing steps that follow the initial pre-processing, specifically atmospheric correction, water column correction, image classification and contextual editing.

Atmospheric correction. The total signal received by satellite sensors is dominated by radiance contributed through atmospheric scattering. Thus, atmospheric correction is essential to retrieve signals from the sea. The simplest such procedure is dark pixel subtraction. Based on an assumption that somewhere in the image is a pixel with zero reflectance, which means the radiance recorded by the sensor is solely from the atmospheric scattering, the minimum pixel value is subtracted from all other pixels to remove the radiance derived from atmospheric scattering. Because images with coral reefs normally contain ocean areas, pixels from deep ocean areas are often used for this correction. Modeling radiative transfer in the atmosphere is another more sophisticated option for correction. Mishra et al. (2005) described a procedure for first-order atmospheric correction for IKONOS imagery based on radiative transfer theory. Radiative transfer codes such as 6S (Vermote et al. 1997) and MODTRAN (http://www.modtran.org) may also be used, and some image processing software has optional modules for atmospheric correction (e.g., FLAASH for ENVI and ATCOR for ERDAS Imagine).

Water column correction. A fundamental challenge for remote sensing of coral reefs is the existence of the water column above the bottom features of interest (i.e., the benthic habitat). Light intensity decreases exponentially with increasing depth. This is known as attenuation, and the degree of attenuation is a function of wavelength, water depth and water properties. Water strongly absorbs wavelengths longer than 700 nm (near infrared), thus visible regions (blue, green and red) are used for mapping bottom features and infrared regions are used for extracting shorelines and emergent vegetation.

Rigorous removal of the water influence on bottom reflectance requires knowledge or estimations of water depth and attenuation characteristics at every pixel. Calculations using these parameters can be effectively used to estimate reef-top reflectance ("reef-up" approach) for substrate mapping (Purkis 2005). In cases where ground truth data (e.g., light attenuation coefficients, water depth, and bottom albedo) are available, wavelength-dependency of light penetration offers high ability to estimate bathymetry of reefs (e.g., Philpot 1989; Stumpf et al. 2003). Recently, Mishra et al. (2005) proposed a method to estimate water depths and attenuation characteristics using IKONOS data, with minimum requirements for ground truth data. They identified a ratio of wavebands (blue and green) that is constant for all bottom types, and developed a polynomial equation to estimate water depths from the ratio. Attenuation characteristics were then estimated by applying radiative transfer theory to deep-water pixels. The same method could be applicable to other multispectral satellite imagery with blue and green bands. In cases where no ground truth data are available, calculating a "depth invariant bottom index", which corrects for water column effect using pairs of multispectral bands instead of calculating bottom reflectance for each band (Lyzenga 1978), may be a practical option. This procedure is known to increase the mapping

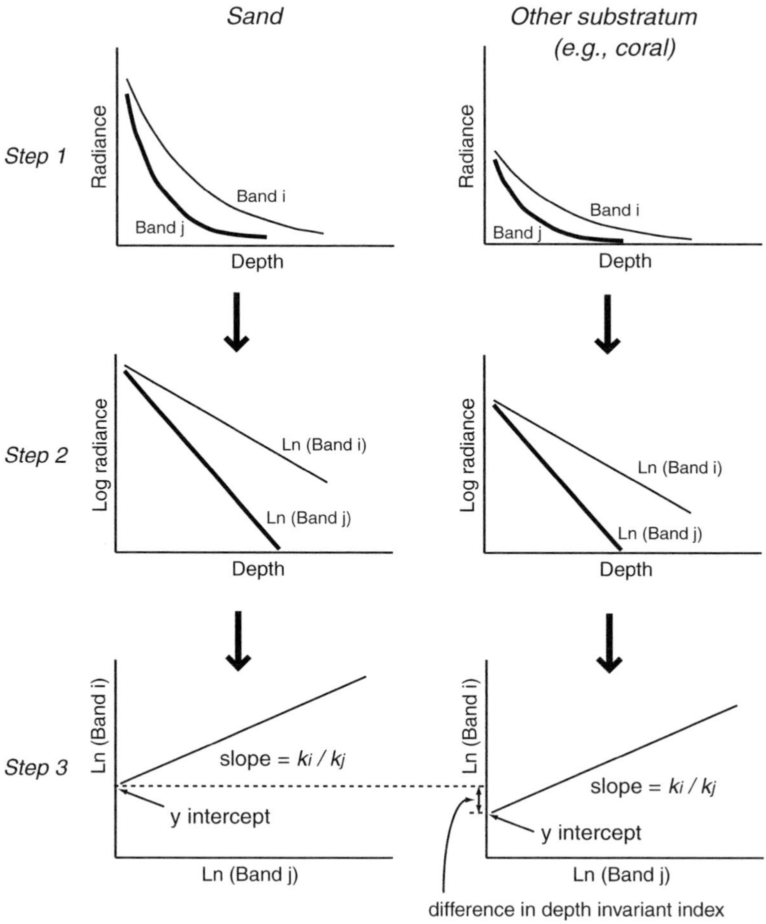

Fig. 3.2 Procedure for water column correction, showing the steps involved in creating depth invariant indices of bottom type for sand and seagrass. Here radiance L denotes atmospherically corrected radiance that is a result of subtracting deep-water (ocean) radiance from pixel radiance. (adapted from Green et al. 2000) (*Step* 1) Exponential attenuation of radiance with depth linearized for bands i and j using natural logarithm. (*Step* 2) Plot of transformed band i against transformed band j for a unique substratum at various depths. Gradient of line represents the ratio of attenuation coefficients, k_i/k_j. The ratio is the same irrespective of bottom type. (*Step* 3) Plot of multiple bottom types. Each bottom type has a unique y-intercept regardless of its depth. The y-intercept therefore becomes a depth invariant index of bottom type

accuracy (Mumby et al. 1998). Figure 3.2 shows an example of the process for invariant bottom index (Green et al. 2000).

Image classification. Benthic habitats can generally be categorized according to either hierarchical geomorphological or ecological classification schemes (Mumby and Harborne 1999; Andréfouët 2011; Table 3.3). Classification is accomplished using spectral, and in some instance textural, techniques, and can be divided into two basic categories, visual interpretation and digital processing.

Visual interpretation has been used since the era of aerial photography (Chap. 2). This approach involves the use of digitizing tablets to draw polygons around different geomorphologic units or habitats based on color (spectra) and texture. Though it suffers from the subjectivity of the operator, it can produce maps with high accuracy (Scopélitis et al. 2010).

Table 3.3 Class hierarchy for the benthic component in Caribbean coral reefs (after Mumby and Harborne 1999)

Level 1	Level 2	Level 3
Coral classes		
	Branching corals	
	Sheet corals	
	Ribbon and fire corals with green calcified algae	
	Massive and encrusting corals	
		Sparse massive and encrusting corals (1–5 % hard coral cover)
		Sparse massive and encrusting corals (1–5 % hard coral cover)
Algal dominated		
	Green algae	
	Fleshy brown algae and gorgonians	
	Lobophora	
	Euchema and *Amphiroa*	
Bare substratum dominated		
	Bedrock/rubble and dense gorgonians	
	Bedrock/rubble and sparse gorgonians	
	Rubble and sparse algae	
	Sand with sparse algae	
	Mud	
	Bedrock	
Seagrass dominated		
	Sparse seagrass	
	Medium density seagrass	
	Dense seagrass	
	Seagrass with distinct coral patches	

Digital processing includes image classification or segmentation based on spectral characteristics (i.e., digital number, radiance, or reflectance) and, in some cases, texture. Spectral processing discriminates features based solely on multispectral signatures, while texture approaches also incorporate spatial variability of the multispectral signatures. Unsupervised classification analyzes images without user input, and then the different segments are assigned to a given benthic category, or class, according to expert knowledge of the user. In supervised classification, ground-truth for each class (i.e., user supplied input) is used to train the classification scheme and identify these classes throughout the image. Classification using texture considers spatial patterns as a function of spectral variation within a particular area. Some substrate types (e.g., corals and macroalgae) may be spectrally similar but retain distinct texture properties by virtue of a systematic variance in spectral characteristics at a discrete spatial scale. Therefore, the boundary of ecotones can be detected (Andréfouët and Roux 1998), and analysis of the texture may thus enhance detection or classification accuracy (LeDrew et al. 2004; Purkis et al. 2006; Lim et al. 2009). Further, high spatial resolution is preferable to observe texture, and thus pan-sharpening of the image (Hanaizumi et al. 2008) may be helpful for improving classification output using texture.

Contextual editing. Some misclassification of habitat categories is inevitable in the digitally classified image. Accordingly, classification results can be edited to take account of known patterns of habitat distribution (Mumby et al. 1998). For example, pixels classified as seagrass, but present on a forereef slope where it is known that seagrass is absent, can be reclassified to the appropriate reef categories (Fig. 3.1). In addition to this type of a posteriori contextual editing, a priori contextual editing is another attractive option (Andréfouët 2008). As indicated in Bouvet et al. (2003) and Andréfouët et al. (2003), accuracy is enhanced if different areas containing different habitats are processed separately. In these situations, it is recommended that this a priori segmentation be based on geomorphology, because geomorphological zonation is related to depth and wave exposure, which also affects the distribution of habitats.

3.2.3 Time-Series Analysis

The recent rapid decline of coral reefs (e.g., Gardner et al. 2003) raises the importance of time-series analysis of historical images to examine the timing and extent of this decline. Two approaches have been proposed for change detection using multi-temporal images (Lunetta and Elvidge 1998), post-classification and pre-classification. The post-classification approach involves the analysis of differences among multi-temporal output products (e.g., habitat maps). The change detection accuracy between the multi-temporal maps is a product of the accuracies of the two maps. Therefore, availability of in situ data for past images is essential to achieve high accuracy for change detection. If only recent in situ data is available, an effective method is to perform calibration and normalization of the images in order to use the training of recent in situ data to assist with classification for all the images (Palandro et al. 2008). The pre-classification approach involves the analysis of change in the actual spectral signatures or spectral indices (Matsunaga et al. 2000; Dustan et al. 2001; Yamano and Tamura 2004). Most spectral change identification techniques require image calibration or normalization to detect spectral differences between pixels in multi-temporal images. For example, normalization using the depth invariant index (Lyzenga 1978) is often insensitive to variations in overall intensity since the index is based on log-transformed values (Fig. 3.2). Direct comparison of spectral values can be achieved by normalizing the effects of changes in the atmosphere, incident light, water depth (tide) and sensor response, and can be accomplished using models and/or known values for pixels of shallow sand, deep-water or the object of interest (e.g., coral) (Yamano and Tamura 2004).

In order to detect long-term changes (>10 years), it is preferable to select images taken in the same season, because macroalgae, which is often difficult to be distinguished from corals due to similar reflectance characteristics, exhibits seasonal changes in distribution and abundance. Additionally, because pixel-based change detection has high sensitivity to spatial mis-registration of the pixels, the RMS errors in geometric correction should be smaller than 0.5 pixels.

3.3 Example Applications

For effective management of coral reefs, Phinn et al. (2006) suggests analysis should follow a progression of knowledge: mapping, monitoring and modeling. Mapping provides baseline surveys or inventories. Monitoring can be achieved through comparison of baseline maps against updated information, enabling changes to be mapped and measured. Modeling includes data integration to make statistical or physics-based links between environmental variables and coral reef processes, enabling prediction of system response to certain environmental conditions. In this section, several studies based on this approach are discussed. Table 3.2 presents a summary of other applications relevant to coral reef management.

3.3.1 *Reef Mapping*

Maps produced using multispectral imagery include: geologic, geomorphologic and sedimentary features (Rankey 2002; Naseer and Hatcher 2004; Purkis et al. 2010); and ecological habitats (Mumby et al. 1997; Andréfouët et al. 2003). These maps not only serve as base maps to examine reef structure and resource inventory but also allow estimation of ecological functions and biodiversity. An overview of two different categories of reef mapping applications are presented below, reef-scale mapping and regional to global-scale mapping.

Reef-scale mapping. Reef-scale maps provide information on habitat structure and composition at a local scale (Fig. 3.1). These maps can be used to assess effectiveness of marine protected areas (MPAs) (Rioja-Nieto and Sheppard 2008), as well as provide estimates of habitat diversity (Mumby 2001; Harborne et al. 2006). Although species distribution and biodiversity cannot be measured directly, such parameters can be inferred from habitat maps (Mumby et al. 2008). Maps with the highest habitat thematic complexity provide better surrogates for inferred parameters than simpler maps (Dalleau et al. 2010), and are more robust to changes in spatial scales. The emergence of high spatial resolution sensors also allows more detailed mapping of reef habitats and associated parameters.

While reef-scale maps for geological and ecological features are commonly produced to estimate resource inventory, maps are not limited to just habitat estimations (e.g., coral, seagrass and macroalgae). Successful applications also include stock assessment of giant clam (Andréfouët et al. 2009b) and invasive brown algae (Andréfouët et al. 2004). They are also used to assess species-habitat correlations, with varying degrees of success. A review of nine studies showed significant relationships between fish parameters (e.g., species richness, total abundance and biomass) and habitats derived from remote sensing images (e.g., geomorphology, benthic habitat, rugosity and depth) (Mellin et al. 2009). However, the studies suggested no clear conclusions or generic rules for fish assemblages across different scales.

Regional to global-scale mapping. This level of mapping enables large-scale, multi-site mapping of coral reefs (Andréfouët et al. 2006; Purkis et al. 2007). Such maps can be used to characterize the landscape structure and composition of reefs, which can in turn be linked to environmental and human impacts. Andréfouët et al. (2001a) classified regions in the Tuamotu Archipelago based on landscape parameters derived from maps produced by SPOT HRV images. It was shown that reef structure was related primarily to exposure to ocean swells. Yamano et al. (2006b) followed the same procedure, but using Landsat ETM+, for the atoll rim in the Marshall Islands, and found human settlements were correlated to reef characteristics, suggesting the approach could be used to assess vulnerability of the islands to environmental change.

Currently the largest-scale mapping product is the Millennium Coral Reef Mapping Project (MCRMP) (Andréfouët et al. 2006; Fig. 3.3), which mapped global geomorphic units of coral reefs using Landsat ETM+ data (Arvidson et al. 2001) (eol.jsc.nasa.gov/reefs/Overview2003/mill.htm). Other examples of large-scale habitat maps include the Japanese Ministry of the Environment coral reef maps (coralmap.coremoc.go.jp/sangomap_eng/index.html), and NOAA benthic map products (www.soest.hawaii.edu/pibhmc/). Data from these large-scale maps have been used to examine regional and global conservation status of coral reefs. Mora et al. (2006) compared MCRMP products with the location of marine protected areas (MPAs), and revealed that only 2 % of the world's coral reefs are within MPAs. Reef geomorphology maps derived from Landsat ETM+ have also been integrated into reserve selection software to identify priority sites for conservation to set up regional MPAs (Beger et al. 2006).

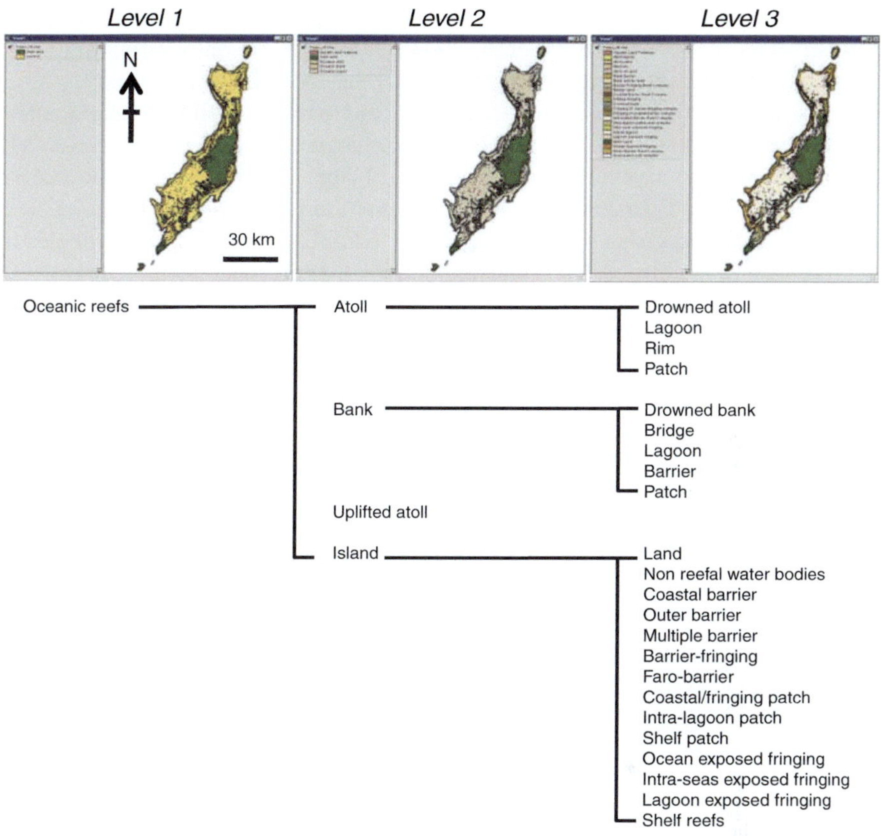

Fig. 3.3 Example of hierarchical geomorphic mapping using Landsat ETM+ images for Palau (7°24′N, 134°32′E), provided by Millennium Coral Reef Mapping Project. Three levels are presented (see Andréfouët et al. 2006 and Andréfouët 2011 for details)

3.3.2 Change Detection

Because decadal-scale degradation of coral reefs has been primarily observed using in situ data (Gardner et al. 2003), scaling up these observations to understand the extent of reef decline at local, regional and global scales is an urgent task. Analysis of aerial photography is one technique that has been used to detect decadal changes of coral reefs back to the 1930s (Chap. 2). Photography provides high spatial resolution and high thematic accuracy can be achieved, however the acquisition interval is typically long (>5 years) and the ability to simultaneously assess associated environmental change is limited. Multispectral satellite remote sensing overcomes these issues, since reefs can be monitored at frequent intervals (i.e., less than monthly if clouds do not obscure observation of target area) and the same data can be used to investigate other environmental parameters (Table 3.2).

Although there are some examples of change detection on coral reefs, (Loubersac et al. 1988; Elvidge et al. 2004), decadal-scale observation and analysis is still scarce. The Landsat program, which has been collecting imagery since 1972, is the best suite of sensors to achieve this purpose (Table 3.1). Radiative transfer simulation to assess the feasibility for change detection by Landsat ETM+ showed that the assessment of the rates of change in three ubiquitous classes 'sand', 'background' (including rubble, pavement, and heavily grazed dead coral structure), and 'foreground' (including living corals and macroalgae) emerges as the most reproducible and feasible application (Andréfouët et al. 2001b). For example, Dustan et al. (2001) explored the use of Landsat TM to detect changes in Florida. Palandro et al. (2008) extended their study, calculating absolute reflectance at the bottom, and performing classification using four classes (sand, bare hard-bottom, covered hard-bottom, and coral) to ensure high classification accuracy. Results confirmed consistent degradation of coral reefs in Florida (Fig. 3.4).

In addition to observing changes in coral habitats, time-series imagery can also be utilized to detect coral reef bleaching. Analysis of aerial photographs taken during the 1998 bleaching event show that information on bleached corals can be obtained only by sensors with high (<2 m) spatial resolution (Andréfouët et al. 2002a). A further complication is that analysis of a single image might not correctly identify pixels that contain both healthy and bleached corals, because pixels that contain other substrate features, such as sand, can have reflectance characteristics similar to those pixels showing partially bleached corals. Comparison of multi-temporal images thus allows more accurate assessment for detecting bleached corals in a pixel, because an increase in reflectance will be recorded where bleaching has taken place. Yamano and Tamura (2004) used 16 normalized Landsat TM images from 1984–2000 to document the extent of severe bleaching in 1998. Analysis of the same imagery was also used to validate radiative transfer simulations that established quantitative limits for detecting coral reef bleaching by satellite sensors.

3.3.3 Reef Modeling

Modeling is an effective tool to understand key processes and to make predictions and forecasting possible. The contribution of remote sensing to modeling application can be divided into two categories: (1) validation and (2) input.

Remote sensing results can provide validation of modeling outputs and can thereby be used to improve modeling accuracy. Examples include: estimation of population size parameters in population dynamics modeling using an IKONOS habitat map (Riegl and Purkis 2009); assessing habitat change using IKONOS imagery for validation of cyclone trajectory modeling (Scopélitis et al. 2007); validation of sediment transport modeling based on accreted and eroded areas in atoll islands identified by comparing maps and IKONOS data (Yokoki et al. 2006); and tuning a local erosion rate coefficient for sediment transport modeling based

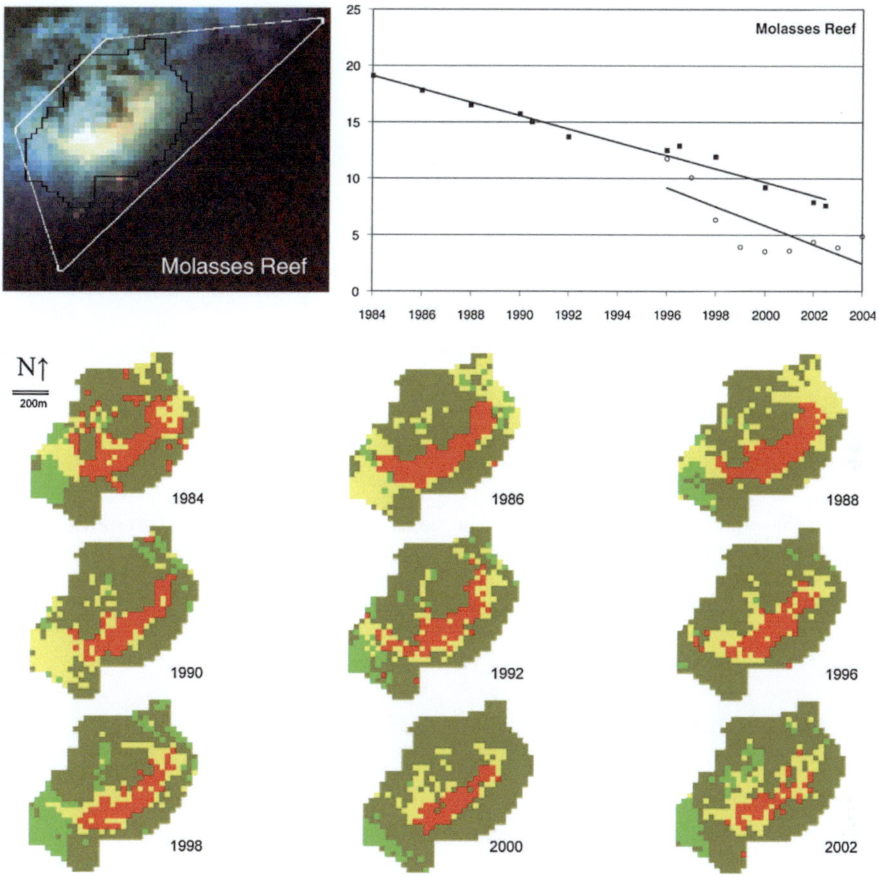

Fig. 3.4 Classified dataset using Landsat TM/ETM+ images for Molasses Reef, Florida Keys (25°00′N, 80°24′W) from 1984 to 2002. Graph trendlines of percent coral cover estimated from the classified images (■) and ground-truth data (¤) showed high consistency. Classification color codes are: *red* coral habitat, *brown* covered hardbottom, *yellow* bare hardbottom and *green* sand (Palandro et al. 2008)

on suspended sediment distribution derived from Landsat ETM+ imagery (Ouillon et al. 2004).

Variables measured or derived through remote sensing can also be input directly to modeling efforts and can also be used to estimate correlation coefficients in statistics-based modeling. For example, oceanographic and climatologic variables measured by NOAA AVHRR, OrbView-2 and SeaWiFS (SST, Chl-a concentration, etc.) have been used to predict bleaching and assess the susceptibility to bleaching (Maina et al. 2008; Fig. 3.5).

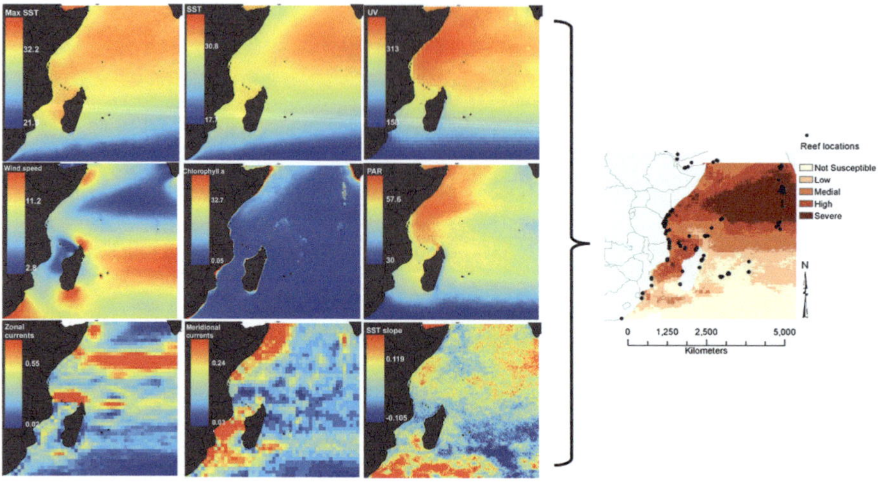

Fig. 3.5 Susceptibility map of coral reefs in the eastern Africa region (*right*) estimated from environmental parameters derived/aggregated from satellite images (*left*). Units for the respective layers are: sea surface temperature (SST) (°C); ultraviolet radiation (UV) (milli-watts/m2); chlorophyll concentration (mg/m3); wind speed (m/s); photosynthetically active radiation (PAR) (Einstein/m2/day); currents (m/s); and SST slope (°C/year) (Maina et al. 2008)

3.4 Conclusions and Future Directions

Multispectral sensors are the only current remote sensing option that provides information on mapping and monitoring of coral reefs at long time scales (>40 years) and large spatial scales (reef to global). Multispectral sensors that contain the important blue band can map basic benthic features with reasonable accuracy (>70 %), which can be improved by addition of texture information, water column correction and contextual editing.

Multispectral satellite sensors are still evolving. Newer instruments offer higher spatial resolution, higher spectral resolution, and greater data acquisition capacity. For example, WorldView-2, which was launched on 8 October 2009, has 1.85-m spatial resolution, 8 multispectral bands (coastal, blue, green, yellow, red, red edge, and two near infrared bands; Table 3.1), and daily data acquisition capacity of 950,000 km^2. As sensors evolve, efforts will continue towards evaluating the new sensors for different mapping applications (e.g., Mumby et al. 1997; Mumby and Edwards 2002; Andréfouët et al. 2003; Yamano et al. 2006a), and analysis techniques will also correspondingly progress and improve.

3.4.1 Integration with Other Sensor Modalities

In addition to using image-based techniques to enhance classification accuracy (e.g., texture information and contextual editing; see Sect. 3.2), integrating multispectral with other mapping technologies is also a promising option. For example, acoustic and optical instruments provide distinct but potentially complementary data regarding the nature of benthic communities (Riegl and Purkis 2005). Acoustic remote sensing can provide seabed roughness (rugosity), seabed hardness, and water depth (Chaps. 8–9). An evaluation in Glovers Atoll, Belize indicated that the accuracy of maps based on the depth invariant index using IKONOS was enhanced when three acoustic measures were added into the analysis (Bejarano et al. 2010). LiDAR remote sensing can also provide measures of seabed rugosity and water depth (Chaps. 5–7), suggesting another strong avenue for data integration. Further, acoustic and LiDAR remote sensing can extend observations of the seabed into more turbid and deeper waters, beyond where observation becomes limited with just optical remote sensing. For example, IKONOS imagery and acoustic remote sensing data were integrated to map sedimentary structure in a high-latitude reef-like setting in Cabo Pulmo, Mexico (Riegl et al. 2007). Similarly, in New Caledonia, reef-associated geomorphology was mapped using a combination of Landsat ETM+ multispectral remote sensing and acoustic multibeam observations (Andréfouët et al. 2009a).

Another benefit of data integration is the ability to utilize depth information to perform a water column correction of the optical bands (Purkis 2005; Bejarano et al. 2010). Bathymetry data can also be utilized as an additional data layer in spatial modeling. Garza-Perez et al. (2004) predicted reef bottom features using spatial modeling based on a combination of environmental data, IKONOS imagery and a digital topographic model. The maps generated by this procedure showed higher classification accuracy than maps generated using only traditional unsupervised classification of the IKONOS image. In all cases of data integration, the proper spatial alignment of different data sources is important to further enhance classification accuracy (Andréfouët and Clareboudt 2000; Andréfouët 2008).

These studies indicate the effectiveness of integrating data from multiple data sources. Additionally, the emerging importance of coral reefs at marginal (e.g., high-latitude, turbid, and moderately deep) settings as refugia for corals under stress from climate change will further enhance the need for the use of such integrated mapping techniques.

3.4.2 Integration with Field Monitoring

While in situ monitoring can cover small areas in superb detail, the measurements and observations can be unrepresentative when extended over larger areas. Linking remote sensing, which provide spatially extensive surveys, with in situ measurements can be very powerful. Consider Reef Check, which is one of the internationally recognized field assessment protocols. Substrate surveys using this protocol include cover types of live hard coral, dead coral, soft coral, fleshy seaweed, sponge, rock, rubble, sand, silt/clay and other. One approach for linking remote sensing data with such protocols is to classify the imagery into comparable cover types. For example, Joyce et al. (2004) classified Landsat ETM+ imagery of the southern Great Barrier Reef, Australia using the Reef Check substrate types, but with accuracies ranging from 12 to 74 % for individual reefs. Scopélitis et al. (2010) further explored the issue, showing that the combination of in situ and satellite data is best suited for mapping dominant coral morphologies and substrate types.

In addition to substrate type monitoring, Leiper et al. (2009) showed the possibility of linking remote sensing with CoralWatch data, which compares field observations of coral color with species-specific Coral Health Charts. The study showed bleached, medium and dark coral could be discriminated with 72.41 % overall accuracy, which extends the capabilities of previous studies focused on just analyzing bleached versus non-bleached corals (Andréfouët et al. 2002a; Elvidge et al. 2004).

These studies suggest the importance of integrating in situ monitoring protocols with remote sensing analysis, but reveal that research is still needed to better establish such links.

3.4.3 Integration with Modeling

The considerable advances in mapping, monitoring and modeling applications derived from multispectral remote sensing (Table 3.2) encourages not just integration with other sources of data but also interaction among the different applications. For example, a link between in situ monitoring, direct remote sensing measuring coral decline (Palandro et al. 2008) and indirect remote sensing measuring anomalous environmental variables (Shinn et al. 2000; Abram et al. 2003; Hu et al. 2003) could be used to identify the cause-and-effect of environmental change and perturbations (Purkis and Riegl 2005). Integration with susceptibility modeling and risk assessments is also possible. For instance, incorporating physical parameters and in situ observations with habitat change trajectories and monitoring information from direct remote sensing can be used to improve susceptibility modeling (Maina et al. 2008; Fig. 3.5).

Modeling efforts using remote sensing data can also be used to assess biodiversity (Fig. 3.6). Because habitat variability can be assumed a surrogate for biodiversity in coral reefs (Mumby et al. 2008; Dalleau et al. 2010), habitat decline may cause loss of biodiversity in coral reefs. For example, Sano et al. (1987) showed significant decrease in fish species numbers after the destruction of staghorn coral (*Acropora cervicornis*) communities due to *Acanthaster planci* infestation. This indicates that mapping habitat change using remote sensing,

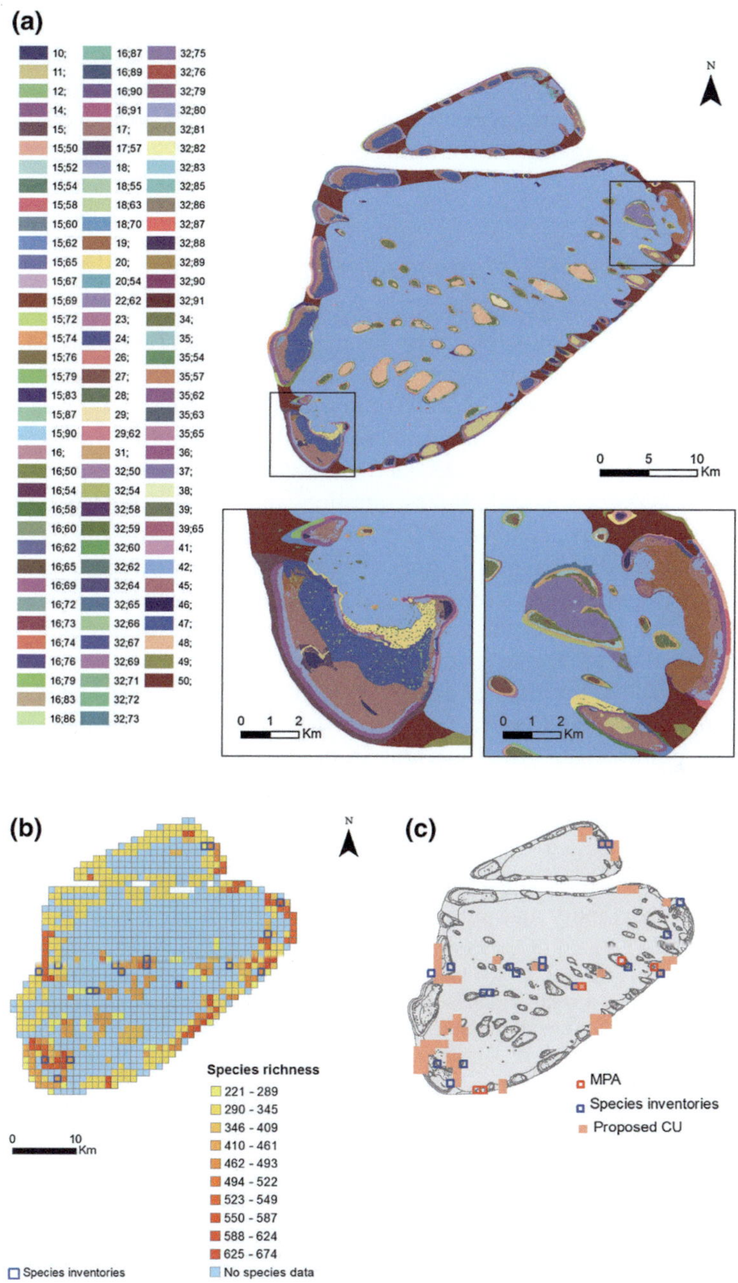

Fig. 3.6 **a** QuickBird-derived habitat maps for Baa Atoll, Republic of Maldives (5°09′N, 73°08′E) for geomorphology and benthic characteristics (Andréfouët et al. 2012). Each color corresponds to a different habitat. **b** Spatial distribution of the estimated species richness derived using combination of point biological census data (cells with a *dark blue border*) and the QuickBird habitat maps. **c** Proposed Conservation Units (*CU*) in Baa Atoll obtained with MPA selection software from the generalized maps of species richness (Hamel and Andrefouët 2012)

particularly loss of habitats such as coral reefs, can be linked with changes in fish diversity. The surrounding physical environment (e.g., SST) is also a good predictor of marine biodiversity (Tittensor et al. 2010). Therefore, integrating habitat surrogacy, habitat change trajectory, and physical variables can contribute to assessments of biodiversity on coral reefs.

3.4.4 Integration with Management

Remote sensing is an important tool in MPA planning (Dalleau et al. 2010). Mapping, in conjunction with recent advance in modeling development for MPA selection (e.g., Possingham et al. 2000) can contribute to more effective management of coral reefs (Hamel and Andréfouët 2012; Fig. 3.6). Furthermore, the integration of multiple, multi-scale data sources, including multispectral remote sensing data, is needed to address key questions in coral reef management (Andréfouët et al. 2005; Chabanet et al. 2005). One outstanding example is the "Reefs at Risk" program, which assesses the risks to global reefs by integrating data from multiple sources, including information on coastal development, watershed-based pollution, marine based pollution and damage, overfishing and destructive fishing and past thermal stress (Burke et al. 2011). In this program, remote sensing provides the reef location (Andréfouët et al. 2006), land use information and sea surface temperatures (SST), and ReefBase (www.reefbase.org) serves as the GIS platform to integrate all these data. ReefBase now includes other data sources such as MPA boundaries and simple habitat maps derived from ALOS AVNIR2. Expanded integration with other sources of data (Robinson et al. 2006), such as the Ocean Biogeographic Information System (OBIS) data for marine biodiversity (www.iobis.org/), is needed to continue improving our ability to assess ecosystem services and biodiversity and plan future management of coral reefs.

Acknowledgments I thank J.A. Goodman, S.J. Purkis and S.R. Phinn for invitation to this book project. This chapter is a contribution to the Biodiversity Priority Program at National Institute for Environmental Studies and to the Strategic Research and Development Project S-9 supported by the Environment Research & Technology Development Fund of the Ministry of the Environment, Japan.

Suggested Reading

Green EP, Mumby PJ, Edwards AJ, Clark CD (2000a) Remote sensing handbook for tropical coastal management. UNESCO, Paris

Miller RL, Del Castillo CE, McKee BA (2005) Remote sensing of coastal aquatic environments. Springer, Dordrecht

Mumby PJ, Skirving W, Strong AE, Hardy JT, LeDrew EF, Hochberg EJ, Stumpf RP, David LT (2004a) Remote sensing of coral reefs and their physical environment. Mar Pollut Bull 48:219–228

Palandro DA, Andréfouët S, Hu C, Hallock P, Müller-Karger FE, Dustan P, Callahan MK, Kranenburg C, Beaver CR (2008a) Quantification of two decades of shallow-water coral reef habitat decline in the Florida Keys National Marine Sanctuary using Landsat data (1984–2002). Remote Sens Environ 112:3388–3399

Richardson LL, LeDrew EF (2006) Remote sensing of aquatic coastal ecosystem processes. Springer, Dordrecht

References

Abram NJ, Gagan MK, McCulloch MT, Chappell J, Hantoro S (2003) Coral reef death during the 1997 Indian Ocean Dipole linked to Indonesian wildfires. Science 301:952–955

Ahmad W, Neil DT (1994) An evaluation of Landsat Thematic Mapper (TM) digital data for discriminating coral reef zonation: Heron Reef (GBR). Int J Remote Sens 15:2583–2597

Andréfouët S (2008) Coral reef habitat mapping using remote sensing: a user vs producer perspective. Implications for research, management and capacity building. J Spatial Sci 53:113–129

Andréfouët S (2011) Reef typology. In: Hopley D (ed) Encyclopedia of modern coral reefs—structure form and process. Springer, Dordrecht

Andréfouët S, Payri C (2000) Scaling-up carbon and carbonate metabolism of coral reefs using in situ data and remote sensing. Coral Reefs 19:259–269

Andréfouët S, Roux L (1998) Characterisation of ecotones using membership degrees computed with a fuzzy classifier. Int J Remote Sens 19:3205–3211

Andréfouët S, Clareboudt M (2000) Objective class definitions using correlation of similarities between remotely sensed and environmental data. Int J Remote Sens 21:1925–1930

Andréfouët S, Riegl B (2004) Remote sensing: a key tool for interdisciplinary assessment of coral reef processes. Coral Reefs 23:1–4

Andréfouët S, Zubia M, Payri C (2004) Mapping and biomass estimation of the invasive brown algae *Turbinaria ornata* (Turner) J. Agardh and *Sargassum mangarevense* (Grunow) Setchell on heterogeneous Tahitian coral reefs using 4-meter resolution IKONOS satellite data. Coral Reefs 23:26–38

Andréfouët S, Cabioch G, Flamand B, Pelletier B (2009a) A reappraisal of the diversity of geomorphological and genetic processes of New Caledonia coral reefs: a synthesis from optical remote sensing, coring and acoustic multibeam observations. Coral Reefs 28:691–707

Andréfouët S, Friedman K, Gilbert A, Remoissenet G (2009b) A comparison of two surveys of invertebrates at Pacific Ocean islands: the giant clam at Raivavae Island, Australes Archipelago, French Polynesia. ICES J Mar Sci 66:1825–1836

Andréfouët S, Clareboudt M, Matsakis P, Pagès J, Dufour P (2001a) Typology of atoll rims in Tuamotu Archipelago (French Polynesia) at landscape scale using SPOT HRV images. Int J Remote Sens 22:987–1004

Andréfouët S, Mumby PJ, McField M, Hu C, Muller-Karger FE (2002a) Revisiting coral reef connectivity. Coral Reefs 21:43–48

Andréfouët S, Hochberg EJ, Chevillon C, Muller-Karger FE, Brock JC, Hu C (2005) Multi-scale remote sensing of coral reefs. In: Miller RL, Del Castillo CE, McKee BA (eds) Remote sensing of coastal aquatic environments. Springer, Dordrecht

Andréfouët S, Muller-Karger FE, Robinson JA, Kranenburg CJ, Torres-Pulliza D, Spraggins SA, Murch B (2006) Global assessment of modern coral reef extent and diversity for regional science and management applications: a view from space. In: Proceedings of 10th international coral reef symposium, 1732–1745

Andréfouët S, Kramer P, Torres-Pulliza D, Joyce KE, Hochberg EJ, Garza-Pérez R, Mumby PJ, Riegl B, Yamano H, White WH, Zubia M, Brock JC, Phinn SR, Naseer A, Hatcher BG, Muller-Karger FE (2003) Multi-site evaluation of IKONOS data for classification of tropical coral reef environments. Remote Sens Environ 88:128–143

Andréfouët S, Muller-Karger FE, Hochberg EJ, Hu C, Carder KL (2001b) Change detection in shallow coral reef environments using Landsat 7 ETM+ data. Remote Sens Environ 78:150–162

Andréfouët S, Berkelmans R, Odrizola L, Done T, Oliver J, Müller-Karger F (2002b) Choosing the appropriate spatial resolution for monitoring coral bleaching events using remote sensing. Coral Reefs 21:147–154

Andréfouët S, Rilwan Y, Hamel MA (2012) Habitat mapping for conservation planning in Baa Atoll, Republic of Maldives. Atoll Res Bull 590:207–221

Arvidson T, Gasch J, Goward SN (2001) Landsat 7's long-term acquisition plan—an innovative approach to building a global imagery archive. Remote Sens Environ 78:13–25

Beger M, Jones GP, Possingham HP (2006) A method of statistical modelling of coral reef fish distribution: can it aid conservation planning in data poor regions? In: Proceedings of 10th international coral reef symposium, pp 1445–1456

Bejarano S, Mumby PJ, Hedley JD, Sotheran I (2010) Combining optical and acoustic data to enhance the detection of Caribbean forereef habitats. Remote Sens Environ 114:2768–2778

Bouvet G, Ferraris J, Andréfouët S (2003) Evaluation of large-scale unsupervised classification of New Caledonia reef ecosystems using Landsat 7 ETM+ imagery. Oceanol Acta 26:281–290

Brock J, Yates K, Halley R (2006) Integration of coral reef ecosystem process studies and remote sensing. In: Richardson LL, LeDrew EF (eds) Remote sensing of aquatic coastal ecosystem processes. Springer, Dordrecht

Burke L, Reytar K, Spalding M, Perry A (2011) Reefs at risk revisited. World Resources Institute, Washington DC

Call KA, Hardy JT, Wallin DO (2003) Coral reef habitat discrimination using multivariate spectral analysis and satellite remote sensing. Int J Remote Sens 24:2627–2639

Capolsini P, Andréfouët S, Rion C, Payri C (2003) A comparison of Landsat ETM+, SPOT HRV, Ikonos, ASTER, and airborne MASTER data for coral reef habitat mapping in South Pacific islands. Can J Remote Sens 29:187–200

Chabanet P, Adjeroud M, Andréfouët S, Bozec YM, Ferraris J, Garcìa-Charton JA, Schrimm M (2005) Human-induced physical disturbances and their indicators on coral reef habitats: a multi-scale approach. Aquat Living Resour 18:215–230

Dalleau M, Andréfouët S, Wabnitz CCC, Payri C, Wantiez L, Pichon M, Friedman K, Vigliola L, Benzoni F (2010) Use of habitats as surrogates of biodiversity for efficient coral reef conservation planning in Pacific Ocean islands. Conserv Biol 24:541–552

Dustan P, Dobson E, Nelson G (2001) Landsat Thematic Mapper: detection of shifts in community composition of coral reefs. Conserv Biol 15:892–902

Elvidge CD, Dietz JB, Berkelmans R, Andréfouët S, Skirving W, Strong A, Tuttle B (2004) Satellite observation of Keppel Islands (Great Barrier Reef) 2002 coral bleaching using IKONOS data. Coral Reefs 23:123–132

Gardner TA, Côté IM, Gill JA, Grant A, Watkinson AR (2003) Long-term region-wide declines in Caribbean corals. Science 301:958–960

Garza-Perez JR, Lehmann A, Arias-Gonzalez JE (2004) Spatial prediction of coral reef habitats: integrating ecology with spatial modeling and remote sensing. Mar Ecol Prog Ser 269:141–152

Green EP, Mumby PJ, Edwards AJ, Clark CD (1996) A review of remote sensing for the assessment and management of tropical coastal resources. Coast Manag 24:1–40

Green EP, Mumby PJ, Edwards AJ, Clark CD (2000b) Remote sensing handbook for tropical coastal management. UNESCO, Paris

Hamel MA, Andréfouët S (2012) Biodiversity-based propositions of conservation areas in Baa Atoll, Republic of Maldives. Atoll Res Bull 590:223–235

Hanaizumi H, Akiba M, Yamano H, Matsunaga T (2008) A pan-sharpened method for satellite image-based coral reef monitoring with higher accuracy. In: Proceedings of 11th international coral reef symposium, pp 626–630

Harborne AR, Mumby PJ, Zychaluk K, Hedley JD, Blackwell PG (2006) Modeling beta diversity of coral reefs. Ecology 87:2871–2881

Hochberg EJ, Atkinson MJ (2003) Capabilities of remote sensing to classify coral, algae, and sand as pure and mixed spectra. Remote Sens Environ 85:174–189
Hochberg EJ, Atkinson MJ (2008) Coral reef benthic productivity based on optical absorptance and light-use efficiency. Coral Reefs 27:49–59
Hochberg EJ, Atkinson MJ, Andréfouët S (2003) Spectral reflectance of coral reef bottom-types worldwide and implications for coral reef remote sensing. Remote Sens Environ 85:159–173
Holden H, LeDrew E (1998) Spectral discrimination of healthy and non-healthy corals based on cluster analysis, principal components analysis, and derivative spectroscopy. Remote Sens Environ 65:217–224
Hu C, Hackett KE, Callahan MK, Andréfouët A, Wheaton JL, Porter JW, Muller-Karger FE (2003) The 2002 ocean color anomaly in the Florida Bight: a cause of local coral reef decline? Geophys Res Lett 30:1151
Jones AT, Thankappan M, Logan GA, Kennard JM, Smith CJ, Williams AK, Lawrence GM (2006) Coral spawn and bathymetric slicks in Synthetic Aperture Radar (SAR) data from the Timor Sea, north–west Australia. Int J Remote Sens 27:2063–2069
Joyce KE, Phinn SR, Roelfsema CM, Neil DT, Dennison WC (2004) Combining Landsat ETM+ and reef check classifications for mapping coral reefs: a critical assessment from the souther Great Barrier Reef, Australia. Coral Reefs 23:21–25
Jupp DLB, Mayo KK, Kuchler DA, Claasen DVR, Kenchington RA, Guerin PR (1985) Remote sensing for planning and managing the Great Barrier Reef of Australia. Photogrammetria 40:21–42
Kuchler DA, Bina RT, van Classen DR (1988) Status of high-technology remote sensing for mapping and monitoring coral reef environments. In: Proceedings of 6th international coral reef symposium 1:97–101
Kutser T, Jupp DLB (2006) On the possibility of mapping living corals to the species level based on their optical signatures. Estuar Coast Shelf Sci 69:607–614
LeDrew EF, Holden H, Wulder MA, Derksen C, Newman C (2004) A spatial statistical operator applied to multidate satellite imagery for identification of coral reef stress. Remote Sens Environ 91:271–279
Leiper IA, Siebeck UE, Marshall NJ, Phinn SR (2009) Coral health monitoring: linking coral colour and remote sensing techniques. Can J Remote Sens 35:276–286
Lim A, Hedley JD, LeDrew E, Mumby PJ, Roelfsema C (2009) The effects of ecologically determined spatial complexity on the classification accuracy of simulated coral reef images. Remote Sens Environ 113:965–978
Loubersac L, Populus J (1986) The applications of high resolution satellite data for coastal management and planning in a Pacific coral island. Geocarto Int 2:17–31
Loubersac L, Dahl AL, Collotte P, Lemaire O, D'Ozouville L, Grotte A (1988) Impact assessment of Cyclone Sally on the almost atoll of Aitutaki (Cook Islands) by remote sensing. In: Proceedings of 6th international coral reef symposium 2:455–462
Lunetta RS, Elvidge CD (1998) Remote sensing and change detection environmental monitoring methods and applications. Ann Arbor Press, Chelsea, Michigan
Lyzenga DR (1978) Passive remote sensing techniques for mapping water depth and bottom features. Appl Opt 17:379–383
Maeder J, Narumalani S, Rundquist DC, Perk RL, Schalles J, Hutchins K, Keck J (2002) Classifying and mapping general coral-reef structure using IKONOS data. Photogram Eng Remote Sens 68:1297–1305
Maina J, Venus V, McClanahan TR, Ateweberhan M (2008) Modelling susceptibility of coral reefs to environmental stress using remote sensing data and GIS models. Ecol Model 212:180–199
Masiri I, Nunez M, Weller E (2008) A 10-year climatology of solar radiation for the Great Barrier Reef: implications for recent mass coral bleaching events. Int J Remote Sens 29:4443–4462
Matsunaga T, Hoyano A, Mizukami Y (2000) Monitoring of coral reefs on Ishigaki Island in Japan using multitemporal remote sensing data. Proc SPIE 4154:212–222

Mellin C, Andréfouët S, Kulbicki M, Dalleau M, Vigliola L (2009) Remote sensing and fish-habitat relationships in coral reef ecosystems: review and pathways for systematic multi-scale hierarchical research. Mar Pollut Bull 58:11–19

Ministry of the Environment (2008) Report on evaluation of coral reef mapping methods (FY H19). Ministry of the Environment, Tokyo

Mishra DR, Narumalani S, Rundiquist D, Lawson M (2005) High-resolution ocean color remote sensing of benthic habitats: a case study at the Roatan Island, Honduras. IEEE Trans Geosci Remote Sens 43:1592–1604

Mishra D, Narumalani S, Rundiquist D, Lawson M (2006) Benthic habitat mapping in tropical marine environments using QuickBird multispectral data. Photogram Eng Remote Sens 72:1037–1048

Mora C, Andréfouët S, Costello MJ, Kranenburg C, Rollo A, Veron J, Gaston KJ, Myers RA (2006) Coral reefs and the global network on marine protected areas. Science 312:1750–1751

Moses CS, Palandro DA, Andréfouët S, Muller-Karger F (2008) Remote sensing of changes in carbonate production on coral reefs: The Florida Keys. In: Proceedings of 11th international coral reef symposium, pp 62–66

Moses CS, Andréfouët S, Kranenburg C, Muller-Karger FE (2009) Regional estimated of reef carbonate dynamics and productivity using Landsat 7 ETM+, and potential impacts from ocean acidification. Mar Ecol Prog Ser 380:103–115

Mumby PJ (2001) Beta and habitat diversity in marine systems: a new approach to measurement, scaling and interpretation. Oecologia 128:274–280

Mumby PJ, Harborne AR (1999) Development of a systematic classification scheme of marine habitats to facilitate regional management and mapping of Caribbean coral reefs. Biol Conserv 88:155–163

Mumby PJ, Edwards AJ (2002) Mapping marine environments with IKONOS imagery: enhanced spatial resolution can deliver greater thematic accuracy. Remote Sens Environ 82:248–257

Mumby PJ, Green EP, Edwards AJ, Clark CD (1997) Coral reef habitat mapping: how much detail can remote sensing provide? Mar Biol 130:193–202

Mumby PJ, Clark CD, Green EP, Edwards AJ (1998) Benefits of water column correction and contextual editing for mapping coral reefs. Int J Remote Sens 19:203–210

Mumby PJ, Skirving W, Strong AE, Hardy JT, LeDrew EF, Hochberg EJ, Stumpf RP, David LT (2004b) Remote sensing of coral reefs and their physical environment. Mar Pollut Bull 48:219–228

Mumby PJ, Broad K, Brumbauch DR, Dahlgren CP, Harborne AR, Hastings A, Holmes KE, Kappel CV, Micheli F, Sanchirico JN (2008) Coral reef habitats as surrogates of species, ecological functions, and ecosystem services. Conserv Biol 22:941–951

Naseer A, Hatcher BG (2004) Inventory of the Maldives' coral reefs using morphometrics generated from Landsat ETM+ imagery. Coral Reefs 23:161–168

Newman CM, LeDrew EF, Lim A (2006) Mapping of coral reefs for management of marine protected areas in developing nations using remote sensing. In: Richardson LL, LeDrew EF (eds) Remote sensing of aquatic coastal ecosystem processes. Springer, Dordrecht

Ouillon S, Douillet P, Andréfouët S (2004) Coupling satellite data with in situ measurements and numerical modeling to study fine suspended-sediment transport: a study for the lagoon of New Caledonia. Coral Reefs 23:109–122

Palandro DA, Andréfouët S, Hu C, Hallock P, Müller-Karger FE, Dustan P, Callahan MK, Kranenburg C, Beaver CR (2008b) Quantification of two decades of shallow-water coral reef habitat decline in the Florica Keys National Marine Sanctuary using Landsat data (1984–2002). Remote Sens Environ 112:3388–3399

Palandro D, Andréfouët S, Muller-Karger FE, Dustan P, Hu C, Hallock P (2003) Detection of changes in coral reef communities using Landsat-5 TM and Landsat-7 ETM+ data. Can J Remote Sens 29:201–209

Philpot WD (1989) Bathymetric mapping with passive multispectral imagery. Appl Opt 28:1569–1578

Phinn S, Joyce K, Scarth P, Roelfsema C (2006) The role of integrated information acquisition and management in the analysis of coastal ecosystem change. In: Richardson LL, LeDrew EF (eds) Remote sensing of aquatic coastal ecosystem processes. Springer, Dordrecht

Possingham HP, Ball IR, Andelman S (2000) Mathematical methods for identifying representative reserve networks. In: Ferson S, Burgman M (eds) Quantitative methods for conservation biology. Springer, New York

Purkis SJ (2005) A "reef-up" approach to classifying coral habitats from IKONOS imagery. IEEE Trans Geosci Remote Sens 43:1375–1390

Purkis SJ, Riegl B (2005) Spatial and temporal dynamics of Arabian Gulf coral assemblages quantified from remote-sensing and in situ monitoring data. Mar Ecol Prog Ser 287:99–113

Purkis SJ, Myint SW, Riegl BM (2006) Enhanced detection of the coral *Acropora cervicornis* from satellite imagery. Remote Sens Environ 101:82–94

Purkis SJ, Kohler KE, Riegl BM, Rohmann SO (2007) The statistics of natural shapes in modern coral reef landscapes. J Geol 115:493–508

Purkis SJ, Rowlands GP, Riegl BM, Renaud PG (2010) The paradox of tropical karst morphology in the coral reefs of the arid Middle East. Geology 38:227–230

Rankey EC (2002) Spatial patterns of sediment accumulation on a Holocene carbonate tidal flat, northwest Andros Island, Bahamas. J Sediment Res 72:591–601

Riegl B, Purkis SJ (2005) Detection of shallow subtidal corals from IKONOS satellite and QTC View (50, 200 kHz) single-beam sonar data (Arabian Gulf; Dubai, UAE). Remote Sens Environ 95:96–114

Riegl B, Purkis SJ (2009) Model of coral population response to accelerated bleaching and mass mortality in a changing climate. Ecol Model 220:192–208

Riegl BM, Halfar J, Purkis SJ, Godinez-Orta L (2007) Sedimentary facies of the eastern Pacific's northernmost reef-like setting (Cabo Pulmo, Mexico). Mar Geol 236:61–77

Rioja-Nieto R, Sheppard C (2008) Effects of management strategies on the landscape ecology of a marine protected area. Ocean Coast Manag 51:397–404

Robinson JA, Andréfouët S, Burke L (2006) Data synthesis for coastal and coral reef ecosystem management at regional and global scales. In: Richardson LL, LeDrew EF (eds) Remote sensing of aquatic coastal ecosystem processes. Springer, Dordrecht

Rowlands GP, Purkis SJ, Riegl BM (2008) The 2005 coral bleaching event Roatan (Hondulas): use of pseudo-invariant features (PIFs) in satellite assessments. J Spatial Sci 53:99–112

Sano M, Shimizu M, Nose Y (1987) Long-term effects of destruction of hermatypic corals by Acanthaster planci infestation on reef fish communities at Iriomote Island, Japan. Mar Ecol Prog Ser 37:191–199

Schuyler Q, Dustan P, Dobson E (2006) Remote sensing of coral reef community change on a remote coral atoll: Karang Kapota, Indonesia. In: Proceedings of 10th international coral reef symposium, 1763–1770

Scopélitis J, Andréfouët S, Largouët C (2007) Modelling coral reef habitat trajectories: evaluation of an integrated timed automata and remote sensing approach. Ecol Model 205:59–80

Scopélitis J, Andréfouët S, Phinn S, Arroyo L, Dalleau M, Cros A, Chabanet P (2010) The next step in shallow coral reef monitoring: combining remote sensing and in situ approaches. Mar Pollut Bull 60:1956–1968

Sharma S, Bahuguna A, Chaudhary NR, Nayak S, Chavan S, Pandey CN (2008) Status and monitoring the health of coral reef using multi-temporal remote sensing—a case study of Pirotan Coral Reef Island, Marine National Park, Gulf of Kachchh, Gujarat, India. In: Proceedings of 11th international coral reef symposium, pp 647–651

Shinn EA, Smith GW, Prospero JM, Betzer P, Hayes ML, Garrison V, Barber RT (2000) African dust and the demise of Caribbean coral reefs. Geophys Res Lett 27:3029–3032

Smith EV, Rogers RH, Reed LE (1975) Automated mapping and inventory of Great Barrier Reef zonation with Landsat data. In: IEEE Ocean '75, 1:775–780

Stumpf RP, Holderied K, Sinclair M (2003) Determination of water depth with high-resolution satellite imagery over variable bottom types. Limnol Oceanogr 48:547–556

Tittensor DP, Mora C, Jetz W, Lotze HK, Ricard D, Vanden Berghe E, Worm B (2010) Global patterns and predictors of marine biodiversity across taxa. Nature 466:1098–1101

Vercelli C, Gabrie C, Ricard M (1988) Utilization of SPOT-1 data in coral reef cartography Moorea Island & Takapoto Atoll, French Polynesia. In: Proceedings of 6th international coral reefs symposium 2:463–468

Vermote EF, Tanré D, Deuzé JL, Herman M, Morcette JJ (1997) Second simulation of the satellite signal in the solar spectrum, 6S: an overview. IEEE Trans Geosci Remote Sens 35:675–686

Webb AP, Kench PS (2010) The dynamic response of reef islands to sea-level rise: evidence from muti-decadal analysis of island change in the central Pacific. Global Planet Change 72:234–246

Willis BL, Oliver JK (1990) Direct tracking of coral larvae: implications for dispersal studies of planktonic larvae in topographically complex environments. Ophelia 32:145–162

Yamano H (2007) The use of multi-temporal satellite images to estimate intertidal reef-flat topography. J Spatial Sci 52:73–79

Yamano H, Tamura M (2004) Detection limits of coral reef bleaching by satellite remote sensing: simulation and data analysis. Remote Sens Environ 90:86–103

Yamano H, Shimazaki H, Matsunaga T, Ishoda A, McClennen C, Yokoki H, Fujita K, Osawa Y, Kayanne H (2006a) Evaluation of various satellite sensors for waterline extraction in a coral reef environment: Majuro Atoll, Marshall Islands. Geomorphology 82:398–411

Yamano H, Yamaguchi T, Chikamori M, Kayanne H, Yokoki H, Shimazaki H, Tamura M, Watanabe S, Yoshii S (2006b) Satellite-based typology to assess stability and vulnerability of atoll islands: a comparison with archaeological data. In: Proceedings of 10th international coral reef symposium, 1556–1566

Yokoki H, Yamano H, Kayanne H, Sato D, Shimazaki H, Yamaguchi T, Chikamori M, Ishoda A, Takagi H (2006) Numerical calculations of longshore sediment transport due to wave transformation in the lagoon of Majuro Atoll, Marshall Islands. In: Proceedings of 10th international coral reef symposium, 1570–1576

Zainal AJM, Dalby DH, Robinson IS (1993) Monitoring marine ecological changes in the east coast of Bahrain with Landsat TM. Photogram Eng Remote Sens 59:415–421

Chapter 4
Hyperspectral Applications

John D. Hedley

Abstract Hyperspectral approaches are at the technological forefront of optical remote sensing of coral reef environments. Currently most hyperspectral data acquisition employs instruments mounted on airplanes, but in the coming years several planned satellite instruments will increase data availability for hyperspectral analysis of reefs. At the simplest level, hyperspectral data permits classification techniques to derive greater number of classes at higher accuracy than multispectral data can support. Alternatively, full spectral reflectance profiles at each pixel allow band-ratio or derivative approaches to look for features of benthic types that occur at specific wavelengths. But while the feasibility of this approach is supported by in situ data, there have been relatively few successfully demonstrated image analyses. Beyond this, working with full spectral reflectance profiles has stimulated exciting new model-based methods that aim to tease apart depth, benthic type and water quality parameters simultaneously. These methods can also incorporate uncertainty propagation, so that error bars can be placed on each derived parameter at every image pixel. Working with hyperspectral data takes coral reef remote sensing to the edge of what can be achieved by per-pixel optical analysis. Natural variations in the reflectance of benthic types and water column properties become limiting and fundamentally confound some objectives. This prompts future developments to look at analyzing spatial patterns and also to establish the cost-benefit ratio of the integration of other data, such as sonar and LiDAR data.

J. D. Hedley (✉)
ARGANS Ltd., Tamar Science Park, Derriford PL6 8BT Plymouth, Devon, UK
e-mail: jhedley@argans.co.uk

J. A. Goodman et al. (eds.), *Coral Reef Remote Sensing*,
DOI: 10.1007/978-90-481-9292-2_4,

4.1 Introduction

4.1.1 Relevance to Coral Reef Management

Coral reefs are typified by colorful assemblages of organisms, and beyond the red, green and blue multispectral capability of our eyes there is potential rich discriminating power in the spectral details of the color of reef benthos. This possibility is suggested by the reef dwelling mantis-shrimp, the eyes of which contain more than ten different wavelength sensitivities over the 400–700 nm range (Cronin and Marshall 1989). Clearly, evolutionary pressures have directed the mantis shrimp to make use of the spectral details of the reef environment, and by exploiting hyperspectral remotely sensed data we may do so too.

Viewed from above, the structure of a reef includes a number of components of optical significance, the air–water interface, the water column and a bottom reflectance that is dependent on the benthic types present. From a benthic mapping perspective the optical contribution of the water column and surface reflectance is just 'noise' that we effectively want to discard, leaving only the benthic reflectance for analysis. Conversely, from the perspective of those only interested in water constituents a variable bottom reflectance is the 'noise' and a confounding factor for analysis. The benthic types that we wish to distinguish have spectral profiles dependent on the pigments they contain. In some cases these pigments are the same across different types, for example chlorophyll occurs in corals and algae; but others are distinct, for example the colors of red, green and brown macroalgae differ because they contain some distinct pigments (Hedley and Mumby 2002).

The premise of hyperspectral analysis is that the optical contributions of a coral reef system's components and differing benthic types have, to some extent, spectrally distinct shapes with respect to wavelength. Multispectral sensors with broadband responses have insufficient resolution to capture detailed peaks and troughs in spectral absorption and scattering, but hyperspectral, or narrow-band, sensors can reveal this detail (Fig. 4.1). The spectral absorption of light by colored dissolved organic matter (CDOM) in the water has a highly characteristic curved shape, as does absorption by chlorophyll in phytoplankton and the pure water itself (Fig. 4.2). Similarly, spectral reflectances of benthic types (Fig. 4.3), while highly variable, are to some extent distinct between types of interest (Hochberg et al. 2003a; Holden and LeDrew 1999). Hyperspectral data offers the promise of separating the contributions of these different components by means of their characteristic spectral shapes, hence separating the 'noise' from the 'signal' for a given application. Nevertheless, the inherent variability of the optical properties of individual components, and in particular the complex benthic spatial structure of coral reefs, is a limiting factor to what can be achieved. It is important to appreciate that components that are spectrally separable may not be in alignment with the components that are desired to be separable from a scientific or management point of view. Corals and macroalgae are significantly different in an ecological sense, but can have very similar spectral reflectances. Both contain

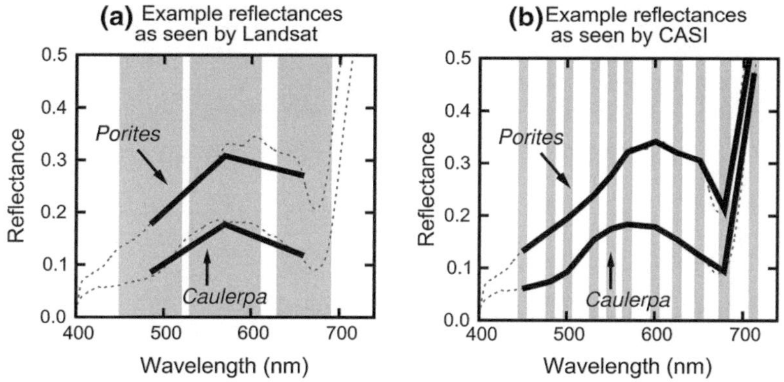

Fig. 4.1 Example coral and macroalgal reflectances resampled to (**a**) Landsat TM blue, green and red bands and (**b**) typical band configuration of the Compact Airborne Spectrographic Imager (CASI), with >10 bands. *Dashed lines* show full resolution reflectance spectra of a coral (*Porites* sp.) and alga (*Caulerpa* sp.), *grey bars* show band locations and widths, and *solid lines* are spectra resampled to the instrument bands

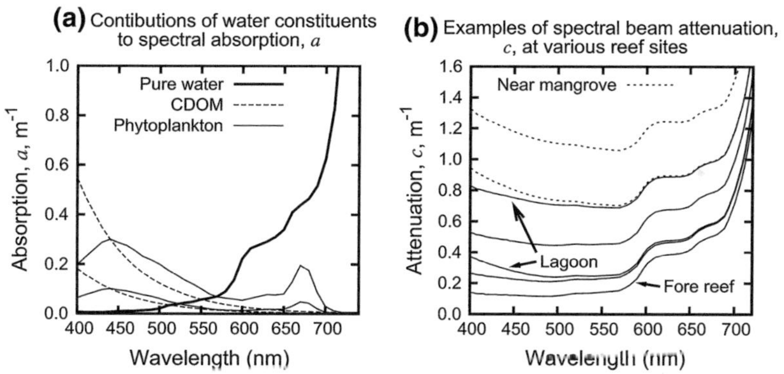

Fig. 4.2 Spectral absorption and attenuation properties of reef waters: **a** example absorption profiles of water constituents in different concentrations (Lee et al. 1998; Pope and Fry 1997); **b** several attenuation (absorption plus scattering) measurements performed with a transmissometer at a variety of reef sites. Note how variation in CDOM affects the shape of the *left* (<500 nm) edge of the attenuation spectra. Other spectral features in attenuation are primarily due to pure water itself as phytoplankton levels are typically low in reef waters. The vertical shift in attenuation across sites (**b**) is due to scattering by differing amounts of suspended particulate matter

chlorophyll and in addition the main accessory pigments for coral symbionts and brown algae, peridinin and fucoxanthin respectively, have similar spectral profiles (Jeffery et al. 1997).

Different reef management objectives require benthic mapping at different levels of descriptive detail. At the most ambitious end of the scale are objectives such as quantification of live coral cover, discrimination of live coral versus dead coral

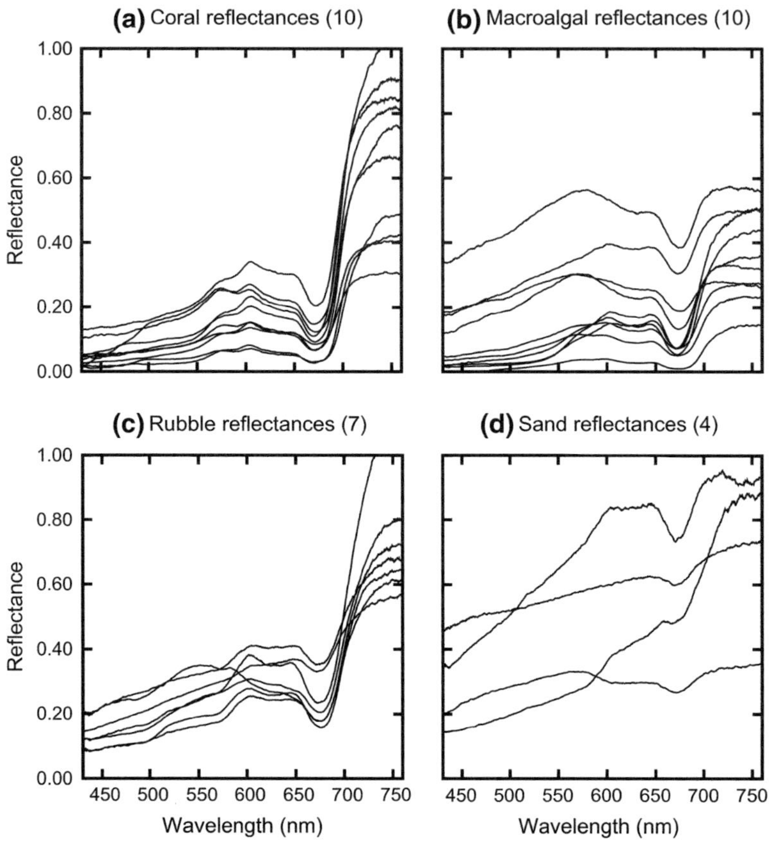

Fig. 4.3 Example spectral reflectances of reef benthic types and sand, as measured in situ with a spectroradiometer in an underwater housing (Roelfsema et al. 2006). Note that chlorophyll is almost ubiquitous and is responsible for the absorption feature (dip) at 680 nm. The reason that coral reflectances go very high above 700 nm (**a**) is often suggested to be due to chlorophyll fluorescence. In fact chlorophyll fluorescence is a small contribution, an alternative interpretation is that the peak occurs because the tissue is transparent at those wavelengths and the underlying coral skeleton is very reflective (Enríquez et al. 2005)

(Mumby et al. 2001, 2004), live coral versus macroalgae (Goodman and Ustin 2007), or detection of coral bleaching events (Elvidge et al. 2004). Although the cited studies and others have indicated positive results, at the time of writing these kinds of objectives are not routinely achieved using hyperspectral data. Habitat level classifications benefit greatly from airborne hyperspectral data in terms of the number of classes that can be accurately estimated (Mumby et al. 1997). However, this success may also be a function of higher spatial resolution and/or scale issues of ground truth surveys, whose effects can co-vary and be hard to isolate (Caplosini et al. 2003). The latest methods for shallow water mapping are based on developing a model for the spectral light received at the instrument that is then 'inverted' to simultaneously extract depth, water optical properties and benthic composition

(Lee et al. 1998, 1999, 2001; Mobley et al. 2005; Brando et al. 2009; Hedley et al. 2009a; Dekker et al. 2011). These methods are primarily designed for hyperspectral data and can routinely achieve excellent results for bathymetric extractions (Hedley et al. 2009a). Results for water column optical properties and benthic composition are more variable but can be well estimated in some environments (Lee et al. 2001; Mobley et al. 2005; Goodman and Ustin, 2007).

Cross-comparing image analysis methods is difficult as there is no universally applicable classification scheme that allows direct inter-comparison of methods for benthic mapping. Different methods may return fundamentally different types of information. Mumby and Harbourne (1999) designed a hierarchical habitat classification scheme for use in Caribbean reefs. A hierarchical scheme has the benefit that classes can be merged for cross comparison between methods of different 'descriptive resolutions' (Green et al. 1996). For coral bleaching surveys Siebeck et al. (2006) have promoted the use of a color reference card. In a field survey or remote sensing context time is well spent in the planning stage to devise a scheme that will allow a meaningful merging of classes if the initial analysis proves too ambitious.

4.1.2 Design and Operational Considerations

The term hyperspectral does not have a rigidly defined meaning, but conveys the sense of data sources with numerous wave bands that are spectrally narrow. The majority of published hyperspectral analyses on coral reefs have been achieved with sensors mounted on airplanes such as the Compact Airborne Spectrographic Imager (CASI) or the Airborne Visible and Infrared Imaging Spectrometer (AVIRIS) (Mumby et al. 2004; Goodman and Ustin 2007). Only a few satellite sensors that could be classed as hyperspectral exist and have been used in shallow water applications, the Hyperion sensor being the most notable to date (Lee et al. 2007). However, the situation regarding available hyperspectral sensors is dynamic, upcoming sensors such as the German EnMAP scheduled for launch 2013 and NASA HyspIRI in 2015 (Table 4.1) may completely change the landscape of hyperspectral reef applications in the near future.

The distinction between multispectral and hyperspectral is sometimes vague and becomes increasingly blurred with the multitude of sensor designs (Table 4.1). For example, in a fairly successful 'hyperspectral' demonstration Mumby et al. (2004) used only 6 bands of a CASI dataset. Next-generation satellite sensor families traditionally seen as 'multispectral' increasingly have hyperspectral-like features, such the extra bands available in DigitalGlobe's WorldView 2 (Table 4.1), or the narrow bands of the European Space Agency's (ESA) upcoming Sentinel 2 (Table 4.1; Hedley et al. 2012a). Hyperspectrally-orientated techniques are likely to see increased use on a wider range of sensors in the future. Most techniques are equally applicable to multispectral and hyperspectral data, the accuracy of estimations may vary but this is not necessarily tied to the number of

Table 4.1 Examples of operational and planned sensors with hyperspectral or relatively narrow-band design philosophy

Name	Bands	Range (nm)	Width (nm)	Pixel size (m)	Example of coral reef or shallow water application
Airborne					
Ocean PHILLS	128	400–1,000	4.6	≥1	Mobley et al. (2005)
AVIRIS	224	400–2,400	~9	≥4	Lee et al. (2001) Goodman and Ustin (2007)
CASI-2	18–288	405–950	~9 for 18 bands	≥1	Mumby et al. (2004) Hedley et al. (2009a)
HyMap	128	450–2,500	15–20	3–10	Heege et al. (2007
AISA Eagle	60–488	400–970	1–10	≥1	Mishra et al. (2007)
Satellite or International Space Station					
Hyperion	220	430–2,500	10	30	Lee et al. (2007) Kutser et al. (2006)
HICO (ISS)	102	380–960	5.7	92	Not yet published
WorldView 2	8	400–1,040	40–180	2	Not yet published
Sentinel 2	13	439–2,280	15–180	10–60	Launch expected 2014
VENμS SSC	12	415–910	16–40	5	Launch expected 2013
PRISMA VNIR	66	400–1,010	≤12	30	Launch expected 2013
EnMAP	94	420–1,000	~6	30	Launch expected 2015
HyspIRI	~212	380–2,500	10	60	Launch expected ~2015

The list does not include sensors with low spatial resolutions (>60 m)

bands alone. Spatial resolution, band narrowness, radiometric accuracy and specific image quality in terms of minimal atmospheric haze and surface reflectance (sunglint) are also important factors. Image revisit times and cost of data are also extremely important factors for change detection applications in particular.

To orientate oneself in the plethora of operational and planned image acquisition systems it is useful to understand the design constraints of an optical remote sensing instrument. To achieve a certain radiometric accuracy in a passive optical imaging system there is a design trade-off between spatial and spectral resolution. Inside the instrument, the photons collected over the area of one pixel are separated into bins for quantifying the response in each spectral band. The rate of photons collected in each bin must be sufficient to ensure a good signal above the internal instrument noise. Collecting from a smaller spatial area implies a smaller field-of-view, which means less photons and hence reduced capability to separate those photons into numerous wavelength bands. As sensor technology improves, overall achievable spatial and spectral resolutions increase, but for any point in time the consequences of this design limitation are reflected in the range of available instrumentation. For example, the ESA MERIS ocean color sensor provides high radiometric accuracy in 15 narrow bands of approximately 10 nm

Table 4.2 Example CASI band configuration used in a reef application (de Vries 1994)

Centre (nm)	Width (nm)	Eye sensitivity	Suggested purpose
449.6	8.8	blue	Water depth of penetration
481.4	8.8		Chlorophyll
500.9	8.9		Ocean chlorophyll reference
530.8	8.9		Scattering
550.6	8.9	green	Reference
568.4	8.9		Phycoerythrin
600.6	8.9	red	Reference
625.7	9.0		Phycocyanin
650.8	9.0		Reference
678.8	10.8		Chlorophyll-a
712.1	9.0		Red Edge 1
751.0	10.9		Red Edge 2
804.4	9.1		Near Infrared 1
848.0	9.1		Near Infrared 2

Also indicated are the peak wavelengths of sensitivity of the color vision of the human eye

width, but at a spatial resolution of 300 m. At the other end of the design spectrum the Ikonos sensor provides 4 m pixels but at lower radiometric accuracy and in only four spectral bands of 65–100 nm width. This design trade-off represents a data level decision for the application of coral reef remote sensing methods. The research indicates that all factors should be optimized, in that coral reef applications benefit from high spatial resolution, high spectral resolution and high radiometric accuracy (Caplosini et al. 2003; Mumby et al. 1997; Mumby and Edwards 2002). However, all three factors cannot be simultaneously maximized, so in practice there must always be a trade-off when choosing a data source for a given application. Note also that bands in the near and far infrared are largely unusable for sub-surface mapping as water is practically opaque in wavelengths above 700 nm (Fig. 4.2). Therefore, although an instrument such as Hyperion has 224 bands, only around 30 carry sufficient information for sub-surface applications.

Airborne sensors are the 'exception that proves the rule' of the above described sensor-design constraint since they fly closer to the Earth's surface. A smaller pixel therefore appears equivalent to a larger one in the sensor field of view, and it is possible to achieve relatively high spatial and spectral resolution simultaneously. However, airborne data introduces numerous processing challenges. Airborne hyperspectral sensors typically operate in a 'push-broom' design, where the sensor records a line of pixels moved across the water surface by the forward motion of the plane. The roll and yaw of the plane must be minimized in flight but still require specialized image geo-correction. Closer proximity to the water surface introduces 'cross track' variation, where the view angle is different onto the water surface to the left and the right of the aircraft travel. Careful flight line design can minimize this (discussed later) but special image corrections are generally required. In all cases the quality of airborne imagery can be substantially affected by the experience of the pilot and operator in collecting this type of data, and on the wind conditions. Another factor to be aware of is that instrument maintenance and deployment is more variable than with satellite platforms. If the instrument is not maintained correctly dust on the sensor and other factors can cause vertical striping in the data, and electrical noise on-board can cause horizontal striping.

For benthic mapping the instrument should have its radiometric sensitivity adjusted to be optimal for the relatively dark sub-surface reflectance. It can be difficult to achieve unsaturated terrestrial data from a sensor tuned for below water applications, especially in the tropics where terrestrial surfaces may include high reflectance coral sand. This impacts the ability to use terrestrial targets as references for atmospheric correction. Operational satellites suffer from these problems less. Their mission requirements and instrument operational characteristics are clearly defined and maintained with long term and multiple users in mind, rather than for ad hoc site deployments for a single customer.

Some airborne sensors such as CASI allow for configurable band wavelengths. In this case it is worth reviewing the literature and features of the study site to position bands in wavelengths that are likely to provide useful information. Table 4.2 shows a published justification of CASI band selection used in a coral reef application (de Vries 1994). However, Table 4.2 should not be taken as definitive and is slightly out-dated; for example near infrared bands are unlikely to show red edge chlorophyll features due to the high absorption by water (Fig. 4.2). The features of the site of interest and recent results on spectral discrimination should also be considered (Hochberg et al. 2003a). Hedley and Mumby (2002) review reef pigment spectral features and their relation to hyperspectral remote sensing.

4.2 Hyperspectral Planning and Preprocessing

A number of distinct approaches exist for mapping or quantifying benthic composition and other biophysical parameters in shallow waters by optical remote sensing. Almost all the described methods are equally applicable to multispectral or hyperspectral data in a practical sense, however the quality of outputs will vary and some approaches have been designed with hyperspectral data in mind. When choosing a method to use the first consideration must be what data is required to be extracted from the imagery for the given application. This must be tempered with what is likely to be actually possible and the practical challenges in applying a specific method. Published methods differ significantly in how complicated they are to apply: from methods that can be applied with little or no image preprocessing and using standard software, to methods that require rigorous atmospheric corrections and custom code. It is important to bear in mind that published results tend to show a positive bias, results where a method performed poorly are far less likely to be published than those where a method worked well. In addition a successful demonstration for a specific image in the scientific literature may be the end result of a long process of methodological development tested on that specific image, but not presented as part of the published paper. Therefore the potential transferability of methods to other sites must always be assessed critically.

4.2.1 Data and Processing Requirements

Subsequent sections give an overview and examples of different image processing techniques for mapping coral reef biophysical parameters. Emphasis will be given not only to the type of data that can be extracted, but also to what is required in a practical sense to apply each method. In general, factors to consider when evaluating the potential of published methods are:

Is the method appropriate for the imagery that is available?

- Does the method rely on hyperspectral imagery? What are the number of bands, spectral width of bands and radiometric accuracy of instruments used in successful demonstrations?
- Is the spatial resolution of the imagery appropriate? What is the scale of benthic heterogeneity at the study site and can the method handle this level of within-pixel mixing?
- If the imagery is already acquired, is it of good quality? An image with substantial sunglint or atmospheric haze may not give as good results as the imagery used in a published paper. Again, there is likely to be a bias as the best published demonstrations will likely have used the highest quality images (see Brando et al. 2009 as a rare exception).
- If the imagery is not already acquired, what is the risk that it will be of insufficient quality to support the analysis? For example, commercial satellite data providers do not always provide quality assurance for sunglint. Is an airborne campaign subject to time, personnel or cost constraints that may compromise the quality of data?

What are the processing requirements and how will these be achieved?

- Is atmospheric correction required? If so, which method or software should be used and what data will be required (see Sect. 4.2.3)?
- Is an air–water interface correction required, either for sunglint removal or to align the magnitude or type of reflectance with modeled or in situ data (e.g., above-surface or below-surface reflectance)?
- What level of geo-correction is required to align with ground truth data? Is alignment even feasible given the spatial scale of the image and available or potential ground truth data?
- Is the analysis procedure a routine that can be found in standard image processing software, or does it require custom code?

- Are adequate personnel resources available to process the data? Bear in mind simply acquiring the data may be a small fraction of the work that needs to be done.
- What long term resources can be allocated to the dataset? For the very best value, datasets should be treated like 'living' entities that receive constant use and attention. The longer a dataset is left unused the less likely it will ever be used again.

What other data beyond the imagery itself will be required or useful?

- For spatial ground truth of benthic composition, how much, what spatial sampling strategy and what level of 'descriptive resolution' is needed?
- Are spectral libraries of benthic type reflectance needed? These are required as input for some methods, but of little practical use for others.
- Are in-water measurements of light fields, constituents or optical properties needed?
- If atmospheric correction is required, above water in situ reflectances measured concurrently with image acquisition for vicarious calibration may be very useful.

What is the status of the method in terms of acceptance and use by practitioners?

- How 'tried and tested' is the technique? Will the wider community find the results easy to accept and interpret? Is this a 'standard procedure' that has been used many times and is well understood?
- How is accuracy reported? Is this easy to interpret and appropriate for the intended application?
- What is the transferability of a method? A method may have been successfully demonstrated at one site but will it work elsewhere if a site has different composition or bio-optical properties?

4.2.2 Preprocessing Considerations

To understand what types of preprocessing are required for different analysis techniques it is helpful to distinguish between two types of image processing algorithms. The first, 'data manipulation', simply performs calculations on the image data without incorporating additional information that varies from pixel-to-pixel. For example, subtracting a dark pixel spectrum from every pixel in an image to perform a basic atmospheric correction, or generating a new data layer as the ratio of the values in two other bands. Clearly this type of processing does not add any information to the image on a pixel-by-pixel basis. The new data values are simply a transform of the existing data at each pixel. In fact, information may be lost, for example a band ratio transform discards information about the differing magnitudes of the values at the two pixels. So the question arises: how can a processing step that does not increase information, or even discards information, be useful? The answer is that the purpose of these image based preprocessing steps (i.e., those that operate on image data without combining any 'external' auxiliary data) is to transform the data into a form more acceptable to a subsequent processing step. Calculating depth invariant indices, for example (described below), attempts to remove information about depth so a classification algorithm is not 'confused' by information that in a benthic cover mapping context we have no interest in.

The situation is different if an external source of data is incorporated into the analysis, such as acoustic bathymetry (Bejarano et al. 2010). In this case information is increased and, in theory, if the information is of good quality, the accuracy of an analysis based on the new information should only increase. If after adding data layers the accuracy remains the same or goes down, either the new data is erroneous or the analysis technique is not appropriate.

These perspectives are very useful when evaluating different processing and preprocessing algorithms. In many cases alternative algorithms are fundamentally similar and by definition operate on the same data source (the image). They differ only in the methods used to remove unnecessary information and the amount and source of potentially useful external information incorporated. Subsequent sections briefly mention a number of preprocessing steps that may be useful in different contexts, to rearrange the existing data or to incorporate external data to facilitate improved operation of the subsequent image analysis.

4.2.3 Atmospheric Correction

The passage of light from its source to the surface and subsequently from the surface of the Earth, or the sea surface, to the sensor is affected by the atmosphere in two ways: (1) light received at the sensor is reduced by absorption and scattering of photons out of the 'beam' that travels from the surface to the sensor; and (2) conversely, the received light is increased by photons that are scattered into the path of the beam. The latter contribution is termed the 'path radiance'. Both processes are dependent on wavelength, so that the apparent spectral reflectance at the sensor deviates from that of the material at the surface. To clarify, a simplified relationship between the radiance received at the sensor and that originating from the surface can be expressed as (Lee et al. 2007):

$$L_t(\lambda) = L_a(\lambda) + t(\lambda)L_w(\lambda) \tag{4.1}$$

where $L_t(\lambda)$ is the radiance received at the sensor, $L_a(\lambda)$ represents the in-scattered component (path radiance), $t(\lambda)$ is the transmission from the surface to the sensor and $L_w(\lambda)$ is the radiance leaving the water surface. A major component of the atmospheric radiance distribution, especially for clear conditions, is Rayleigh scattering. This is scattering that occurs at the molecular level and is responsible for the blue sky when looking upwards from the Earth's surface. Similarly, when looking downwards from a sensor through the atmosphere, the apparent reflectance of the surface tends to be excessively blue. Water vapor and other aerosols also contribute to absorption and scattering and can be highly variable across sites and times.

Not all approaches require atmospheric correction, in particular atmospheric correction methods that affect every pixel in the same way across an image do not affect the information content of the image. However, time series analysis may require atmospheric effects be removed to align images radiometrically, and simple visual interpretation may benefit from the removal of path radiance (haze). Below we review the different approaches that can be taken and what is involved in a practical sense.

Empirical line and dark pixel subtraction—Strictly speaking, Eq. 4.1 is an approximation as it ignores the more complex scattering pathways and spatial variation in surface reflectance. Nevertheless Eq. 4.1 is the basis of a number of atmospheric correction schemes. Importantly Eq. 4.1 establishes a linear relationship at each wavelength between the surface reflectance, $L_w(\lambda)$, (what we want) and the at-sensor reflectance, $L_t(\lambda)$, (what we have). The two components that must be deduced in each band are the scaling factor and offset, $t(\lambda)$ and $L_a(\lambda)$. If reference ground truth reflectances are available then these parameters can be estimated by linear regression for each band and this is the basis of the so-called 'empirical line' atmospheric correction. Smith and Milton (1999) review the requirements to accurately perform this correction, in particular reference targets must be homogenous and substantially larger than the pixel size to avoid adjacency or point-spread function (PSF) effects (Milovich et al. 1995). In addition reference targets should encompass the range of reflectances in each band, to avoid extrapolation. Empirical line is also suitable for high spatial resolution satellite sensors (Karpouzli and Malthus 2003). A basic form of empirical line is dark pixel subtraction (Mather 1999), which assumes the darkest band values in the image represent a surface reflectance of zero. This is not recommended for most processing algorithms that require atmospheric correction (e.g., inversion methods, Sect. 4.3.5) and will have little or no effect for most classification approaches (Caplosini et al. 2003).

Cloud shadow method—Lee et al. (2007) present a convenient method that can be applied if an image contains deep water areas, clouds, and cloud shadows on deep water. The method also requires an estimation of the ratio of direct to total irradiance on the surface, which can be obtained from freely available and simple to use radiative transfer models such as SBDART (Ricchiazzi et al. 1998) or libRadtran (Mayer and Kylling 2005). The method is relatively insensitive to this estimation so while it requires specifying atmospheric constituents a 'standard atmosphere' assumption may be sufficient.

Radiative transfer modeling—A more sophisticated approach is to use a radiative transfer model, either to evaluate $t(\lambda)$ and $L_a(\lambda)$ or to parameterize a more complicated inversion that more fully captures multiple scattering photon paths.

MODTRAN 4 for example allows for pixel adjacency effects, and is the basis of the commercially available FLAASH (Fast Line-of-sight Atmospheric Analysis of Spectral Hypercubes) plugin for the image processing software ENVI (Adler-Golden et al. 1999; Exelis VIS 2012). A number of research level atmospheric codes that are used in the literature have limited availability or no support. Two that occur regularly in the shallow water mapping literature are TAFKAA (used in: Goodman and Ustin 2007; Mobley et al. 2005; Lesser and Mobley 2007) and c-WOMBAT-c (used in Brando et al. 2009). TAFKAA is documented in Gao et al. (2000) and Montes et al. (2001) whereas the basis of c-WOMBAT-c is described in de Haan et al. (1997) and Brando and Dekker (2003). Freely available codes such as the above mentioned SBDART and libRadtran can also be used to parameterize a correction, as can the 6SV code which includes polarization (Kotchenova et al. 2006; Kotchenova and Vermote 2007). It should be noted that almost all use of radiative transfer models for atmospheric correction requires some estimating of unknown parameters. For this reason, despite the demonstrable success of model based approaches (Ferrier and Trahair 1995), highly accurate atmospheric corrections are rarely achieved simply by 'turning the handle' on a model based approach.

Vicarious calibration—Vicarious calibration refers to the process of taking in situ above or below water reflectances for refining an atmospheric correction. In principle it operates identically to the empirical line correction, but is typically performed after a radiative transfer model based atmospheric correction and uses actual above or below-water reflectances over the area of interest. Specific instruments exist for collecting boat based in situ reflectances, such as the gimbal mounted DALEC instrument (Slivkoff 2010). Given the expense of acquiring airborne hyperspectral data and the difficulty of performing accurate corrections by modeling alone, collection of vicarious calibration data is always recommended.

4.2.4 Cross Track Variation and Correction

Imagery from both satellite and airborne sensors can be affected by variations in the view angle onto the Earth's surface across the image. This is particularly true for aquatic targets as the air–water interface can be highly reflective for certain solar-view geometries. The extent of variation across an image is a function of the altitude of the sensor and the width of imaged area. The satellite sensor IKONOS, for example, orbits at around 700 km and has a 0.9° field of view imaging 11 km on the surface. For IKONOS the variation in view angle is therefore extremely small, less than 0.5° from nadir, and cross-image view angle effects are negligible. Push-broom airborne sensors have a wider field of view to accommodate a sufficiently wide swath at low altitudes. For example the current CASI-550 has 40.4° across track field of view (to the left and right of aircraft travel) (ITRES 2008) imaging a swath of width approximately equal to the altitude of flight. Therefore the variation in view angle from left to right side of the swath is ±20.2° off nadir, greater if the aircraft rolls. Cross track effects are greatest when the direction of flight is perpendicular to the solar azimuth, since then the instrument cross-track direction is in the solar plane and direct solar reflectance is the greatest (Mobley 1994; Kay et al. 2011).

In an aquatic application several processes will cause the detected radiance to vary with view angle: (1) at steeper angles the path through the atmosphere is greater; (2) reflectance of the upper side of the air–water interface is highly directionally dependent (Kay et al. 2011); (3) at steeper angles the transmitted path through the water is greater; and (4) the benthos may exhibit non-Lambertian bi-directional reflectance function (BRDF) (Hedley and Enríquez 2010). Of these, the effect of air–water interface reflectance is by far the most apparent (Fig. 4.2). While atmospheric path can be corrected by atmospheric correction codes, the in-water path and benthic BRDF in aquatic applications are rarely corrected. Image processing software packages such as ENVI (Exelis VIS 2012) may contain cross-track correction algorithms, but typically these have been developed with terrestrial applications in mind and may be parameterized based on canopy BRDFs only (Kennedy et al. 1997). Since the primary cross-track effect in an aquatic image is reflection from the air water interface, a sunglint correction procedure may be used provided there is a deep water area across the entire track from which to parameterize the visible band-NIR relationship (see Sect. 4.2.5 and Kay et al. 2009).

4.2.5 Sunglint Correction

A major source of pixel-to-pixel variation in high spatial resolution images is the reflection of the sun from the upper side of the water surface. For pixels smaller than surface waves, undulations in the water surface introduce bright speckle or wave shaped 'sunglint' patterns (Fig. 4.4). As spatial resolution decreases below that of waves, the effect tends to become a gradual cross-image effect (Sect. 4.2.4). These patterns obscure benthic features and will confound classification algorithms. For model inversion algorithms (Sect. 4.3.5) the surface reflectance is a complicating factor. The reflected component of the radiance has never penetrated the water surface, so spectrally it carries no information about sub-surface features and it simplifies analysis to remove it. Although note that other remote sensing methods can usefully infer wave energy and bathymetry from the effect on wave glint patterns (Cureton et al. 2007; Splinter and Holman 2009).

A number of sunglint removal algorithms have been proposed (Joyce 2004; Hochberg et al. 2003b; Hedley et al. 2005; Lyzenga et al. 2006) which all rely on the assumption that the spatial variation in a near-infrared (NIR) band is solely due to glint. Kay et al. (2009) demonstrated that these algorithms are essentially functionally identical and differ only in what is assumed to be the 'base level' NIR. Using the depth of the oxygen absorption feature at 760 nm is an alternative approach that has been suggested by Kutser et al. (2009), assuming the imagery

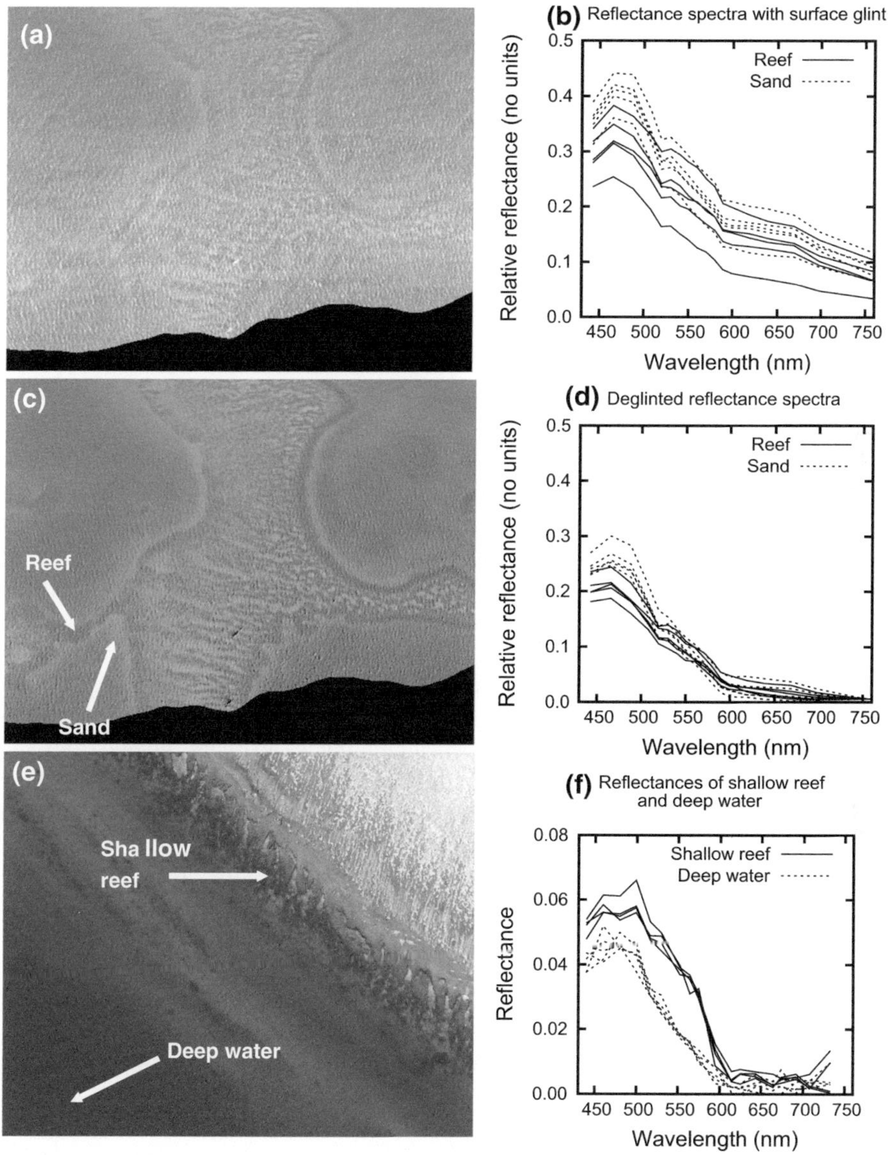

Fig. 4.4 Examples of airborne hyperspectral data of reefs: **a**, **c** CASI image of reefs at St. John, U.S Virgin Islands before and after deglint procedure (reproduced with permission from Hedley et al. 2005), red green and blue are 508, 488, 467 nm, respectively; and (**e**) CASI image of Heron Reef, Australia, red, green and blue are 523, 508, 470 nm, respectively. Each plotted type depicts five reflectance spectra from adjacent pixels. Note that deglint decreases the spread of the data but does not necessarily increase the spectral separation between differing types (**b**, **d**). **f** Example where reef and deep water have similar reflectance profiles but are separable up to 600 nm despite sources of variation

has the required wavelength bands to elucidate this feature. However, this method is currently demonstrated as anecdotally successful only in a single test case.

For airborne campaigns it is sensible to design image acquisition strategy to minimize glint. Flight paths towards or away from the sun with solar zenith angles of 30°–60° have been recommended (Mustard et al. 2001; Dekker et al. 2003). In one specific example, Lesser and Mobley (2007) acquired imagery with solar zenith 40°–55° and flight direction aligned with the sun.

4.2.6 Depth Correction

Water depth has a major effect on the spectral reflectance measured over a reef. The range of the absorption coefficient of water over the 400–700 nm wavelengths is high and the spectral shape of absorption is also affected by water constituents such as CDOM (Fig. 4.2). Hence the same bottom reflectance at different depths will give rise to very different above-water reflectances, and this is clearly a complication in identifying the bottom type. Two simple methods can be used to remove the effect of variable depth: (1) 'depth correction' requires that depth be known across the image either from acoustic (Bejarano et al. 2010; Chaps. 8–10) or LiDAR data (Chaps. 5–7), or for some reef topographies a simple stratification into depth zones may be possible (Mumby et al. 2004); or (2) calculating 'depth invariant indices' using an image based pre-processing method that aims to remove the effect of variable depth by calculating new image layers from the logarithm of pairs of bands (Lyzenga 1981; Green et al. 2000). Both depth correction and depth invariant indices require regions of homogenous benthos with variable depth within the image. In coral reef applications sand is ideal and has the advantage of providing a bright relatively strong signal over the visible wavelength range. Both methods rely on the concept of approximate exponential attenuation of light as expressed by the diffuse attenuation coefficient, k_i, in each band i, such that for depth z, the above water reflectance in band i is proportional to $\exp(-2k_i z)$. For depth correction the k_i values are estimated by regression for areas of sand at a range of depths, the exponential relationship can then adjust other pixels to a fixed depth (Bejarano et al. 2010). For classification approaches the uniform depth does not have to be zero. For depth invariant indices new bands are created using $r = \ln(r_i) - (k_i / k_j) \times \ln(r_j)$, where r_i and r_j are the reflectance in bands i and j, and k_i and k_j are the diffuse attenuation coefficients at the wavelengths of those bands. In this case it is not necessary to calculate the actual k values, the ratio (k_i / k_j) can be derived by regression between bands i and j over variable depth sand even when actual depth is unknown (Lyzenga 1981; Green et al. 2000). Depth invariant indices are typically used with classification (Sect. 4.3.1; Green et al. 2000) but can also be incorporated into band difference methods (Sect. 4.3.2; Isoun et al. 2003).

4.3 Hyperspectral Algorithms

4.3.1 Classification

Currently the majority of practical coral reef mapping from remote sensing is performed to habitat-level using classification algorithms, such as the k-means or maximum likelihood algorithms, as implemented in software packages like ENVI (Exelis VIS 2012) or ERDAS IMAGINE (ERDAS 2011) (Bertels et al. 2007; Caplosini et al. 2003; Harborne et al. 2006). Classification algorithms seek to group pixels in an image into a number of classes based on their spectral similarity. The approach can be unsupervised, the user tells the algorithm how many classes are desired and the algorithm will automatically try to group spectrally similar pixels, or supervised, a set of pixels are identified as being of known classes and the algorithm proceeds on that basis. In either case it is necessary to have some a priori knowledge of the classes of some pixels (i.e., 'ground truth' or 'calibration' data). For the unsupervised classification normally two to three times as many classes as required are requested and the user subsequently identifies what the classes represent, merging down to the required number. In the absence of field survey data it is possible for 'ground truth' to simply be visual interpretation of the image combined with local knowledge. It is recommended to retain some ground truth data for a subsequent independent accuracy assessment (sometimes called 'validation'). However, caution should be exercised in the selection and structure of the calibration and validation data. Having more points is not necessarily better if the structure is such that data from the same areas is simply duplicated across the calibration and validation data sets, which will inflate the reported accuracy with respect to uncharacterized areas of the imagery. Contextual editing may also be used to improve map accuracy; this simply means changing the class of pixels where visual interpretation indicates they are clearly in error (Mumby et al. 1998).

The level of detail that classification can provide from different imagery sources has been investigated a number of times (Mumby et al. 1997; Caplosini et al. 2003) and the conclusions are fairly consistent. Multispectral broad band data such as Landsat TM/ETM+ data cannot derive much more than three or four classes of coarse scale reef structures, while more and narrower bands may permit up to ten or more habitat level classes (Fig. 4.5a). 'Habitat level classes' means areas described by phrases such as 'sparse corals with high algal cover' or 'sand with occasional branching red algae' (Mumby et al. 1997; Harborne et al. 2006). Appropriate choices of classes will be site-dependent. Classification techniques have not been demonstrated as able to map individual reef components such as 'live coral' or 'macroalgae'. This distinction is important to realize; it is often assumed coral reef remote sensing can map 'coral cover' but what can realistically be achieved is mapping of 'habitats containing coral'. Nevertheless the level of detail obtainable from hyperspectral data can be used to derive interesting products, such as the beta-diversity map of Fig. 4.5b. This was produced from a 19-band CASI image of U.S. Virgin Islands reefs, which was first classified to 19

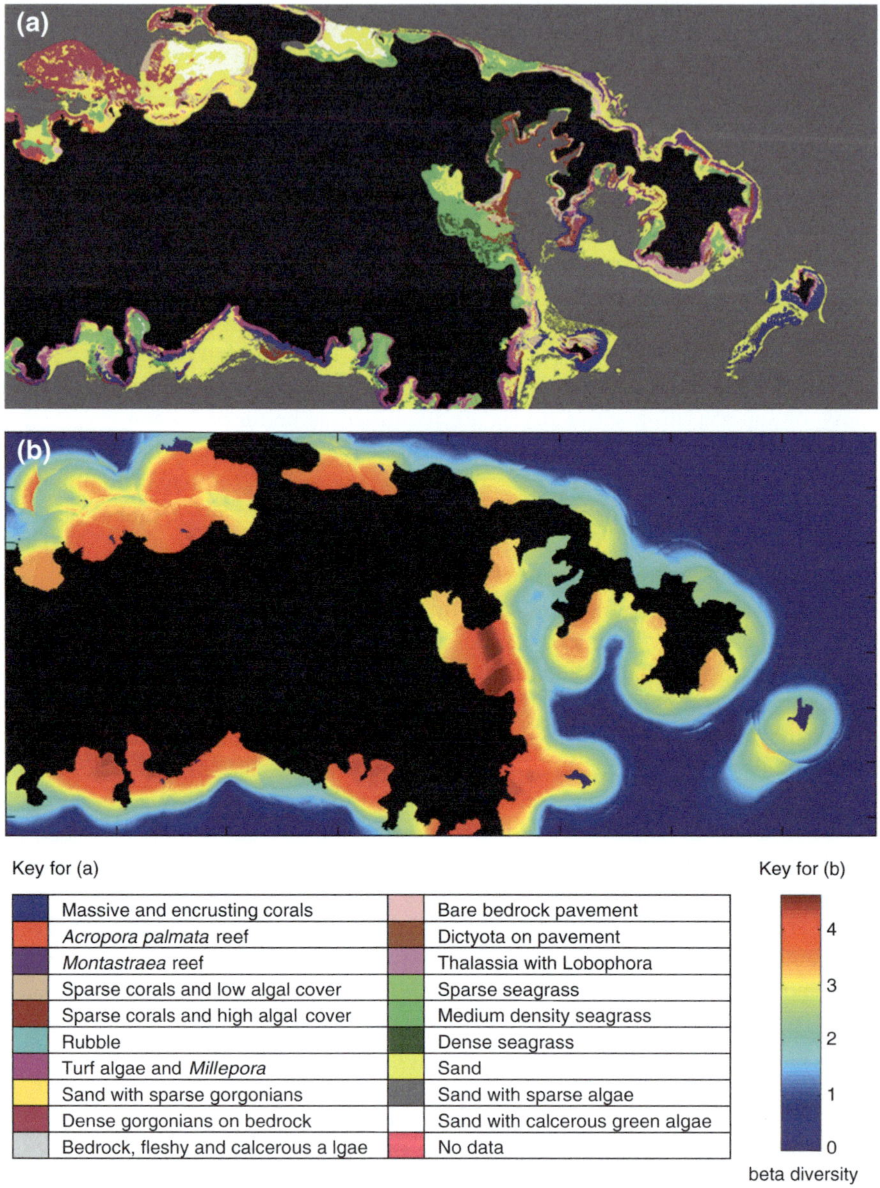

Fig. 4.5 CASI derived benthic maps of St. John in the U.S. Virgin Islands: **a** 19-class habitat map and **b** beta diversity. Map width is 15.7 km (Harborne et al. 2006)

benthic classes (Harborne et al. 2006). Hyperspectral data is required to support this relatively high number of classes.

For a practitioner, the main decision to be made is the number of habitat classes to aim to identify and to establish an appropriate and realistic classification scheme that ideally will allow class merging (Mumby and Harbourne 1999). To some extent this will depend on available ground truth data. For classification approaches it is important to perform sunglint correction if glint patterns are visible in the imagery (Sect. 4.2.5) but the majority of atmospheric correction schemes will make little or no difference, and in practice atmospheric correction is often omitted. Calculation of depth invariant indices (Sect. 4.2.6) can increase accuracy, but strictly speaking if ground truth data is available across the full range of depths this should be unnecessary and in fact the depth may carry habitat information. Note that working with more classes demands more detailed ground truth data for calibration and validation. So there may be associated costs in fully exploiting hyperspectral data that extend beyond the cost of data acquisition alone.

4.3.2 Band-Specific Analysis

A number of coral reef remote sensing studies have investigated the concept of mapping benthic types such as live coral or algae by evaluating band differences, ratios, or 'derivatives', the slope of the spectral profile (Tsai and Philpot 1998), at specific wavelengths. These 'wavelength feature' approaches offer the potential to map reefs to the level of individual reef components, rather than that of habitats as typically elucidated by classification. While actual remote sensing applications that have successfully used wavelength features for mapping benthic type are scarce and have used differing methodologies, when successful they have produced results at the fore-front of coral reef remote sensing capability (Hochberg and Atkinson 2000; Isoun et al. 2003; Mumby et al. 2001, 2004).

The basis of using wavelength features is informed by analysis of libraries of in situ reflectance profiles, where the spectral reflectances of benthic types are taken using a spectroradiometer either underwater or exposed (i.e., 'field spectroscopy' as it is termed in terrestrial environments). For example, Holden and LeDrew (1999) took a number of spectral reflectance measurements of live and bleached coral and identified three wavelength regions between 500 and 650 nm that could be used for discrimination. Hochberg and Atkinson (2000, 2003c) and Hochberg et al. (2003a) have extensively studied in situ spectral libraries and identified wavelength regions where discrimination of benthic types may be possible (see also Wettle et al. 2003). Hedley and Mumby (2002) review a number of previous studies on the same theme and attempt to establish a causal biological link to the observed spectral features in terms of pigments.

To use wavelength specific features for mapping requires hyperspectral data, first because a free choice is required of which wavelengths to use, and secondly because narrow bands are necessary to elucidate the spectral features. Effectively, data in wavelength regions that are not of interest are discarded. In ocean color remote sensing, important wavelength regions for phytoplankton pigments have long been established; so operational satellite sensors such as MERIS have narrow bands located at these wavelengths. Coral reef remote sensing is a dynamic field still under development; hence sensors with reef specific bands are unlikely to be launched in the near future, or ever. For now hyperspectral data facilitates wavelength feature analysis even when only a few wavelengths are required.

One of the most successful demonstrations of measuring a proxy for 'reef health' by a remote sensing study was that of Mumby et al. (2001, 2004) where derivatives from CASI data were used to discriminate live and dead *Porites* coral in a shallow French Polynesian atoll and to quantify live coral among several other benthic types. In particular the spectral slope 506–565 nm was key to the discrimination of live from dead coral. Isoun et al. (2003) mapped the cover of living coral at a Hawaiian site using three narrow wavelength bands at 488, 551 and 577 nm. In another Hawaiian example, Hochberg and Atkinson (2000) used an automated procedure to establish optimal separating wavelengths from in situ reflectance measurements and then applied them to map coral, algae, and sand. For these methods depth correction and atmospheric correction is recommended, especially if a direct comparison to in situ spectra is part of the method. Again there is little consistency between publications. The site for Mumby et al. (2004) consisted of very shallow reef (< 4 m) and a basic depth correction was applied whereas Isoun et al. (2003) used a form of depth invariant indices. Due to the scant availability of validated applications wavelength feature methods should be considered the most experimental reef mapping approach and with the most questionable transferability to other sites.

4.3.3 Spectral Unmixing

Reefs are spatially heterogeneous at scales smaller than even the highest resolution sensors. Spectrally and functionally diverse benthos and substrates such as corals, macroalgae, rubble and sand can co-exist at sub-meter scales. This mixing clearly presents a challenge for analysis techniques that attempt to characterize each pixel as a single class. Spectral unmixing is an approach that attempts to tease apart the hyperspectral reflectance measurement of a pixel to quantify the proportions of the individual sub-pixel constituents. Unmixing relies on having a spectral reflectance library of the 'pure endmembers' and the mathematics of the approach assumes the pixel reflectance is the sum of the reflectances of the components, weighted by their proportion in the pixel.

The unmixing approach has been successful in mineralogical applications (Adams et al. 1986) but in mineralogy the endmember reflectances are well defined and the physical mixing often occurs on a fine scale or on relatively flat surfaces. In contrast for coral reefs endmember spectra are not well defined (Fig. 4.3), the overlying water column complicates the reflectance, and the three-dimensional structure of the reef may invalidate the linear mixing assumption (Hedley 2008). Nevertheless modeling and experimental studies indicate unmixing has potential (Hedley and Mumby 2003; Hedley et al. 2004) and only a few published image analyses have included an unmixing component (Goodman et al 2007;

Hamylton 2011). However, with recent developments unmixing approaches can now be thought of a subset of model inversion based techniques, which apply similar reasoning but are more flexible in their formulation (Sect. 4.3.6)

4.3.4 Bathymetry

Depth variation across a coral reef can display strong and characteristic features, from relative shallow back-reef lagoons and emergent fore-reef features and slopes, to spur and groove zones and upstanding isolated coral heads. Mapping bathymetry is an important objective for navigation or evaluating benthic light levels, and published methods using optical remotely sensed data are numerous and among the earliest shallow water remote sensing applications (Lyzenga 1978, 1981). While bathymetry is a distinct objective from mapping benthic type, the effect of depth on the above-water spectral reflectance means the two are intimately tied. Indeed, the latest semi-empirical or 'physics-based' model inversion methods (Sect. 4.3.6) extract both depth and benthic reflectance simultaneously. Lyzenga's depth invariant indices technique (Lyzenga 1978, 1981) can also produce a depth estimate but requires the bottom to be classified first in order to factor out variation due to bottom reflectance. However, while variable bottom reflectance and variation in water column optical properties are considered the biggest weakness of bathymetry extraction algorithms (Dekker et al. 2011), depth estimation can be surprisingly robust if multispectral or hyperspectral data are used (Lyzenga et al. 2006). This is because the light absorption by pure water is approximately exponential with depth and has a wide range over visible wavelengths (Fig. 4.2; Maritorena et al. 1994).

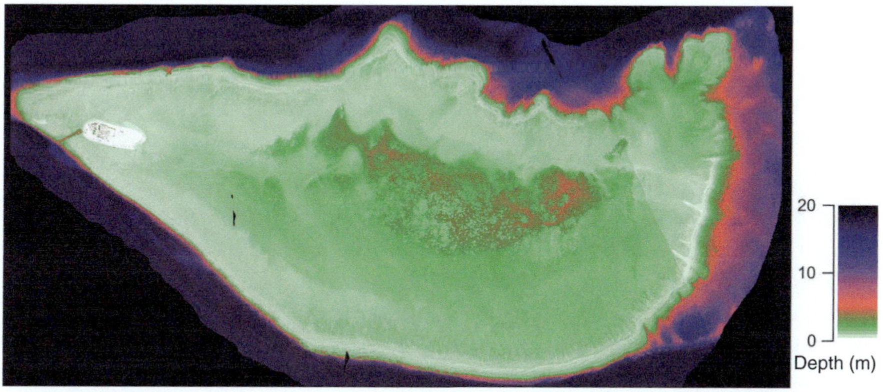

Fig. 4.6 Bathymetric map of Heron Reef, Australia, derived by radiative transfer model inversion applied to 19-band CASI data. The reef length is 11 km and the full image is at a resolution of 1 m with around 50 million pixels. The linear discontinuity on the right is due to tidal change between adjacent flight lines (Hedley et al. 2009a, the data set is available online, Hedley et al. 2012c)

Increasing depth modifies the above-water spectral reflectance across all wavelengths in a characteristic way that presents a strong 'signal' even in the presence of water column constituents or variable bottom reflectance. Another consequence of the exponential attenuation with depth is that small depth changes in shallow areas are equally resolvable as large changes in deeper water; hence the sensitivity of bathymetric extraction methods is typically sensibly structured. If bathymetric extraction from hyperspectral data is desired the very best methods are the multi-parameter algorithms described in Sect. 4.3.6 (Fig. 4.6). Although these methods do not require any a priori bathymetry data, without any it is difficult to assess if the resultant map can be trusted. However, the implementation of model inversion methods is technically challenging and at the time of writing no off-the-shelf processing software exists. Hence the simpler bathymetry-only methods (Lyzenga et al. 2006) may be more practical. The strength of many of these latter methods is that if bathymetry is known for some image pixels, then parameterizing the methods directly from image pixels ameliorates the effects of imperfect atmospheric correction or other data bias.

4.3.5 Change Detection

A distinct approach to coral reef remote sensing analysis that warrants further development is multi-image change detection. The principle is simple, in that two or more images at different time points are spatially aligned by geo-rectification and regions of change are identified on a pixel-by-pixel basis. However, there are substantial practical challenges:

- *Radiometric alignment* - different sensor characteristics, and illumination or atmospheric conditions must be removed or factored out.
- *Spatial alignment* - benthic features on reefs occur at sub-pixel scales even for the highest available image resolutions.
- *Acquisition of two or more suitable images* - costs may be prohibitive or satellite acquisitions may not be conveniently scheduled.

A change detection approach was the basis of the demonstration of the detection of the Keppel Islands 2002 bleaching event using IKONOS data by Elvidge et al. (2004), which remains at the time of writing probably the only peer-review published demonstration of bleaching detection from optical satellite data. In another example Dadhich et al. (2011) assessed post-bleaching change in coral cover using two QuickBird images. Due to low availability of hyperspectral satellite data, and the cost and complications of airborne data, no hyperspectral multi-image change detection appears to have been attempted in reef environments. However, the launch of future hyperspectral satellite sensors such as EnMAP, with its 4-day revisit time, will provide a new impetus for multi-image approaches. In addition to detecting change, using multiple images provides more information and could help factor out variations due to atmospheric conditions and sea state. Fully understanding the confounding affect of environmental fluctuations is a prerequisite of image based change detection that is not trivial. For example, sediment resuspension on a reef could be mistaken for bleaching. Relatively little is known about the optical effects of reef sediments in general (Hedley 2011a).

In the earliest days of Landsat applications on reefs, Bina and Ombac (1979) proposed the use of multiple time point data to minimize the effect of tidal variation. Still to this date the use of multiple time series data on coral reefs has largely been restricted to Landsat (Andréfouët et al. 2001; Dustan et al. 2002; Phinney et al. 2002; Schuyler et al. 2006), which is clearly a function of data availability. Despite the well-established advantages of airborne hyperspectral data for one-off mapping (Mumby et al. 1997) operational satellite data will continue to have a role to play. Dustan et al. (2002) and Phinney et al. (2002) made the promising observation of reflectance changes over 20 Landsat images consistent with known coral to algal shifts in the Caribbean. ESA's upcoming Sentinel 2 is often posited as a Landsat and SPOT continuity mission but in fact this instrument has five narrow bands that are 'hyperspectral' in character (Table. 4.1). Models suggest it will outperform Landsat in reef applications (Hedley et al. 2012a) and combined with 10 m spatial resolution, 5-day revisit for coastal areas, the capability for reef change detection using Sentinel 2 will surely be a priority for future investigation.

4.3.6 Inversion Methods

The most recent developments in shallow water remote sensing are in radiative transfer model inversion methods, which are variously referred to as 'physics-based' or 'semi-analytical' and largely stem from key publications by Lee et al. (1998, 1999) and Mobley et al. (2005). The idea behind these methods is to construct a 'forward model' for above-water spectral reflectance that takes a number of input parameters including depth, the concentration of various water column constituents such as CDOM and phytoplankton, and the choice of bottom material, which determines the bottom spectral reflectance. For each pixel in an image an inversion algorithm effectively runs the model backwards, to find the best combination of input parameter values to give the closest spectral match between the model output and the measured pixel reflectance. These methods were developed for hyperspectral data, since numerous wavelength bands are necessary to tease apart the spectral influence of all the different components (Fig. 4.2 and 4.3). The methods can be equally applied to multispectral data but the inherent uncertainty will increase (see Sect. 4.4).

Figures 4.7 and 4.8 illustrate the kind of products that can be derived from these inversion methods. The raw outputs are image layers for each of the model input parameters: bathymetry, benthic type, and values that express the relative concentrations of CDOM or phytoplankton and the water column backscatter. From these raw parameters it is easy to calculate secondary outputs, the bottom reflectance can be reconstructed, water column optical properties can be recombined to give

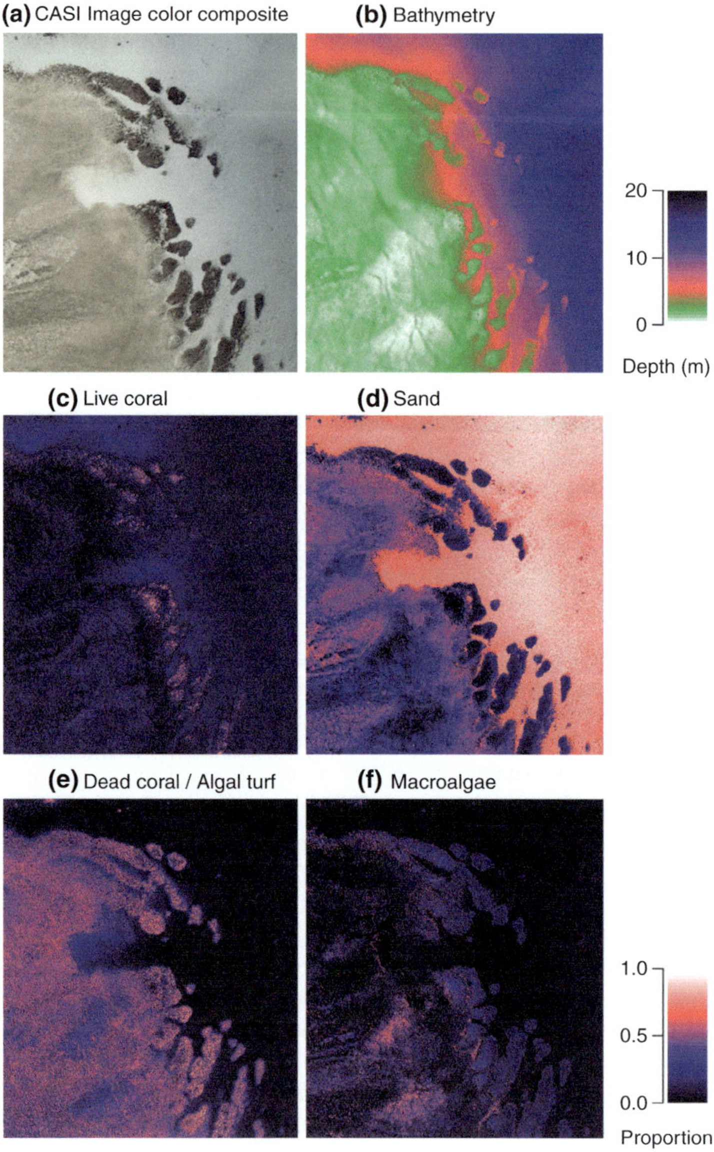

Fig. 4.7 Output layers from an inversion method applied to (**a**) CASI data of Heron Reef, GBR, Australia. Output includes: (**b**) estimated bathymetry; (**c**–**f**) estimated cover of four benthic classes, which should be interpreted with caution. Sand (**d**) is reasonable but coral (**c**), dead coral (**e**) and algae (**f**) are spectrally similar so may be confused (Fig. 4.2). Meaningful benthic validation is extremely challenging due to the spatial complexity and scales involved. The image subset is from the *upper right* of Fig. 4.4 and represents an area approximately 500 × 600 m in pixels of 1 m

spectral absorption coefficients, or depth and attenuation can be combined to produce a relative benthic light product. However, the accuracy of the raw individual layers may differ widely. Depth estimation is fairly robust, and in reef environments is always the most accurate retrieval (Dekker et al. 2011, 2009a, 2009b, 2010; Mobley et al. 2005; Lesser and Mobley 2007). In reef environments where the water is often very clear there is little 'signal' to establish the optical properties of the water column, and hence CDOM, phytoplankton and backscatter retrievals are often invalid and simply correlate with bottom reflectance. The situation is very different in waters with higher phytoplankton or CDOM levels (Lee et al. 2001). In waters that are not 'optically shallow', conversely, the bottom reflectance cannot be determined. A further potential disadvantage of spectral matching approaches is that the measure of spectral fit employed usually favors retrieval of factors, such as depth, which affect the whole spectrum. Narrow wavelength features like those discussed in Sect. 4.3.2 only weakly affect the matching process. Benthic type mapping may benefit from a hybrid approach where matching is weighted in wavelength regions of known pigment features, but this has not yet been tried. In all cases basic inversion methods offer no 'quality assurance' and simply rely on user interpretation or validation to deduce which estimations are reliable. The next section discusses refinements that can quantify or automate quality assurance.

With respect to choosing a methodological approach, the methods of Lee et al. (1998, 1999) and Mobley et al. (2005) differ substantially in the practical sense of the forward model and inversion method used, but fundamentally the methods are very similar. Mobley et al. (2005) populates a look-up table of reflectances using the commercial software Hydrolight (or the associated Ecolight). Hydrolight is a full numerical integration based model for light propagation in water, sometimes referred to as an 'exact' model in that it embodies the physical theory of radiative transfer. Lee et al. (1998, 1999) uses a more approximate forward model that gives above-water reflectance directly as the output of a series of simple equations. This model can be evaluated sufficiently fast to apply successive approximation algorithm such as Levenberg–Marquardt (Wolfe 1978) at each pixel. However, several key parameters in Lee et al. (1998) were derived by multiple runs of Hydrolight, hence with the same parameterization the models should give almost identical results. The majority of differences are due to implementation details and discretization or local optima problems from inversion via look up tables versus successive approximation (Dekker et al. 2011; Hedley et al. 2009a, unpublished data). Since no off-the-shelf software currently exists for these algorithms the look-up table approach is the most straightforward solution for practitioners. While Hydrolight (Mobley and Sundman 2000) is a commercial package, a free validated open-source implementation of the same solution algorithm exists, PlanarRad (Hedley 2011b). PlanarRad is functionally similar to Hydrolight but lacks chlorophyll fluorescence. Incorporating chlorophyll fluorescence may be important in high phytoplankton environments (Tote et al. 2011) but is an insignificant factor in most reef environments.

Since the Lee et al. (1998, 1999) original publications numerous other workers have devised their own variants of the semi-analytical algorithm. Wettle and Brando (2006) use 'specific inherent optical properties' (SIOPs), where water samples from the site are required to characterize the concentration-dependent, or 'specific', spectral absorptions of water constituents. The suggested advantages are two-fold: (1) the method is calibrated for local water constituents, and (2) the outputs are actual constituent concentrations whereas the original formulation of Lee et al. (1999) returned concentration proxies that evade direct interpretation. Wettle and Brando (2006) also introduced a linear mixing model for substrate reflectance; this concept has been used by several other workers (Klonowski et al. 2007; Hedley et al. 2009a). A number of these method variants were tested in a cross-comparison exercise with a Caribbean and Australian data set (Dekker et al. 2011). It should be noted that these methods require high quality atmospheric correction (Goodman et al. 2008).

4.4 Conclusions

The remote sensing community in general increasingly recognizes the need to provide confidence intervals on remotely sensed products. For classification approaches on reefs the normal practice is to describe overall accuracy in terms of misclassification rates, but this does not express confidence at the pixel level and is based on regions where ground truth data is available. In reality inaccuracies may be dependent on the location in the image. For example, determination of benthic type will be increasingly uncertain in deeper water. Inaccuracies also result from the various assumptions and simplifications inherent to the calibration, correction and image processing workflow (e.g., atmospheric correction, sunglint correction, radiative transfer modeling, etc.). It is important to gain an understanding of how these inaccuracies are manifest in the classification and mapping output. Analysis methods based on wavelength features or model inversion will always produce 'an answer' at each pixel, even if there is almost no information to extract. Confidence in the output values of any method may vary widely across an image, and this is especially true in coral reef applications where the environment displays heterogeneity in multiple aspects: depth, water constituents, and sea surface state (e.g., inside and outside of a lagoon).

One way to characterize uncertainty is to consider the noise-equivalent delta-reflectance, NEΔR (Brando and Dekker 2003). In shallow water applications, perturbation in pixel-to-pixel spectral reflectance caused by the top of the water surface, atmospheric turbulence, and sensor noise can be grouped to 'environmental noise' and evaluated from the band-wise standard deviation over an area of homogenous deep water in an image (Brando and Dekker 2003). Brando et al. (2009) used this measure directly to provide pixel quality assurance in a shallow water model inversion, excluding pixels where depth was such that the reflectance was below a NEΔR determined threshold. However, this requires absolute depth to be correct and excludes other uncertainties, such as confusing deep waters for shallow dark benthos.

A more general approach to quantify uncertainty is to apply a noise-perturbed repeated analysis to determine multiple model solutions in every pixel (Hedley et al. 2009b, 2010). Figure 4.8 illustrates uncertainty propagation in a physics-based inversion algorithm applied to CASI and QuickBird data of Heron Reef, Australia. In this case environmental noise has been characterized as the covariance matrix of reflectance over a deep water area. For each pixel in the image the model has been inverted 20 times, with the pixel reflectance perturbed by a random noise term each time. Hence for every pixel there are 20 estimates of depth, 20 of benthic composition etc., from this 90 % confidence intervals can be calculated giving error bars on every parameter for every pixel.

Figure 4.8 illustrates the value of 19-band CASI data over 4-band QuickBird data. Both datasets are capable of supporting bathymetric estimations (Fig. 4.8f), although the uncertainty for QuickBird starts to increase below 5 m. However, only the CASI data can support any level of benthic type mapping (Fig. 4.8b and c). Uncertainty for sand is low and visual interpretation indicates the estimated values are sensible (corresponds to Fig. 4.8d). While CASI estimated coral cover is reasonable (Figs. 4.8b and 4.7c), QuickBird has extremely high uncertainty for benthic type (Fig. 4.8b and c) and the mean estimations are clearly in error. Note that uncertainty applies to all components of the system, CASI has high uncertainty for water column absorption where the water is shallow and benthos is heterogeneous (left of Fig. 4.7d), but water properties over deep substrate that is clearly identifiable as sand are more certain (right of Fig. 4.8d).

Therefore the limit of what can be achieved in a particular remote sensing objective is a function of both environmental variation or 'noise' and sensor configuration and sensor noise (Fig. 4.9). Mapping benthic type by QuickBird is a 'sensor-limited' scenario (Fig. 4.9b) because it's possible to do better with the same analysis with hyperspectral CASI data (Figs. 4.8b and c). If perfect hyperspectral data with zero sensor noise were available, environmental variations would be the limiting factor (Fig. 4.9c). The key question is, how far along the scale with current sensor technologies are we to the 'environmentally limited' scenario? That is, could a new sensor with more or narrower bands, and better signal to noise characteristics, lead to better or more consistent results in reef mapping than have been demonstrated to date? In fact, modeling experiments (Hedley et al. 2012b) suggest we are very close to the environmental limit. So the next significant developments in coral reef remote sensing will most likely not be in sensor technology, but in techniques that minimize uncertainties by exploiting multiple sources of data, image time series and spatial patterns.

One way to reduce uncertainties and increase accuracy is to constrain the possibilities embodied in the analysis. For example, certain benthic types have known depth ranges, and so the solution that they occur in combination with certain depths is unlikely. Fearns et al. (2008) used a Bayesian analysis in this way to improve a habitat map derived from HyMap data of Ningaloo Reef in Western Australia. Statistical approaches to combine ancillary data such as sonar (Bejarano

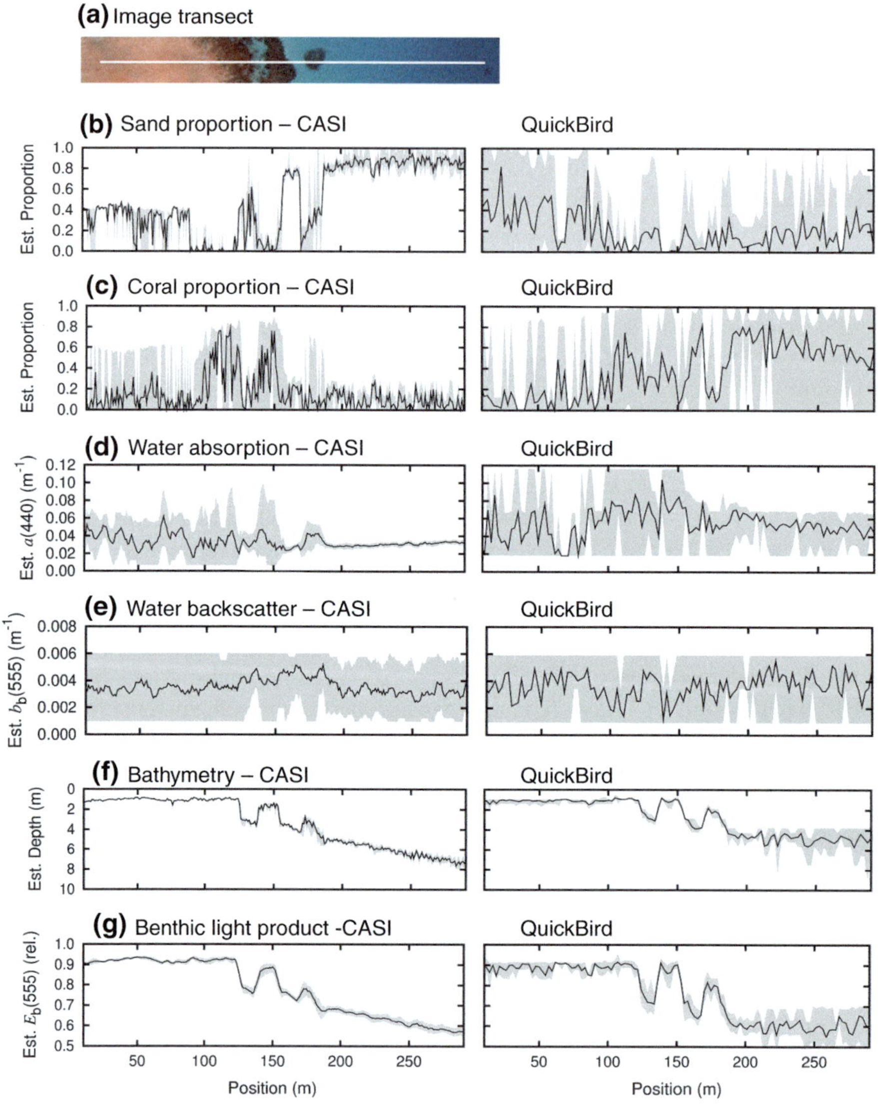

Fig. 4.8 Uncertainty propagation through a physics-based inversion model incorporating six benthic types represented by thirteen spectral reflectances (see Hedley et al. 2009a). Plots show an image transect from shallow lagoon (*left*) through high coral cover fore reef (~80 % cover) to deep sand area (*right*). Grey regions show 90 % confidence intervals on estimations from 17-band CASI data and QuickBird

et al. 2010) or other a priori knowledge are certain to play a key role in future developments in coral reef remote sensing.

As a final note, this chapter started by highlighting the capabilities of the mantis shrimp's 'hyperspectral' eye. However, looking at the simple red–green–blue reef

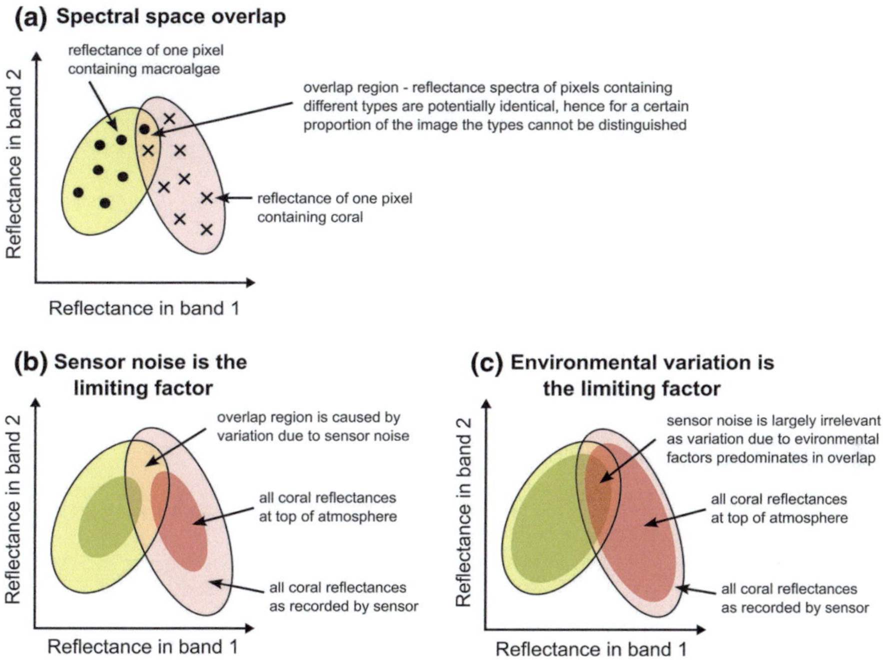

Fig. 4.9 Achievable remote sensing objectives, such as mapping coral versus macroalgae, may be limited by the sensor or by sources of environmental 'noise'. **a** Benthic types cannot be distinguished in some cases if there is spectral space overlap between pixels that contain the types of interest. Spectral variation across pixels of the same benthic type is the product of multiple environmental and sensor noise contributions. **b**, **c** Processes that contribute the most to spectral space overlap are the primary limiting factor

image in Fig. 4.7a, or even a monochrome version, it is fairly easy for a person familiar with coral reefs to identify sand regions, fore reef and back reef, and where coral is likely to be found. Spatial pixel-to-pixel patterns of light and dark are enough for the human eye (and brain) to do almost as good a job as hyperspectral remote sensing. It is clear that treating the mapping problem as one in which each single pixel is examined in isolation runs counter to the methods evolution has shaped in our own image processing system. This observation points the way to future developments: to mirror the capabilities of human vision is technically challenging but offers rich potential in an environment that is so strongly characterized by spatial patterns.

Acknowledgments Reflectance spectra in Fig. 4.3 were collected by Chris Roelfsema. Heron Island imagery was collected by Stuart Phinn and funded by the Australian Research Council, and image pre-processing was conducted by Karen Joyce. Inherent optical properties in Fig. 4.2b were collected with instrumentation held by the UK's NERC Field Spectroscopy Facility and during field work part-funded by the World Bank/Global Environment Facility Coral Reef Targeted Research Program. Figure 4.4a and b were derived from a figure previously published in Hedley et al. 2005 and are reproduced with permission from Taylor and Francis. Figure 4.5 is reproduced from Harborne et al. 2006 with permission from the Ecological Society of America.

Suggested Reading

Dekker AG, Phinn SR, Anstee J, Bissett P, Brando VE, Casey B, Fearns P, Hedley J, Klonowski W, Lee ZP, Lynch M, Lyons M, Mobley C, Roelfsema C (2011a) Inter-comparison of shallow water bathymetry, hydrooptics, and benthos mapping techniques in Australian and Caribbean coastal environments. Limnol Oceanogr Methods 9:396–425

Hedley JD, Mumby PJ (2002a) Biological and remote sensing perspectives of pigmentation in coral reef organisms. Adv Marine Biol 43:277–317

Kirk JTO (2010) Light and photosynthesis in aquatic ecosystems. Cambridge University Press, Cambridge

Lesser MP, Mobley CD (2007a) Bathymetry, water optical properties, and benthic classification of coral reefs using hyperspectral remote sensing imagery. Coral Reefs 26:819–829

Mobley CD (1994a) Light and water. Academic, San Diego

References

Adams JB, Smith MO, Johnson PE (1986) Spectral mixture modelling: a new analysis of rock and soil types at the Viking Lander I site. J Geophys Res 91:8098–8112

Adler-Golden SM, Matthew MW, Bernstein LS, Levine RY, Berk A, Richtsmeier SC, Acharya PK, Anderson GP, Felde JW, Gardner JA, Hoke ML, Jeong LS, Pukall B, Ratkowski AJ, Burke HK (1999) Atmospheric correction for short-wave imagery based on MODTRAN 4. SPIE Proc 3753:61–69

Andréfouët S, Muller-Karger FE, Hochberg EJ, Hu C, Carder KL (2001) Change detection in shallow coral reef environments using Landsat 7 ETM + data. Remote Sens Environ 78:150–162

Bejarano S, Mumby PJ, Hedley JD, Sotheran I (2010) Combining optical and acoustic data to enhance the detection of Caribbean forereef habitats. Remote Sens Environ 114:2768–2778

Bertels L, Vanderstraete T, Collie SV, Knaeps E, Sterckx S, Goossens R, Deronde B (2007) Mapping of coral reefs using hyperspectral CASI data; a case study: Fordata, Tanimbar, Indonesia. Int J Remote Sens 29:2359–2391

Bina RT, Ombac ER (1979) Effects of tidal fluctuations on the spectral patterns of Landsat coral reef imageries. In: Proceedings of the 13th international symposium of the remote sensing of environment. University of Michigan, Ann Arbor, pp 2051–2070

Brando VE, Dekker AG (2003) Satellite hyperspectral remote sensing for estimating estuarine and coastal water quality. IEEE Trans Geosci Remote Sens 41:1378–1387

Brando VE, Anstee JM, Wettle M, Dekker AG, Phinn SR, Roelfsema C (2009) A physics based retrieval and quality assessment of bathymetry from suboptimal hyperspectral data. Remote Sens Environ 113:755–770

Caplosini P, Andréfouët S, Rion C, Payri C (2003) A comparison of Landsat ETM+, SPOT HRV, Ikonos, ASTER, and airborne MASTER data for coral reef habitat mapping in South Pacific islands. Canadian J Remote Sens 29:187–200

Cronin TW, Marshall NJ (1989) A retina with at least ten spectral types of photoreceptors in a mantis shrimp. Nature 339:137–140

Cureton GP, Anderson SJ, Lynch MJ, McGann BT (2007) Retrieval of wind wave elevation spectra from sunglint data. IEEE Trans Geosci Remote Sens 45:2829–2836

Dadhich AP, Nadaoka K, Yamamoto T, Kayanne H (2011) Detecting coral bleaching using high-resolution satellite data analysis and 2-dimensional thermal model simulation in the Ishigaki fringing reef, Japan. Coral Reefs. doi:10.1007/s00338-011-0860-1

de Haan JF, Kokke JMM, Hoogenboom HJ, Dekker AG (1997) An integrated toolbox for processing and analysis of remote sensing data of inland and coastal waters—atmospheric correction. In: 4th international conference on remote sensing for marine and coastal environments, Orlando, Florida de Vries DH (1994) Imaging spectroscopy: CASI operations in Australia during summer 1992/93. In: 7th Australasian remote sensing conference proceedings, pp 136–140, ARSC, Melbourne, Australia

Dekker A, Byrne G, Brando V, Anstee J (2003) Hyperspectral mapping of intertidal rock platform vegetation as a tool for adaptive management. CSIRO Land and Water, Canberra, Australia

Dekker AG, Phinn SR, Anstee J, Bissett P, Brando VE, Casey B, Fearns P, Hedley J, Klonowski W, Lee ZP, Lynch M, Lyons M, Mobley C, Roelfsema C (2011b) Inter-comparison of shallow water bathymetry, hydrooptics, and benthos mapping techniques in Australian and Caribbean coastal environments. Limnol Oceanogr Methods 9:396–425

Dustan P, Dobson E, Nelson G (2002) Landsat Thematic Mapper: detection of shifts in community composition of coral reefs. Conserv Biol 15:892–902

Elvidge CD, Dietz JB, Berkelmans R, Andréfouët S, Skirving W, Strong AE, Tuttle BT (2004) Satellite observation of Keppel Islands (Great Barrier Reef) 2002 coral bleaching using IKONOS data. Coral Reefs 23:123–132

Enriquez S, Mendez ER, Iglesias-Prieto R (2005) Multiple scattering on coral skeletons enhances light absorption by symbiotic algae. Limnol Oceanogr 50:1025–1032

ERDAS (2011). ERDAS IMAGINE. http://www.erdas.com

Exelis VIS, Exelis Visual Information Solutions (2012) ENVI—Environment for visualizing images, Version 4.8

Fearns P, Rodrigo G, Klonowski W (2008) Combining hyperspectral and environmental knowledge using probabilistic methods to produce shallow water habitat maps. In: Proceedings of ocean optics XIX, Ciocco, Tuscany, Italy

Ferrier G, Trahair NS (1995) Evaluation of apparent surface reflectance estimation methodologies. Int J Remote Sens 16:2291–2297

Gao B-C, Montes MJ, Ahmad Z, Davis CO (2000) Atmospheric correction algorithm for hyperspectral remote sensing of ocean color from space. Appl Opt 39:887–896

Goodman JA, Ustin SL (2007) Classification of benthic composition in a coral reef environment using spectral unmixing. J Appl Remote Sens 1:011501

Goodman JA, Lee ZP, Ustin SL (2008) Influence of atmospheric and sea-surface corrections on retrieval of bottom depth and reflectance using a semi-analytical model: a case study in Kaneohe Bay, Hawaii. Appl Optics 47:F1–F11

Green EP, Mumby PJ, Edwards AJ, Clark CD (1996) A review of remote sensing for tropical coastal resources assessment and management. Coastal Manage 24:1–40

Green EP, Mumby PJ, Edwards AJ, Clark CD (2000) Remote sensing handbook for tropical coastal management. UNESCO, Paris

Hamylton S (2011) Estimating the coverage of coral reef benthic communities from airborne hyperspectral remote sensing data: multiple discriminant function analysis and linear spectral unmixing. Int J Remote Sens. doi:10.1080/01431161.2011.574162

Harborne AR, Mumby PJ, Zychaluk K, Hedley JD, Blackwell PG (2006) Modeling the beta diversity of coral reefs. Ecology 87:2871–2881

Heege T, Hausknecht P, Kobryn H (2007) Hyperspectral seafloor mapping and direct bathymetry calculation using HyMap data from the Ningaloo Reef and Rottnest Island areas in Western Australia. In: Proceedings of the 5th EARSel Workshop on imaging spectroscopy, Bruges, 23–25 April, pp 1–8

Hedley JD (2008) A three-dimensional radiative transfer model for shallow water environments. Opt Express 16:21887–21902

Hedley JD (2011a) Modelling the optical properties of suspended particulate matter of coral reef environments using the finite difference time domain (FDTD) method. Geo Marine Letters. doi:10.1007/s00367-011-0265-8

Hedley JD (2011b) PlanarRad user manual. http://www.planarrad.com

Hedley JD, Enríquez S (2010) Optical properties of canopies of the tropical seagrass *Thalassia testudinum* estimated by a three-dimensional radiative transfer model. Limnol Oceanogr 55:1537–1550

Hedley JD, Harborne AR, Mumby PJ (2005) Simple and robust removal of sun glint for mapping shallow water benthos. Int J Remote Sens 26:2107–2112
Hedley JD, Mumby PJ (2002b) Biological and remote sensing perspectives of pigmentation in coral reef organisms. Adv Marine Biol 43:277–317
Hedley JD, Mumby PJ (2003) A remote sensing method for resolving depth and subpixel composition of aquatic benthos. Limnol Oceanogr 48:480–488
Hedley JD, Mumby PJ, Joyce KE, Phinn SR (2004) Spectral unmixing of coral reef benthos under ideal conditions. Coral Reefs 23:60–73
Hedley JD, Roelfsema C, Phinn SR (2009a) Efficient radiative transfer model inversion for remote sensing applications. Remote Sens Environ 113:2527–2532
Hedley JD, Roelfsema C, Phinn S (2009b) Uncertainty propagation in a physics-based shallow water mapping algorithm applied to CASI and QuickBird imagery of Heron Reef, GBR. In: Proceedings of RSPSoc conference, Leicester
Hedley JD, Roelfsema C, Phinn S (2010) Propagating uncertainty through a shallow water mapping algorithm based on radiative transfer model inversion. In: Proceedings of ocean optics XX, Anchorage
Hedley JD, Roelfsema C, Koetz B, Phinn S (2012a) Capability of the Sentinel 2 mission for tropical coral reef mapping and coral bleaching detection. Remote Sens Environ. doi:10.1016/j.rse.2011.06.028
Hedley JD, Roelfsema C, Phinn S, Mumby PJ (2012b) Environmental and sensor limitations in optical remote sensing of coral reefs: implications for monitoring and sensor design. Remote Sens 4:271–302. doi:10.3390/rs4010271
Hedley JD, Roelfsema C, Phinn SR (2012c). Bathymetric map of Heron Reef, Australia, derived from airborne hyperspectral data at 1 m resolution. doi: 10.1594/PANGAEA.779522
Hochberg EJ, Atkinson MJ (2000) Spectral discrimination of coral reef benthic communities. Coral Reefs 19:164–171
Hochberg EJ, Atkinson MJ, Andrefouet S (2003a) Spectral reflectance of coral reef bottom-types worldwide and implications for coral reef remote sensing. Remote Sens Environ 85:159–173
Hochberg E, Andrefouet S, Tyler M (2003b) Sea surface correction of high spatial resolution Ikonos images to improve bottom mapping in near-shore environments. IEEE Trans Geosci Remote Sens 41:1724–1729
Hochberg EJ, Atkinson MJ (2003) Capabilities of remote sensors to classify coral, algae and sand as pure and mixed spectra. Remote Sens Environ 85:174–189
Holden H, LeDrew E (1999) Hyperspectral identification of coral reef features. Int J Remote Sens 13:2545–2563
ITRES (2008) CASI-550 airborne hyperspectral solutions. www.itres.com/assets/pdf/CASI-550.pdf
Isoun E, Fletcher C, Frazer N, Gradie J (2003) Multi-spectral mapping of reef bathymery and coral cover; Kailua Bay, Hawaii. Coral Reefs 22:68–82
Jeffery SW, Mantoura RFC, Bjørnland T (1997) Data for the identification of 47 key phytoplankton pigments. In: Jeffery SW, Mantoura RFC, Wright SW (eds) Phytoplankton pigments in oceanography: guidelines to modern methods. UNESCO, Paris, pp 449–559
Joyce KE (2004) A method for mapping live coral cover using remote sensing. Ph.D. thesis, University of Queensland, Brisbane, Australia
Karpouzli E, Malthus T (2003) The empirical line method for atmospheric correction of IKONOS imagery. Int J Remote Sens 24:1143–1150
Kay S, Hedley JD, Lavender S (2009) Sun glint correction of high and low spatial resolution images of aquatic scenes: a review of methods for visible and near-infrared wavelengths. Remote Sensing 1:697–730
Kay S, Hedley JD, Lavender S, Nimmo-Smith A (2011) Light transfer at the ocean surface modeled using high resolution sea surface realizations. Opt Express 19:6493–6504
Kennedy RE, Cohen WB, Takao G (1997) Empirical methods to compensate for a view-angle-dependent brightness gradient in AVIRIS imagery. Remote Sens Environ 62:277–291

Klonowski WM, Fearns PRCS, Lynch MJ (2007) Retrieving key benthic cover types and bathymetry from hyperspectral imagery. J Appl Remote Sens 1:011505
Kotchenova SY, Vermote EF, Matarrese R, Klemm FJ Jr (2006) Validation of a vector version of the 6S radiative transfer code for atmospheric correction of satellite data Part I: path radiance. Appl Optics 45:6726–6774
Kotchenova SY, Vermote EF (2007) Validation of a vector version of the 6S radiative transfer code for atmospheric correction of satellite data. Part II: homogeneous Lambertian and anisotropic surfaces. Appl Opt 46:4455–4464
Kutser T, Miller I, Jupp D (2006) Mapping coral reef benthic substrates using hyperspectral space-borne images and spectral libraries. Estuar Coast Shelf Sci 70:449–460
Kutser T, Vahtmäe E, Praks J (2009) A sun glint correction method for hyperspectral imagery containing areas with non-negligible water leaving NIR signal. Remote Sens Environ 113:2267–2274
Lee Z, Carder KL, Mobley CD, Steward RG, Patch JS (1998) Hyperspectral remote sensing for shallow waters I. A semianalytical model. Appl Optics 37:6329–6338
Lee Z, Carder KL, Mobley CD, Steward RG, Patch JS (1999) Hyperspectral remote sensing for shallow waters. 2. Deriving bottom depths and water properties by optimization. Appl Opt 38:3831–3843
Lee Z, Carder KL, Chen RF, Peacock TG (2001) Properties of the water column and bottom derived from airborne visible infrared imaging spectrometer (AVIRIS) data. J Geophys Res 106:11639–11651
Lee Z, Casey B, Arnone R, Weidemann A, Parsons R, Montes MJ, Gao B-C, Goode W, Davis CO, Dyef J (2007) Water and bottom properties of a coastal environment derived from Hyperion data measured from the EO-1 spacecraft platform. J Appl Remote Sens 1:011502
Lesser MP, Mobley CD (2007b) Bathymetry, water optical properties, and benthic classification of coral reefs using hyperspectral remote sensing imagery. Coral Reefs 26:819–829
Lyzenga DR (1978) Passive remote sensing techniques for mapping water depth and bottom features. Appl Opt 17:379–383
Lyzenga DR (1981) Remote sensing of bottom reflectance and water attenuation parameters in shallow water using aircraft and Landsat data. Int J Remote Sens 2:71–82
Lyzenga D, Malinas N, Tanis F (2006) Multispectral bathymetry using a simple physically based algorithm. IEEE Trans Geosci Remote Sens 44:2251–2259
Maritorena S, Morel A, Gentili B (1994) Diffuse reflectance of oceanic shallow waters: Influence of water depth and bottom albedo. Limnol Oceanogr 39:1689–1703
Mather PM (1999) Computer processing of remotely sensed images, 2nd edn. Wiley, Chichester
Mayer B, Kylling A (2005) Technical note: the libRadtran software package for radiative transfer calculations—description and examples of use. Atmos Chem Phys 5:1855–1877
Milovich JA, Frulla LA, Gagliardini DA (1995) Environmental contribution to the atmospheric correction for Landsat-MSS images. Int J Remote Sens 16:2515–2537
Mishra DR, Narumalani S, Rundquist D, Lawson M, Perk R (2007) Enhancing the detection of coral reef and associated benthic habitats: a hyperspectral remote sensing approach. J Geophys Res 112: CO8014
Mobley CD (1994b) Light and water. Academic, San Diego
Mobley CD, Sundman L (2000) Hydrolight 4.1 user's guide. Sequoia Scientific. http://www.sequoiasci.com/products/Hydrolight.aspx
Mobley CD, Sundman LK, Davis C, Bowles JH, Downes TV, Leathers RA, Montes MJ, Bissett WP, Kohler DDR, Reid RP, Louchard EM, Gleason A (2005) Interpretation of hyperspectral remote-sensing imagery by spectrum matching and look-up tables. Appl Opt 44:3576–3592
Montes MJ, Gao B-C, Davis CO (2001) A new algorithm for atmospheric correction of hyperspectral remote sensing data. Proc SPIE-Int Soc Opt Eng SPIE 4383:23–30
Mumby PJ, Green EP, Edwards AJ, Clark CD (1997) Coral reef habitat-mapping: how much detail can remote sensing provide? Mar Biol 130:193–202
Mumby PJ, Clark CD, Green EP, Edwards AJ (1998) Benefits of water column correction and contextual editing for mapping coral reefs. Int J Remote Sens 19:203–210

Mumby PJ, Chisholm JRM, Clark CD, Hedley JD, Jaubert J (2001) A bird's-eye view of the health of coral reefs. Nature 413:36
Mumby PJ, Edwards AJ (2002) Mapping marine environments with IKONOS imagery: enhanced spatial resolution can deliver greater thematic accuracy. Remote Sens Environ 82:248–257
Mumby PJ, Harbourne AR (1999) Development of a systematic classification scheme of marine habitats to faciltate regional management and mapping of Caribbean coral reefs. Biol Conserv 88:155–163
Mumby PJ, Hedley JD, Chisholm JRM, Clark CD, Ripley H, Jaubert J (2004) The cover of living and dead corals from airborne remote sensing. Coral Reefs 23:171–183
Mustard J, Staid M, Fripp W (2001) A semianalytical approach to the calibration of AVIRIS data to reflectance over water application in a temperate estuary. Remote Sens Environ 75:335–349
Pope RM, Fry ES (1997) Absorption spectrum (380–700 nm) of pure water II. Integrating cavity measurements. Appl Optics 36:8710–8723
Phinney J, Muller-Karger F, Dustan P, Sobel J (2002) Using remote sensing to reassess the mass mortality of *Diadema antillarum* 1983–1984. Conserv Biol 15:885–891
Ricchiazzi P, Yang S, Gautier C, Sowle D (1998) SBDART: a research and teaching software tool for plane-parallel radiative transfer in the earth's atmosphere. Bull Am Meteorol Soc 79:2101–2114
Roelfsema CM, Marshall J, Hochberg E, Phinn S, Goldizen A, Joyce KE (2006) Underwater spectrometer system 2006 (UWSS04). http://ww2.gpem.uq.edu.au/CRSSIS/publications/UW %20Spec %20Manual %2029August06.pdf
Schuyler Q, Dustan P, Dobson E (2006) Remote sensing of coral reef community change on a remote coral atoll: Karang Kapota, Indonesia. In: Proceedings of 10th international coral reef symposium, Okinawa, Japan. 28 June– 2 July, 2004, pp 1763–1770
Siebeck UE, Marshall NJ, Kluter A, Hoegh-Guldberg O (2006) Monitoring coral bleaching using a colour reference card. Coral Reefs 25:453–460
Slivkoff M (2010) Dynamic above-water radiance and irradiance collector (DALEC). http://wwwinsitumarineoptics.com
Smith GM, Milton EJ (1999) The use of the empirical line method to calibrate remotely sensed data to reflectance. Int J Remote Sens 20:2653–2662
Splinter KD, Holman RA (2009) Bathymetry estimation from single-frame images of nearshore waves. IEEE Trans Geosci Remote Sens 47:3151–3160
Tsai F, Philpot W (1998) Derivative analysis of hyperspectral data. Remote Sens Environ 66:41–51
Tote C, Sterckx S, Knaeps E, Raymaekers D (2011) Remote sensing of shallow water bodies: effect of the bottom substrate on water leaving reflectance. In: 7th EARSeL workshop of the special interest group in imaging spectroscopy, 11–13th April 2011, Edinburgh
Wettle M, Ferrier G, Lawrence AJ, Anderson K (2003) Fourth derivative analysis of Red Sea coral reflectance spectra. Int J Remote Sens 24:3867–3872
Wettle M, Brando VE (2006) Sambuca: semi-analytical model for bathymetry un-mixing and concentration assessment. CSIRO Land and Water Science Report 22/06
Wolfe MA (1978) Numerical methods for unconstrained optimization. Van Nostrand Reinhold Company, New York

Section II
LiDAR

Chapter 5
LiDAR Overview

Samuel J. Purkis and John C. Brock

Abstract LiDAR is rapidly gaining prominence in coral reef research and management due to its ability to represent complex topographic structures with very high precision. The technology is now moving out of the research arena into operational use with an accompanied dramatic drop in survey cost. Typically aircraft mounted, LiDAR instruments are capable of auditing large areas more rapidly than could be achieved using more traditional vessel-based techniques. LiDAR, in contrast to passive optical remote sensing data, which rely on inference using some radiance measurement or reflectance index, provides direct measurements of elevation from which both terrestrial and marine topography can be easily derived. The combination of high point density scanning laser altimetry data coupled with high precision GPS provide very detailed three-dimensional information. While typically used as the basis for mapping, these data are also well poised for the derivation of coral reef biophysical properties, such as seabed geomorphology, rugosity, texture and bed-form geometry. In turn, such parameters can be related to both the biology and geology of the audited ecosystem.

S. J. Purkis (✉)
National Coral Reef Institute, Oceanographic Center, Nova Southeastern University,
8000 N. Ocean Drive, Dania, FL 33004, USA
e-mail: purkis@nova.edu

J. C. Brock
U.S. Geological Survey, Coastal and Marine Geology Program,
USGS National Center, Mail Stop 915-B 12201 Sunrise Valley Drive,
Reston, VA 20192, USA
e-mail: jbrock@usgs.gov

J. A. Goodman et al. (eds.), *Coral Reef Remote Sensing*,
DOI: 10.1007/978-90-481-9292-2_5,

5.1 Introduction

Laser systems used in measuring distances are called LiDAR (Light Detection and Ranging). As with radar, LiDAR is an 'active' remote sensing technology since it provides its own energy. The sensors considered in Chaps. 1–4 are 'passive', because they measure natural electromagnetic energy, typically provided by the sun. LiDAR instruments are unique in that they confine the coherent light energy within a very narrow beam, providing pulses of very high peak intensity. This enables blue-green wavelength LiDAR to penetrate clear to moderately turbid coastal waters for bathymetric measurements and allows mainly near-infrared LiDAR to probe gaps in forest canopies to provide topographic data for digital elevation models (DEMs) (Brock and Sallenger 2000; Brock and Purkis 2009). For these measurements, LiDAR systems employ an accurate clock that records the round-trip travel time between the transmission and reception of a reflected laser pulse.

The development of LiDAR technology commenced in the 1970s, with early systems built in the USA and Canada (Ackermann 1999). However, the technology was not implemented aboard aircraft until the late 1980s, at which point it started to be used for the accurate determination of terrain models (Baltsavias 1999). The application in these early years was limited by its complexity, cost effectiveness, and poor georeferencing. These limitations were gradually overcome following the availability of more reliable electronics, more advanced lasers, and the increased geo-accuracy afforded by the Global Positioning System (GPS). In 1988, the U.S. Army Corps of Engineers constructed an operational LiDAR system which was subsequently developed commercially by Optech Inc. (LaRocque and West 1990; Irish and Lillycrop 1999). Around this time the Australian Laser Airborne Depth Sounder (LADS) system was also developed and later commercialized by the Tenix LADS Corporation (Irish and White 1998). In the last decade, LiDAR has overcome many of the hurdles that plagued early setups.

Current systems use commercial lasers that can pulse at rates as high as 200 kHz and inertial measurement units (IMUs) that provide pointing precision that enables georeferencing accuracy to the sub-metre level. An IMU is an electronic device that measures and reports on an aircraft's velocity, orientation, and gravitational forces using a combination of digital accelerometers and gyroscopes. Good positional accuracy of LiDAR postings demands precise information on the aircraft location at both the time the laser sounding is dispatched groundward, and the time the reflection is subsequently received (Latypov 2002). GPS is only sufficiently accurate for this purpose when differentially corrected against a base-station (DGPS). Here, the position of the aircraft is acquired by GPS, and differential-GPS on-the-fly algorithms are applied at post processing (not real time). The GPS reference station should be positioned within the survey area and the rover unit (aboard the survey-aircraft) should not depart by more than 25 km. Note that at extremely stable (but rare) conditions of the troposphere and ionosphere this distance might exceed 100 km. With the differential correction robustly applied to the survey GPS data during post-processing, one can expect that the position is accurate within 0.05 m in all directions (Katzenbeisser 2003). Such advances in the use of GPS, alongside the fine precision afforded by IMU units, have served to decrease the cost and increase the accuracy of LiDAR as a survey technology.

LiDAR sensors can be categorized into three types: (1) profiling, (2) discrete return, and (3) waveform. Profiling sensors are the simplest and the sensor records only one return at fairly course sample densities along a narrow swath. Discrete return instruments are more advanced and employ a type of laser transceiver that records multiple returns (typically ~ 5) for each pulse of light sent towards the target, while waveform instruments record a digitized profile of the full return pulse. The addition of a scanning component (which may be zig-zag, parallel, elliptical, etc.) greatly increases the ground covered beneath each flight line and is typical for more modern systems. The mode of delivery of the laser light can also be used to characterize the technology. Most commonly used for ranging applications, 'pulsed' LiDAR consists of a pulsed laser transmitter, an optical telescope receiver that amplifies the backscatter, and photomultiplier receiver tube to convert optical energy into electrical impulses. Range to object is determined by recording the time taken by the transmitted pulse to the target and back. By contrast, 'continuous wave' LiDAR ranges by modulating the intensity of the laser light. Here, travel time is directly proportional to the phase difference between the received and transmitted sinusoidal laser signal. It is important to appreciate that while advanced scanning LiDAR can yield data which, if processed, can be viewed as an image; more simple profilers only deliver widely spaced range postings along a narrow corridor. In order that LiDAR data not be misinterpreted, it is paramount that the user be fully informed as to the specifics of the instrument used to acquire the soundings and the associated caveats with the dataset.

5.2 Physical Principles

5.2.1 Aircraft-Deployed LiDAR

Bathymetric LiDAR instruments are typically aircraft-mounted, operate on the transmission of green laser light (typically 532 nm), and record the intensity of back-scattered energy. In addition to a green laser, most bathymetric LiDAR systems also operate a separate near-infrared laser pulse. This is used to find the range to the water surface, but some designs employ the same green laser pulse to range to the sea surface as well as the sea bottom. Only systems with short laser pulses can follow this strategy without impacting the accuracy and precision of the data. Common to all of these designs is that the green laser pulse propagates through the water column and reflects off the sea bottom. The water depth is calculated from the time-lapse between the water surface return and the bottom return (Fig. 5.1). Dividing the travel-time of a reflected laser pulse by two, and

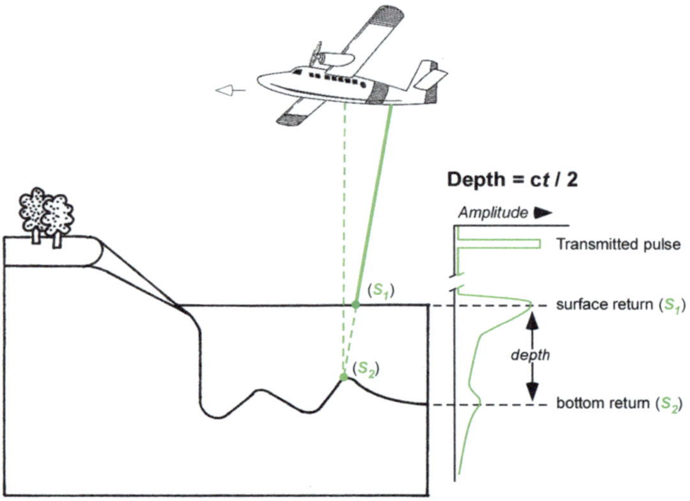

Fig. 5.1 Principles of operation of a LiDAR bathymeter. The water depth can be calculated from the travel time difference (t) between the water surface (S_1) and bottom (S_2) pulse returns. Here c represents the velocity of the laser light pulse

multiplying the result by the speed of light, yields the distance between an overhead LiDAR unit and a surface target. It should be recognised that since the laser energy is travelling at the speed of light, the timing mechanism of the sensor must at least be accurate to within a few nanoseconds, but ideally less than one nanosecond. As an example, a mistiming of 1 ns will yield a vertical error in the range of 30 cm.

Laser energy is lost due to refraction, backscattering, and absorption at the water surface, the sea bottom, and inside the water column. These effects all serve to diminish the strength of the bottom return and limit the maximum detectable depth. There is a distinction between bathymetric LiDAR that best measures water depth versus sensors that can capture submerged topography. The latter needs a more accurate aircraft trajectory and does not require tidal or swell correction. One also has to take into account that the speed of light depends on the density of the atmosphere, which means it varies with pressure, humidity and temperature. Considering that survey flights with a LiDAR will only be conducted under clear atmospheric conditions, one can neglect humidity. But pressure has to be considered, specifically if one is flying at various altitudes. For example, assume two survey flights, one at a coastline (0 m MSL) and one at a high elevation area (2,000 m MSL), both 2,000 m above ground. Taking the speed of light valid at the coast and applying it for the high region will lead to calculated distances which are about 12 cm too short, about twice the error that would be expected to arise from just positioning inaccuracies (Katzenbeisser 2003).

Each sounding must be corrected for water level fluctuations using either vertical aircraft positioning derived from GPS, or by referencing the LiDAR

Table 5.1 Typical LiDAR flight parameters

Flying height	200–500 m (400 m typical)
Vertical accuracy	±15 cm
Horizontal accuracy	dGPS = 3 m; kGPS = 1 metre
Max mapping depth	60 m (exceptionally clear water)
Typical *kd* product	4
Typical coastal *k* range	0.2–0.8 (d = 5–20 m)
Typical estuarine *k* range	1.0–4.0 (d = 1–4 m)
Sounding density	3–15 m
Sun angle	18–25° (to minimize glare)
Swath width	Typically in the range of 250 m
Sea state	Low (0–1 beaufort on the scale)
Water penetration	Blue-green LiDAR (532 nm) used
Aircraft altitude relative to the water or land surface	NIR LiDAR (1,064 nm) used

dGPS differential GPS mode; *kGPS* kinematic GPS mode

measurements of water surface location with water level gauge measurements. When coupled with robust GPS information, repeated postings within designated flight-line swaths provide a highly accurate characterization of bathymetry. Typical flight parameters for airborne LiDAR used in bathymetry are shown in Table 5.1.

Laser depth sounding techniques have proven most effective in clear, shallow waters. Since optical water clarity is the most limiting factor for depth detection, it is important to conduct the overflights during tidal and current conditions that minimize the water turbidity due to sediment re-suspension and river inflow. The LiDAR system must have a *kd* factor large enough to accommodate the water depth and water turbidity at the study site (k = attenuation coefficient; d = water depth). For instance, if a given LiDAR system has a $kd = 4$ and the turbid water has an attenuation coefficient of $k = 1$, the system will be effective only to depths of approximately 4 m. Beyond that depth, one may have to use acoustic echo-sounding techniques or side scanning sonar systems (Chaps. 8–10). By virtue of the laser intensity, a LiDAR pulse will travel deeper into the water than diffuse sunlight, commonly reaching 2–3 Secchi depths (Cecchi et al. 2004; Wang and Philpot 2007; Mohammadzadeh and Valadan Zoej 2008), which could be as deep at 60 m in clear waters over coral reefs (Fig. 5.2 - white rectangle). This is far superior to the depth penetration of passive optical systems which are generally limited to no better than 1.5 Secchi depths (Sinclair 1999).

The reason bathymetric LiDAR commonly employ a 532 nm blue-green laser to range the distance to the seabed is to maximize water penetration. Providing that the water is blue, and with the near exponential attenuation of electromagnetic energy by water that increases with increasing wavelength through the visible to near infrared spectrum, a pure blue laser with a wavelength shorter than 500 nm would offer greater penetration than the blue-green lasers that are typically employed. However, there are both engineering and physical constraints why shorter wavelength instruments are not used and blue-green lasers perform better

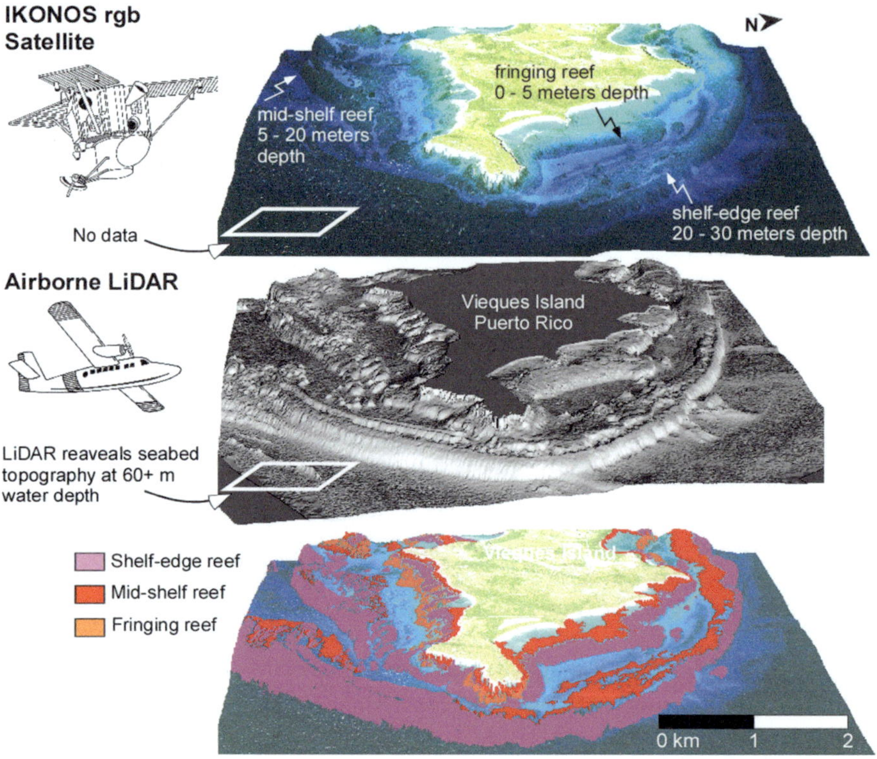

Fig. 5.2 Combined reef mapping (*bottom*) using IKONOS multispectral satellite data (*top*) and airborne bathymetric LiDAR soundings (*middle*). While this LiDAR lacks any multispectral capability, over exceptionally clear waters it is able to resolve the seabed to depths exceeding 60 m (*white rectangle*). The resulting thematic map product (*bottom*) is both highly accurate and 3-D. These data were acquired over the eastern point of the Island of Vieques, Puerto Rico. Satellite image: GeoEye

than blue in turbid coastal waters. From an engineering standpoint, by placing a nonlinear medium, usually a crystal, into the laser beam, the 1,064 nm output from a Nd:YAG laser can be converted to visible light with wavelengths of 532 nm. This process, termed 'frequency-doubling', is a relatively inexpensive way to produce a visible-light laser and hence the 532 nm adopted by many LiDAR systems has arisen from the easy availability and low cost of Nd:YAG 1,064 nm lasers. From a practical standpoint, wavelengths shorter than 532 nm are also problematic since they interact more strongly with the atmosphere than longer wavelengths, and secondly, creating a high-intensity blue laser is energetically significantly less efficient than the 532 nm blue-green laser. For all of these reasons, combined with the fact that blue lasers suffer from temperature problems at high powers, explains why 532 nm is the preferred wavelength for LiDAR sensors. Nonetheless, a constraint of adopting visible-wavelength lasers for LiDAR is that they are easily absorbed by the eye and hence the maximum power of the system is limited by the need to make them eye-safe.

Terrestrial topographic LiDAR systems typically utilize near-infrared (NIR) lasers with a wavelength of 1,064 nm, generated using the same Nd:YAG lasers as previously discussed, but not frequency-doubled into the visible spectrum. As with the 532 nm blue-green laser, this NIR wavelength is focused by the eye and similarly the power must be limited for safety. Although less accurate, military instruments often utilize lasers with infrared wavelengths as long as 1,550 nm. This holds the dual advantage of being eye-safe at much higher power levels and the beam is not visible using night-vision goggles.

Both terrestrial and bathymetric lasers suffer null or poor returns from dark substrates. For the former, typical culprits are surfaces such as water, asphalt and tar, while for bathymetric lasers, dark patches of dense seagrass and algal meadows may be problematic. Fog and clouds also absorb both NIR and visible wavelengths. Clouds tend to be more problematic for terrestrial surveys as they are flown at an altitude of 2,000 m, higher than that adopted for marine work.

Given that they do not penetrate water, NIR topographic lasers cannot be used to assess bathymetry, though they can be used to detect the water surface in bathymetric applications. By contrast, blue-green hydrographic lasers do reflect off of terrestrial targets and can be used to measure emergent terrain elevations. By employing temporal waveform digitizing, hybrid systems such as the Experimental Advanced Airborne Research LiDAR (EAARL) have demonstrated the capability of measuring both topography and bathymetry from sounding by a single blue-green laser (Bonisteel et al. 2009; McKean et al. 2009; Nayegandhi et al. 2009; Wright and Brock 2002). Figure 5.3 shows such a dataset. This experimental instrument may signal a future move towards commercial implementation of multiple-application, single blue-green or NIR wavelength instruments. The SHOALS and LADS systems have this capability too, but to a lesser accuracy because of footprint size. The single laser technique is also implemented in HawkEyeII and the Coastal Zone Mapping and Imaging LiDAR (CZMIL), with the addition of a segmented detector approach to increase the density of the measurements. The CZMIL is a new sensor development effort within the National Coastal Mapping Program which is based around the CHARTS (Compact Hydrographic Airborne Rapid Total Survey) sensor suite, the main components of which are a SHOALS-3000 topo-hydro LiDAR and a CASI 1500 hyperspectral imager.

Dual-wavelength LiDAR provides both bathymetric and topographic LiDAR mapping capability by carrying both a NIR and a blue-green laser. Access to a harmonised and consistent elevation model containing both bathymetry and topography is valuable in reef research because the architecture of terrestrial watersheds imparts considerable control over the health of corals (Rogers 1990; Lapointe and Clark 1992), an idea that will be further investigated later in this chapter. The NIR laser of a dual-wavelength LiDAR is not redundant over water, because it reflects from the water surface and can be used to acquire the range to the air–water interface, as well as to distinguish dry land from water using the

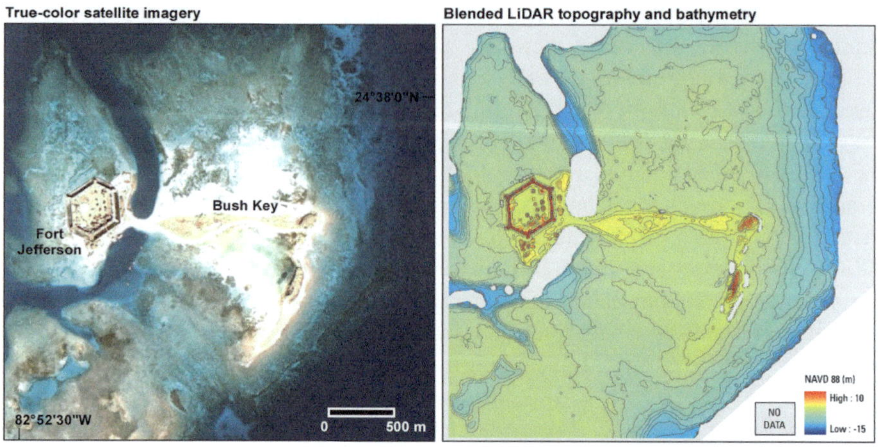

Fig. 5.3 On the *left* is a true-color satellite image, while the *right* plots seamless terrestrial topography and submarine bathymetry for the Dry Tortugas, an offshore coral reef system in South Florida, USA, acquired using the EAARL LiDAR. The elevated structure of both Fort Jefferson and the mangroves on Bush Key are clearly captured in the laser returns. The data are located within UTM Zone 17 and North is top. Credit: USGS

signal polarisation (Guenther 2007). And since the speed of light is different in air and water, knowledge of the distance to the water surface improves the overall ranging accuracy of the LiDAR system. In addition, specific LiDAR instruments, for example the Optech SHOALS system, record the red wavelength water Raman signal (647 nm). The water Raman signal derives from interactions between the blue-green laser and H_2O molecules that cause part of the energy to be back-scattered while changing wavelength (Guenther et al. 1994). Recording the Raman signal provides another means of identifying the water surface and can be used to identify incorrect surface detections that may arise due to land reflection or the presence of unexpected targets, such as birds.

As demonstrated by both SHOALS and the EAARL, one of the key advances in the development of bathymetric LiDAR systems has been the ability to survey across the littoral zone through the seamless acquisition of near-shore topography and shallow bathymetry with a single system on a single survey (LaRocque and West 1990; Irish and Lillycrop 1999; Guenther et al. 2000; Wozencraft 2003). As discussed, this objective can now be achieved through use of single- or dual-laser technology. Acquisition of terrestrial and bathymetric data using a single system considerably reduces the cost and increases the utility of LiDAR.

While both the EAARL and a dual-wavelength LiDAR offer near-unbroken profiles between seabed topography and terrestrial terrain, neither system can acquire dependable bathymetric data in the near-shore or surf zone. When white-caps are present, the laser does not penetrate the water column. Even when the water surface is clear, if the depth is less than 2 m, systems that employ relatively long transmitted pulse lengths (>10 ns) are not capable of separating the peak in the returning laser-waveform that corresponds to the water surface from that of the seabed. For coastal mapping, this problem may be resolved through the use of a temporally shortened transmitted pulse (<2 ns), or may be obviated by combining successive flights at low-tide with a topographic LiDAR, and at high-tide with a bathymetric LiDAR (Pastol et al. 2007). Modern LiDAR systems address this problem by employing "shallow-water" algorithms and waveform deconvolution techniques. Though experimental, the shallow-water problem can alternatively be tackled using a statistical relationship that links water depth to the Raman waveform shape (Pe'eri and Philpot 2007). Since reef-tops tend to be considerably shallower than 2 m, these advances are very relevant for the successful application of LiDAR as a complete-system for coral reef mapping.

State-of-the-art LiDAR systems are calibrated and have the capability to capture reflectance data from the returning pulse, in addition to the three-dimensional coordinates of the laser returns (Lillesand et al. 2004; Tuell and Park 2004; Tuell et al. 2005). Like the strength of radar returns, the intensity of LiDAR "echoes" varies with the wavelength of the source energy and the reflectance spectrum of the material that reflects the downwelling laser. For bathymetric applications, such radiometric information, termed LiDAR "intensity", can aid in the identification of seabed character. Intensity values are extracted from the returning laser waveform and represent the magnitude of the bottom return. Via interpolation, an intensity image of the seafloor can be created from spot soundings. To capture a true reflectance image, it is necessary to employ a calibrated system and all environmental and system response parameters must be appropriately modelled. The reflectance image is captured at the wavelength of the transmitting laser within the LiDAR system. The interpolated image, while still only single-band (i.e., monochromatic), can be considered spectrally distinct. Accordingly, there exists the opportunity of seabed mapping based upon blue-green reflectance. Bathymetric sensor-packages that collect LiDAR intensity include the Optech SHOALS system (Fig. 5.4), the Tenix LADS ADS Mk II (Fig. 5.5), and Hawk-EyeII. For SHOALS and CZMIL, first-generation versions of these systems did not return a spectrally calibrated signal, and therefore the resulting soundings could not be used in a classification that relied upon spectral albedo. It was therefore common that the data were interpolated manually, with a user digitizing reef features (Walker et al. 2008), or alternatively, a classifier was devised that operates on the uncalibrated laser backscatter values (Filin 2004; Arefi and Hahn 2005; Collin et al. 2008). In either case, good ground-truth was demanded in the form of seabed descriptions, photographs, or video.

Neither airborne topographic nor bathymetric LiDAR can be collected through clouds or dense haze/smoke; however, the data are acquired at relatively low altitudes, often below cloud-level. This is unlikely to remain the case as the next-generation LiDAR, which now boasts much higher pulse repetition frequencies than today's instruments, and are able to acquire data at altitudes in excess of 5,000 m, while still maintaining acceptable point spacing. Providing that visible-spectrum imagery is not being captured concurrently and that aircraft safety considerations can be adequately addressed, LiDAR surveys can also be flown at

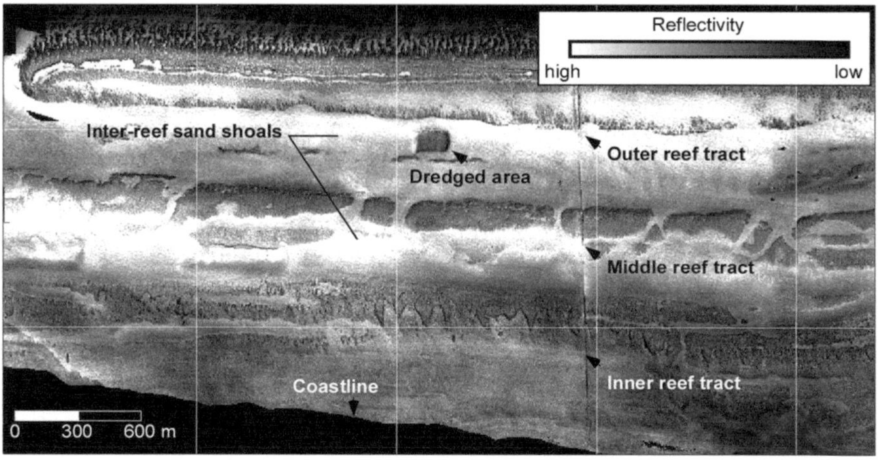

Fig. 5.4 Intensity image of a section of the Florida reef tract (offshore Dania Beach) obtained at 532 nm from the SHOALS-3000 LiDAR instrument, a component of the CHARTS instrument array. In addition to water depth, SHOALS is able to provide two additional products that have been unavailable with previous LiDAR instruments: seafloor reflectance and water column attenuation. Credit: Optech International

night when cloudy conditions are less pervasive. Night flights confer an added benefit when turbidity conditions are high by reducing solar noise in the returned waveforms.

5.2.2 Field-Deployed LiDAR

Beyond its more common airborne application, LiDAR can also be used to considerable effect in lower-cost field-mounted instruments that are relevant for coral reef research. In their 500 million year tenure on Earth, reefs have built vast carbonate edifices, through tectonics and sea-level change, many ancient examples of which now sit on land (Wood 1999). The vertical faces of outcrops can be read to reveal information on, for example, climate and sea-level cyclicity over geological time-scales. Outcrops have traditionally been examined through intensive fieldwork, but more recently LiDAR range finders have been used to produce a 3-D geological computer model of vertical rock formations (Bellian et al. 2005). Analysis of time- and spectral-resolved laser signals permits the 3-D positioning of the cliff face and, if not obscured by vegetation, classification based on the reflectance characteristics of the rock. In this case, the LiDAR is tripod-mounted and can aim its laser beam in a wide range; its head rotates horizontally, and a mirror flips vertically. The laser measures the distance to the first object in its path. As with airborne instruments, ground-based green LiDAR also have power constraints due to eye-safety issues.

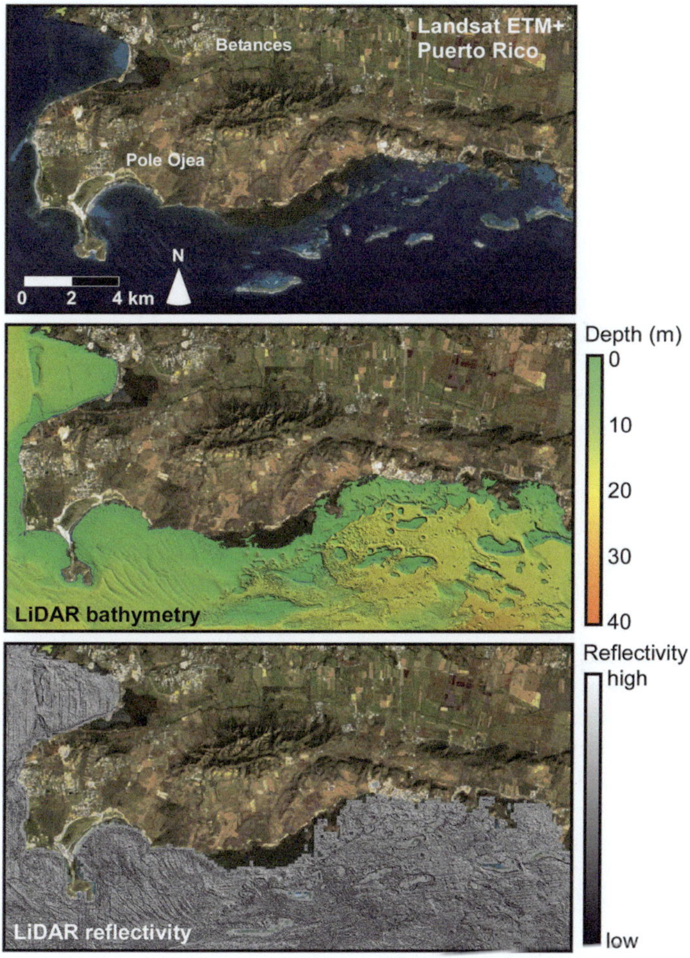

Fig. 5.5 At the *top* is a Landsat Enhanced Thematic Mapper image of southwestern Puerto Rico. The *middle* and *bottom* depict LiDAR bathymetry and LiDAR seafloor intensity, respectively, acquired in 2006 using an ADS Mk II Airborne System operated by Tenix LADS. Landsat image: NASA. LiDAR Credit: NOAA

Laser range-finders are similarly used in the marine realm as a payload on manned or remotely operated submersibles. When 'flown' in close proximity to the seabed, the LiDAR can be used to measure topography, and if spectrally calibrated, also quantify seabed character based on reflectance and fluorescence (Reuter et al. 1995; Harsdorf et al. 1997). Such applications may be accompanied by the capture of seabed video and sonar. Another well investigated application of both airborne and submersible laser remote sensing is for fisheries surveys (Squire and Krumboltz 1981; Krekova et al. 1994; Churnside and Hunter 1997; Churnside et al. 2003). Submarine fluorescence LiDAR is also used for the detection of oil spills and floating chemicals on the sea surface (Reuter et al. 1995), as well as to estimate the amount of yellow substance and phytoplankton pigment within the water column (Babichenko and Poryvkina 1992; Kopilevich et al. 2005; Tuell et al. 2005).

5.2.3 Cost and Application

Coupling a high pulse-rate with low aircraft speed and altitude, allows LiDAR to cover the ground or seabed with a high density of sample points. This density is critical in coral reef environments where the terrain is highly variable. The cost of conducting laser profiling surveys varies greatly with the mapping density desired. Rohman and Monaco (2005) quoted costs ranging from approximately $375 USD per sq. km at 5 × 5 m resolution, to $2,000 USD per sq. km at 2 × 2 m resolution; however, since the quotes are now outdated, rather than take these prices as absolute, it is best to just consider the relevant difference for the two spatial resolutions. To deliver data that can be used for marine charting, 200 % coverage of the seabed is demanded to ensure that all obstacles to marine navigation have been captured and to eliminate artefacts such as fish and flotsam, which will not be present in both acquisitions. The necessity for double coverage understandably elevates the cost of a LiDAR survey. For purposes of mapping seabed geomorphology, only a single overpass may be necessary to achieve the desired sounding spatial density, but typically swath overlap of up to 30 % is used to determine flightline-to-flightline bias for purposes of quality control and to eliminate data gaps arising from excessive aircraft roll, yaw, etc.

For surveys <100 sq. km with water depths <50 m, airborne LiDAR is deemed to be more cost effective than multi-beam acoustic surveys, and faster to acquire (Rohman and Monaco 2005). As highlighted by Costa et al. (2009), these higher efficiencies for LiDAR are due to the system's distinct acquisition geometry, wider swath-widths, and faster survey speeds. In particular, the average acquisition speed is much faster for LiDAR, approximately 140 knots, while the average speed of a survey ship is only 8 knots. Swath-width varies as a function of scan angle and aircraft altitude, but is nearly independent of water depth (Stephenson and Sinclair 2006). Conversely, the relationship between swath-width and water depth is proportional for multi-beam systems (i.e., the shallower the water, the narrower the swath and the less area mapped on a single survey line), an undesirable trait when working in shallow water and/or areas of high bathymetric relief. The fact that LiDAR is an airborne technology also means that remote areas can be accessed more rapidly than could be achieved by vessel. However, depending on the range of the aircraft in relation to the study site and appropriate airfields, LiDAR surveys over very remote oceanic reef provinces are problematic. It should also be born in mind, though, that boats may also be excluded from sensitive coral reef areas.

Due to the inherent fine resolution and high density of postings, even a moderately sized LiDAR survey generates vast quantities of data (Table 5.2). This necessitates both reliable electronic storage and advanced software for processing and manipulation. These programs and the computers needed to run them are more

Table 5.2 Approximate file sizes for raw x, y, z point LiDAR data in ASCII-txt format (i.e., not including waveforms and associated digital-camera images)

Survey area	1-m resolution (MB)	2-m resolution (MB)	3-m resolution (MB)	4-m resolution (MB)	5-m resolution (MB)
1 sq. mile	77	19	8.5	5	3
1 sq. km	30	7.5	3	2	1

specialised and costly than, for example, needed to process satellite imagery. An example of a high-end 3-D visualization software is Fledermaus, manufactured by IVS 3D. The cost and demanded expertise may render the processing of raw LiDAR data at a level beyond the typical coral reef manager or scientist. An exception to this trend is the USGS distributed software 'ALPS' (Airborne LiDAR Processing System), which has been developed in an open-source programming environment on a Linux platform. ALPS supports the exploration and processing of LiDAR data acquired by the EAARL system in an interactive or batch mode. Available at no-cost, ALPS encompasses the processing workflow to convert raw LiDAR data to marine and terrestrial DEMs, including bare-earth and canopy structure models. Manual interaction and interrogation of the raw EAARL laser waveforms is also facilitated. The United States' Naval Oceanographic Office (NAVOCEANO) produce no-cost software which contain area-based-editing (ABE) tools, but these focus only on visualization of the processed depth data and associated waveforms, as well as down-look imagery.

In the United States, LiDAR technology is frequently used throughout the remote sensing community. One reason for this being that a great deal of data are acquired by government agencies and therefore must reside in the public domain. The Center for LiDAR Information Coordination and Knowledge (CLICK) website (http://lidar.cr.usgs.gov/) serves as a portal to information for accessing both terrestrial and marine LiDAR data. NASA, NOAA, and USGS acquired soundings can all be obtained via CLICK and the site includes a viewer to visualise the LiDAR data held within the database. Further, the NOAA Digital Coast (http://csc.noaa.gov/digitalcoast/) serves up both topographic and bathymetric LiDAR collected along the United States coastline, and the NOAA National Geophysical Data Center (NGDC) serves DEMs created from combined sonar and LiDAR data (http://ngdc.noaa.gov/).

5.3 Image Products and Environmental Variables

5.3.1 Bathymetric Products

LiDAR is not the only solution for remotely deriving water depth over coral reef areas. Methods do exist for extracting bathymetry from visible-spectrum passive remote sensing data (Chaps. 1–4). For multispectral imagery, these are based on the differential spectral attenuation of light with wavelength by water (Lyzenga 1981;

Stumpf et al. 2003; Lyzenga et al. 2006), else derived empirically or via optimisation when hyperspectral information is captured (Gordon and Morel 1983; Lee et al. 1999; Dekker 1993; Durand et al. 2000; Hedley and Mumby 2003). However, the relationships of attenuation on which these methods are based are often tenuous and even under optimum conditions, such techniques cannot match the accuracy or depth of penetration delivered by LiDAR. This is demonstrated in Fig. 5.6, where for areas of dark substrate (seagrass), the optically-derived DEM consistently underestimates water depth, while overestimating depth in the bright areas. LiDAR, by comparison, is not confounded in this way. The disparity in performance between passive visible-band imagery and LiDAR becomes more pronounced as depth increases.

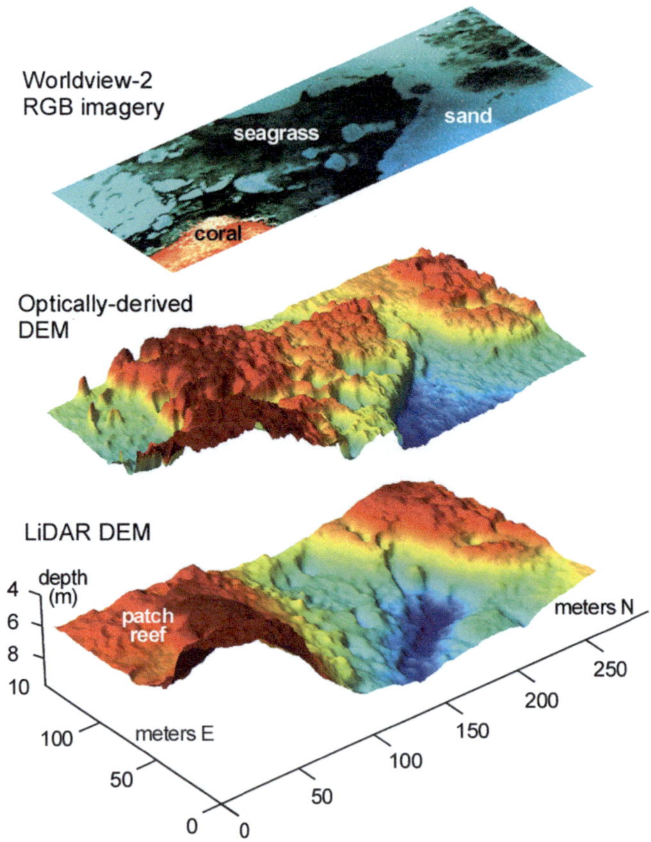

Fig. 5.6 Comparison of remotely derived DEMs acquired passively from Worldview-2 (WV2) and actively via LiDAR. The *top* is WV2 imagery from the northern Florida Keys, which has been processed according to Stumpf et al. (2003) to yield an optically derived DEM (*middle*). For comparison, the *bottom* shows a DEM derived from NASA-EAARL LiDAR for the same area. While the first-order trend of both DEMs is the same, there are important differences. The color-scale for the DEMs is the same. Credit: Jeremy Kerr. Satellite imagery: DigitalGlobe. LiDAR: NASA-EAARL

Extraction of bathymetry from synthetic aperture radar (SAR) is also a mature strategy, competitively priced, and since the launch of TerraSAR-X, also of high resolution. Marine DEM-derivation is, however, a complex undertaking and has limitations. The caveat to retrieving water depth from SAR is that the image must be acquired under favourable meteorological and hydrodynamic conditions; moderate winds of 3-10 m/s and significant currents of about 0.5 m/s (Alpers and Hennings 1984; Vogelzang et al. 1997). These criteria are necessary to ensure that the waterbody is in motion over the seabed. This flow (typically tidally driven) interacts with the bottom topography causing modulation of the surface current velocity, which in turn give rise to local variations in surface wave patterns. An over-flying SAR senses such variations through modulations in the backscattered radar signal (Lyzenga 1991). As with the previously discussed optical derivation of water depth, but unlike LiDAR, reference soundings are required to tune the SAR extraction model. Suitable conditions for SAR imaging can be elusive and have to date mitigated widespread use of the technology for resolving reef geomorphology. Nonetheless, in cases where favorable conditions are found, the potential of SAR is heightened by its ability to image the sea-surface through cloud cover at any time of the day or night. On land, and if the data have already been acquired, extraction of topography from stereo-paired satellite images or aerial photographs yields an accurate DEM at a cost many orders cheaper than a LiDAR survey. For shallow water at least, this technique is transferable to the marine realm, but due to refraction at the air–water interface, the calculations are complex and the methodology has so far not been applied extensively (Murase et al. 2008).

Given the lack of accuracy and necessity of considerable ground-control for deriving DEMs from satellite imagery, be it SAR or visible-spectrum, LiDAR has become the technology of choice for bathymetric mapping. It delivers a reliable, fast, and accurate representation of a reefscape through geo-referenced 3-D point clouds. Whether terrestrial or submarine, LiDAR-based measurements are commonly used for three families of information: (1) the production of bare-earth models, (2) DEM generation, and (3) geomorphology studies. The latter two have most relevance for coral reef research, but for completeness all three will be covered.

Many applications, for example contouring, require a bare-earth digital terrain model (DTM). Unfortunately, the raw data points captured by LiDAR do not constitute a bare-earth DTM. Even though most LiDAR systems can measure "last return" data points, in terrestrial environments these often measure ground parameters like shrubbery, cars, buildings, and the canopy of dense foliage. For some applications, raw LiDAR points must therefore be post-processed to remove these returns as they are considered undesirable. In others cases, such as terrestrial bare earth mapping, the last returns are valuable and desired by engineers, scientists, and researchers interested in buildings, infrastructure, and forest canopy, among other land-cover mapping applications. In the marine realm, the problem of surface clutter is less pronounced since submerged aquatic vegetation has less vertical relief than terrestrial foliage. Even so, if the data are to be used for

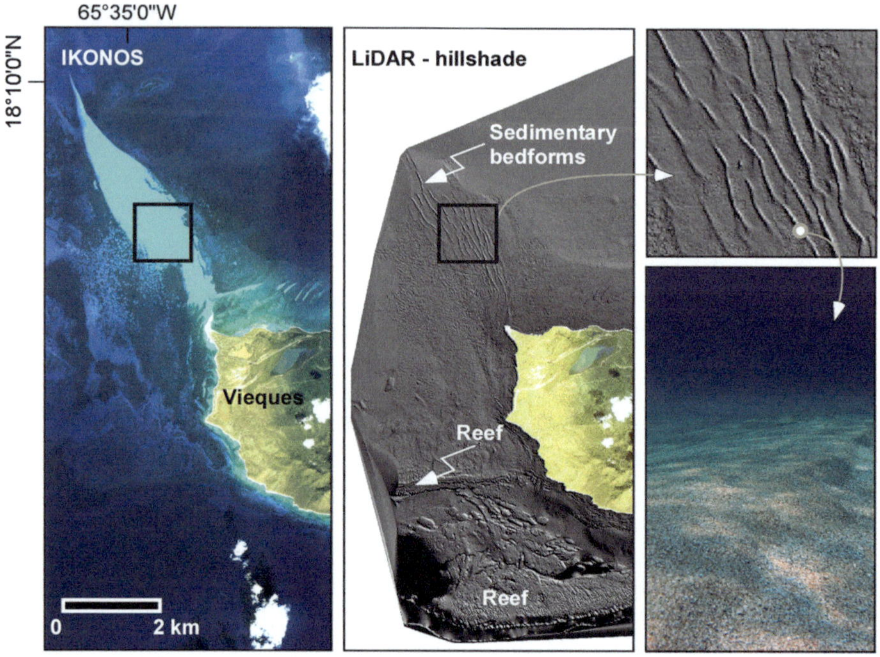

Fig. 5.7 The ability of IKONOS versus LiDAR to resolve reef features and sedimentary bedforms. The former are visible in both, the latter only in LiDAR. These data were acquired off the western coast of Vieques, Puerto Rico, a mixed carbonate depositional environment in the Caribbean

charting, some effort must be made to separate the bare-earth seabed model from the clutter which may represent a hazard to navigation.

In its native form, a LiDAR dataset is composed of discrete topographic soundings. Contour maps, DEMs, and TINs (triangulated area networks) can be extracted from the point-cloud to yield a topographic surface. This information has considerable utility when combined with passive satellite or aircraft imagery, because, by offering an independent measure of bathymetry, it can be used as input to a spectral water column correction (see Chap. 7).

While not a pixel-based imaging technology, bathymetric LiDAR soundings can be interpolated to a raster and viewed as a single-band image (Fig. 5.5). Processed in this way, the data reveal a picture of seabed topography and can be used as the basis for mapping, much like a satellite image (Storlazzi et al. 2003; Brock et al. 2004, 2006; Tuell et al. 2005; Collin et al. 2008; Walker et al. 2008; Nayegandhi and Brock 2008; Purkis and Kohler 2008; Costa et al. 2009). Once rasterized, derivative products such as contour maps and hillshade-relief images can be easily created using GIS software (Fig. 5.7). Hillshade is a gray-scale 3-D model of a topographic surface, with the sun's relative position taken into account for shading the image. Application of this technique delivers depth perspective to a flat image and is an intuitive method with which to view a LiDAR grid. More advanced metrics that can be derived from a raster are roughness, rugosity, and fractal dimension (Purkis and Kohler 2008; Zawada and Brock 2009). These are all useful means of distilling complex topographic patterns into relatively simple numerical explanations. The use of geomorphology will be explored in the following sections as a means of interrogating biotic and abiotic aspects of a coral reef, as well as the surrounding marine and terrestrial environment.

5.3.2 Biotic Features

The aggradation of coral reefs creates bottom roughness, resulting in topographic complexity ranging from centimetres to kilometres in spatial scale, which both influences and reflects many ecological variables. LiDAR sensing of benthic topographic complexity shows great promise as a proxy for more comprehensive habitat complexity (Brock et al. 2004; Brock et al. 2008), a fundamental ecological factor on coral reefs that is relevant to species diversity and richness, herbivore shelter, predation, recruitment, metabolic processes, hydrodynamics, and nutrient fluxes (McCormick 1994; Sale 1991; Sebens 1991; Szmant 1997; Purkis et al. 2008). On the basis of seabed roughness alone, for example, reef habitat can reliably be separated from low relief sandy or rocky substrates. Though a simplistic split, this means of classifying reef tracts may be sufficient for the development of a management plan for a reef area. The high spatial resolution that will be offered by the next generation LiDAR will allow more detailed partitions of habitat to be made on the basis of the rich topographic complexity of the data.

Rugosity, a measure of topographic complexity traditionally assessed in the field, is one of the important attributes to describe both biotic and abiotic seafloor features, and is a parameter that can be readily extracted from LiDAR DEMs. Several permutations on this calculation exist that range from an assessment of slope variation within an area of interest (Greene et al. 2004), to ranges of elevation inside a kernel window (Dartnell 2000). A more sophisticated representation of surface roughness presented by Jenness (2002) calculates the ratio between surface and planar (projective) area. This ratio will be equal to 1 for flat planes and increases in value for more complex surfaces (Purkis et al. 2008; Purkis and Kohler 2008).

With LiDAR soundings typically posted with spacings on the order of one meter, the spatial resolution is insufficient to differentiate coral assemblages strictly on the basis of their topographic variation. This renders LiDAR, for now at least, as less capable than, for example, hyperspectral airborne imagery, for the assessment of ecological communities. But note that the highest spatial density now available for bathymetric LiDAR have already shown promise in the discrimination of large boulder coral colonies (Brock et al. 2006), and assemblage-scale differences in LiDAR rugosity at scales of tens of metres can allow the separation of coral-dominated habitat from other seabed types (Foster et al. 2009;

Zawada and Brock 2009). By contrast, seagrass and algal meadows, classes that do have ecological relevance, cannot be identified by topographic roughness alone and their identification demands the use of spectral or photographic data in unison with LiDAR (Chap. 7). Both the EAARL and the CZMIL acquire digital imagery in tandem to laser soundings. Calibrated lasers (see Sect. 5.2.1) may also offer some differentiation capacity for habitats such as seagrass that lack a topographic signature (Tuell et al. 2005).

An ecosystem parameter that is now commonly sought in both field and remote sensing surveys is 'ecological resilience'. This describes the capacity of an ecosystem to cope with disturbances, such as storms, pollution, and global change, without shifting into a qualitatively different state. For coral reefs, this shift is typically from a vital coral-dominated state, to one with low coral cover and an abundance of fleshy macro-algae. When assessed by a diver, ecosystem status will be measured in terms of the prevalence of key assemblages such as: (1) coral cover in percentage, (2) algal cover in percentage, and (3) soft coral, sponges and corallimorphs, quantified as percentages. Further measurements of the community will also be required to assess resilience, and likely focus on parameters such as the number of small corals in the assemblage (a proxy for recruitment), the size-frequency distribution of colonies, a calculation of dominance/evenness/overall diversity of the assemblage, as well as physiological parameters such as bleaching susceptibility. None of these parameters can be derived from LiDAR remote sensing. However, despite these limitations, some biotic parameters relevant to reef resilience can be accessed using LiDAR. Laser surveys do deliver information that can be used to calculate resilience 'indicators', such as the propensity of a water body to resist warming beyond thresholds tolerable to reef growth. This can be summarised as ranked assessments of water movement, distance to deep water, depth of reef base, and exposure, all readily calculable from a DEM.

5.3.3 Abiotic Features

Besides living coral cover, reefs are also classified according to geomorphology. For instance, the evolutionary reef classification scheme of Hopley (1982) and Hopley et al. (2007) adopts a threefold separation into juvenile, mature and senile reefs, on the basis of the relative depth of the lagoon and surrounding rim. These 'architectural' properties of the reef do not pertain to the coverage or diversity of live corals and cannot be appraised without detailed information on the water depth across the system. In contrast to satellite imagery, LiDAR is well poised to provide this overview and can cover the scales of entire reef tracts. As demonstrated by various papers that analyse morphometric attributes of reefs from topographic data (e.g., size, shape, orientation and complexity), reefs can be partitioned mathematically according to their morphology (Purkis et al. 2007; Brock et al. 2008: Purkis and Kohler 2008; Zieger et al. 2009; Purkis et al. 2010; Harris et al. 2011). Though limited to shallow waters, bathymetric LiDAR represents the state-of-the-art for reef-scale seafloor topography and offers an unparalleled means of mathematically quantifying reef geomorphology.

While corals are the architects and builders, coral reefs are not, by any means, made of corals alone. In fact, many other calcareous organisms, both animal and plant, may contribute more to the volume of a reef than do the corals (Blanchon et al. 1997; Wood 1999; Braithwaite et al. 2000; Perry et al. 2009). In near-shore systems, there may also be considerable quantities of siliciclastic sediments within the reef environment. Such settings are commonly termed 'mixed' carbonate systems. As depicted in Fig. 5.7, the bedforms that these sediments create can be appraised via bathymetric LiDAR. In the figure, reef features are easily discerned in the IKONOS satellite imagery because they are spectrally different from the surrounding substrate, as well as by LiDAR because they have considerable topographic relief. By contrast, sedimentary bedforms are not spectrally distinct and are unresolved by the satellite imagery, but visible by virtue of the differences in relief observed in the LiDAR. This same figure demonstrates a further advantage of LiDAR over satellite imagery for reef mapping; in the southern portion of the two images is an expansive reef complex, which due to depth cannot be seen via IKONOS, but is intricately resolved in the LiDAR survey.

Naturally occurring bedforms include sediment waves, submerged aquatic vegetation root masses, and obstruction scour and deposits. The structure of these features can be related to factors such as sediment supply and grain size, as well as serving as a useful proxy for wave, current, and tidal characteristics in the coastal zone (the region depicted in Fig. 5.7 is clearly subjected to very rigorous tidal currents). Furthermore, repeat coverage with time-separated LiDAR surveys allows temporal change of these environments to be assessed. This is a pertinent capability in the context of human modification of the nearshore environment through coastal construction, beach nourishment, and land reclamation (Gares et al. 2006).

5.3.4 *Surrounding Environment*

Examples of LiDAR topographic applications in the vicinity surrounding reefs include regional mapping of changes along sandy coasts due to storms or long-term sedimentary processes (Guenther et al. 1996; Sallenger et al. 1999; Gutierrez et al. 1998; Arens et al. 2002; Woolard and Colby 2002; Bonisteel et al. 2009; Brock and Purkis 2009; Kempeneers et al. 2009; Klemas 2009). Beyond measuring seabed height, the laser return of a LiDAR also has the capacity to be used to interrogate water quality (Babichenko and Poryvkina 1992; Kopilevich et al. 2005; Tuell et al. 2005). This is feasible because the quality and quantity of laser light returned to the overflying instrument is altered by fluorescence, absorption, and scattering within the water column. By quantifying these changes, Hoge (2006) demonstrated that oceanic beam attenuation can be retrieved from airborne laser-induced and depth-resolved colored dissolved organic matter (CDOM)

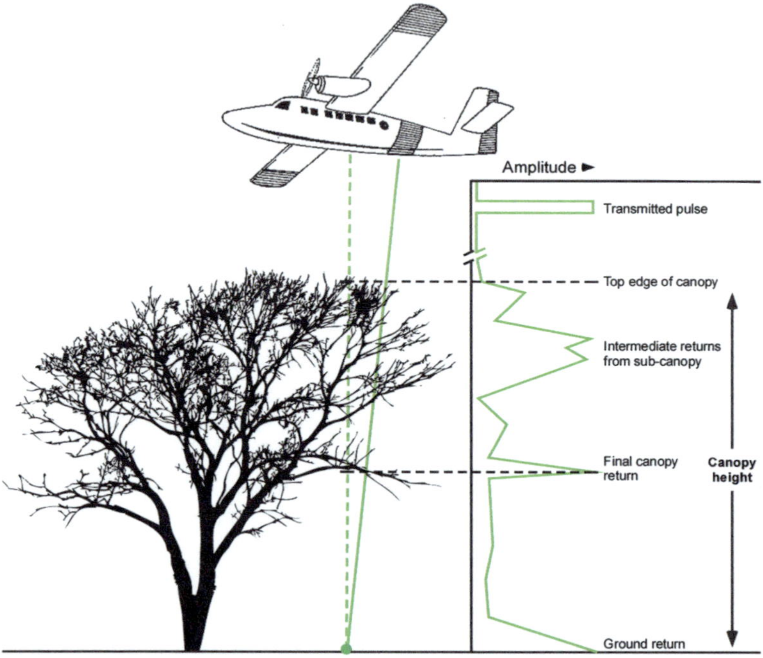

Fig. 5.8 Principles of measurement of canopy-structure using airborne LiDAR. Incident pulses of laser energy reflect off various portions of the canopy, resulting in a return waveform where the amplitude of the pulse at a given height is a function of the canopy architecture. The last large-amplitude spike is the ground return

fluorescence. The attenuation coefficient of oceanic and inland waters can also be remotely sensed with LiDAR (Tuell et al. 2005; Hoge 2006). These applications are relevant to reefs, since their health is strongly linked to water quality and clarity (Rogers 1990; Fabricius et al. 2005).

Depending on the reflective property and structure of the terrestrial target, a transmitted pulse may be returned multiple times to the receiver because it is distended and altered as it interacts with the target. For instance, over a stand of mangroves, if part of the pulse hits a branch in the upper canopy, the rest of the pulse may continue to travel to reach the ground. Both interactions induce reflections. At present, some LiDAR receivers are able to record many such multiple returns, which, through post-processing, can be useful in determining the physical geometry of sparse structures such as tree canopies (Purkis and Klemas 2011). This application has yet to be harnessed in the marine realm, but since mangroves are a key nursery habitat for reef fish (Mumby et al. 2004), it is relevant in the context of LiDAR sensing of the environments that surround reefs. Full-waveform (FW) airborne LiDAR is required for such studies, meaning that the system is capable of recording the entire emitted and backscattered signal of each laser pulse. By contrast, more conventional (i.e., non-FW) instruments only

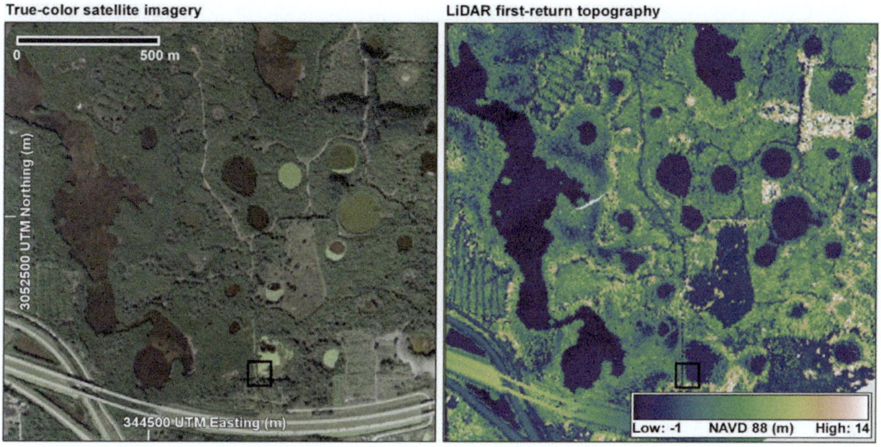

Fig. 5.9 The *left* image shows a region of Tampa Bay, Florida, USA populated by dense stands of mangroves. The *right* image depicts LiDAR first-return topography for the same area. Here, the raw-waveform LiDAR data acquired by the EAARL have been converted to georeferenced spot (x,y,z) returns. The zero crossing of the second derivative has been used to detect the first arrival laser signal, which is the first significant measurable portion of the return pulse. This can be presumed to have reflected from the upper branches of the mangrove crowns. The resulting surface therefore describes canopy height over this highly vegetated area. The data are located within UTM Zone 17 and North is top. Credit: USGS

capture clusters of 3-D points. As the FW LiDAR laser energy penetrates into the vegetation canopy, the returned waveforms are directly linked with the vertical and horizontal structure of the target (Figs. 5.8 and 5.9). For instance, direct relationships exist between full LiDAR waveforms and parameters such as tree height, stem diameter, and above ground biomass (Blair et al. 1999; Dubayah and Drake 2000; Brock et al. 2001; Harding et al. 2001; Lefsky et al. 2002).

A watershed describes the drainage basin or catchment area for a parcel of land. Watersheds drain into other watersheds in a hierarchical form, smaller ones merging into larger ones, with topography determining where the water flows. To reconstruct the geometry of a watershed and its connections, it is vital to have a sound understanding of both the geomorphology and landcover of the area in question. Beyond their importance for quantifying the freshwater hydrological cycle, watershed maps are a precursor to the analysis of many aspects of coral reef health and resilience (Rogers 1990; Lapointe and Clark 1992). In order to survive, coral reefs need specific environmental conditions, such as low nutrient and sediment levels. These conditions can easily be altered by the content and quantity of water that flows through watersheds and into coral reef waters. Human activities, including deforestation, agriculture, coastal development and dam construction have altered the natural flow of watersheds, putting coral reefs at risk. In addition, pollutants, such as sewage and chemical fertilizers, make their way to reefs through watersheds. For these reasons, consideration of watersheds is of particular importance for conservation planning.

As with bathymetry, LiDAR is not the only means of measuring watershed topography, but it is the most accurate. For applications where a regional-scale but coarse-resolution watershed map is sufficient, there are three relevant remote sensing programs; NASA's Shuttle Radar Topography Mission (SRTM), the ASTER Global Digital Elevation Model (GDEM), and the TerraSAR-X/TanDEM-X topography program. The first two of these are accessible at no-cost, while the latter is for sale through the German consortium that operates the mission. If a finer resolution product is required, as is often the case for reef studies, a topographic LiDAR survey is appropriate. To provide maximum return from the financial outlay, mapping of the watershed should be combined with a bathymetric survey through use of an instrument such as SHOALS or EAARL, which can acquire both in a single mission. However, if the topographic area is very large relative to the bathymetric area, it may be more cost effective to use a topo-only sensor for their faster coverage rates and therefore lower cost.

5.4 Processing and Validation Requirements

Discrete-return LiDAR data requires several complex processing steps to develop useable products; the workflow is usually divided into two phases:

- *Preprocessing* is the preparation of the raw data, merging of the GPS/IMU with the laser ranges to produce a point cloud, correcting for errors induced by flight geometry, removing overlaps between flight lines (particularly if the LiDAR system is internally inconsistent), surfacing operations, and accuracy assessment of the raw point locations. Preprocessing is typically undertaken by the LiDAR contractor.
- *Postprocessing* is the development of usable information from the point clouds, including development of TINs, DEMs, and other products like canopy height maps.

Because of the great volume of data acquired during a laser survey, validation is laborious and every effort is taken to reduce error at source, so as to lessen the need for post hoc validation. Indeed, a considerable portion of the error associated with a LiDAR survey can be mitigated by thorough validation of the instrument package before flight. Some of this, such as laser scanner calibration, laser beam alignment, noise reduction of the LiDAR signal, and so forth, are carried out by the instrument design company (Adams 2000; Fang and Huang 2004; Latypov 2005; Wagner et al. 2006). These parameters may drift through time and so the instrument-package must be returned periodically for factory-service. In an effort to further reduce errors, other calibrations are conducted on-site by the operator prior to take-off. These include measurement of the position-shift between laser scanner, IMU, and GPS. This shift distance will be applied to the sounding data during post-processing georectification. Also, the laser scan rate may not be in concert with the data acquisition rate of the GPS and the IMU, and must be synchronised such that every LiDAR posting is assigned a geographic position. To ensure consistent and accurate DGPS positioning throughout the survey, the GPS base-stations on the ground, of which there may be several, must be checked for their broadcast quality. It is also typical that an expert operator accompanies the instrument during flight to provide real-time monitoring of overall data accuracy and quality, which is done in an effort to avoid large and unexpected errors (Mohammadzadeh and Valadan Zoej 2008). The operator is also tasked with monitoring the survey-coverage to avoid data gaps.

Upon landing, the raw data are promptly downloaded and converted to a format readable by LiDAR processing software for further inspection and quality control. At this stage, outliers may be filtered out to deliver a more realistic dataset and the point-cloud is transformed to the desired projection system. Despite rigorous calibration, inaccuracies arise from sources beyond the control of the operator. These include the necessity to fly a survey at a sub-optimum altitude because of air traffic control restrictions and turbulence, or because of unavoidable poor sea-state conditions. To account fully for the influence of all potential error sources, the 1 sigma vertical accuracy of topographic LiDAR is generally quoted to be ± 0.15 m (Wozencraft 2003). For bathymetric data, this sigma is likely to be slightly greater.

Validation of a bathymetric LiDAR survey can be conducted against an ancillary dataset of depth soundings, which is commonly acquired by sonar or multi-beam (Chaps. 8–10). Optical depth-extraction from satellite imagery is not of sufficient accuracy to serve as a validation set. LiDAR, like sonar, delivers spot postings of depth and hence a direct spot-by-spot validation is not feasible since the two surveys will not be exactly coincident. Hence, the vessel-acquired acoustic data must be gridded prior to attempting a validation in which the LiDAR points are compared to the grid. It is typical that multi-beam soundings are of higher density than LiDAR. A further complication is that the ground-verified bathymetry contains a tidal signal. By contrast, LiDAR postings may be collected with reference to the ellipsoid and are not tidally influenced. If the multi-beam data, like the LiDAR, are acquired using post-processed kinematic techniques, the tidal offset is mitigated. Another discrepancy between bathymetric LiDAR and sonar is that, since laser profilers are designed primarily to provide data for hydrographic charting, they typically return the depth of the highest object within the bounds of the laser spot. For this reason, bathymetric LiDAR is referred to as being 'shoal biased' (Quadros et al. 2008). Because the laser spot is broad in comparison to a sonar sounding, the validation of submarine LiDAR with sonar is also problematic when seabed terrain is rough. For dual-laser LiDAR systems that acquire data across the intertidal, there may exist the opportunity for validation between the marine and topographic datasets. Caution must be applied to this approach as the size of the laser spot for a marine LiDAR is generally larger than that of a topographic instrument. Because of the aforementioned problem with shoal biasing, the bathymetric LiDAR will return a terrain height above that of a topographic laser scanner. Despite all of these concerns, acoustic data remains the best source of validation data for verifying LiDAR accuracy.

The spectral calibration of LiDAR sensors that collect reflectance data from the returning laser pulse must also be validated. This is most easily achieved through comparison with optical field measurements acquired using a hand-held spectro-radiometer. These measurements must be made coincident in space and time to the airborne LiDAR survey. For remote sensing coral reefs, such bottom reflectance measurements are also part of the typical work-flow for inversion of hyperspectral radiative transfer modeling and hence there is considerable overlap between the two applications (Tuell and Park 2004). Other LiDAR products, such as habitat characteristics and water quality, must be calibrated utilizing in situ measurements acquired coincident with the LiDAR data acquisition. This too is comparable to efforts undertaken during multispectral and hyperspectral investigations.

As the capabilities of LiDAR sensors and associated analysis tools continue to evolve, so too does the sophistication of the map products that can be derived using this technology. This development trajectory holds great promise for LiDAR to play an important role in future coral reef remote sensing applications.

Acknowledgments Sam Purkis was supported by the National Coral Reef Institute, Nova Southeastern University.

Suggested Reading

Brock JC, Purkis SJ (eds) (2009) Coastal applications of airborne LiDAR remote sensing. J Coast Res 25(6):59–65 (Special issue)
Guenther GC (2007) Digital elevation model technologies and applications: the DEM users manual. In: Maune D (ed) Airborne LiDAR bathymetry, 2nd edn. American Society for Photogrammetry and Remote Sensing, USA, pp 253–320 (Chapter 8)
Lillesand TM, Kiefer RW, Chipman JW (2004) Remote sensing and image interpretation, 5th edn. Wiley, New York
Purkis SJ, Klemas V (2011) Global environmental change and remote sensing. Wiley, New York

References

Ackermann F (1999) Airborne laser scanning: present status and future expectations. ISPRS J Phogrammetry Remote Sens 54:64–67
Adams MD (2000) LiDAR design, use, and calibration concepts for correct environmental detection. IEEE Trans Robot Autom 16:753–761
Alpers W, Hennings I (1984) A theory for the imaging mechanism of underwater bottom topography by real and synthetic aperture radar. J Geophys Res 89:10529–10546
Arefi H, Hahn M (2005) A hierarchical procedure for segmentation and classification of airborne LiDAR images. In: Geoscience and remote sensing symposium, IGARSS '05, Vol 7, pp 4950–4953
Arens JC, Wright CW, Sallenger AH, Krabill WB, Swift RN (2002) Basis and methods of NASA airborne topo-graphic mapper LiDAR surveys for coastal studies. J Coast Res 18:1–13
Babichenko S, Poryvkina L (1992) Laser remote sensing of phytoplankton pigments. LiDAR Remote Sens SPIE 1714:127–131

Baltsavias EP (1999) A comparison between photogrammetry and laser scanning. ISPRS J Photogrammetry Remote Sen 54:83–94
Bellian JA, Kerans C, Jennette DC (2005) Digital outcrop models: applications of terrestrial scanning LiDAR technology in stratigraphic modelling. J Sediment Res 75:166–176
Blair JB, Rabine DL, Hofton MA (1999) The laser vegetation imaging sensor: a medium-altitude, digitization-only, airborne laser altimeter for mapping vegetation and topography. ISPRS J Photogrammetry Remote Sens 54:115–122
Blanchon P, Jones B, Kalbfleisch W (1997) Anatomy of a fringing reef around Grand Cayman: storm rubble, not coral framework. J Sediment Res 67:1–16
Bonisteel JM, Nayegandhi A, Wright CW, Brock JC, Nagle DB (2009) Experimental advanced airborne research LiDAR (EAARL) data processing manual. U.S. Geological Survey Open-File Report 2009-1078, p 38
Braithwaite CJR, Montaggioni LF, Camoin GF, Dalmasso H, Dullo WC, Mangini A (2000) Origins and development of Holocene coral reefs: a revisited model based on reef boreholes in the Seychelles, Indian Ocean. Int J Earth Sci 89:431–445
Brock J, Sallenger A (2000) Airborne topographic mapping for coastal science and resource management. USGS Open-File Report 01-46
Brock J, Wright CW, Hernandez R, Thompson P (2006) Airborne LiDAR sensing of massive stony coral colonies on patch reefs in the Northern Florida reef tract. Remote Sens Environ 104:31–42
Brock JC, Palaseanu-Lovejoy M, Wright CW, Nayegandhi A (2008) Patch-reef morphology as a proxy for Holocene sea-level variability, Northern Florida Keys, USA. Coral Reefs 27:555–568
Brock JC, Purkis SJ (2009b) The emerging role of LiDAR remote sensing in coastal research and resource management. J Coastal Res 53:1–5
Brock JC, Sallenger AH, Krabill WB, Swift RN, Wright CW (2001) Recognition of fiducial surfaces in LiDAR surveys of coastal topography. Photogrammetric Eng Remote Sens 67:1245–1258
Brock JC, Wright CW, Clayton TD, Nayegandhi A (2004) LiDAR optical rugosity of coral reefs in Biscayne National Park, Florida. Coral Reefs 23:48–59
Cecchi G, Palombi L, Mochi I, Lognoli D, Raimondi V, Tirelli D (2004) LiDAR measurement of the attenuation coefficient of natural waters. In: Proceedings of the 22nd international laser radar conference, European Space Agency, Paris, p 827
Churnside J, Hunter J (1997) Laser remote sensing of epipelagic fishes. In: Proceedings of laser remote sensing of natural waters: from theory to practice, SPIE, vol 2964., pp 38–53
Collin A, Archambault P, Long B (2008) Mapping the shallow water seabed habitat with SHOALS. IEEE Trans Geosci Remote Sens 46:2947–2955
Costa BM, Battista TA, Pittman SJ (2009) Comparative evaluation of airborne LiDAR and ship-based multibeam SoNAR bathymetry and intensity for mapping coral reef ecosystems. Remote Sens Environ 113:1082–1100
Churnside JH, Demer DA, Mahmoudi B (2003) A comparison of LiDAR and echosounder measurements of fish schools in the Gulf of Mexico. ICES J Mar Sci 60:147–154
Dartnell P (2000) Applying remote sensing techniques to map seafloor geology/habitat relationships. Master's thesis, San Francisco State University, CA
Dekker AG (1993) Detection of optical water quality parameters for eutrophic waters by high resolution remote sensing. PhD thesis, Vrije Universiteit Amsterdam. ISBN 90-9006234-3
Dubayah RO, Drake JB (2000) LiDAR remote sensing of forestry. J Forest 98:44–46
Durand D, Bijaoui J, Cauneau F (2000) Optical remote sensing of shallow-water environmental parameters: a feasibility study. Remote Sens Environ 73:152–161
Fabricius K, De'ath G, McCook L, Turak E, Williams DM (2005) Changes in algal, coral and fish assemblages along water quality gradients on the inshore Great Barrier Reef. Mar Pollut Bull 51:384–398
Fang H-T, Huang D-S (2004) Noise reduction in LiDAR signals based on discrete wavelet transform. Optics Communications 233:67–76

Filin S (2004) Surface classification from airborne laser scanning data. Comput Geosci 30:1033–1041
Foster G, Walker BK, Riegl BM (2009) Interpretation of single-beam acoustic backscatter using LiDAR-derived topographic complexity and benthic habitat classifications in a coral reef environment. J Coastal Res 53:16–26
Gares PA, Wang Y, White SA (2006) Using LiDAR to monitor a beach nourishment project at Wrightsville Beach, North Carolina, USA. J Coastal Res 22:1206–1219
Gordon HR, Morel AY (1983) Remote assessment of ocean color for interpretation of satellite visible imagery: a review. Springer Verlag, New York (Volume 4 of Lecture notes on coastal and estuarine studies)
Greene HG, Kvitek R, Bizzaro JJ, Bretz C, Iampietro PJ (2004) Fisheries habitat characterization of the California continental margin. California sea Grant College Program, University of California, CA
Guenther G (2007) Airborne LiDAR bathymetry digital elevation. Model technologies and applications. In: Maune D (ed.) The DEM users manual. American Society for Photogrammetry and Remote Sensing, pp 253–320
Guenther G, LaRocque P, Lillycrop W (1994) Multiple surface channels in SHOALS airborne LiDAR. SPIE: Ocean Optics XII 2258:422–430
Guenther GC, Tomas RWL, LaRocque PE (1996) Design considerations for achieving high accuracy with the SHOALS bathymetric LiDAR system. In: Proceedings of laser remote sensing of natural waters: from theory to practice, SPIE, 15:54–71
Guenther GC, Brooks MW, LaRocque PE (2000) New capabilities of the SHOALS airborne LiDAR bathymeter. Remote Sens Environ 73:247–255
Gutierrez R, Gibeaut JC, Crawford MM, Mahoney MP, Smith S, Gutelius W, MacPherson CDE (1998) Airborne laser swath mapping of Galveston Island and Bolivar Peninsula, Texas. In: Proceedings of 5th international conference on remote sensing for marine and coastal environments, San Diego 1:236–243
Harding DJ, Lefsky MA, Parker GG, Blair JB (2001) Laser altimetry height profiles methods and validation for closed-canopy, broadleaf forests. Remote Sens Environ 76:283–297
Harris PM, Purkis SJ, Ellis J (2011) Analyzing spatial patterns in modern carbonate sand bodies from Great Bahama Bank. J Sediment Res 81:185–206
Harsdorf S, Janssen M, Reuter R, Wachowicz B (1997) Design of an ROV-based LiDAR for seafloor monitoring. In: Analysis of water quality and pollutants. Proceedings of SPIE, Vol 3107, pp 288–297
Hedley JD, Mumby PJ (2003) A remote sensing method for resolving depth and subpixel composition of aquatic benthos. Limnol Oceanogr 48:480–488
Hoge FE (2006) Beam attenuation coefficient retrieval by inversion of airborne LiDAR-induced chromophoric dissolved organic matter fluorescence. I Theor Appl Opt 45:2344–2351
Hopley D (1982) The geomorphology of the Great Barrier Reef: quaternary development of coral reefs. Wiley, New York
Hopley D, Smithers SG, Parnell K (2007) The geomorphology of the Great Barrier Reef: development, diversity, and change. Vol xiii., Cambridge University Press, Cambridge, p 532
Irish J, Lillycrop W (1999) Scanning laser mapping of the coastal zone: the SHOALS system. J Photogrammetry Remote Sens 54:123
Irish JL, White TE (1998) Coastal engineering applications of high-resolution LiDAR bathymetry. Coast Eng 35:47–71
Jenness JS (2002) Calculating landscape surface area from digital elevation models. Wildl Soc Bull 32:829–839
Katzenbeisser R (2003) About the calibration of LiDAR sensors. In: 3-D Reconstruction form Airborne Laser-Scanner and InSAR data. ISPRS Workshop, 8–10 Oct, Dresden
Kempeneers P, Deronde B, Provoost S, Houthuys R (2009) Synergy of airborne digital camera and LiDAR data to map coastal dune vegetation. J Coastal Res 53:73–82
Klemas VV (2009) The role of remote sensing in predicting and determining coastal storm impacts. J Coastal Res 25:1264–1275

Kopilevich YI, Feygels VI, Tuell GH, Surkov A (2005) Measurement of ocean water optical properties and seafloor reflectance with scanning hydrographic operational airborne LiDAR system (SHOALS): I. Theoretical Background. In: Proceedings of SPIE. vol 5885

Krekova MM, Krekov GM, Samokhvalov IV, Shamanaev VS (1994) Numerical evaluation of the possibilities of remote laser sensing of fish schools. Appl Opt 33:5715–5720

Lapointe BE, Clark MW (1992) Nutrient inputs from the watershed and coastal eutrophication in the Florida Keys. Estuaries Coasts 15:465–476

LaRocque PE, West GR (1990) Airborne laser hydrography: an introduction. In: ROPME/PERSGA/IHB workshop on hydrographic activities in the ROPME sea area and Red Sea (Kuwait City)

Latypov D (2002) Estimating relative LiDAR accuracy information from overlapping flight lines. ISPRS J Photogrammetry Remote Sens 56:236–245

Latypov D (2005) Effects of laser beam alignment tolerance on LiDAR accuracy. ISPRS J Photogrammetry Remote Sens 59:361–368

Lee Z, Carder KL, Mobley CD, Steward RG, Patch JS (1999) Hyperspectral remote sensing for shallow waters: 2. Deriving bottom depths and water properties by optimization. Appl Opt 38:3831–3853

Lefsky MA, Cohen WB, Parker GG, Harding DJ (2002) LiDAR remote sensing for ecosystem studies. Bioscience 52:19–30

Lyzenga DR (1981) Remote sensing of bottom reflectance and water attenuation parameters in shallow water using aircraft and Landsat data. Int J Remote Sens 2:71–82

Lyzenga DR (1991) Interaction of short surface and electromagnetic waves with ocean fronts. J Geophys Res 93:10765–10772

Lyzenga DR, Malinas NP, Tanis FJ (2006) Multispectral bathymetry using a simple physically based algorithm. IEEE Trans Geosci Remote Sens 44:2251–2259

McCormick MI (1994) Comparison of field methods for measuring surface topography and their associations with a tropical reef fish assemblage. Mar Ecol Prog Ser 112:87–96

McKean J, Nagel D, Tonina D, Bailey P, Wright CW, Bohn C, Nayegandhi A (2009) Remote sensing of channels and riparian zones with a narrow-beam aquatic-terrestrial LIDAR. Remote Sens 1:1065–1096

Mohammadzadeh A, Valadan Zoej MJ (2008) A state of art on airborne LiDAR application in hydrology and oceanography: a comprehensive overview. Int Arch Photogrammetry, Remote Sens Spat Inf Sci 37:315–320 (Part B1. Beijing)

Mumby PJ, Edwards AJ, Arias-González JE, Lindeman KC, Blackwell PG, Gall A, Gorczynska MI, Harborne AR, Pescod CL, Renken H, Wabnitz CC, Llewellyn G (2004) Mangroves enhance the biomass of coral reef fish communities in the Caribbean. Nature 427:533–536

Murase T, Tanaka M, Tani T, Miyashita Y, Ohkawa N, Ishiguro S, Suzuki Y, Kayanne H, Yamano H (2008) A Photogrammetric correction procedure for light refraction effects at a two-medium boundary. Photogrammetric Eng Remote Sens 74:1129–1135

Nayegandhi A, Brock JC (2008) Assessment of coastal vegetation habitats using LiDAR. In: Yang X (ed) Lecture notes in geoinformation and cartography—remote sensing and geospatial technologies for coastal ecosystem assessment and management. Springer, pp 365–389

Nayegandhi A, Brock JC, Wright CW (2009) Small-footprint, waveform-resolving LiDAR estimation of submerged and sub-canopy topography in coastal environments. Int J Remote Sens 30:861–878

Pastol Y, Le Roux C, Louvart L (2007) LITTO3D: a seamless digital terrain model. Int Hydrogr Rev 8:38–44

Pe'eri S, Philpot W (2007) Increasing the existence of very shallow-water LiDAR measurements using the red-channel waveforms. IEEE Trans Geosci Remote Sens 45:1217–1223

Perry CT, Smithers SG, Johnson KG (2009) Long-term coral community records from Lugger Shoal on the terrigenous inner-shelf of the central Great Barrier Reef, Australia. Coral Reefs 28:1432

Purkis SJ, Graham NAJ, Riegl BM (2008) Predictability of reef fish diversity and abundance using remote sensing data in Diego Garcia (Chagos Archipelago). Coral Reefs 27:167–178

Purkis SJ, Kohler KE (2008) The role of topography in promoting fractal patchiness in a carbonate shelf landscape. Coral Reefs 27:977–989
Purkis SJ, Kohler KE, Riegl BM, Rohmann SE (2007) The statistics of natural shapes in modern coral reef landscapes. J Geol 115:493–508
Purkis SJ, Rowlands GP, Riegl BM, Renaud PG (2010) The paradox of tropical karst morphology in the coral reefs of the arid Middle East. Geology 38:227–230
Purkis SJ, Klemas V (2011) Global environmental change and remote sensing. Wiley, New York
Quadros ND, Collier PA, Fraser CS (2008) Integration of bathymetric and topographic LiDAR: a preliminary investigation. Int Arch Photogrammetry, Remote Sens Spat Inf SciPart B 37:315–320 (Part B1. Beijing)
Reuter R, Wang H, Willkomm R, Loquay K, Braun A, Hengstermann T (1995) A laser fluorosensor for maritime surveillance: measurement of oil spills. EARSeL Adv Remote Sens 3:152–169
Rogers CS (1990) Responses of coral reefs and reef organisms to sedimentation. Mar Ecol Prog Ser 62:185–202
Rohmann SO, Monaco ME (2005) Mapping southern Florida's shallow-water coral ecosystems: an implementation plan, NOAA Technical Memorandum NOS NCCOS 19 (Online)
Sale PF (1991) Habitat structure and recruitment in coral reef fishes. In: Bell SS, McCoy ED, Mushinsky HR (eds) Habitat structure: the physical arrangement of objects in space. Chapman and Hall, New York, pp 211–234
Sallenger AH, Krabill WB, Brock JC, Swift RN, Jansen M, Manizade S, Richmond B, Hampto M, Eslinger D (1999) Airborne laser study quantifies El Niño-induced coastal change. American Geophysical Union, EOS Transactions 80:89–93
Sebens KP (1991) Habitat structure and community dynamics in marine benthic systems. In: Bell SS, McCoy ED, Mushinsky HR (eds) Habitat structure: the physical arrangement of objects in space. Chapman and Hall, New York, pp 211–234
Sinclair M (1999) Laser hydrography—commercial survey operations. In: Proceedings of US hydrographic conference, Alabama, USA
Squire JL Jr, Krumboltz H (1981) Profiling pelagic fish schools using airborne optical lasers and other remote sensing techniques. Mar Technol Soc J 15:27–31
Stephenson D, Sinclair M (2006) NOAA LiDAR data acquisition and processing report: Project OPR-I305-KRL-06, NOAA data acquisition and processing report NOS OCS (Online)
Storlazzi CD, Logan JB, Field ME (2003) Quantitative morphology of a fringing reef tract from high-resolution laser bathymetry. Geol Soc Am Bull 115:1344–1355
Stumpf RP, Holderied K, Sinclair M (2003) Determination of water depth with high-resolution satellite imagery over variable bottom types. Limnol Oceanogr 48:547–556
Szmant AM (1997) Nutrient effects on coral reefs: a hypothesis on the importance of topographic and trophic complexity to reef nutrient dynamics. In: Proceedings of the 8th international coral reef symposium, Smithsonian Tropical Research Institute, Panama, pp 1527–1532
Tuell GH, Feygels VI, Kopilevich YI, Cunningham AG, Weidemann AD, Mani R, Podoba V, Ramnath V, Park JY, Aitken J (2005) Measurement of ocean water optical properties and seafloor reflectance with scanning hydrographic operational airborne LiDAR sysem (SHOALS): II. Practical results and comparison with independent data. In: Proceedings of SPIE, vol 5885
Tuell GH, Park JY (2004) Use of SHOALS bottom reflectance images to constrain the inversion of a hyperspectral radiative transfer model. In: Kammerman G (ed) Laser Radar and Technology Applications IX. Proceedings of SPIE, vol 5412, p 185–193
Vogelzang J (1997) Mapping submarine sand waves with multiband imaging radar 1. Model development and sensitivity analysis. J Geophys Res 102:1163–1181
Wagner W, Ullrich A, Ducic V, Melzer T, Studnicka N (2006) Gaussian decomposition and calibration of a novel small footprint full-waveform digitising airborne laser scanner. ISPRS J Photogrammetry Remote Sens 60:100–112
Walker BK, Riegl B, Dodge RE (2008) Mapping coral reef habitats in southeast Florida using a combined technique approach. J Coastal Res 24:1138–1150

Wang C-K, Philpot WD (2007) Using airborne bathymetric LiDAR to detect bottom type variation in shallow waters. Remote Sens Environ 106:123–135

Wood R (1999) Reef evolution. Oxford University Press, Oxford, p 414

Woolard JW, Colby JD (2002) Spatial characterization, resolution, and volumetric change of coastal dunes using airborne LiDAR: Cape Hatteras, North Carolina. Geomorphology 48:269–288

Wozencraft JM (2003) SHOALS airborne coastal mapping: past, present and future. J Coastal Res 38:207–216

Wright CW, Brock J (2002) EAARL: a LiDAR for mapping shallow coral reefs and other coastal environments. In: Paper in the proceedings of the 7th international conference on remote sensing for marine and coastal environments, Miami, 20–22 May 2002

Zawada DG, Brock JC (2009) A multiscale analysis of coral reef topographic complexity using LiDAR-derived bathymetry. J Coastal Res 53:6–15

Zieger S, Stieglitz T, Kininmonth S (2009) Mapping reef features from multibeam sonar data using multiscale morphometric analysis. Mar Geol 264:209–217

Chapter 6
LiDAR Applications

Simon J. Pittman, Bryan Costa and Lisa M. Wedding

Abstract Coral reef ecosystems exhibit biotic complexity and spatial heterogeneity in physical structure at multiple spatial scales. The recent application of LiDAR technology to coral reef ecosystems has vastly improved the mapping and quantification of these physically complex ecological systems. Understanding the geomorphology of coral reefs, from a three-dimensional perspective, using LiDAR, offers great potential to advance our knowledge of the functional linkages between geomorphic structure and ecological processes in the marine environment. The recent application of LiDAR in coral reef ecosystems also demonstrates the depth and breadth of the potential for this technology to support research and mapping efforts in the coastal zone. This chapter builds upon the previous one, which covered the background and principles of LiDAR altimetry, by reviewing coral reef LiDAR applications and providing several case studies that highlight the

S. J. Pittman (✉) · B. Costa · L. M. Wedding
NOAA/NOS/NCCOS/CCMA, Biogeography Branch,
1305 East West Highway, Silver Spring, Maryland, MD 20910, USA
e-mail: simon.pittman@noaa.gov

B. Costa
e-mail: bryan.costa@noaa.gov

L. M. Wedding
e-mail: lisa.wedding@noaa.gov

S. J. Pittman
Marine Science Center, University of the Virgin Islands, 2 John Brewers Bay, St. Thomas VI, Virgin Islands 00802, USA

L. M. Wedding
Institute of Marine Science, University of California at Santa Cruz, 100 Shaffer Rd, Santa Cruz, CA 95060, USA

L. M. Wedding
NOAA/SWFSC, Fisheries Ecology Division, 110 Shaffer Rd, Santa Cruz, CA 95060, USA

J. A. Goodman et al. (eds.), *Coral Reef Remote Sensing*,
DOI: 10.1007/978-90-481-9292-2_6,

utility of this technology. The application of LiDAR for navigational charting, engineering, benthic habitat mapping, ecological modeling, marine geology and environmental change detection are presented. The future directions of LiDAR applications are considered in the conclusion of this chapter, as well as the next steps for expanding the use of this remote sensing technology in coral reef environments.

6.1 Introduction

In tropical marine ecosystems, LiDAR systems have been used predominantly to acquire bathymetric information about the seafloor in order to support navigational charting (Irish and Lillycrop 1999; McKenzie et al. 2001; Wozencraft et al. 2008), coastal engineering (Irish and White 1998; Wozencraft et al. 2000), benthic habitat mapping (Brock et al. 2006; Wang and Philpot 2007; Wozencraft et al. 2008; Walker et al. 2008; Walker 2009), ecological modeling (Wedding et al. 2008b; Pittman et al. 2009, 2011a, b), shoreline extraction (Liu et al. 2007) and change detection (Zhang et al. 2009). Airborne LiDAR has provided accurate seafloor data for shallow coral reefs, as well as seamless, high resolution land-sea coastal terrain models with sufficient vertical resolution for the forecasting of flood impacts from tsunami and sea-level rise (Tang et al. 2009). In addition, vulnerability maps produced from LiDAR data that depict regions prone to flooding have proven to be essential to planners and managers responsible for mitigating the associated risks and costs to both human communities and coral reef ecosystems (Brock and Purkis 2009; Gesch 2009).

6.2 Example LiDAR Applications

This chapter reviews coral reef LiDAR applications and highlights several case studies to demonstrate the utility of this technology. Here we include examples of applications of LiDAR related to: (1) navigational charting, (2) characterization and ecological study of coral reef ecosystems, (3) examination of the geomorphology of coral reefs, (4) coastal engineering and modeling, and (5) understanding and monitoring environmental change. Wherever possible we provide examples of direct applications of LiDAR to coral reef ecosystems. However, due to the limited number of LiDAR surveys specifically addressing coral reefs, and few published studies, some of our examples and applications are focused more broadly in the coastal zone. We also include several applications that highlight the potential for LiDAR to improve our knowledge of the broader scale patterns and processes that influence the structure and function of coastal ecosystems, such as monitoring coastal sedimentary processes across tropical seascapes.

6.2.1 Navigational Charting

LiDAR supports navigational charting by acquiring seafloor depths and identifying possible hazards to navigation. This is particularly important in shallow waters with hardbottom features, such as coral reefs, to avoid potentially hazardous groundings and damage to sensitive and valuable coral reef communities. According to International Hydrographic Organization (IHO) navigational charting standards, LiDAR surveys must not exceed predetermined levels of vertical and horizontal uncertainties at the 95 % confidence level (IHO 2008). The maximum levels of vertical and horizontal uncertainty allowed depend primarily on the depth of the surveyed area. In general, shallower areas (<40 m) are subject to more rigorous standards, where under-keel clearance is critical. Deeper areas (>100 m) are subject to less rigorous uncertainty standards, where a general description of the seafloor is adequate. Given the depth dependent nature of these specifications, bathymetric LiDAR surveys are most often conducted to meet the highest standards of uncertainty (i.e., IHO Special Order or Order 1), since most LiDAR systems on average only penetrate 30 m into the water column (but in clear water typical of many reef environments can penetrate as much as 60–70 m).

In 2006, a LiDAR survey of southwestern Puerto Rico was commissioned by NOAA's Office of Coast Survey (OCS) to map elevations between 50 m above sea level downwards to 70 m below sea level. This survey was conducted using the Laser Airborne Depth Sounder (LADS) Mk II Airborne System (Stephenson and Sinclair 2006), which uses a 900 Hz Nd: YAG (neodymium-doped yttrium aluminum garnet) laser that is split by an optical coupler into infrared (1,064 nm) and blue-green (532 nm) beams. The infrared beam measures the height of the plane above the water surface at nadir, while the green beam oscillates beneath the sortie in a rectilinear pattern to measure depths and elevations. The data were collected with 4 × 4 m sounding densities and 200 % seabed coverage, which thereby dictated the swath width, line spacing and speed of the survey (Table 6.1; Baltsavias 1999). The data collected for this project met IHO Order 1 uncertainty standards, and were used by NOAA to update parts of the nautical charts for the west coast of Puerto Rico (i.e.,

Table 6.1 Scan pattern configuration of the LADS Mk II LiDAR system. Adapted from Stephenson and Sinclair (2006)

Sounding density (m)	Swath width (m)	Line spacing 200 % coverage (m)	Line spacing 100 % coverage (m)	Survey speed (kts)
6 × 6	288	125	250	210
5 × 5	240	100	200	175
4 × 4	192	80	160	140
4a × 4a	150	60	120	175
3 × 3	100	40	80	150
2 × 2	50	20	40	140

Each pattern is available at all of the operational altitudes (e.g., 500–1,000 m in 100 m increments)

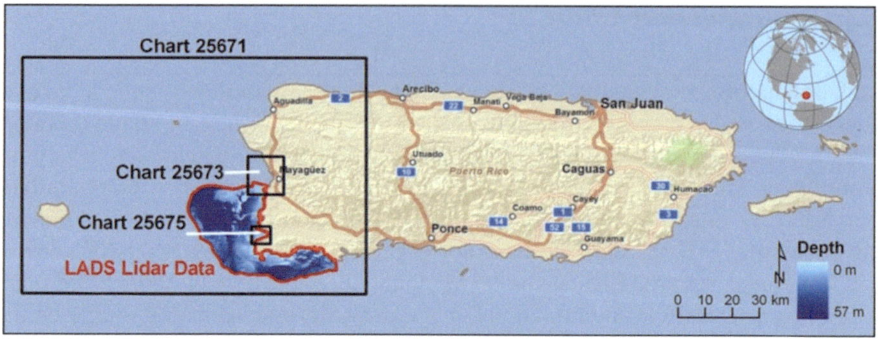

Fig. 6.1 Nautical charts (25671, 25673 and 25675) in western Puerto Rico that were updated using LADS LiDAR data. The *red polygon* denotes the complete spatial extent of the LADS data

charts 25671, 25673 and 25675) (Fig. 6.1). Charts 25671 and 25675 had not been updated since 2003, while chart 25673 had not been updated since 2006. New shoal features and potential hazards to navigation were identified during the survey (Fig. 6.2). These features were incorporated in the new versions of these charts, which were released to the maritime community in 2010. Similar projects were conducted using the LADS sensor in Miami, Florida and on the Alaskan Peninsula (Fugro LADS 2010). In addition, several previously uncharted reefs were identified by a LiDAR survey in the United States Virgin Islands in 2010, a region that was last surveyed in 1924, and where boat groundings frequently occur.

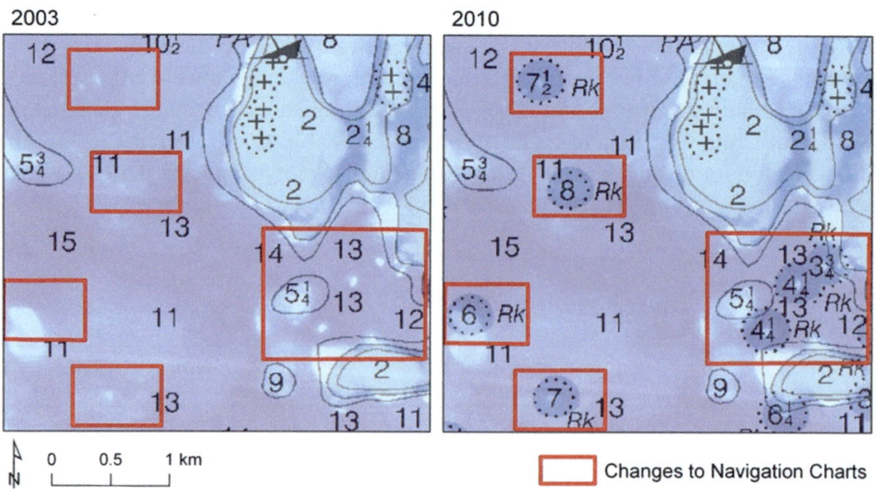

Fig. 6.2 In the U.S. Caribbean, nautical chart 25671 for the west coast of Puerto Rico was updated using the LADS LiDAR system. New shoal features and hazards to navigation (located within the *red squares*) were identified during the survey, and were used to update the 2003 edition of the chart (*left*). The new chart (*right*) was released in 2010. Soundings for both charts are in fathoms (1 fathom = 1.83 m)

6.2.2 Benthic Habitat Mapping

An important goal of benthic habitat mapping is to help resource managers make informed and ecologically relevant decisions, thereby supporting the process of ecosystem-based management and marine spatial planning. Benthic habitat maps have been used to: (1) understand and predict the spatial distribution of resources, (2) detect environmental change, (3) design monitoring sampling strategies, and (4) delineate zones and assess the efficacy of marine protected areas (Ward et al. 1999; Friedlander et al. 2007a, b; Pittman et al. 2011a, b). LiDAR supports benthic habitat mapping by acquiring continuous information about the depth and structural properties of the seafloor in depths reaching 60–70 m under optimal conditions (Stumpf et al. 2003). Seafloor habitats are differentiated from each other based on their geomorphological structure (e.g., their physical composition) and biological cover (i.e., the types and abundance of sessile organisms that colonize those structures). The three-dimensional detail provided by LiDAR offers the potential to develop highly accurate benthic habitat maps even in the absence of other remote sensing data types. In locations with overlapping multispectral or hyperspectral imagery and LiDAR data sets, combining LiDAR derived digital elevation models (DEMs) with spectral data enhances the overall accuracy of the derived benthic habitat maps (Chust et al. 2010; see Chap. 7). In Hawaii, Conger et al. (2006) used LiDAR bathymetry from the USACE SHOALS system (U.S. Army Corps of Engineers Scanning Hydrographic Operational Airborne LiDAR Survey; Irish and Lillycrop 1999; Irish et al. 2000) in conjunction with multispectral QuickBird imagery to develop a simple technique to decorrelate remote sensing color band data from depth in areas of shallow water. The method produced pseudo-color bands that were suitable for direct knowledge-based interpretation, as well as for calibration to absolute seafloor reflectance.

Seamless land topography and marine bathymetry digital elevation models are now becoming available (see Chap. 5) and provide an opportunity for the development of models that quantify land-sea interactions, such as runoff impacts to nearshore coral reef ecosystems. Furthermore, combined bathymetric and topographic LiDAR systems can survey land and seafloor simultaneously, a useful capability for mapping land adjacent to coral reef ecosystems or where emergent features such as cays and intertidal flats exist. LiDAR provides a three-dimensional representation of the seafloor, which has important utility in identifying and mapping habitat types with differing geomorphological characteristics and varying levels of topographic complexity. Three-dimensional surface features are also important in predicting species distribution patterns across coral reef ecosystems (Pittman et al. 2009; Pittman and Brown 2011; see Sect. 6.3.2).

The Experimental Advanced Airborne Research LiDAR (EAARL) (Wright and Brock 2002) developed by the National Aeronautics and Space Administration (NASA) and U.S. Geological Survey (USGS) was used to collect 1×1 m bathymetry for a broad swath of the northern Florida reef tract to map stony coral reefs in Biscayne National Park (Brock et al. 2006). Rugosity, a measure of surface

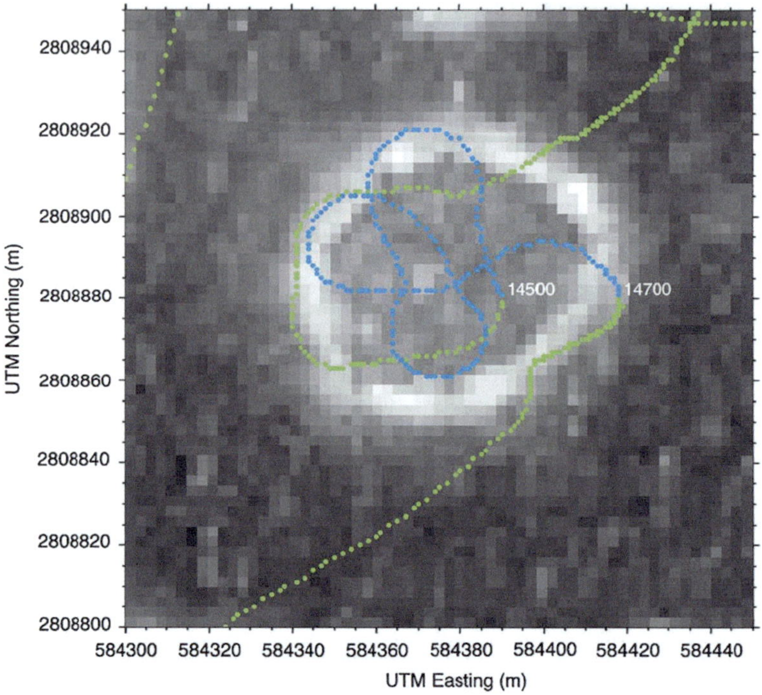

Fig. 6.3 LiDAR derived rugosity surface illustrating a patch reef in Biscayne Bay, Florida. The *green* and *blue* points denote the location of underwater video that was taken of the seafloor (adapted from Brock et al. 2006)

complexity, was calculated as the ratio of planar surface area to actual surface area. Features exhibiting high rugosity were investigated further and correlated with in situ observations using an underwater video camera (Fig. 6.3). This video was manually classified into seven substratum classes having statistically different rugosity values, with live coral having the highest mean rugosity out of the coral colony classes. The EAARL system has also been used to map coral reefs at sub-meter resolution for specific reefs, such as Johnson's Reef in the U.S. Virgin Islands, producing a topographic map with vertical and horizontal uncertainties of 10 and 40 cm, respectively. Given these results, the EAARL system has been shown to have great potential for identifying and mapping stony coral colonies. Other LiDAR systems, such as the SHOALS system (Wang and Philpot 2007; Wozencraft et al. 2008) and LADS system (Walker 2009), have also been applied to map geomorphology of coral reef ecosystems, albeit at broader spatial resolution of 1 acre minimum mapping unit (MMU).

An under-utilized data product, but currently evolving application area, of some LiDAR systems is the intensity surface, which quantifies the amount of laser light energy returned from the seafloor (e.g., seafloor pseudo reflectance or absolute reflectance; see Chap. 7). For acoustic systems, intensity information is indicative of sediment properties, including grain size, roughness and hardness (Hamilton and Bachman 1982; Chaps. 8–10). These types of sediment properties, particularly porosity, are important for benthic habitat mapping, as many tropical marine organisms respond differently to hard bottom and soft bottom habitat types (Friedlander and Parrish 1998; Pittman et al. 2007). Deriving intensity information from LiDAR data is an active area of research. Most recently, intensity information was processed for the Compact Hydrographic Airborne Rapid Total Survey (CHARTS) system, and used to map benthic habitats and different submerged aquatic vegetation types in Plymouth Harbor, MA (Reif et al. 2011). In the future, more LiDAR systems may be capable of producing intensity surfaces similar to acoustic multibeam sensors, as the technology advances and research refines signal processing techniques and algorithms for classifying complex multivariate data (Costa et al. 2009). Nonetheless, fundamental technical differences and data characteristics exist between LiDAR and acoustic mapping systems, which are indicative of different inherent capabilities between these systems.

6.2.3 Morphology and Topographic Complexity

Bathymetric mapping of three-dimensional habitat using remote sensing technology is of great interest to ecologists because the structure and composition of habitat greatly influences marine ecosystems. Coral reef ecosystems exist as topographically complex surfaces varying across a wide range of morphological characteristics that have ecological implications for the distribution of individuals, species and spatial patterns in marine biodiversity (Pratchet et al. 2008; Pittman et al. 2009; Zawada and Brock 2009). Topographic complexity also influences the movement of water across coral reef seascapes (Monismith 2007; Nunes and Pawlak 2008), and also enhances energy dissipation, which thus increases nutrient uptake of benthic communities (Hearn et al. 2001). Very little is known about the causal mechanisms that link bathymetric morphology to biological distributions and ecosystem function, but it is emerging that patterns of topographic complexity quantified across a range of spatial scales provide useful proxies or surrogate variables for predicting spatial distributions of fishes and corals (Pittman et al. 2007; Purkis et al. 2008, 2009; Hearn et al. 2001). Understanding the ecological relevance of structural complexity is increasingly important because human activity in the coastal zone, combined with hurricanes, marine diseases, and thermal stress, have resulted in broad-scale loss and degradation of biogenic structure created by reef forming scleractinian corals, seagrasses and mangroves. Over the past 20 years, for example, coral reefs of the Caribbean region have experienced a significant decline in coral cover (Gardner et al. 2003) resulting in a 'flattening' of the topographic complexity (Alvarez-Filip et al. 2009).

LiDAR-derived bathymetry provides a primary surface from which many morphological derivatives (e.g., slope, aspect, curvature), including topographic complexity, can be modeled and quantified using surface morphometrics from the fields of digital terrain modeling and industrial surface metrology. In these fields, morphometrics are used to quantify geomorphological surface features and irregularities or roughness in engineered surfaces, such as for quality control or examination of damage (Pike 2001a, b). Pittman et al. (2009) examined seven surface morphometrics and found that topographic complexity, particularly the slope-of-slope (a measure of the maximum rate of maximum slope change), emerged as the most useful predictor of faunal diversity and abundance across Caribbean coral reef seascapes. Although some co-linearity existed between morphometrics, the differences between them, even if only subtle, appeared to matter when predicting faunal distributions (Fig. 6.4). Subsequently, Pittman and Brown (2011) examined the interaction between topographic complexity and across-shelf location in SW Puerto Rico and found improved predictive performance in mapped habitat suitability for several key fish species associated with Caribbean coral reef seascapes. LiDAR derived topographic complexity, for example, contributed most to the spatial model of habitat suitability for threespot damselfish (*Stegastes planifrons*), an important indicator species of live coral cover, producing a highly reliable prediction (Fig. 6.5). Studies by Wedding and Friedlander (2008) in Hawaii, and Walker et al. (2009) in Florida, have also found useful predictability between LiDAR topographic complexity and fish metrics. Variance in depth (within a 75 m radius) demonstrated the strongest relationships with fish abundance and species richness, while depth and slope were also found to be useful spatial pattern metrics (Wedding and Friedlander 2008). Walker et al. (2008) reported a depth dependent relationship between topographic complexity and species richness, which was more pronounced in shallow coral reefs, as well as a correlation between topographic complexity and fish abundance, which was strongest in deeper offshore coral reefs. With increasing concern over the structural collapse of coral reefs, studies are now underway using LiDAR bathymetry to forecast the impact of declining reef complexity on habitat suitability for fish species and diversity to provide advance warning on the potential consequences for fish and fisheries that depend on coral reef structure (Pittman et al. 2011b).

Variations in topographic complexity can also be used to characterize differences between benthic habitat classes. Pittman et al. (2009) showed that in SW Puerto Rico aggregated patch reefs had the greatest proportion of high slope-of-slope, followed by spur and groove; whereas the largest areal extent of high slope-of-slope was quantified for the more common class of colonized pavement with sand channels. These habitat classes were correspondingly found to support the highest live coral cover and fish species richness values (Pittman et al. 2009). For the Florida reef tract, Zawada and Brock (2009) quantified topographic complexity using the fractal dimension (D) and found spatial patterns in D were positively correlated with known reef zonation in the area, and consistent with physical processes operating on the reef geomorphology, such as erosion and sea-level dynamics. In similar studies using multibeam data from the Caribbean island of

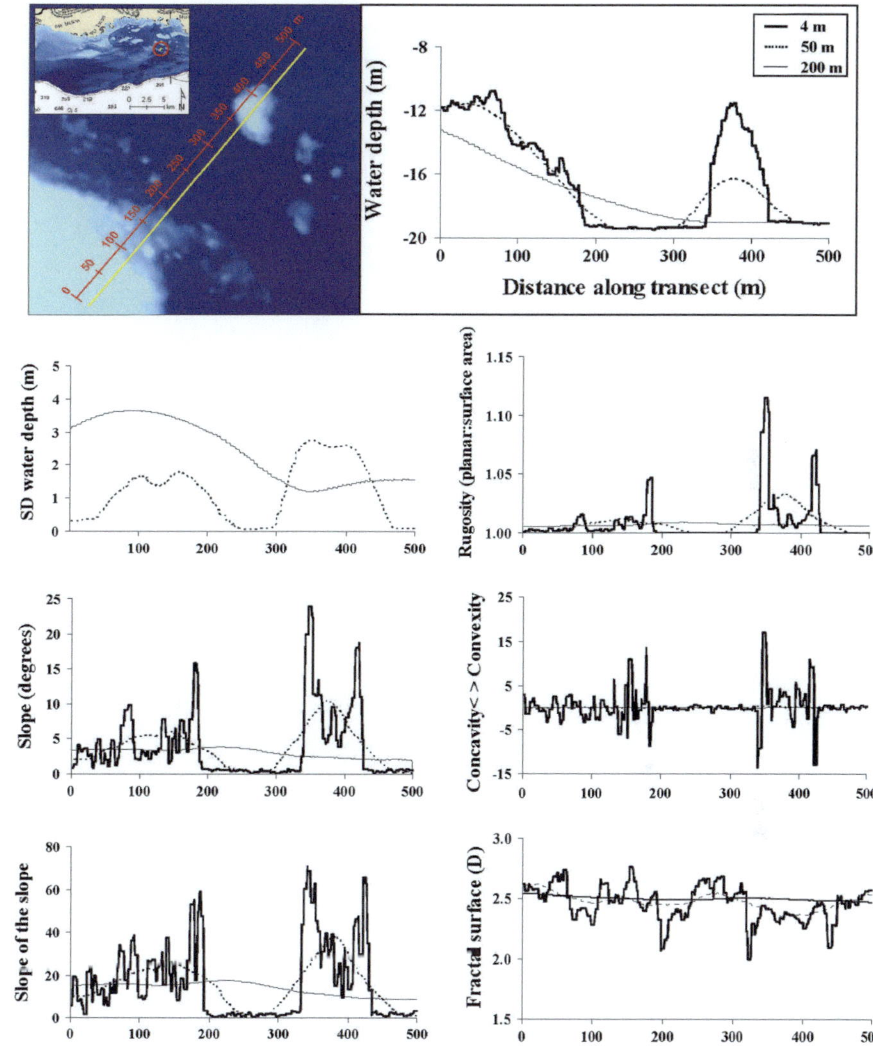

Fig. 6.4 Profiles for individual morphometrics at 1 m intervals along a 500 m transect across a coral reef seascape in the La Parguera region of southwestern Puerto Rico. To examine scale effects the seven morphometrics were calculated at multiple spatial scales using circular neighborhoods of 4, 50 and 200 m radii (adapted from Pittman et al. 2009)

Navassa, the highest fractal dimensions were quantified in areas characterized by highest live coral cover (Zawada et al. 2010).

The high predictability of marine fauna across complex coral reef ecosystems using LiDAR derivatives indicates that LiDAR is a useful tool for rapidly and cost-effectively gathering broad scale data in support of conservation planning,

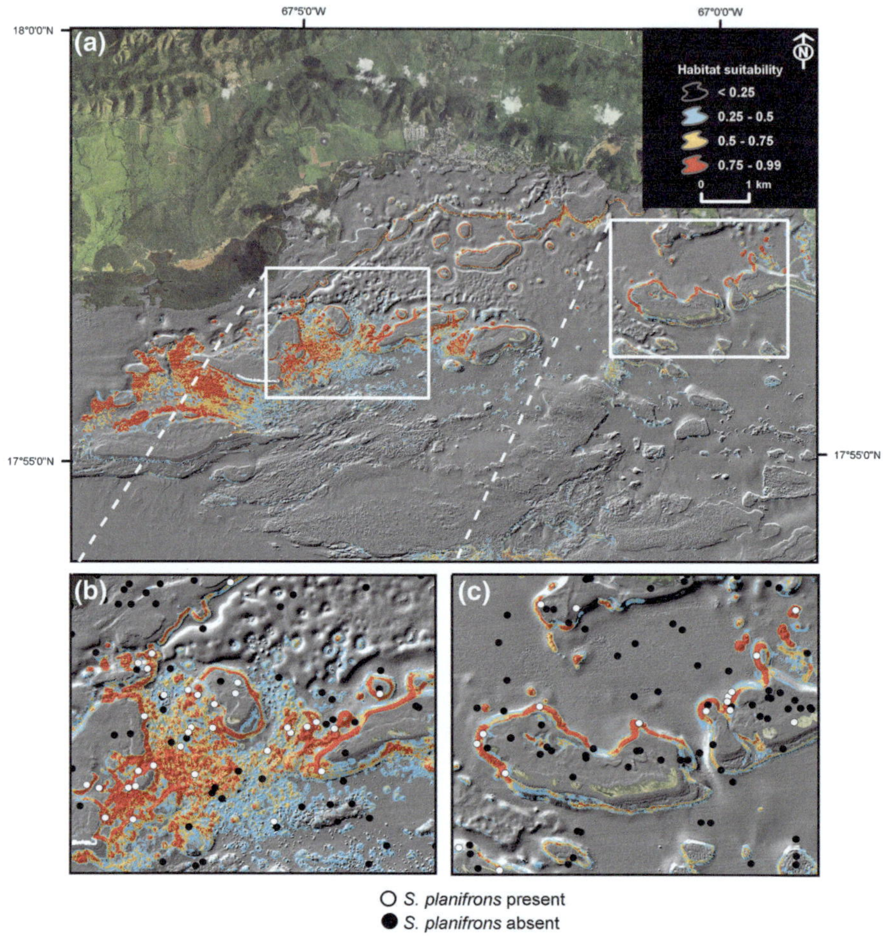

Fig. 6.5 Model of predicted habitat suitability for a potential indicator species of coral health, the threespot damselfish (*Stegastes planifrons*), across the coral reef seascapes of southwestern Puerto Rico. Maximum Entropy Distribution Modeling (MaxEnt) determined that LiDAR derived slope-of-slope together with distance across the shelf were the most important spatial predictors (adapted from Pittman and Brown 2011)

designing targeted monitoring activities, and for improving our ecological understanding of coral reef ecosystems. Nevertheless, a general consensus from these studies is that finer-scale in situ measurements of topographic complexity were more strongly correlated with fish variables than LiDAR-derived variables (Wedding and Friedlander 2008; Pittman et al. 2009; Walker et al. 2009). This suggests that finer resolution LiDAR may be required to boost the predictive power of remotely sensed topographic complexity.

6.2.4 Marine Protected Area Planning

Effective implementation of coastal and marine spatial planning (CMSP) relies on a comprehensive geospatial framework. For example, planning units are typically discrete geographic locations or zones that may have particular characteristics of interest and are considered to be 'place-based' (Norse et al. 2005; Olsen et al. 2010). In the marine environment, marine protected areas (MPAs) are among the most widely implemented forms of place-based management (Lorenzen et al. 2010). One of the critical first steps in CMSP involves mapping and integrating biological and physical datasets (Douvere 2008; Pittman et al. 2011a). This method has been successful in marine planning and spatial conservation prioritization efforts worldwide (Sala et al. 2002; Friedlander et al. 2003; Jordan et al. 2005).

Presented here is an example of marine spatial planning in Hawaii, where LiDAR technology was applied to assist in the spatial characterization of complex habitats to inform marine conservation planning and evaluation. In the Main Hawaiian Islands, SHOALS data was utilized to spatially characterize habitat complexity across a broad range of nearshore coral reef ecosystems. An initial pilot study was first conducted in Hanauma Bay Marine Life Conservation District (MLCD) to determine the utility of LiDAR data to quantify complexity in a contiguous reef environment (Wedding et al. 2008). Digital maps of surface rugosity were produced at 4 × 4 m resolution for the purpose of characterizing fish habitat utilization patterns inside and outside of marine protected areas (Wedding et al. 2008; Friedlander et al. 2007b, 2010; Fig. 6.6). Results indicated that LiDAR-derived rugosity was significantly correlated with in situ chain-tape rugosity, as measured by obtaining the ratio of the length of a chain laid across the bottom along a transect line to the linear distance of the transect line (Wedding et al. 2008). The initial study was also used to examine MPA configuration and design in order to assess the range of habitat characteristics, such as water depth

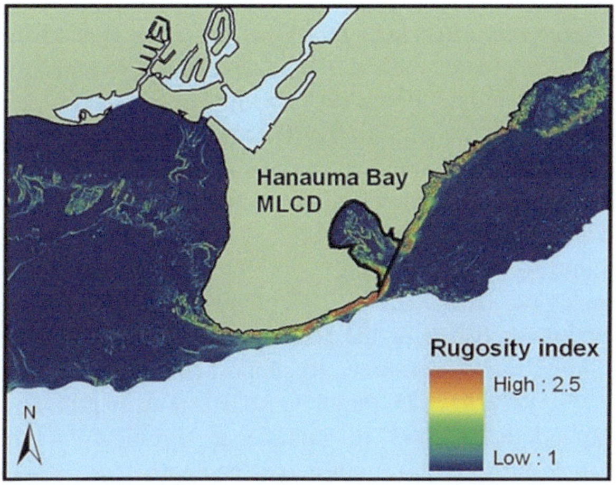

Fig. 6.6 Hanauma Bay Marine Life Conservation District pilot study site for evaluation of USACE SHOALS LiDAR technology for measuring coral reef habitat complexity. Lidar-derived rugosity was calculated by obtaining the ratio of seascape surface area to the planimetric area in a neighborhood analysis

Table 6.2 Summary of LiDAR derived depth and habitat complexity for Marine Life Conservation Districts (MLCDs) on Oahu, Hawaii based on bathymetric grids

MLCD	Established	Depth (m)			Habitat complexity		
		Mean	SD	Range	Mean	SD	Range
Pupukea	1983[a]	8.1	4.2	0.0–16.9	29.9	21.8	0–84.7
Hanauma bay	1967	8.6	6.7	0.1–27.7	18.8	17.6	0–80.3
Waikiki	1988	2.1	1.2	0.0–5.0	7.5	8.6	0–64.6

Habitat complexity represented by slope-of-slope, and table values are percent
[a] Pupukea MLCD was originally established in 1983 and the boundaries were modified in 2003. Data in the above table were calculated based on the 2003 boundary

and habitat complexity, and mosaic of interconnected habitat types present in the MPA. The application of LiDAR was then expanded in Hawaii to assist NOAA in the evaluation of MPAs throughout the State (Friedlander et al. 2010). LiDAR data was used to spatially characterize and quantify the three-dimensional seafloor structure within each MPA (Friedlander et al. 2010). Here we highlight the results from the MLCDs on the island of Oahu, where LiDAR-derived depth and slope-of-slope were summarized to calculate the mean, standard deviation and range of values within each MLCD boundary (Table 6.2; Fig. 6.7).

Waikiki MLCD. The Waikiki MLCD, located on the South Shore of Oahu, has a very small depth range (0–5 m) and relatively low habitat complexity (Friedlander et al. 2010), but Williams et al. (2006) reported fish biomass of target species in the Waikiki MLCD was twice that of the adjacent area. Meyer and Holland (2005) conducted a study of bluespine unicornfish (*Naso unicornis*) movements using acoustic tracking and found the habitat utilization patterns were aligned with topographically complex features on the fringing reef (e.g., the reef crest). So for a large bodied surgeonfish, such as *N. unicornis,* this small (0.34 km^2) MPA provides effective protection because their general home ranges are contained within the MPA boundary (Meyer and Holland 2005). It also suggests that there is an appropriate range of depth and habitat complexity within the MPA boundary for protection of this species.

Hanauma Bay MLCD. In the Hanauma Bay MLCD, the depth range (0–28 m) is much greater than in the Waikiki MLCD and the protected area shelters more diverse benthic habitat types with a wide range of structural complexity (Fig. 6.7; Friedlander et al. 2010). The fish assemblage within Hanauma Bay MLCD boundary was found to harbor eight times the biomass, and shelter a greater number of large-bodied fish species, compared to other adjacent open access areas (Friedlander et al. 2006, 2007a, b). In Hanauma Bay, LiDAR-derived rugosity was found to be a statistically significant predictor of fish biomass at multiple spatial scales (4, 10, 15, 25 m) (Wedding et al. 2008). This MLCD offers physical protection to fishes in the form of structurally complex habitat in the absence of fishing, which combines to support the high fish biomass.

Pupukea MLCD. Pupukea MCLD was originally established in 1983, and later expanded in 2003 to include a significantly greater area of the seascape (>6 × larger area), with a greater depth (e.g., 12–17 m) and habitat range (e.g.,

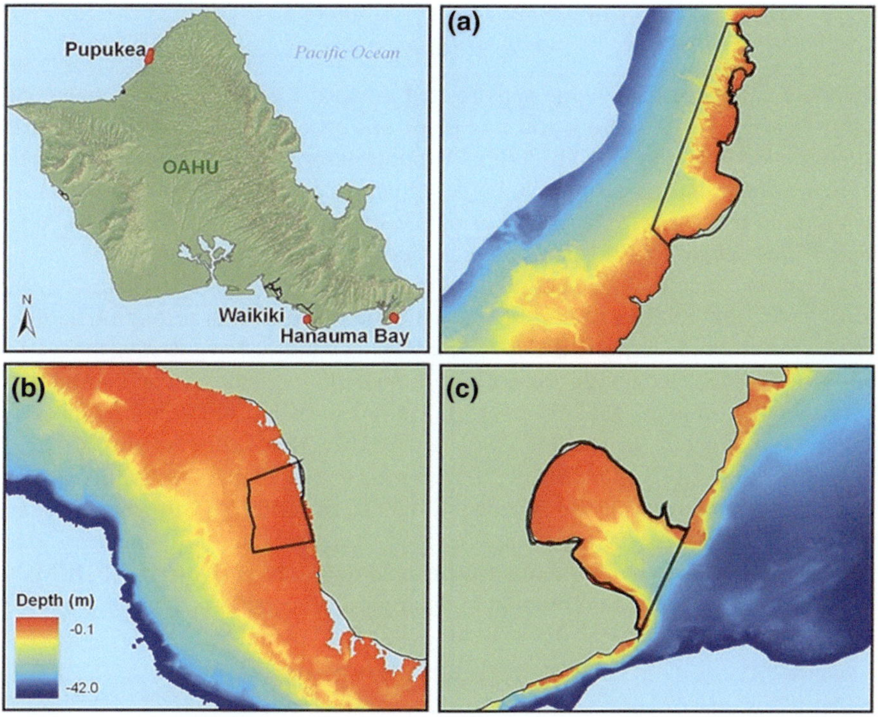

Fig. 6.7 Map of LiDAR derived depth for marine life conservation districts (*MLCDs*) on Oahu, Hawaii: **a** Pupukea MLCD, **b** Waikiki MLCD, **c** Hanauma bay MLCD

inclusion of deeper coral rich habitat and sand channels) (Friedlander et al. 2010; Fig. 6.7). NOAA Biogeography Branch benthic habitat maps were utilized to compare the change in biological cover within the expanded Pupukea MLCD boundary following the MLCD expansion (Friedlander et al. 2010). By coupling these habitat maps with the LiDAR data it was evident that the 1983 MLCD protected a very small depth range that was dominated by macroalgae. After the boundary expansion in 2003, the LiDAR data characterized a greater depth range, and the NOAA benthic habitat maps demonstrated the MLCD now protected deeper coral-rich habitat and large sand channels. With the inclusion of deeper coral habitats in Pupukea MLCD, a NOAA fish-habitat utilization study found that there was a greater diversity and biomass of fishes protected within this new reserve boundary (Friedlander et al. 2010).

These studies indicate that LiDAR data can prove useful towards identifying depth range, habitat complexity, and identify natural borders or corridors for fish movement in order to reduce the possibility of fish home ranges extending outside MPA boundaries. It also reveals that remotely sensed LiDAR data can be effectively combined with acoustic fish tracking (see Chap. 8), and other fish-habitat utilization information, as well as benthic habitat maps, to design boundary alternatives that support the optimal placement of marine protected areas.

6.2.5 Marine Geology

There are extensive knowledge gaps related to marine geomorphology since only approximately 10 % of the world's seafloor has been mapped from air and ship-borne sensors (Sandwell et al. 2003). Airborne laser altimetry has recently been applied to map marine geomorphology and enhance the understanding of coastal geomorphic processes (Sallenger et al. 2003; Brock et al. 2004; Brock and Purkis 2009; Chust et al. 2010). Coral reef geomorphology is a result of the unique oceanographic and geological conditions distinct to each geographic location, and the complex morphology of coral reefs can be mapped at high resolution across a broad spatial extent using LiDAR. A number of studies have demonstrated the utility of LiDAR technology for collecting quantitative data sets on coastal geomorphological systems (Sallenger et al. 2003; Liu et al. 2007) and in mapping geomorphic structure in shallow coral reef environments (Storlazzi et al. 2003; Finkl et al. 2005, 2008; Banks et al. 2007; Purkis and Kohler 2008). In this section, we present a case study of the application of LiDAR technology to understand the processes that shaped a large fringing reef tract in South Molokai (Fig. 6.8). LiDAR technology provided three dimensional data sets in the form of DEMs to allow for the enhanced interpretation of geological processes that shaped coral reef morphological development (Field et al. 2008; Storlazzi et al. 2008).

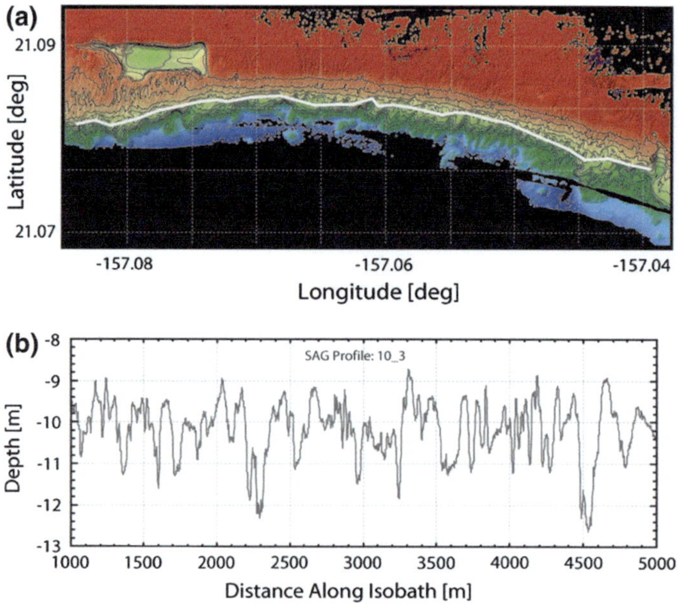

Fig. 6.8 **a** Shaded relief map of SHOALS LiDAR bathymetry overlaid with 2 m contours. **b** Example of shore-parallel bathymetric profile along the 10 m isobath (*bold white line* in **a**) (adapted from Storlazzi et al. 2003, courtesy USGS)

Field et al. (2008) utilized SHOALS LiDAR data combined with NOAA aerial photographs from the island of Molokai to study shallow-water coral reef development and response to sedimentation. The study area included a 40 km fringing coral reef located on the southern coast of the island of Molokai in the main Hawaiian Islands. Molokai's south shore is well protected from storm surge and wave energy and this has allowed for the development of one of the largest continuous fringing reefs in Hawaii. In addition, the steep terrestrial slopes and extensive runoff of upland soils has impacted coral reefs along the south shore. The fusion of aerial photography (2D) and bathymetric LiDAR (3D) were supplemented with in situ observations to infer linkages between the morphological patterns in reef structure and the coastal processes that shaped this reef tract. For instance, the LiDAR data highlighted a pronounced channel in the fringing reef off the coast that was formed from stream erosion during a period of lower sea-level (Fig. 6.9; Field et al. 2008; Storlazzi et al. 2008).

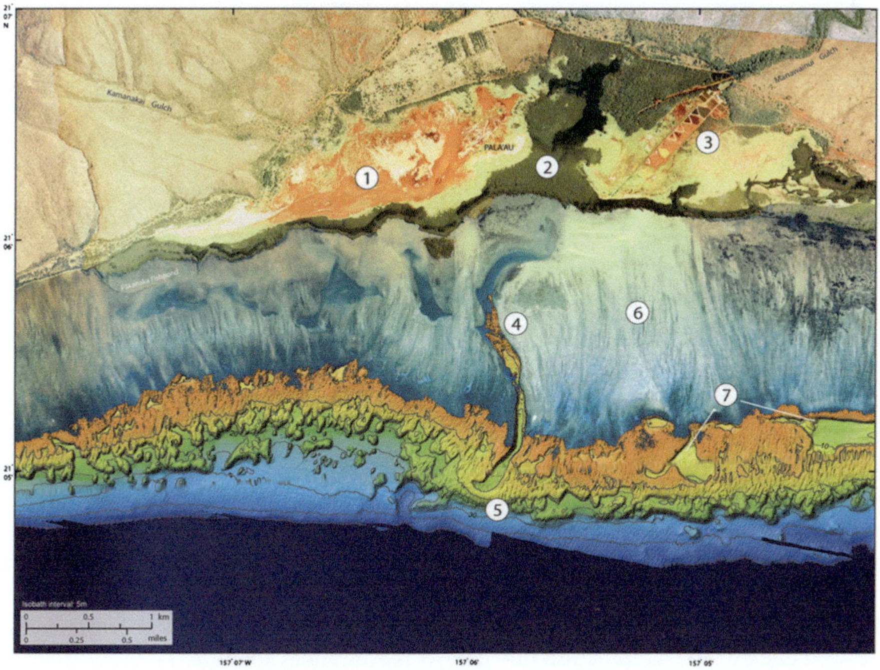

Fig. 6.9 The coastal area at Palaau is characterized by an extensive mud and salt flat (*1*) that formed from heavy flooding and run-off in the early 1900s and an extensive mangrove forest (*2*) that was started in 1903 to curb the heavy sediment run-off. The elongated structure (*3*) east of the mangroves is a shrimp farm. The reef at Palaau is dissected by a meandering channel (*4*) that resulted from erosion during a period of lower sea-level (>12,000 years ago). Note that the reef is not breached at the end of the channel (*5*), possibly because the water flowed through the porous reef rather than over it. East of the channel the reef flat is a broad, barren surface (*6*) covered by thin deposits of muddy sand. The middle part of the reef is characterized by large pits (*7*), which likely result from extensive, long-term karstic dissolution by fresh water flowing through the reef (adapted from Field et al. 2008, courtesy USGS)

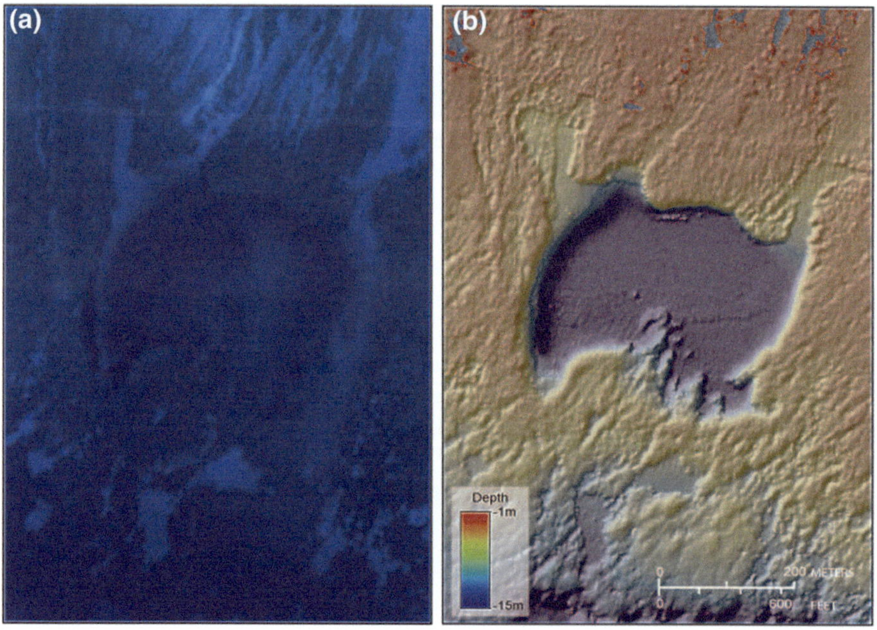

Fig. 6.10 Example of 'blue holes' on the reef flat in Molokai: **a** air photo shows the *dark blue color* of the water in a blue hole off Kakahaia, **b** SHOALS LiDAR bathymetry of the same area (adapted from Storlazzi et al. 2008, courtesy USGS)

The fusion of LiDAR and aerial photography also highlighted an extensive, shallow reef flat (<2 m) with scattered deep, sediment filled pits (i.e., blue holes, <25 m in depth; Fig. 6.10). Many of the blue holes were found to be correlated with onshore drainage and it is hypothesized that these patterns may have been produced during sea-level low-stands from either freshwater-induced (karst) dissolution, or stream incision. The morphology of spur-and-groove structures on the fringing reef was defined from the LiDAR DEMs and a series of depth profiles taken along transects running perpendicular to shore and used to quantify the broader scale (1–10 km) morphology of the reef structure. The LiDAR depth profiles identified extensive reef flats (extending >1,200 m offshore) along the well protected, central portion of the fringing reef complex, but along the eastern and western ends of the south shore no shallow reef flat was identified (Storlazzi et al. 2008, 2003).

Beyond the Molokai case study, the application of LiDAR technology has supported the identification and mapping of coral reef geomorphology in a number of other locations (Brock et al. 2006, 2008; Banks et al. 2007; Finkl et al. 2005, 2008). EAARL LiDAR in the Florida Keys, for example, was utilized to quantify morphologic differences in patch reef systems and to interpret fluctuating sea-level conditions in the Holocene based on two stages of reef accretion (Brock et al. 2008). The LiDAR-derived DEMs assisted in identifying two morphologically different patch-reef populations, and infer differences from changing sea-level regimes during the Early versus Late Holocene (Brock et al. 2008).

Other types of active remote sensors (i.e., acoustic systems; Chaps. 8–10) are available to map coral reef geomorphology, and may be the only viable option for mapping the seafloor in turbid and/or deep (>30 m) water. However, in other situations (e.g., in clear, shallow waters), LiDAR can be more time and cost efficient at certain spatial resolutions ($\geq 4 \times 4$ m), allowing for large areas of shallow and emergent seafloor to be rapidly mapped (Costa et al. 2009). With the increasing construction of LiDAR sensors and the lowering cost of data acquisition combined with opportunities for data fusion (e.g., hyperspectral; Chap. 7), LiDAR is becoming a viable technology for a wide range of geomorphological studies. For instance, in the Molokai case study, the fusion of LiDAR and aerial imagery provided enhanced information about marine geomorphology in the coastal environment (Field et al. 2008; Storlazzi et al. 2008, 2003). Walker et al. (2008) similarly combined aerial photography and laser bathymetry to map coral reefs, but also integrated acoustic ground discrimination and sub-bottom profiling into a GIS environment to support mapping efforts. In addition, bathymetric DEMs produced using the Hawk Eye LiDAR system in Spain were combined with multispectral imagery to enhance coastal habitat classification and mapping efforts (Chust et al. 2010). Understanding the geomorphology of coral reefs from a three-dimensional perspective, and across a range of spatial scales, offers great potential to advance our knowledge of the functional linkages between geomorphic structure and ecological processes.

6.2.6 Coastal Sediment Management

LiDAR supports engineering projects by acquiring seamless topographic elevations and seafloor depths, which can be used to calculate relative sediment area for regional sediment management. The goals of coastal sediment management are to increase efficiency of dredging operations through an understanding of coastal processes, and to provide a regional context for coastal projects so that they can be managed as a system of projects, rather than individual projects (Wozencraft and Millar 2005). The Regional Sediment Management Demonstration Program (RSMDP) has provided opportunities to show how broad scale, high resolution, bathymetric and topographic data can be used to identify sediment transport pathways and to reliably calculate spatial distribution of relative sediment volumes for regional sediment budgets (Wozencraft and Irish 2000). The RSMDP encompasses 360 km of shoreline in the Gulf of Mexico, stretching from Dauphin Island, Alabama east to Apalachicola Bay, Florida. In this area, approximately five million topographic and bathymetric LiDAR soundings were collected using the SHOALS system from 1995 to 2000. The SHOALS system was developed by USACE in the early 1990s as a tool for monitoring near-shore marine environments and later for coastal terrestrial environments. The SHOALS system is made up of two parts: the airborne system and the ground-processing system. The airborne system uses a 400 Hz Nd: YAG infrared (1,064 nm) and a blue-green (532 nm) laser transmitter with five receiver channels. The infrared frequency measures the sea surface distance at nadir, while the blue-green frequency scans below the sortie to measure marine depths and/or terrestrial elevations. SHOALS can be mounted on a variety of aircraft, and is usually operated at an altitude of 200–400 m and speed of 117–140 knots. This configuration allows for data collection with a horizontal spot spacing of 4 m in a 100–300 m swath below the aircraft.

In support of the RSMDP, several SHOALS surveys near Destin, Okaloosa County, Florida were analyzed (Wozencraft and Irish 2000). In Destin, a navigable depth of 4.3 m is authorized by the federal government for the tidal inlet of East Pass, which connects Choctawhatchee Bay and the Gulf of Mexico. The first surveys followed Hurricane Opal in 1995, which caused significant sediment infilling throughout the entire inlet system. The LiDAR survey detected this infilling, and illustrated the need to dredge sand from the navigation channel, nourish eroded adjacent beaches, and use it to repair breaches of Norriego Point. The subsequent surveys occurred in 1996 and later, in 1997, to document the repair of jetties along the mouth of the inlet. Additional rock was used to rebuild these jetties, which were washed away by the storm surges of Hurricane Opal. This survey also detected additional breaches of Norriego Point, despite previous efforts to restore it using dredged material. By comparing the different depth surfaces through time, the USACE was able to understand the morphological changes that were taking place in this dynamic environment (Fig. 6.11). These depth surfaces were also used to compute sediment volumes that were lost and gained during this two year time period, allowing engineers to quantify the sediment budget of the inlet and begin to explain the transport mechanisms (e.g., waves, tides, currents, wind, etc.) driving this exchange of material.

The USACE has invested in data collection to support regional sediment management by establishing the National Coastal Mapping Program (Wozencraft and Lillycrop 2006). Using the NAVOCEANO CHARTS system, topographic lidar, bathymetric lidar, aerial photography, and hyperspectral imagery are collected around the coast of the U.S. on a re-occuring schedule to provide the repeat, high-resolution, high-accuracy data needed to perform these analyses for all USACE coastal projects (Reif et al. 2012).

6.2.7 Risk Assessment and Environmental Change

Climate change threatens coral reef ecosystems in several ways. Rising ocean temperatures and increasing ocean acidification levels, in particular, may lead to mass coral bleaching events and disease epidemics (Hoegh-Guldberg 2007). Climate change also threatens the livelihoods of communities that depend on coral reef ecosystems, by altering the capacity to provide ecosystem goods and services,

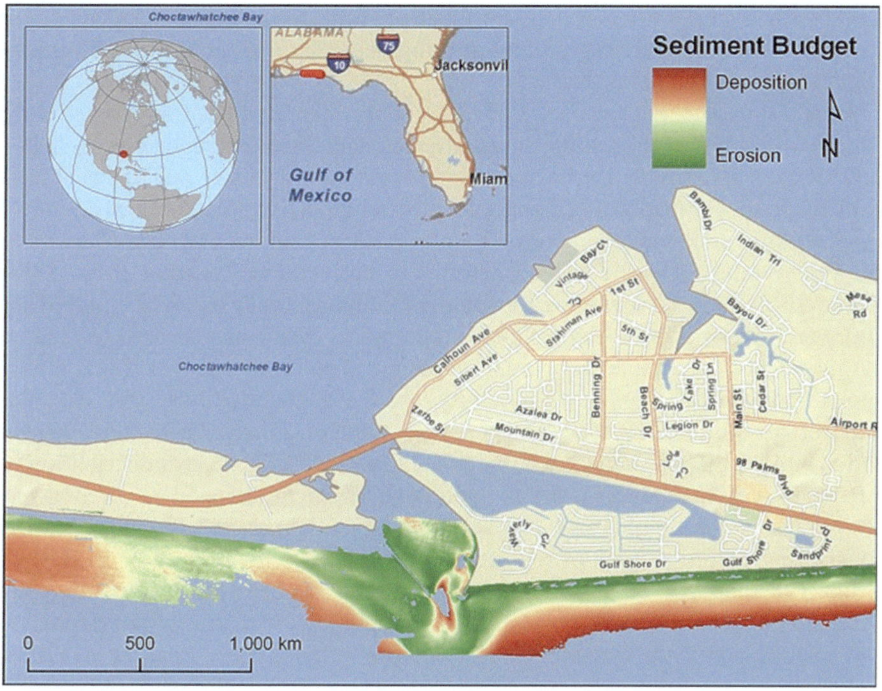

Fig. 6.11 LiDAR collected in Destin, Okaloosa County, Florida by the U.S. Army Corps of Engineers. This dataset was used to describe the sediment budget (i.e., erosion and accretion of sand) to inform dredging operations in the East Pass navigable waterway

as well as by threatening to inundate low lying areas as sea levels rise and storm events intensify. Technologies such as LiDAR can help assess the risks of flooding in the coastal zones by allowing governments to design, plan, implement and evaluate climate change mitigation and adaptation strategies.

One such LiDAR project is the Future Coasts Program in Australia conducted by the Victoria State Government Department of Sustainability and Environment (VicDSE) (www.climatechange.vic.gov.au/index.html) to prepare Australia's coasts for the effects of climate change as well as manage and mitigate the long term risks to coastal communities and natural environments (Sinclair and Quadros 2010). High resolution topographic and bathymetric information was needed to assess the effects of rising sea levels which could lead to significant changes to the coastline of Australia. This topographic and bathymetric information was collected at 2.5–5 m horizontal resolution using two LiDAR sensors (LADS Mk II and Hawk Eye II). The LADS Mk II system mapped the entire coastline 100 m inland from the vegetation line offshore to the 20 m isobath. The Hawk Eye II system mapped certain small bays and inlets to about 10 m in depth. The datasets from the two systems were later integrated to create a seamless topographic/bathymetric surface for the entire Victorian coastline. This seamless surface is currently being used by the VicDSE to model coastal flooding from storm surge events, assess the areas that are at risk, manage future development along the coasts and determine effective prevention measures.

In addition to sea-level rise, LiDAR products can also be used to assess the effects of tsunamis and storm surges (Brock and Purkis 2009b; Gesch 2009). LiDAR systems provide the accurate, high-resolution data sets that are necessary to evaluate the vulnerability of coastal areas to inundation (Stockdon et al. 2009). For example, dune elevations have been extracted from LiDAR data to evaluate the vulnerability of barrier island beaches to hurricanes (Stockdon et al. 2009). Recurrent LiDAR surveys support volumetric change analysis (White and Wang 2003) and repeat coastal surveys after major storm events can be used to monitor the magnitude of coastal change and evolution (Liu et al. 2010). LiDAR is also applied to subtidal regions to quantify change in habitat type and calculate transport of sediment or sand. Conger et al. (2009a) utilized QuickBird imagery and SHOALS LiDAR data to identify and characterize sand deposit distribution on a fringing reef in Oahu (Fig. 6.12). Sand is an important component of coral reef ecosystems and is a highly dynamic substrate type (Conger et al. 2009a) especially considering accretion rates of reef building corals (e.g., 0–2 mm/year in Hawaii; Grigg 1982, 1998). This study found that sand deposits in the fringing reef environment were strongly controlled by morphology and to a lesser degree by wave action and hydrodynamic energy (Conger et al. 2009b). Finkl et al. (2005)

Fig. 6.12 LiDAR map of sand distribution on the South shore of Oahu, Hawaii. Sand bodies are denoted by *red polygons*

has similarly inferred linkages between coastal processes (e.g., wave transformation patterns and beach morphodynamics) and geomorphic pattern in the seabed morphology in southeast Florida. Identifying this relationship between coastal processes and geomorphic patterns using high resolution LiDAR data is an important step in the field of marine geology.

Tsunami modeling predicts which coastal areas will be inundated in the event of a tsunami. LiDAR data provides high resolution continuous seafloor depths and topographic elevations, which can be used to simulate tsunami propagation and inundation along the coastline. These high resolution surfaces are needed in order to realistically model the non-linear wave dynamics of coastal inundation (González et al. 2005; Venturato 2005), because even small variations in nearshore depths, coastlines and topography can affect the behavior of a tsunami (Tang et al. 2006). In the United States, tsunami inundation predictions and evacuation planning fall under the responsibility of NOAA's two Tsunami Warning Centers. The West Coast and Alaska Tsunami Warning Center (WC/ATWC) is located in Palmer, Alaska and is responsible for issuing tsunami warnings for the west and east coasts of North America. The Pacific Tsunami Warning Center (PTWC) is located in Honolulu, Hawaii and is responsible for issuing warnings for most of the countries bordering the Pacific Ocean (under the auspices of the UNESCO/IOC International Coordination Group for the Pacific Tsunami Warning System). In 2006, a new site was proposed for the PTWC on Ford Island in Pearl Harbor. Before the center's relocation, the vulnerability of the site to inundation by a tsunami was assessed using a seamless topographic/bathymetric digital elevation model (Tang et al. 2006). Several datasets were used to create this DEM, including two LiDAR datasets. One LiDAR dataset was collected by the Joint Airborne LiDAR Bathymetry Technical Center of Expertise (JALBTCX) at 1–5 m horizontal resolution using the SHOALS system. The other LiDAR dataset was collected by NOAA's Coastal Services Center (CSC) at a 3 m horizontal resolution using the Leica ALS-40 Aerial LiDAR system. Together, these surfaces (and several acoustic datasets) were combined to create a 10 m resolution digital elevation model for Pearl Harbor in Honolulu. Tsunami waveforms were modeled at 16 distinct points (Fig. 6.13) in order to evaluate the potential impacts on Pearl Harbor. Tang et al. (2006) concluded that none of the 18 modeled tsunami scenarios, or the past recorded tsunami events, have caused inundation at the proposed NOAA site on Ford Island, Oahu. The NOAA building site on Ford Island is situated at 3.0 m above mean high water level (MHW) and all of the modeled tsunami scenarios were less than 1.5 m above MHW.

Airborne LiDAR systems have also been widely applied to map shorelines, understand coastal geomorphology, and support change detection (Brock and Purkis 2009). Shoreline information is critical for coastal geomorphologists to quantify coastal erosion, accretion and estimate sediment transport budgets (Liu et al. 2007). Traditionally, shoreline extraction for accurate maps was done using in situ surveys and aerial photography interpretation (Morton et al. 2005). The LiDAR-derived shorelines, however, can be explicitly referenced to the tidal datum surface and therefore represent a great improvement from using the beach

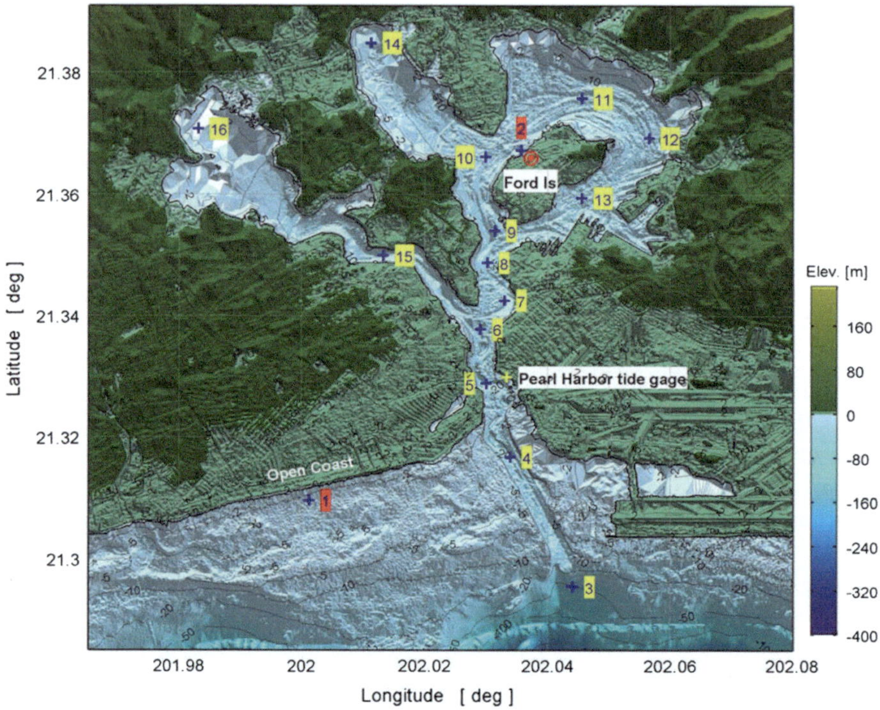

Fig. 6.13 Map denoting the 16 tsunami inundation modeling locations overlaid on a digital elevation model generated partly from LiDAR depths and elevations (adapted from Tang et al. 2006)

line on aerial photographs as the shoreline proxy (Liu et al. 2007). Beyond shoreline extraction, DEMs support the three dimensional visualization of coast habitat and volumetric change analysis in these systems (Zhang et al. 2009). For instance, DEMs produced from LiDAR data have been used to study geomorphological change in coastlines and barrier islands (White and Wang 2003). Further, LiDAR-derived metrics have been applied to establish a relationship between coastal erosion and accretion with beach morphology. Saye et al. (2005), for example, found that LiDAR characterized eroding dunes commonly located in association with steep-sloping, narrow beaches and that accreting dunes were found adjacent to low-sloping, wide beaches.

6.3 Future Directions in LiDAR

6.3.1 Integration with Other Sensors

In the last decade, research in data fusion and integration techniques has grown with access to multi-resolution, multi-temporal and multi-frequency datasets (Pohl and Van Genderen 1998). Remotely sensed imagery collected using different sensors can be fused into integrated analysis approaches to glean additional information than otherwise could be extracted from the individual images on their own. LiDAR data has been integrated with a variety of sensors, including multispectral (Cochran-Marquez 2005; Chust et al. 2008; Walker 2009) and hyperspectral sensors (Lee 2003; Chap. 7), in order to improve the classification of nearshore coral reefs and improve hydrographic surveying (Smith et al. 2000). In addition to multispectral and hyperspectral sensors, LiDAR data has also been integrated with imagery from acoustic sensors (Tang et al. 2009; Walker et al. 2008). In particular, in the Walker et al. (2008) study, shallow-water (<35 m) benthic habitat maps were developed for areas offshore of Broward County, Florida by integrating LiDAR with aerial photography and two types of acoustic information: acoustic ground discrimination systems (AGDS) and sub-bottom profilers. Habitats were defined by their geographic location, geomorphologic characteristics and biological communities. The LiDAR data, collected using the LADS system, was used primarily to map the location and geomorphology of seafloor features. The final habitat map had an overall thematic accuracy of 89.6 %. Given the importance of habitat maps, it is essential to extract as much information about the seafloor as possible from the imagery. The fusion and integration of LiDAR with different sensors offers new ways for extracting this information, and ultimately, to better understand the benthic marine environment.

6.3.2 Deployment on Different Platforms

In addition to being mounted on piloted airplanes, LiDAR systems can also be mounted on ground vehicles, unmanned aerial vehicles (UAVs) or integrated with satellites. For example, the Ice, Cloud, and Land Elevation Satellite (ICESat) collected laser altimetry data that was used primarily to describe ice sheet mass balance until it went out of operation in 2009. It is scheduled to be replaced in 2016 by ICESat-2. Such LiDAR systems are also used to measure chemical concentrations (e.g., ozone, water vapor and other pollutants; Fig. 6.14; Engel-Cox et al. 2006) as well as wind speeds at different altitudes in the atmosphere (Gentry et al. 2000) based on the backscattered return and the Doppler shift effect (Baker et al. 1995). For instance, the Cloud-Aerosol LiDAR Infrared Pathfinder Satellite Observations (CALIPSO) is providing new opportunities to study clouds and aerosols, which are important because they have direct effects on the radiation balance of the Earth (Ramanathan et al. 2001), making them relevant to coral bleaching studies and the future of coral reef ecosystems. If cloud cover were to decrease during the summer months, shallow-water corals would be at higher risk for bleaching, as was the case with the 1983 bleaching event in Indonesia, which followed windless and cloudless conditions (Brown and Suharzono 1990). Consequently, space-based LiDAR systems may prove to be a valuable tool in a

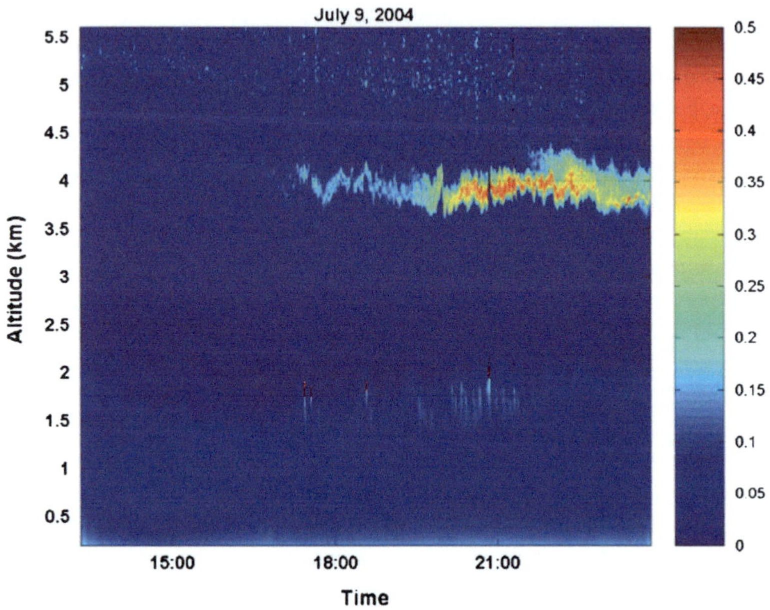

Fig. 6.14 LiDAR image depicting high-level (~4 km) smoke in the atmosphere (adapted from Engel-Cox et al. 2006)

resource manager's toolbox for predicting and responding to bleaching events that will affect the health of the coral reef ecosystems.

6.4 Conclusion

This chapter highlighted LiDAR applications that have successfully integrated this remote sensing technology for navigational charting, engineering, benthic habitat mapping, ecological modeling, marine geology and environmental change detection in coral reef ecosystems. These LiDAR applications demonstrated the depth and breadth of applications to support research and mapping efforts on coral reefs and surrounding ecosystems. Several case studies were described in greater detail to demonstrate the utility of LiDAR technology to address specific research goals and to illustrate the potential for wider applications. Understanding the geomorphology of coral reefs from a three-dimensional perspective using LiDAR offers great potential to advance our knowledge of the functional linkages between geomorphic structure and ecological processes in the marine environment. Further, seamless land topography and marine bathymetry DEMs are now becoming available, providing a valuable opportunity for the development of models that quantify land-sea interactions. The future directions of LiDAR applications involve mounting LiDAR sensors on alternative platforms, fusing LiDAR with other high resolution imagery to further enhance the information on coral reef structure, and exploiting the information that can be derived from LiDAR-derived seafloor intensity surfaces. In the future, as the technology advances, and research efforts continue to refine signal processing techniques and algorithms, the capabilities and products that can be derived from LiDAR will similarly improve and expand.

Acknowledgments This chapter was made possible with contributions from Tim Battista (NOAA Biogeography Branch), Alan M. Friedlander (University of Hawaii/USGS), Curt D. Storlazzi (USGS), Michael E. Field and (USGS) and Christopher L. Conger. Support for the authors was provided by NOAA's Coral Reef Conservation Program.

Suggested Reading

Brock JC, Purkis SJ (2009) The emerging role of LiDAR remote sensing in coastal research and resource management. J Coast Res SI 53:1–5

Conger CL, Fletcher CH, Hochberg EH, Frazer N, Rooney J (2009) Remote sensing of sand distribution patterns across an insular shelf: Oahu, Hawaii. Mar Geo 267:175–190

Costa BM, Battista TA, Pittman SJ (2009) Comparative evaluation of airborne LiDAR and ship-based multibeam sonar bathymetry and intensity for mapping coral reef ecosystems. Remote Sens Environ 113:1082–1100

Pittman SJ, Costa BM, Battista TA (2009) Using LiDAR bathymetry and boosted regression trees to predict the diversity and abundance of fish and corals. J Coast Res 53(SI):27–38

Pittman SJ, Brown KA (2011) Multiscale approach for predicting fish species distributions across coral reef seascapes. PLoS ONE 6(5):e20583. doi:10.1371/journal.pone.0020583

Storlazzi CD, Logan JB, Field ME (2003) Quantitative morphology of a fringing reef tract from high-resolution laser bathymetry: Southern Molokai, Hawaii. Geol Soc Am Bull 115:1344

References

Alvarez-Filip L, Dulvy NK, Gill JA, Cote IM, Watkinson AR (2009) Flattening of Caribbean coral reefs: region-wide declines in architectural complexity. Proc Roy Soc Ser B 276:3019–3025

Baker WE, Emmitt GD, Robertson F, Atlas RM, Molinari JE, Bowdle DA, Paegle J, Hardesty M, Menzies RT, Krishnamurti TN, Brown RA, Post MJ, Anderson JR, Lorenc AA, McElroy J (1995) LiDAR-measured winds from space: a key component for weather and climate prediction. Bull Am Meteorol Soc 76(6):869–888

Baltsavias EP (1999) Airborne laser scanning: basic relations and formulas. ISPRS J Photogrammetry Remote Sens 54:214–1999

Banks KW, Riegl BM, Shinn EA, Piller WE, Dodge RE (2007) Geomorphology of the Southeast Florida continental reef tract (Miami-Dade, Broward, and Palm Beach Counties, USA). Coral Reefs 26:617–633

Brock JC, Wright WC, Clayton TD, Nayegandhi A (2004) LiDAR optical rugosity of coral reefs in Biscayne National Park, Florida. Coral Reefs 23:48–59

Brock JC, Wright CW, Kuffner IB, Hernandez R, Thompson P (2006) Airborne LiDAR sensing of massive stony coral colonies on patch reefs in the Northern Florida reef tract. Remote Sens Environ 104:31–42

Brock J, Palaseanu-Lovejoy M, Wright CW, Nayegandhi A (2008) Patch-reef morphology as a proxy for Holocene sea-level variability, Northern Florida Keys, USA. Coral Reefs 27:555–568
Brock JC, Purkis SJ (2009) The emerging role of LiDAR remote sensing in coastal research and resource management. J Coast Res SI 53:1–5
Brown BE, Suharsono (1990) Damage and recovery of coral reefs affected by El Nino related seawater warming in the Thousand Islands, Indonesia. Coral Reefs 8:163–170
Chust G, Galparsoro I, Borja Á, Franco J, Uriarte A (2008) Coastal and estuarine habitat mapping, using LiDAR height and intensity and multi-spectral imagery. Estuar Coast Shelf Sci 78:633–643
Chust G, Grande M, Galparsoro I, Uriarte A, Borja A (2010) Capabilities of the bathymetric Hawk Eye LiDAR for coastal habitat mapping: a case study within a Basque estuary. Estuar Coast Shelf Sci 89(3):200–213
Cochran-Marquez SA (2005) Moloka'i benthic habitat mapping. US geological survey open file report 2005-1070. Available online: http://pubs.usgs.gov/of/2005/1070/index.html. Visited 27 Sep 2010
Conger CL, Hochberg EJ, Fletcher CH, Atkinson MJ (2006) Decorrelating remote sensing color bands from bathymetry in optically shallow waters. Trans Geosci Remote Sens 44(6): 1655–1660
Conger CL, Fletcher CH, Barbee M (2009a) Artificial neural network classification of sand in all visible submarine and subaerial regions of a digital image. J Coastal Res 21(6):1173–1177
Conger CL, Fletcher CH, Hochberg EH, Frazer N, Rooney J (2009b) Remote sensing of sand distribution patterns across an insular shelf: Oahu, Hawaii. Mar Geol 267:175–190
Costa BM, Battista TA, Pittman SJ (2009) Comparative evaluation of airborne LiDAR and ship-based multibeam sonar bathymetry and intensity for mapping coral reef ecosystems. Remote Sens Environ 113:1082–1100
Douvere F (2008) The importance of marine spatial planning in advancing ecosystem-based sea use management. Mar Policy 32:762–771
Engel-Cox JA, Hoff RM, Rogers R, Dimmick F, Rush AC, Szykman JJ, Al-Saadi J, Chu DA, Zell ER (2006) Integrating LiDAR and satellite optical depth with ambient monitoring for 3-dimensional particulate characterization. Atmos Environ 40:8056–8067
Field ME, Logan JB, Chavez Jr PS, Storlazzi CD, Cochran SA (2008) Views of the South Moloka'i Watershed-to-Reef System. In: Field ME, Cochran SA, Logan JB, Storlazzi CD (eds) The coral reef of South Moloka'i, Hawai'i: portrait of a sediment-threatened fringing reef. U.S. Geological survey scientific investigation report 2007-5101, pp 17–32
Finkl CW, Benedet L, Andrews JL (2005) Submarine geomorphology of the continental shelf off Southeast Florida based on interpretation of airborne laser bathymetry. J Coastal Res 21:1178–1190
Finkl CW, Becerra JE, Achatz V, Andrews JL (2008) Geomorphological mapping along the Upper Southeast Florida Atlantic Continental Platform. J Coastal Res 1388–1417
Friedlander AM, Parrish JD (1998) Habitat characteristics affecting fish assemblages on a Hawaiian coral reef. J Exp Mar Biol Ecol 224:1–30
Friedlander A, Sladek-Nowlis J, Sanchez JA, Appeldoorn R, Usseglio P, McCormick C, Bejarno S, Mitchel-Chui A (2003) Designing effective marine protected areas in Seaflower Biosphere Reserve, Columbia, based on biological and sociological information. Conserv Biol 17:1–16
Friedlander AM, Brown E, Monaco ME, Clark A (2006) Fish habitat utilization patterns and evaluation of the efficacy of marine protected areas in Hawaii: integration of NOAA digital benthic habitat mapping and coral reef ecological studies. NOAA Technical Memorandum NOS NCCOS 23:217
Friedlander AM, Brown E, Monaco ME (2007a) Defining reef fish habitat utilization patterns in Hawaii: comparisons between marine protected areas and areas open to fishing. Mar Ecol Prog Ser 351:221–233
Friedlander AM, Brown EK, Monaco ME (2007b) Coupling ecology and GIS to evaluate efficacy of marine protected areas in Hawaii. Ecol Appl 17:715–730

Friedlander AM, Wedding LM, Brown E, Monaco ME (2010) Monitoring Hawaii's marine protected areas: examining spatial and temporal trends using a seascape approach. NOAA Technical Memorandum NOS NCCOS 117. Prepared by the NCCOS Center for Coastal Monitoring and Assessment's Biogeography Branch. Silver Spring, MD, p 130

Fugro LADS (2010) Capabilities: nautical charts, oil and gas, coastal management, climate change and seabed classification. Available online: www.fugrolads.com/capabilities.htm. Visited 22 Sep 2010

Gardner TA , Côté IM, Gill JA, Grant A, Watkinson AR (2003) Long-term region-wide declines in Caribbean corals. Science 301(5635):958–960

Gentry BM, Chen H, Li SX (2000) Wind measurements with 355-nm molecular Doppler LiDAR. Opt Lett 25(17):1231–1233

Gesch DB (2009) Analysis of LiDAR elevation data for improved identification and delineation of lands vulnerable to sea-level rise. J Coastal Res 10053:49–58

González FI, Titov VV, Mofjeld HO, Venturato AJ, Simmons RS, Hansen R, Combellick R, Eisner RK, Hoirup DF, Yanagi BS, Yong S, Darienzo M, Priest GR, Crawford GL, Walsh TJ (2005) Progress in NTHMP hazard assessment. Nat Hazards 35(1):89–110

Grigg RW (1982) Darwin point: a threshold for atoll formation. Coral Reefs 1:29–34

Grigg RW (1998) Holocene coral reef accretion in Hawaii: a function of wave exposure and sea level history. Coral Reefs 17:263–272

Hamilton EL, Bachman RT (1982) Sound velocity and related properties of marine sediments. J Acoust Soc Am 72(6):1891–1904

Hearn CJ, Atkinson MJ, Falter JL (2001) A physical derivation of nutrient-uptake rates in coral reefs: effects of roughness and waves. Coral Reefs 20:347–356

Hoegh-Guldberg O, Mumby PJ, Hooten AJ, Steneck RS, Greenfield P, Gomez E, Harvell CD, Sale PF, Edwards AJ, Caldeira K (2007) Coral reefs under rapid climate change and ocean acidification. Science 318:1737

IHO (International Hydrographic Organization) (2008) IHO standards for hydrographic surveys: special publication, vol 44, 5th edn. International Hydrographic Bureau, Monaco, France, pp 1–36

Irish JL, White TE (1998) Coastal engineering applications of high-resolution LiDAR bathymetry. Coast Eng 35(1–2):47–71

Irish JL, Lillycrop WJ (1999) Scanning laser mapping of the coastal zone: the SHOALS system. ISPRS J Photogrammetry Remote Sens 54:123–129

Irish JL, McClung JK, Lillycrop WJ (2000) Airborne LiDAR bathymetry: the SHOALS system. Int Navig Assoc PIANC Bull 103:43–53

Jordan A, Lawler M, Halley V, Barrett N (2005) Seabed habitat mapping in the Kent Group of islands and its role in marine protected area planning. Aquat Conserv Mar Freshw Ecosyst 15:51–70

Lee M (2003) Benthic mapping of coastal waters using data fusion of hyperspectral imagery and airborne laser bathymetry. PhD dissertation, University of Florida. Gainsville, Florida, U.S.A, p 119

Lorenzen K, Steneck RS, Warner RR, Parma AM, Coleman FC, Leber KM (2010) The spatial dimensions of fisheries: putting it all in place. Bull Mar Sci 86:169–177

Liu H, Sherman D, Gu S (2007) Automated extraction of shorelines from airborne light detection and ranging data and accuracy assessment based on Monte Carlo simulation. J Coastal Res 23:1359–1369

Liu H, Wang L, Sherman D, Gao Y, Wu Q (2010) An object-based conceptual framework and computational method for representing and analyzing coastal morphological changes. Int J Geogr Inform Sci 24(7):1015–1041

McKenzie C, Gilmour B, Van Den Ameele EJ, Sinclair M (2001) Integration of LiDAR data in CARIS HIPS for NOAA charting. Fugro Pelagos white papers. Fugro Pelagos, San Diego, CA, pp 1–16

Meyer CG, Holland KN (2005) Movement patterns, home range size and habitat utilization of the bluespine unicornfish, *Naso unicornis* (Acanthuridae) in a Hawaiian marine reserve. Environ Biol Fishes 73:201–210

Monismith SG (2007) Hydrodynamics of coral reefs. Annu Rev Fluid Mech 39:37–55

Morton RA, Miller T, Moore L (2005) Historical shoreline changes along the US Gulf of Mexico: a summary of recent shoreline comparisons and analyses. J Coastal Res 21:704–709

Norse EA, Crowder LB, Gjerde K, Hyrenbach D, Roberts C, Safina C, Soulé ME (2005) Place-based ecosystem management in the open ocean. Mar Conserv Biol Sci Maintaining Sea's Biodivers, pp 302–327

Nunes V, Pawlak G (2008) Observations of bed roughness of a coral reef. J Coastal Res 24(2 Suppl. B):39–50

Olsen E, Kleiven A, Skjoldal H, von Quillfeldt C (2010) Place-based management at different spatial scales. J Coastal Conserv, pp 1–13

Pike RJ (2001a) Digital terrain modelling and industrial surface metrology—converging crafts. Int J Mach Tools Manuf 41(13–14):1881–1888

Pike RJ (2001b) Digital terrain modeling and industrial surface metrology: converging realms. Prof Geogr 53(2):263–274

Pittman SJ, Christenson J, Caldow C, Menza C, Monaco M (2007) Predictive mapping of fish species richness across shallow-water seascapes in the Caribbean. Ecol Model 204:9–21

Pittman SJ, Costa BM, Battista TA (2009) Using LiDAR bathymetry and boosted regression trees to predict the diversity and abundance of fish and corals. J Coastal Res 53(SI):27–38

Pittman SJ, Brown KA (2011) Multiscale approach for predicting fish species distributions across coral reef seascapes. PLoS ONE 6(5):e20583. doi:10.1371/journal.pone.0020583

Pittman SJ, Connor DW, Radke L, Wright DJ (2011a) Application of estuarine and coastal classifications in marine spatial management. In: Wolanski E, McLusky DS (eds) Treatise on estuarine and coastal science, vol 1. Academic Press, UK, pp 163–205

Pittman SJ, Costa BM, Jeffrey CFG, Caldow C (2011b) Importance of seascape complexity for resilient fish habitat and sustainable fisheries. In: Proceedings of the 63rd Gulf and Caribbean Fisheries Institute meeting, San Juan, Puerto Rico, 2010

Pratchett MS, Munday MS, Wilson SK, Graham NAJ, Cinner JE, Bellwood DR, Jones GP, Polunin NVC, McClanahan TR (2008) Effects of climate-induced coral bleaching on coral-reef fishes: ecological and economic consequences. Oceanogr Mar Biol Annu Rev 46:251–296

Pohl C, Van Genderen JL (1998) Multisensor image fusion in remote sensing: concepts, methods and applications. Int J Remote Sens 19(5):823–854

Purkis SJ, Kohler KE (2008) The role of topography in promoting fractal patchiness in a carbonate shelf landscape. Coral Reefs 27:977–989

Purkis SJ, Graham NAJ, Riegl BM (2008) Predictability of reef fish diversity and abundance using remote sensing data in Diego Garcia (Chagos Archipelago). Coral Reefs 27:167–178

Ramanathan V, Crutzen PJ, Kiehl JT, Rosenfeld D (2001) Aerosols, climate and the hydrological cycle. Science 294:2119–2124

Reif M, Dunkin L, Wozencraft J, Macon C (2011) Sensor fusion trends. Earth Imaging J 8(2):32–35

Reif MK, Wozencraft JM, Dunkin LD, Sylvester CS, Macon CL (2012) U.S. Army Corps of Engineers airborne coastal mapping in the Great Lakes. J Great Lakes Res (in press)

Sala E, Aburto-Oropeza O, Paredes G, Parra I, Barrera J, Dayton P (2002) A general model for designing networks of marine reserves. Science 298:1991–1993

Sallenger AH, Krabill WB, Swift RN, Brock J, List J, Hansen M, Holman RA, Manizade S, Sontag J, Meredith A (2003) Evaluation of airborne topographic LiDAR for quantifying beach changes. J Coastal Res 125–133

Sandwell D, Gille S, Orcutt J, Smith W (2003) Bathymetry from space is now possible. Eos Trans AGU 84:37

Saye SE, Van Der Wal D, Pye K, Blott SJ (2005) Beach-dune morphological relationships and erosion/accretion: an investigation at five sites in England and Wales using LiDAR data. Geomorphology 72:128–155

Sinclair M, Quadros N (2010) Airborne bathymetric LiDAR survey for climate change. FIG Congress: facing the challenges and building capacity, 11–16 April, Sydney, Australia, p 17

Smith RA, JL Irish, Smith MQ (2000) Airborne LiDAR and airborne hyperspectral imagery: a fusion of two proven sensors for improved hydrographic surveying. In: Proceedings of Canadian hydrographic conference, Montreal, Canada, May 15–19

Stephenson D, Sinclair M (2006) NOAA LiDAR data acquisition and processing report: Project OPR-I305-KRL-06, NOAA Data acquisition and processing report NOS OCS (Online)

Stockdon HF, Doran KS, Sallenger Jr AH (2009) Extraction of LiDAR-based dune-crest elevations for use in examining the vulnerability of beaches to inundation during hurricanes. J Coastal Res 53(SI):59–65

Storlazzi CD, Logan JB, Field ME (2003) Quantitative morphology of a fringing reef tract from high-resolution laser bathymetry: Southern Molokai, Hawaii. Geol Soc Am Bull 115:1344

Storlazzi CD, Logan JB, Field ME (2008) Shape of the South Moloka'i fringing reef: trends and variation. In: Field ME, Cochran, SA, Logan JB, Storlazzi CD (eds) The coral reef of South Moloka'i, Hawai'i-portrait of a sediment-threatened fringing reef. U.S. Geological Survey Scientific Investigation Report 2007-5101, pp 33–36

Stumpf RP, Holderied K, Sinclair M (2003) Determination of water depth with high-resolution satellite imagery over variable bottom types. Limnol Oceanogr 48(1):547–556

Tang L, Chamberlin CD, Tolkova E, Spillane M, Titov VV, Bernard EN, Mofjeld HO (2006) Assessment of potential tsunami impact for Pearl Habor, Hawaii, Tech. Memo. OAR PMEL-131. Seattle, Wash, p 36

Tang L, Titov VV, Chamberlain CD (2009) Development, testing, and applications of site-specific tsunami inundation models for real-time forecasting. J Geophy Res 114, C12025, pp 22

Venturato AJ (2005) A digital elevation model for seaside Oregon: procedures, data sources and analysis. NOAA Technical Memorandum OAR PMEL-129. Seattle, WA, p 21

Wang C-K, Philpot WD (2007) Using airborne bathymetric LiDAR to detect bottom type variation in shallow waters. Remote Sens Environ 106:123–135

Walker BK, Riegl B, Dodge RE (2008) Mapping coral reef habitats in Southeast Florida using a combined technique approach. J Coastal Res 24(5):1138–1150

Walker BK (2009) Benthic habitat mapping of Miami-Dade County: visual interpretation of LADS bathymetry and aerial photography. Florida DEP report #RM069. Miami Beach, FL, p 47

Walker BK, Jordan LKB, Spieler RE (2009) Relationship of reef fish assemblages and topographic complexity on Southeastern Florida coral reef habitats. J Coastal Res SI 53:39–48

Ward TJ, Vanderklift MA, Nicholls AO, Kenchington RA (1999) Selecting marine reserves using habitats and species assemblages as surrogates for biological diversity. Ecol Appl 9:691–698

Wedding LM, Friedlander AM, McGranaghan M, Yost R, Monaco M (2008) Using bathymetric LiDAR to define nearshore benthic habitat complexity: implications for management of reef fish assemblages in Hawaii. Remote Sens Environ 112:4159–4165

Wedding LM, Friedlander AM (2008) Determining the influence of seascape structure on coral reef fishes in Hawaii using a geospatial approach. Mar Geodesy 31:246–266

White SA, Wang Y (2003) Utilizing DEMs derived from LiDAR data to analyze morphologic change in the North Carolina coastline. Remote Sens Environ 85:39–47

Williams ID, Walsh WJ, Miyasaka A, Friedlander AM (2006) Effects of rotational closure on coral reef fishes in Waikiki-Diamond head fishery management area, Oahu, Hawaii. Mar Ecol Prog Ser 310:139–149

Wozencraft JM, Irish JL, Wiggins CE, Stupplebeen H, Chavez PS (2000) Regional mapping for coastal management, Maui and Kauai, Hawaii. In: Proceedings of national beach preservation conference 2000, Maui, Hawaii

Wozencraft JM, Irish JL (2000) Airborne LiDAR surveys and regional sediment management. In: Proceedings 20th EARSeL symposium: workshop on LiDAR remote sensing of land and sea, European association of remote sensing laboratories, June 16–17, Dresden, Germany, p 11

Wozencraft JM, Lillycrop WJ (2006) JALBTCX coastal mapping for the USACE. Int Hydrogr Rev 7(2):28–37
Wozencraft JM, Macon CL, Lillycrop WJ (2008) High resolution coastal data for Hawaii. Am Soc Civ Eng, 6–8 Nov, Pittsburgh, PA
Wozencraft JM, Millar D (2005) Airborne LiDAR and integrated technologies for coastal mapping and charting. Mar Technol Soc J 39(3):27–35
Wright CW, Brock JC (2002) EAARL: a LiDAR for mapping shallow coral reefs and other coastal environments. In: Proceedings of the 7th international conference on remote sensing for marine and coastal environments, May 20–22, 2002, Miami, FL
Zawada DG, Brock JC (2009) A multiscale analysis of coral reef topographic complexity using LiDAR-derived bathymetry
Zawada DG, Piniak GA, Hearn CJ (2010) Topographic complexity and roughness of a tropical benthic seascape. Geophys Res Lett 37:L14604
Zhang K, Whitman D, Leatherman S, Robertson W (2009) Quantification of beach changes caused by Hurricane Floyd along Florida's Atlantic coast using airborne laser surveys

Chapter 7
Integrated LiDAR and Hyperspectral

Jennifer M. Wozencraft and Joong Yong Park

Abstract Integrating LiDAR data and hyperspectral imagery is an area of active research in remote sensing, inclusive of application for coastal and coral reef mapping. These two technologies can be combined in a number of different ways, and at a number of stages of processing to produce benthic classification maps. This chapter introduces the concept of data fusion, presents a data fusion model, and describes the different ways in which LiDAR and hyperspectral data can be integrated for benthic mapping. Examples are presented to first demonstrate data fusion during the preprocessing stage prior to classification, followed by data fusion performed during processing and classification. The chapter concludes with examples of how classification maps derived from LiDAR data and hyperspectral imagery individually can be combined in a postprocessing high-level fusion approach to produce an integrated benthic classification map.

7.1 Introduction

The concept of integrating LiDAR data with hyperspectral imagery for benthic mapping has been percolating in the coastal remote sensing community for nearly two decades. Early attempts primarily took advantage of the water depth

J. M. Wozencraft (✉)
Joint Airborne lidar Bathymetry Technical Center of Expertise, Coastal and Hydraulics Laboratory, U.S. Army Corps of Engineers, Engineer Research and Development Center, 7225 Stennis Airport Road, Suite 100, Kiln, MS 39556, USA
e-mail: jennifer.m.wozencraft@usace.army.mil

J. Y. Park
Optech, Inc, 7225 Stennis Airport Road, Suite 300, Kiln, MS 39556, USA
e-mail: joongyongpark@optech.com

J. A. Goodman et al. (eds.), *Coral Reef Remote Sensing*,
DOI: 10.1007/978-90-481-9292-2_7,

information provided by the bathymetric LiDAR to perform the "depth-correction" step of hyperspectral imagery preprocessing (Lillycrop and Estep 1995; Bissett et al. 2005), as described in Sect. 4.2.6. The capability for expanding the integration of these two data types was accomplished by Lee and Tuell (2003), who developed a similar methodology for depth-correcting the "pseudo-reflectance" data derived from bathymetric LiDAR, producing an estimate of seafloor reflectivity at the laser wavelength of 532 nm. Lee (2003) used maximum likelihood algorithms to derive habitat maps from both depth-corrected hyperspectral imagery and seafloor pseudo-reflectance imagery, and then integrated the output in a high-level data fusion approach described by Park (2002). This high-level integration yielded a final classification map of greater accuracy than either of the individual classifications alone. Tuell and Park (2004) extended the utility of LiDAR data in the preprocessing of hyperspectral imagery by using LiDAR pseudo-reflectance images to identify the homogeneous seafloor areas required to implement a hyperspectral depth-correction scheme.

More recently, advanced modeling of environmental and sensor response functions have provided the capability to extract water column attenuation and perform radiometric calibration of LiDAR-derived reflectance data (Kopilevich et al. 2005; Tuell et al. 2005a). These improvements in the processing of bathymetric LiDAR allow for the estimate of absolute seafloor reflectance, as opposed to the more empirical pseudo-reflectance. Tuell et al. (2005b) demonstrated the use of LiDAR-derived depth as a fixed constraint, and the LiDAR-derived water column attenuation and absolute seafloor reflectance as weak constraints, for implementing the radiative transfer inversion model to solve for absolute bottom reflectance using hyperspectral imagery (expanding on hyperspectral processing concepts presented in Sect. 4.3.6). Tuell and Lohrenz (2006) then introduced a data fusion model to encapsulate the procedures for integrating LiDAR data with hyperspectral imagery for benthic mapping. Wozencraft et al. (2007) reports that LiDAR-derived water-leaving reflectance has also been used in a radiometric balancing step for the hyperspectral imagery.

Current research focuses on using LiDAR-derived parameters to constrain a combined atmospheric-oceanographic spectral optimization model for a variety of seafloor and water column properties (Kim et al. 2010). Additionally, Park et al. (2010) describe ongoing work to extract textural features from the LiDAR data. All of the LiDAR-derived parameters (e.g., depth, LiDAR seafloor reflectance, hyperspectral seafloor reflectance, and texture) can then be used as input to a combined analysis of the receiver operating characteristic (ROC) and linear discriminant functions to identify which LiDAR and hyperspectral features best inform the benthic classification process. Similar applications on land are using the fusion of LiDAR data and hyperspectral imagery to improve classification of landcover types and give new insights into changes in landcover (Reif et al. 2011).

The following sections describe a data fusion model for benthic mapping, explain how LiDAR-derived parameters can be used to inform hyperspectral preprocessing, and finally present the integration of hyperspectral imagery and LiDAR to produce benthic classification maps. Because many of the techniques

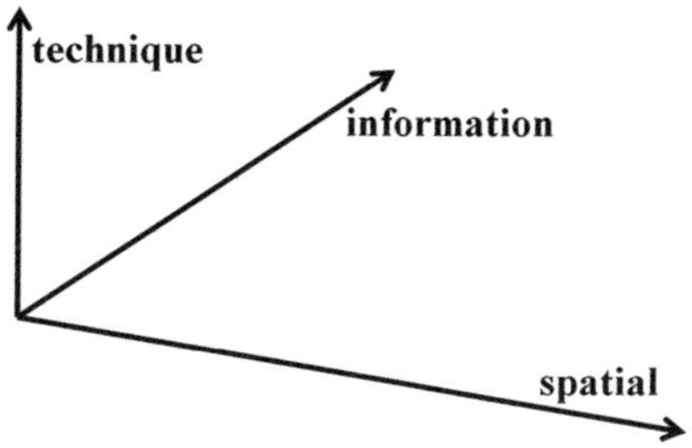

Fig. 7.1 The SIT data fusion model, named for its axes: spatial, information and technique, was created by modifying and combining two general data fusion models into a single model specifically applicable to the integration of remote sensing data for benthic mapping (after Park et al. 2010)

described in this chapter are fairly new, there are not many specific examples of their application to coral reef remote sensing. Nonetheless, a number of the primary datasets for developing these techniques were collected in areas with coral reefs because of their diverse bottom features, and the examples provided herein are from these areas: Ft. Lauderdale, FL, Looe Key, FL, and Hilo Bay, HI.

7.2 LiDAR/Hyperspectral Processing

7.2.1 SIT Data Fusion Model

The SIT data fusion model was created by modifying and combining two existing general data fusion models (Abidi and Gonsalves 1992; Hall 1992) to create a single model that is specifically applicable to fusion of remote sensing data for benthic mapping (Tuell and Lohrenz 2006). The model is named for the constituent axes that make up its domain: spatial, information and technique (Fig. 7.1). The axes increase in abstraction from raw data with unknown properties to objects and features with known identities. For example, along the spatial axis, LiDAR or hyperspectral data transitions from raw data with no geometric or geographic information into georeferenced pixels or point clouds with explicit geographic positions. Along the information axis, pixels transition from raw digital numbers into environmental and habitat parameters, such as depth, water column attenuation, and habitat type. Along the technique axis, algorithms increase in complexity (Table 7.1) from straightforward extraction methods to artificial intelligence techniques like neural nets.

Data fusion procedures can then be defined as a function of their relative position on the three axes. Figure 7.2 shows where each of the examples in this chapter are positioned in the spatial and information domains. From these examples, it is evident that the data fusion techniques devised so far for benthic mapping occur close to the raw sensor data on the spatial domain, but span the entire information domain. This suggests that the benthic maps derived from data

Table 7.1 Example processing methods for the 'Technique' axis of the SIT data fusion model, with increasing complexity from simple models to complex cognition approaches

Technique	Examples
Model	Simulation, estimation (least squares)
Extraction	Signal, pixel, segment
Inference	Parametric (clustering), non-parametric (neural nets)
Cognition	Templates, fuzzy set theory, knowledge systems (rules, Dempster-Shafer)

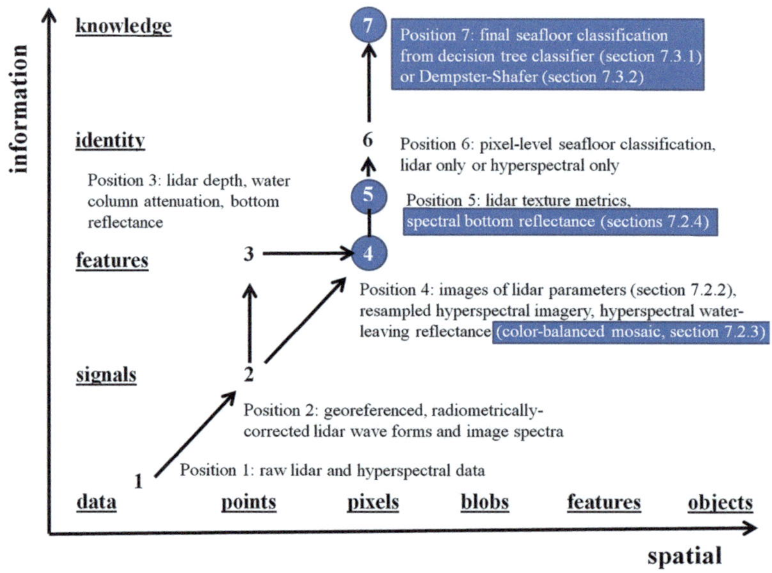

Fig. 7.2 The spatial and information axes of the SIT data fusion model, where each numbered position represents a step in the processing of LiDAR data and hyperspectral imagery for benthic mapping: where 1 is raw sensor data, and 7 is classified pixels

fusion techniques can contain rich levels of information at the pixel level. New developments are underway to advance fusion in the spatial domain, so that identification and classification occurs at the object level in the spatial domain (i.e., processing groups of pixels rather than pixel-by-pixel methods). Current data fusion methods for benthic mapping include techniques from all of the levels identified in Table 7.1.

7.2.2 LiDAR-Derived Parameters

The first steps toward integrating LiDAR data with hyperspectral imagery are the extraction of the various LiDAR-derived parameters, such as water depth, seafloor reflectance, water column attenuation, water column volume reflectance, and water

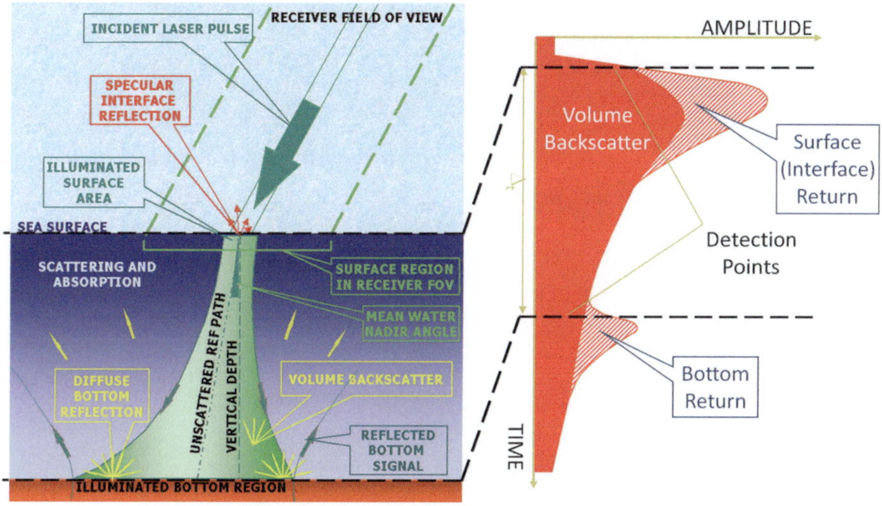

Fig. 7.3 Interactions of the sea-surface, water column, and seafloor with the bathymetric laser pulse (*left*) (after Wozencraft and Millar 2005). Bathymetric LiDAR waveform (*right*). The surface return captures the interaction of the laser pulse with the sea surface, while the bottom return captures its interaction with the seafloor. The volume backscatter captures photons scattered from water molecules and particulates suspended in the water column

leaving reflectance (Fig. 7.3). As introduced in Chap. 5, full waveform LiDAR measures returned light at discrete intervals (on the order of nanoseconds), thereby recording a "waveform" for each emitted pulse. The first peak of the waveform corresponds to the return of photons from the water surface. The second peak corresponds to the return of photons from the seafloor, where the seafloor reflectance is captured in the magnitude of this second peak. The return of photons from water molecules, entrained sediments, and organic matter occurs throughout the transit of the laser pulse through the water and is called the volume backscatter. Water column attenuation is determined as the log slope of the volume backscatter. Water column volume reflectance is the integral of the water column attenuation through the water column, and water leaving reflectance is the sum of water column volume reflectance and the seafloor reflectance. Examples of each of these parameters are shown in Fig. 7.4.

The images depicted in Fig. 7.4 were generated using data from the SHOALS system following the approach described in Kopilevich et al. (2005), which is one of the most rigorous approaches currently available. The approach first estimates seawater inherent optical properties (IOPs) by fitting the measured LiDAR waveforms with those generated from a multiple-forward-single-backscattering model of the LiDAR signal for different water column and bottom properties. Bottom reflectance is then derived using these IOPs as constraints on modeling bottom returns and seafloor geometry. This approach requires radiometric calibration of the bathymetric LiDAR, which relates radiance to the measured voltage response of the LiDAR receivers (Tuell et al. 2005a). In the bathymetric LiDAR

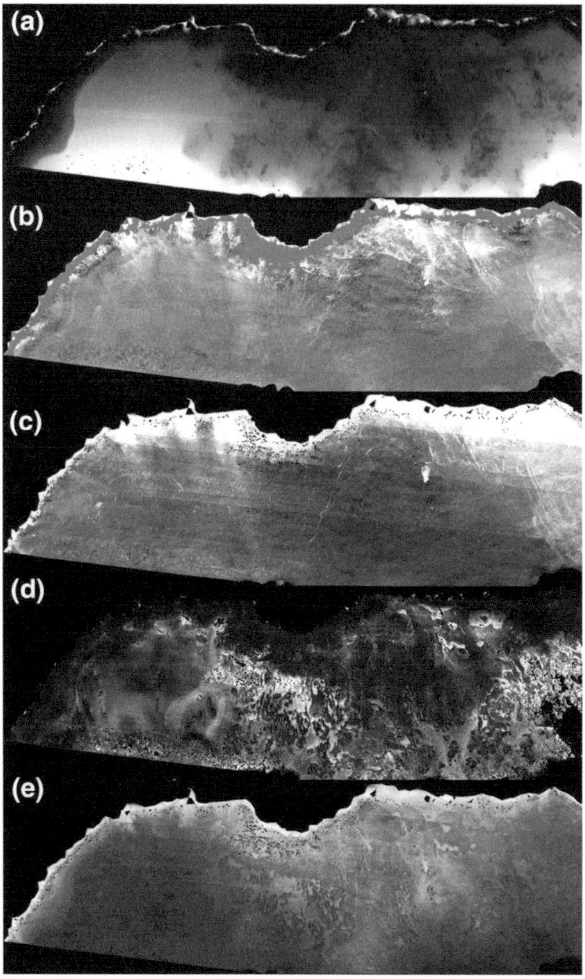

Fig. 7.4 Parameters extracted from SHOALS bathymetric LiDAR waveforms for Hilo Bay, Hawaii: (**a**) SHOALS depth from 0–40 m, where lighter areas are deeper; (**b**) $a + b_b$, water column absorption and backscattering, a measure of water clarity, where lighter areas are more turbid; (**c**) water column volume reflectance, which is $a + b_b$ integrated over the entire water column, where lighter areas are more reflective; and (**d**) bottom reflectance, where lighter areas are more reflective, in this case sand appears light and corals appear dark; and (**e**) water surface reflectance, which is the sum of water column volume reflectance and bottom reflectance

community, there are several different meanings of reflectance: reflectivity, relative reflectance, pseudo-reflectance and absolute reflectance. Reflectivity is the same as the intensity of laser energy return. Relative reflectance represents differences in reflectivity between pixels or clouds of pixels. Pseudo-reflectance and absolute reflectance are derived using physics-based equations applied to full waveform analysis or simple empirical estimation. Pseudo-reflectance can be

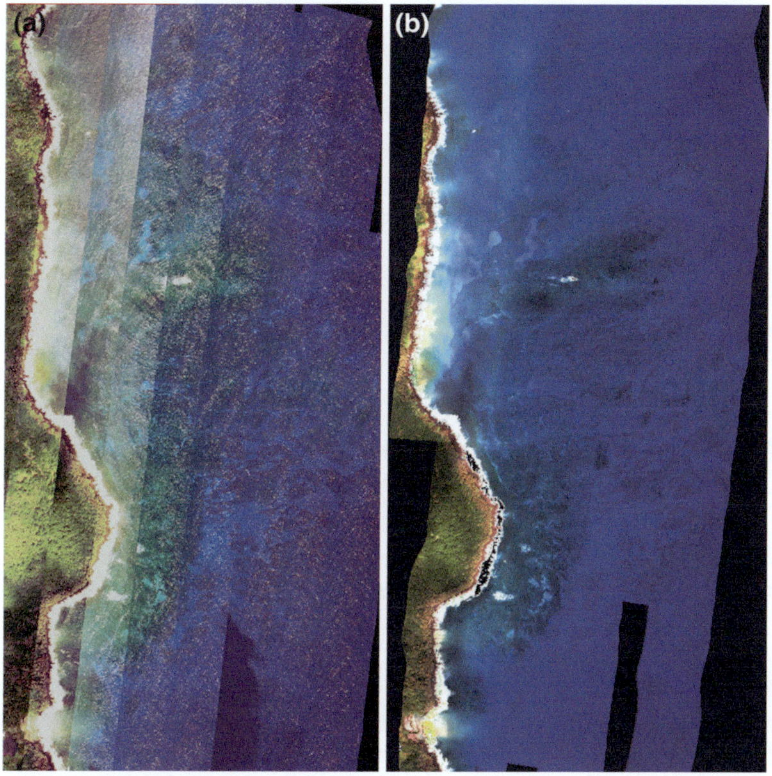

Fig. 7.5 True-color images of CASI-1500 hyperspectral imagery for Hilo Bay, HI: (**a**) mosaic of radiometrically and geometrically corrected images; (**b**) color-balanced mosaic showing the combined result of path length radiance bias correction, atmospheric correction, sunglint removal, and color-balancing using the SHOALS LiDAR-derived water-leaving reflectance

computed as a function of just the digital counts of laser return energy at the sensor, whereas absolute reflectance requires radiometric calibration to convert laser return energy to radiance. Of the bathymetric LiDAR systems in operation today, only the SHOALS systems are radiometrically calibrated.

The value of LiDAR-derived bathymetry and seafloor reflectance to coral reef mapping was discussed in Chaps. 5 and 6. The next section will describe how these parameters are used in conjunction with additional LiDAR-derived water column information in the processing of hyperspectral imagery for benthic habitat mapping.

7.2.3 Hyperspectral Color Balancing

Following typical image preprocessing steps of radiometric, geometric, atmospheric and sunglint corrections (see Chap. 4), LiDAR-derived water-leaving reflectance values can be used to normalize radiometric differences in adjacent hyperspectral images. As defined in the previous section, LiDAR-derived water-leaving reflectance is calculated by adding together the LiDAR-derived water column volume reflectance and seafloor reflectance. This water leaving reflectance is then used to scale the hyperspectral water-leaving reflectance across all wavelengths based on the ratio between the LiDAR-derived water-leaving reflectance at 532 nm and water-leaving reflectance at the closest band in the hyperspectral imagery. The resulting image is referred to as a "color-balanced" mosaic of hyperspectral imagery (Wozencraft et al. 2008). Figure 7.5 shows a mosaic of at-sensor radiance images compared with a color-balanced mosaic of water-leaving reflectance for Hilo Bay, Hawaii, HI. The LiDAR data were collected by a SHOALS sensor and the hyperspectral imagery were collected by a CASI-1500.

7.2.4 Constrained Optimization Modeling

LiDAR-derived water column information can be used to constrain the inversion of the hyperspectral radiative transfer equation. A spectral seafloor reflectance image is defined here as bottom reflectance across all viable wavelengths of the hyperspectral imagery (typically limited to just the visible portion of the spectrum). One method for deriving this parameter using just hyperspectral imagery is by inverting the radiative transfer equations to simultaneously solve for seafloor reflectance, spectral water column volume reflectance, and spectral water column attenuation using an iterative non-linear least-squares approach (Lee 2003). As an alternative, LiDAR data can be used to constrain this inversion, thereby increasing the accuracy of the derived reflectance. For example, Tuell et al. (2005a) identify homogenous areas in the LiDAR-derived reflectance image, and calculate water column attenuation in these areas such that the difference between hyperspectral-derived depth and LiDAR-derived depth is minimized. Tuell and Park (2004) suggested the creation of LiDAR water column layers by interpolating among the homogenous areas, which Tuell et al. (2005b) refined by using the LiDAR-derived water column attenuation to scale hyperspectral-derived water column attenuation. This approach yields spatially-varying water column attenuation at every wavelength of the hyperspectral data for use in the inversion of the radiative transfer equation. The resulting spectral seafloor reflectance image is improved by reducing the impact of assumptions regarding constant water column attenuation throughout a hyperspectral scene. Figure 7.6 shows example reflectance products generated from SHOALS LiDAR and CASI-1500 hyperspectral data collected in Fort Lauderdale, FL. The water-leaving reflectance image on the left is an example of the color-balanced mosaic described in Sect. 7.2.3, while the image on the right is the derived hyperspectral seafloor reflectance.

LiDAR-derived water depth, water column attenuation, and bottom reflectance can also be used to constrain the number of parameters in a combined atmospheric-oceanographic spectral optimization model for deriving water column properties and seafloor reflectance (Kim et al. 2010). The oceanographic portion of

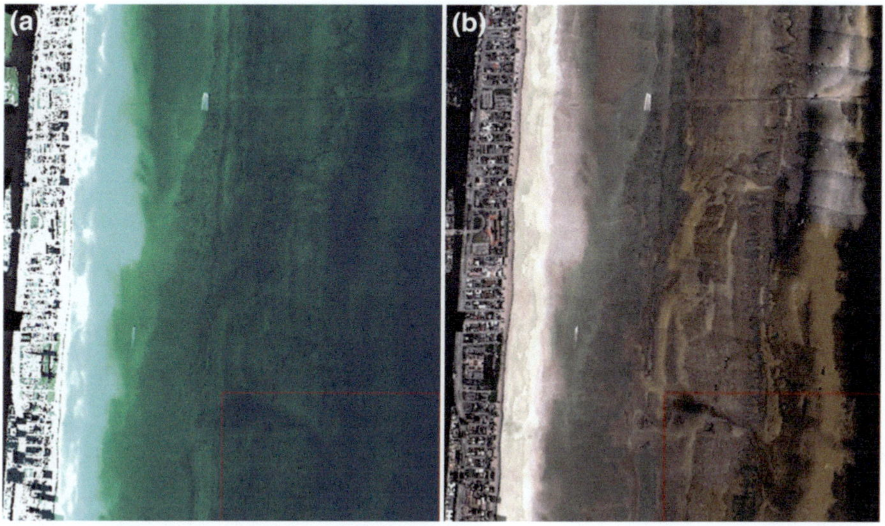

Fig. 7.6 True-color images built from *red*, *green*, and *blue bands* of the CASI-1500 hyperspectral imagery for Ft. Lauderdale, FL: **a** color-balanced mosaic of water-leaving reflectance; and **b** seafloor reflectance

this spectral optimization model follows similar inversion methods as described above and in Chap. 4; however, in this coupled model, the atmospheric correction is also performed as part of the inversion process. Using this method, the atmosphere, water column, and seafloor are decomposed into their component constituents, and incorporated into the radiative transfer equation inversion process through a series of analytical and empirical relationships, all well-established in the ocean optics community. The results of this approach are a series of output data layers, including spectral water-leaving reflectance, water column attenuation, *Chl a* and colored dissolved organic matter absorption, spectral seafloor reflectance, and abundance images depicting the proportionate contributions of seafloor constituents in each pixel based on input bottom spectra (e.g., see spectral unmixing in Chap. 4). Figure 7.7 shows examples of spectral seafloor reflectance and three abundance images (sand, coral, and sea grass) generated from SHOALS LiDAR data and CASI-2 hyperspectral imagery collected near Looe Key, FL.

7.3 Applications of LiDAR/Hyperspectral Fusion

The techniques introduced for integrating LiDAR data with hyperspectral imagery have focused predominantly on improving the quality of spectral seafloor reflectance and spectral water column information. This section will demonstrate how information extracted from the LiDAR data can be combined with spectral information in a decision-tree classifier to improve a semi-automated pixel-level

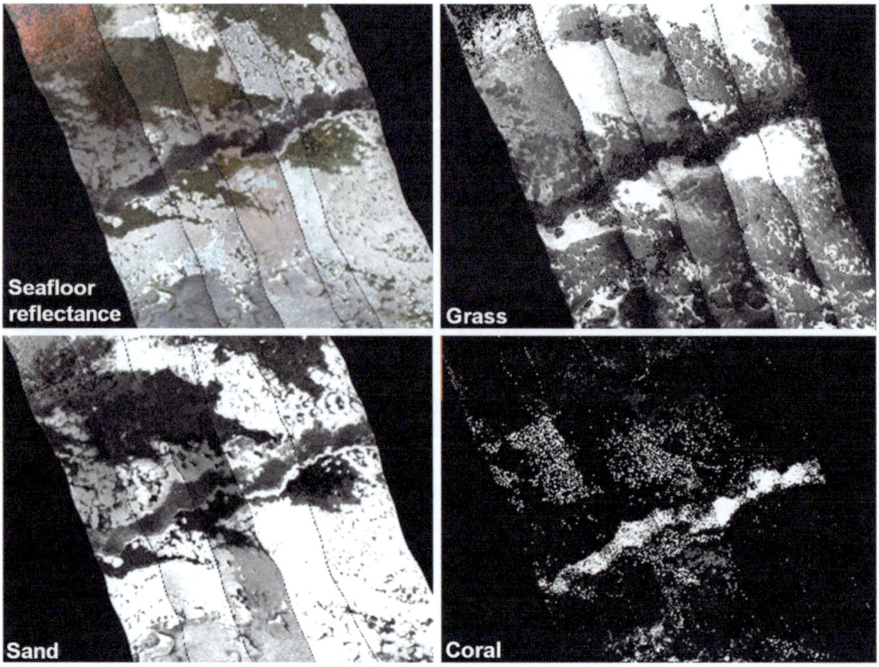

Fig. 7.7 Spectral optimization output generated from SHOALS LiDAR and CASI-2 hyperspectral imagery of Looe Key, FL. The spectral seafloor reflectance image is true-color of spectral seafloor reflectance, where the reddish area in the *upper left* is an artifact of the spectral optimization processing. The remaining images are abundance images of seagrass, sand, and coral. In these images, the brighter pixels are those with greatest similarity to the input seafloor spectra for that type (i.e., higher abundance), and the darker pixels are the least similar (i.e., lower abundance)

landcover classification. Although the analysis is based on land, the conceptual approach can be extended to submerged ecosystems as bathymetric LiDAR and hyperspectral data fusion becomes more prevalent. An advanced approach for combining seafloor classifications generated from LiDAR data and hyperspectral imagery is also presented.

7.3.1 Decision-Tree Classification

A relatively straightforward technique for integrating information derived from LiDAR data and hyperspectral imagery is a decision-tree classifier. In the example presented here, the height above ground, or surface height, was derived from topographic LiDAR data, and combined with select hyperspectral bands for classification into basic landcover classes on a pixel-by-pixel basis (Reif et al. 2011). The land-cover types and decision parameters are described in Table 7.2. The schematic in Fig. 7.8 demonstrates the decision process from preprocessed imagery

Table 7.2 Landcover classes and general descriptions used in the decision-tree classification of Hilo Bay, HI

Land-cover class	Class description
Unclassified/saturated	Includes "no data" pixels and saturated pixels (undiscerned bright image objects)
Bare ground/road	Includes non-vegetation pixels with height <1 m
Structures	Includes non-vegetation pixels with height >1 m
Low vegetation	Includes vegetation pixels defined by NDVI value > 0.3 and height < 0.5 m (i.e., grasses)
Medium vegetation	Includes vegetation pixels defined by NDVI value > 0.3 and height 0.5–6 m (i.e., small trees/shrubs)
Tall vegetation	Includes vegetation pixels defined by NDVI value > 0.3 and height > 6 m (i.e., trees)

NDVI normalized difference vegetation index

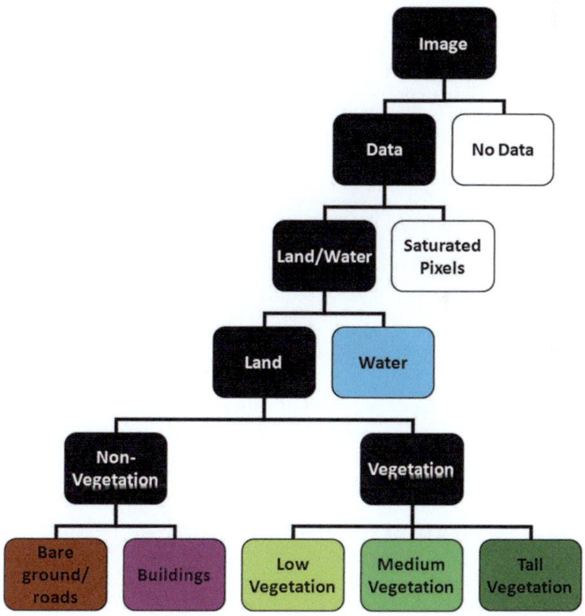

Fig. 7.8 Schematic demonstrating the decision process for combining LiDAR-derived above-ground height with hyperspectral indices for a basic landcover classification

(i.e., after radiometric, geometric, and atmospheric corrections) to final classification. An example classification resulting from this technique using SHOALS LiDAR and CASI-1500 hyperspectral imagery is shown in Fig. 7.9 for Hilo Bay, HI.

Chapter 3 discussed the utility of landcover classifications, in particular changes in landcover, to help assess the impact of anthropogenic processes on nearby coral reef. The decision-tree approach is widely applicable to combining LiDAR data with hyperspectral imagery, and could be used for seafloor classification to support coral reef mapping, for example using LiDAR-derived reflectance, a texture-based metric, and hyperspectral-derived seafloor reflectance to derive pixel-based seafloor classification.

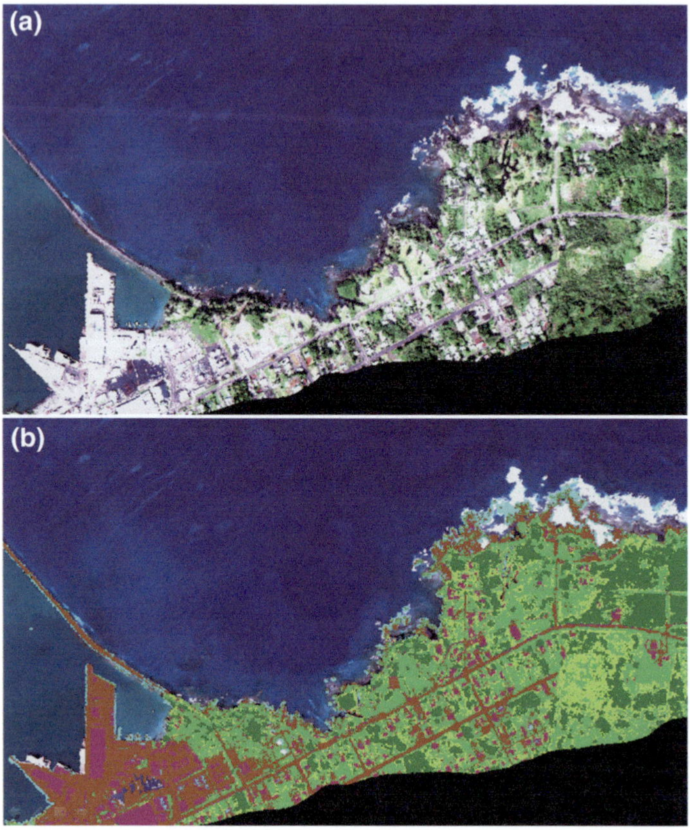

Fig. 7.9 Basic landcover classification generated using decision-tree classifier with SHOALS LiDAR derived above-ground height and CASI-1500 hyperspectral vegetation indices for Hilo Bay, HI: **a** hyperspectral image; **b** hyperspectral image with landcover classification overlay

7.3.2 Dempster-Shafer Method

The Dempster-Shafer (D-S) method is an example of decision-level data fusion. In this method, Dempster and Shafer (Shafer 1976) present a generalization of the Bayesian theory of uncertainty (Lowrance and Garvey 1982). The D-S method is based on the concept that humans assign measures of belief according to combinations of all available evidence (i.e., according to multiple rather than single events) (Hall 1992). In the D-S method, probability and uncertainty intervals are used to determine the likelihood of a hypothesis based on a combination of several decision variables. In remote sensing, for example, one sensor may be able to provide information that can be used to distinguish the height of objects, whereas information from another sensor may be able to distinguish only the shape of objects. In the Bayesian approach, all unknown propositions (e.g., objects in the environment) are assigned an equal a priori probability. When the number of

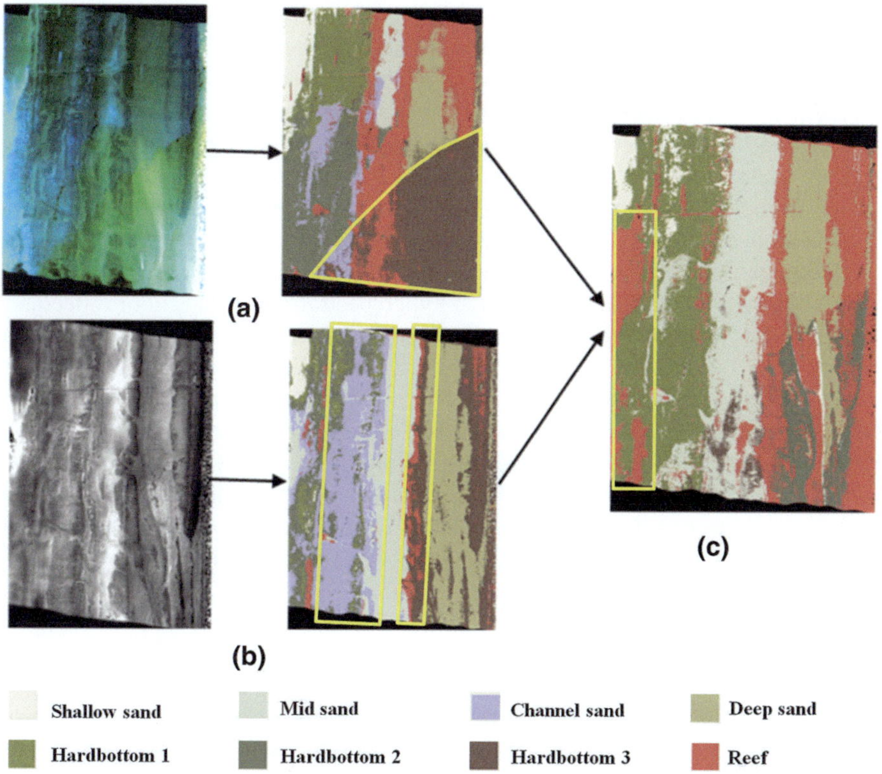

Fig. 7.10 Data fusion classification procedure: **a** maximum likelihood classification image from hyperspectral seafloor image; **b** maximum likelihood classification image from LiDAR seafloor reflectance image; and **c** final bottom classification using Dempster-Shafer method. Areas of misclassification are indicated by *yellow polygons*

unknown propositions is relatively large compared with the number of known propositions, the main problem of the Bayesian approach is that the probabilities of the known propositions become unstable, leading to questionable results (Abidi and Gonzales 1992). The D-S method was developed in an attempt to overcome such limitations. Many researchers have explored the application of the D-S method to multisensor target identification, military command and control, and land-cover classification (Bogler 1987; Waltz and Buede 1989; Park 2002).

As an example of the D-S method in practice, a maximum likelihood classifier (MLC) is applied to both LiDAR-derived and hyperspectral-derived seafloor reflectance images. The output probabilities for each class from the MLC are then used as a priori probabilities in the D-S method. The relevant input images and processed MLC images are shown in Fig. 7.10. Each MLC image has eight classes: shallow sand, mid sand, channel sand, deep sand, hard bottom type 1, hard bottom type 2, hard bottom type 3, and reef. In some instances areas are classified as the same class in both images, but in many areas the classifications differ, which is where the statistically-based D-S method becomes useful. The D-S method utilizes two fundamental components: hypothesis and proposition. An hypothesis is a fundamental statement about nature (e.g., a pixel or object is a reef). A proposition may be either a hypothesis or a combination of hypotheses, which in turn may contain overlapping or conflicting hypotheses. For example, 'proposition 1' = the seafloor class is sand, 'proposition 2' = the seafloor class is sand or reef, and 'proposition 3' = the seafloor type is reef. When these propositions are input to the D-S method, along with probabilities for each proposition, output consists of a single integrated classification image (Fig. 7.10c).

Examining the MLC image from the hyperspectral-derived seafloor reflectance in Fig. 7.10a, there is a large patch of deeper water that is misclassified as 'hard bottom type 3'. Similarly, there are 'mid sand' areas in the south misclassified as 'reef', and there are 'hard bottom type 1' areas in the south misclassified as 'hard bottom type 2'. The MLC image from the LiDAR-derived seafloor reflectance in Fig. 7.10b appears to have more detail and sharper boundaries between classes, but nonetheless it also has several misclassified areas. For example, most 'hard bottom type 1' areas were misclassified as 'channel sand', and some 'reef areas' were misclassified as 'hard bottom type 3'. Areas of misclassification in Figure 7.10 are indicated by *yellow polygons*.

In the fusion classification image in Fig. 7.10c, many areas were re-assigned different classes based on the propositions used in the D-S method. For example, the 'hard bottom type 2' areas in the hyperspectral MLC image and the 'channel sand' areas in the LiDAR MLC image were both correctly re-classified as 'hard bottom type 1'. The large area that was classified as 'hard bottom type 3' in the hyperspectral MLC image was re-classified as 'deep sand', 'reef', or 'hard bottom type 2.' However, the 'shallow sand' areas in the hyperspectral MLC image and some 'mid sand' and 'channel sand' areas in the LiDAR MLC image were incorrectly re-classified as 'reef' (as again indicated by the *yellow polygons*), suggesting there is still room for improvement. Nevertheless, the D-S method produces better overall results compared with either of the MLC images alone.

7.4 Summary and Discussion

This chapter introduced the concept of data fusion and presented a model developed specifically for integrating LiDAR data and hyperspectral imagery to improve benthic classification. Examples demonstrated the extraction of seafloor reflectance and water column attenuation from LiDAR waveforms, and then how this information can be used to improve the processing of hyperspectral data to spectral seafloor reflectance using modified depth-correction and model inversion techniques. An approach for combining LiDAR-derived information with spectral data in a decision tree classifier was also presented, along with a higher-level data fusion technique known as the Dempster-Shafer method for combining independent classification images into an integrated product.

There is ongoing research in both hardware and software for integrating LiDAR data with hyperspectral imagery. Many researchers are looking at the possibility of extending the techniques presented here with other LiDAR sensors. Specifically, a new sensor development effort, the Coastal Zone Mapping and Imaging LiDAR (CZMIL), is currently in the flight-testing stage. CZMIL was designed as an imaging system, and incorporates many hardware changes to improve implementation of data fusion at all levels presented in this chapter. The CZMIL data processing system, developed in tandem with the airborne hardware, is essentially a LiDAR/hyperspectral data fusion processing system.

Other areas of research are conversely investigating how hyperspectral information might inform LiDAR processing. One example is using hyperspectral water column attenuation to aid in processing LiDAR seafloor reflectance imagery in shallow water, where estimation of water column attenuation is confounded due to convolution of the surface and bottom returns. Techniques are also under development for selecting the most information-rich features from the LiDAR data and hyperspectral imagery, like various texture metrics and spectral indices, for inclusion in decision-tree, maximum likelihood, and other classifiers.

The types of information provided by integrating hyperspectral imagery with LiDAR data are not different from those provided by the other remote sensing techniques described throughout this book. The examples presented here are specific to LiDAR and hyperspectral, but there is every reason to expect similar data fusion techniques to be applied to other combinations of remote sensing products. The advantage and ultimate goal of data fusion is the improved accuracy of the final coral reef and environmental data products derived from the imagery.

Acknowledgments The data collection, data processing, and data fusion technique development summarized in this chapter were funded by the Joint Airborne Lidar Bathymetry Technical Center of Expertise (JALBTCX) through the Naval Oceanographic Office's Adding Hyperspectral to CHARTS Project, the U.S. Army Corps of Engineers National Coastal Mapping Program, and the National Ocean Partnership Program's High-level Data Fusion Software for SHOALS-1,000TH project; and by the U.S. Naval Research Laboratory's Countermine Lidar UAV-Based System Project. The data collection, data processing, and data fusion technique development summarized in this chapter were accomplished by personnel at JALBTCX, Optech, Inc. (USA,formerly OptechInternational), and the University of Southern Mississippi.

Suggested Reading

Lee M (2003) Benthic mapping of coastal waters using data fusion of Hyperspectral Imagery and Airborne Laser Bathymetry. Ph.D. dissertation. University of Florida. Gainsville, Florida, p 119

Park JY, Ramnath V, Feygels V, Kim M, Mathur A, Aitken J, Tuell GH (2010) Active-passive data fusion algorithms for seafloor imaging and classification from CZMIL data. In: Lewis PE (eds) Proceedings SPIE, 7,695. Shen SS, Algorithms and technologies for multispectral, hyperspectral, and ultraspectral imagery 16

Reif M, Macon CL, Wozencraft JM (2011) Post-katrina land-cover, elevation, and volume change assessment along the south shore of lake pontchartrain, Louisiana. J Coast Res Appl Lidar Tech [Pe'eri, Long] USA 62:30–39

Tuell GH, Park JY, Aitken J, Ramnath V, Feygels VI, Guenther GC, Kopilevich YI (2005) SHOALS-enabled 3-D benthic mapping. In: Chen S, Lewis P, (eds) Algorithms and technologies for multispectral, hyperspectral, and ultraspectral imagery 11, Proceedings of SPIE 5806:816–826

Wozencraft JM, Macon CL, Lillycrop WJ (2008) High resolution coastal data for Hawaii. Proceedings of sessions of the conference: solutions to coastal disasters, Am Soc Civ Eng, pp 422–431

References

Abidi M, Gonzalez R (1992) Data fusion in robotics and machine intelligence. Academic, San Diego

Bogler PL (1987) Shafer-dempster reasoning with applications to multisensor target identification systems. IEEE Trans Syst Man Cybern SMC 17(6):968–977

Bissett WP, DeBra S, Kadiwala M, Kohler DDR, Mobley CD, Steward RG, Weidemann AD, Davis CO, Lillycrop J, Pope RL (2005) Development, validation, and fusion of high-resolution active and passive optical imagery. In: Kadar I (ed) Signal processing, sensor fusion, and target recognition XIV. Proceedings of SPIE, vol 5809. SPIE, Bellingham, WA, pp 341–349

Hall D (1992) Mathematical techniques in multisensor data fusion. Artech House, Boston

Kim M, Park JY, Tuell G (2010) A constrained optimization technique for estimating environmental parameters from CZMIL Hyperspectral and Lidar Data. In: Shen SS, Lewis PE (eds) Proceedings SPIE, 7695, algorithms and technologies for multispectral, hyperspectral, and ultraspectral imagery XVI

Kopilevich Y, Feygels V, Tuell G, Surkov A (2005) Measurement of ocean water optical properties and seafloor reflectance with scanning hydrographic operational airborne lidar survey (SHOALS): I Theoretical background. In: Proceedings remote sensing of the coastal oceanic environment, SPIE, vol 5885, pp 106–114

Lowrance JD, Garvey TD (1982) Evidential reasoning: a developing concept. Proceedings of IEEE International conference on cyberbetics society, pp 6–9

Lee M (2003) Benthic mapping of coastal waters using data fusion of hyperspectral imagery and airborne laser bathymetry. Ph.D. dissertation, University of Florida. Gainsville, Florida, p 119

Lee M, Tuell G (2003) A technique for generating bottom reflectance images from SHOALS data, presented at U.S. Hydro 2003 hydrographic conference, Biloxi, Mississippi, pp 24–27

Lillycrop WJ, Estep LL (1995) Generational advancements in coastal surveying. Mapping Sea Technol 36(6):10–16

Park JY (2002) Data fusion techniques for object space classification using airborne laser data and airborne digital photographs. Ph.D. dissertation, University of Florida, Department of civil and coastal engineering, Gainesville, Florida

Park JY, Ramnath V, Feygels V, Kim M, Mathur A, Aitken J, Tuell GH (2010) Active-passive data fusion algorithms for seafloor imaging and classification from CZMIL Data. In: Shen SS, Lewis PE (eds) Proceedings SPIE, 7695, algorithms and technologies for multispectral, hyperspectral, and ultraspectral imagery XVI

Reif M, Macon CL, Wozencraft JM (2011) Post-katrina land-cover, elevation, and volume change assessment along the south shore of Lake Pontchartrain, Louisiana, J Coast Res. Special issue, applied lidar techniques [Pe'eri, Long], vol 62, pp 30–39

Shafer G (1976) A mathematical theory of evidence. Princeton University Press, Princeton

Tuell G, Park JY (2004) Use of SHOALS bottom reflectance images to constrain the inversion of a hyperspectral radiative transfer model. In: Kammerman G (ed) Proceedigs SPIE vol 5412, laser radar and technology applications IX. pp 185–193

Tuell G, Lohrenz S (2006) High level data fusion for SHOALS-100th, Annual Report for FY 2006, National ocean partnership program

Tuell GH, Park JY, Aitken J, Ramnath V, Feygels VI, Guenther GC, Kopilevich YI (2005a) SHOALS-enabled 3-d benthic mapping. In: Chen S, Lewis P (eds) Proceedings SPIE vol 5806, algorithms and technologies for multispectral, hyperspectral, and ultraspectral imagery XI. pp 816–826

Tuell GH, Feygels V, Kopilevich Y, Weidemann AD, Cunningham AG, Mani R, Podoba V, Ramnath V, Park JY, Aitken J (2005b) Measurement of ocean water optical properties and seafloor reflectance with scanning hydrographic operational airborne lidar survey (SHOALS): II. Practical results and comparison with independent data. In: Proceedings remote sensing of the coastal oceanic environment, SPIE vol 5885. pp 115–127

Waltz EL, Buede DM (1989) Data fusion and decision support for command and control. IEEE Trans Syst Man Cybern SMC 16(6):865–879

Wozencraft JM, Macon CL, Lillycrop WJ (2008) High resolution coastal data for Hawaii. Proceedings of sessions of the conference: solutions to coastal disasters, Am Soc Civ Eng, pp 422–431

Wozencraft JM, Macon CL, Lillycrop WJ (2007) CHARTS-Enabled data fusion for coastal zone characterization. Proceedings of the 6th international symposium on coastal engineering and science of coastal sediment processes 3, Am Soc Civ Eng, Reston, VA. ISBN-0-7844-0926-9, pp 1827–1836

Wozencraft JM, Millar D (2005) Airborne lidar and integrated technologies for coastal mapping and charting. Mar Technol Soc J. 39(3):27–35

Section III
Acoustic

Chapter 8
Acoustic Methods Overview

Bernhard Riegl and Humberto Guarin

Abstract Acoustic methods are widely used for the production of physical, environmental and biological data required for the responsible management of marine resources, such as coral reefs. Here, we review the basic physical properties of sound in water that can be harnessed for active or passive acoustic remote sensing systems. Sound, by assessing the return characteristics of emitted sound waves, can be used to derive information on seafloor topography via depth (obtained by measuring travel time), on seafloor makeup (obtained by measuring backscatter intensity), or on water column characteristics (obtained by measuring Doppler shifts). Sound is also used to track organisms such as fish or even to create images by harnessing natural sound sources to "illuminate" objects like fish. Acoustic methods have a place in the toolbox of every coral reef manager.

8.1 Introduction

Acoustic methods used in the ocean are frequently subsumed under the term (SONAR) Sound Navigation And Ranging, but this is in reality only a subset of the full array of available techniques. Acoustic methods are extremely versatile and find many applications as 'active' or 'passive' technologies. SONAR, an active technology, is similar to radar or (LiDAR) Light Detection And Ranging,

B. Riegl (✉)
National Coral Reef Institute, Oceanographic Center, Nova Southeastern University, 8000 N. Ocean Drive, Dania, FL 33004, USA
e-mail: rieglb@nova.edu

H. Guarin
Bert Instruments Inc, 2646 Sherman St, Hollywood, FL 33020, USA
e-mail: hguarin@bertinst.com

J. A. Goodman et al. (eds.), *Coral Reef Remote Sensing*,
DOI: 10.1007/978-90-481-9292-2_8,

providing its own pulse energy and thus differing from passive sensors that measure energy emanating from the environment (e.g., listening arrays). SONAR technologies are largely based on measurements of the speed of sound in water (i.e., the travel time between a transmitted and received sound pulse), as well as the characteristics of the scattered sound. Dependent on the strength, width and orientation of the pulse, the acoustic signal can be used to investigate features of the water body, the sediment surface, or the sediment (or bedrock) interior. SONAR methods are therefore among the most versatile and frequently used oceanographic remote sensing tools in the open ocean, as well as in coastal and reef environments. Listening arrays are also increasingly used in fisheries science. Fish and/or marine mammals are either tracked by using their own vocalizations (in particular whales lend themselves very well to such applications) or acoustic emitters are attached to animals and then tracked by passive listening arrays.

Absorption of sound in water is dependent mostly on the frequency of the acoustic pulse. At the high-frequency end, above 1 Hz (Hz = cycles per second), absorption by seawater is important and applications are largely limited to acoustic imaging and side-looking SONAR. At the low end, below 1 Hz, the generation of sound is technically challenging, needing small earthquakes or large explosions to create measurable events. Thus, ocean acoustics generally concerns itself with a band of frequencies between 1 Hz and several hundred KHz (Tolstoy and Clay 1966).

SONAR technology is not new and has served many purposes. Early in its development, it was mainly dedicated to shipping safety and military use. Soon after the Titanic struck an iceberg and sunk under great loss of life in 1912, patents were filed to use the echo of sound waves to detect large objects underwater (Medwin and Clay 1998). The technology was based on the idea that once speed of sound in water is precisely known, measurement of its travel time from a transmitter source and back to a receiver would allow an equally precise measurement of the scattering object's distance from the sensor. The earliest measurements of sound velocity in water were performed by Colladon and Sturm (1827) in Switzerland's Lake Geneva. In 1916 Chilowsky and Langevin (1916) obtained echoes from the shallow seafloor and from an iron plate at 200 m depth. They used an electrostatic sound source as transmitter and a carbon button microphone as receiver. During WWI, the Canadian physicist Boyle and the British scientist Wood used quartz piezoelectric crystals to build the first SONAR (Medwin and Clay 1998), which became known as Anti-Submarine Detection Investigations Committee (ASDIC). Langevin used quartz piezoelectric transducers both as projector and receiver to obtain sound transmission up to 8 km. In 1919, Marti patented an 'echosounder' that was capable of producing a continuous visual record of the seafloor, and in 1925, the German ship Meteor ran echosounding lines across the South Atlantic. With this, a scientific revolution began in the understanding of the deep seafloor following the detection of the mid-oceanic ridges and abyssal plains. By 1935, acoustic bathymetry was developed enough to allow soundings of ocean depth, and backscatter began to be used for the detection of fish schools (Medwin and Clay 1998). The onset of WWII gave great impetus to the use of physics for the detection of remote objects, and both radar and SONAR methods were well advanced (Jones 1999).

Modern applications of acoustic remote sensing techniques persist in part within the military realm, but civilian application has also become increasingly important. As agencies entrusted with the management of marine resources strive to better understand the systems put into their charge, habitat mapping has become of prime importance. Fisheries management, conservation management, and marine spatial planning rely increasingly on large-scale habitat maps that can, in depths beyond the resolution of optical sensors, only be produced by acoustic means. Thus, acoustic applications directly related to the needs of habitat mapping, as well as biological and fisheries assessment, are proliferating.

8.2 Physical and Technical Principles

Sound waves exhibit different physical characteristics from electromagnetic waves (e.g., visible and infrared wavelengths) that make them well suited for use in underwater investigations. Radio, radar and other electromagnetic waves propagate at the speed of light for long distances in the atmosphere, but as soon they enter the water, these properties become rapidly limited due to attenuation. Water, especially salt water, has high conductivity and is highly dissipative. Assessment of objects and surfaces at greater depths thus requires another means of transmitting energy, such as sound, that utilizes different physical properties. The mechanical propagation of a disturbance underwater can travel great distances (thousands of kilometers in the case of a whale song). Sound waves are described by the wave equation and compensate for lower propagation speeds with lower transmission loses (e.g., lower speeds and losses than light).

8.2.1 The Sound Wave

Sound is a mechanical disturbance that travels through a medium, in our case seawater (Fig. 8.1). The propagating disturbance is identified as an incremental acoustic pressure, the magnitude of which is smaller than ambient pressure (Medwin and Clay 1998). Sound is transmitted through gases and liquids as a longitudinal, or compressional, wave characterized by local regions of compression and rarefaction (Fig. 8.1). The term "longitudinal" refers to the displacements in the medium occurring along the direction of wave propagation. In contrast, sound transmission in solids can also be expressed as a "transverse" wave characterized by waves of alternating shear stress (i.e., particle displacement) at right angles to the direction of propagation. These types of waves have about half the velocity of longitudinal waves and cannot pass through liquids, since they cannot sustain the shear forces.

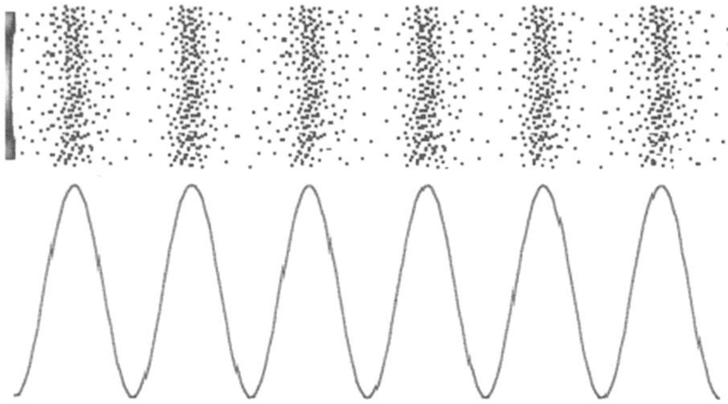

Fig. 8.1 Sound waves are areas of higher and lower particle density in the medium, which can be mathematically expressed as a sine wave

Basic to all acoustic techniques is the physics of the sound wave as an energy carrier. Waves are defined as propagations of a vibrational state that is periodic in space and time, in which energy is transported without simultaneous mass transport (Benenson et al. 2002). The vibrational state of a wave is described by its phase. Sound waves can exist as plane waves, in which the wave fronts are planes perpendicular to the propagation vector, or as spherical waves, in which the wave fronts are surfaces of concentric spheres around the center. The Huygens-Fresnel principle (Medwin and Clay 1998) states that each point of an advancing wave front is, in fact, the center of a fresh disturbance and the source of a new train of waves; the entire advancing wave is the sum of all the secondary waves arising from points in the medium already traversed. Application of this principle allows the explanation of diffraction, because every point on an object encountered by a sound wave becomes the source of a new one. It also allows convenient visualization and explanation of wave propagation (Fig. 8.2).

8.2.2 Sound in Water

The speed of sound in a medium is dependent on its density, and since water can have varying physical properties, the speed of sound also varies among and even within distinct water bodies (Fig. 8.3). Speed is determined by the water's bulk modulus and mass density. The earliest measurements were taken in 1827 by Colladon and Sturm in Lake Geneva. They suspended a bell underwater from a boat and struck it by means of a lever. That same lever also ignited a charge of powder that set off a light, allowing a distant observer to exactly mark the time of origination of the underwater sound. Using a listening tube from a second boat, the time differential between flash and arriving sound could be measured. Thus, a value of 1,435 m/s was obtained in this study. Soon after, it was realized that water temperature, density, and salinity are also important variables in sound speed. Since these variables are commonly horizontally stratified in the ocean, this results in the speed of sound being similarly horizontally stratified. The speed of sound in water also varies as a function of depth and environmental conditions, including

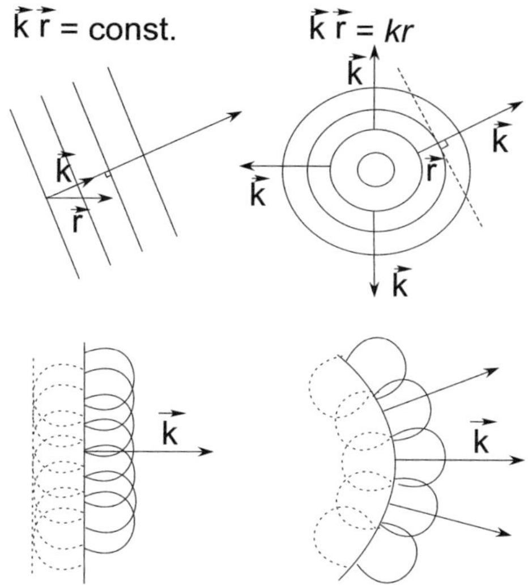

Fig. 8.2 Wave fronts of a plane wave (*top left*) and of a spherical wave (*top right*); k is the wave (propagation) vector with unit (1/m) and r is the position vector with unit m. In a plane wave, the wave fronts are perpendicular to the propagation vector, while in the spherical wave, the wave fronts are surfaces of concentric spheres around the source at r = 0. Propagation of wave fronts according to Huygen's principle for a plane wave (*bottom left*) and a spherical wave (*bottom right*). Any point of a wave front serves as starting point of an elementary wave. The wave front at a later instant is the envelope of the superposition of all elementary waves from a given wave front

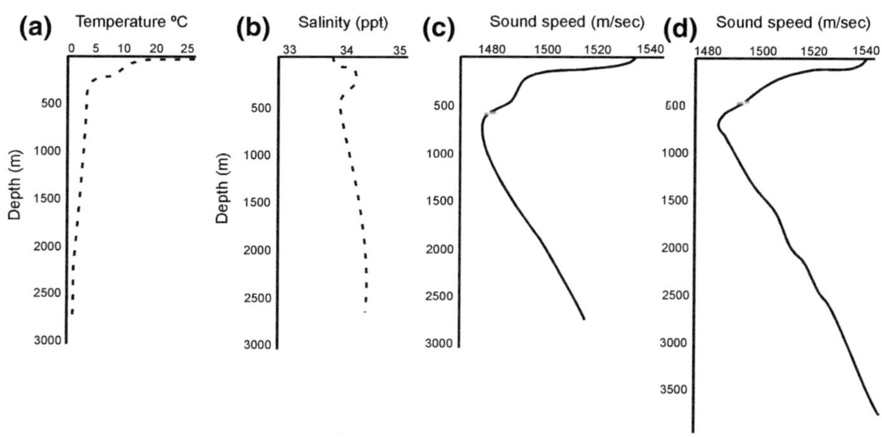

Fig. 8.3 (**a**, **b**, **c**) Variation of sound speed with temperature and salinity in the central Pacific, where a sound speed minimum occurs at 650 m. **d** Variation of sound speed in the equatorial Atlantic. Due to different properties of the respective water columns, the sound speed profiles differ between the world's oceans (modified from Jones 1999, by permission of Wiley)

seasonal and geographic variations, currents, tides, swells, internal waves, and even time of the day. The average velocity of a sound-wave travelling through seawater is approximately 1,500 m/s, but changes in water temperature, pressure and salinity cause important deviations (Wilson 1960; Medwin and Clay 1998).

The sound speed plotted as function of depth for a particular location is called a sound speed profile and provides a visual indication of location-specific sound speed variations (Fig. 8.3). It is observed that close to the surface, the controlling variable for the speed of sound in water is temperature, while at larger depths the controlling factor is depth. Salinity has only a small influence within the range from 34 to 35 parts per thousand (34–35 Practical Salinity Units, PSU). As a practical approximation, per every degree centigrade of temperature increase, sound speed close to the surface increases by 3 m/s; per every increase of one PSU increase in salinity, sound speed increases 1.4 m/s; and per every kilometer of depth increase, sound speed increases by 17 m/s.

Localized sound speed can be measured using an instrument such as a Time of Flight Probe, which uses an acoustic transducer transmitting a pulse that is reflected by a plate located at a fixed distance. It measures the two way travel time, divides it by two, and thereby computes the sound speed. For acoustic remote sensing, however, it is necessary to measure the sound speed profile for the entire water column. The sound speed profile can be obtained using instruments such as a CTD, which is lowered from the surface while periodically measuring the conductivity "C" (salinity), temperature "T" and depth "D" (pressure) through the water column. The United Nations Educational and Scientific Organization (UNESCO) algorithm to compute the sound speed profile based on CTD data, based on the Chen and Millero (1977) equation, is considered a universal standard in oceanography. Alternatively, an eXpendable Bathy Thermograph (XBT), which is less accurate but cheaper and easier to use, can similarly be used to compute the sound speed profile. While free falling at a known rate through the water column, the XBT sends temperature information to the surface using two very thin wires, before breaking loose. For a vessel that is underway, the Moving Vessel Profiler (MVP) utilizes a computer controlled winch and a custom 'fish' (a general term for a hydrodynamically shaped instrument) to measure sound velocity without the need to stop the vessel.

An important factor for acoustic remote sensing is attenuation in the water column, where active SONAR suffers two-way attenuation and passive SONAR only one-way attenuation. Since sound spreads geometrically from its source and then again upon reflection, it is attenuated by this spreading. The classic example to visualize a wave spreading is to throw a rock into a calm pond. A circular wave is produced when the rock hits the pond, where circles with increasing circumferences, but decreasing amplitudes, are seen. The total amount of energy in a wave remains constant as it spreads out from a source, but as the circle gets bigger, the energy spreads to fill it. So the energy per unit length of the surface wave gets smaller, decreasing the height of the wave. If a disturbance is created in the water column and the waves spread out in all directions, they get smaller even more rapidly than surface waves; this is called spherical spreading. Sooner or later, the spreading wave may fill the entire available depth, hit the surface and bottom of the pond and therefore only be able to spread sideways. This is called cylindrical spreading. With sound spreading in a cylindrical shape, sound intensity (power transmitted through a unit area in a specified direction) decreases more slowly than in spherical spreading. We therefore see, that the chemical and physical make-up, as well as the morphology of the waterbody (or any other body that sound travels in) influences propagation and properties of sound. Different types of sound propagation can therefore be harnessed for the study of the ocean and for remote-sensing purposes.

Sound is also impacted by absorption and scattering. When sound travels through a medium, it interacts with the molecules of the medium. If it has enough power to overcome the molecules' resistance to movement, these will start vibrating and thus absorbing some energy from the acoustic wave. The molecules get the energy to vibrate from the acoustic wave, and the higher the frequency of the wave, the faster the molecules in the medium will vibrate, thus taking more energy from the acoustic wave. This is why, under the same conditions, a high frequency wave will not travel as far as a low frequency wave. The ocean also has dissolved salts that absorb sound and convert it to heat, thus decreasing the amplitude of the acoustic wave. Further, when a sound wave interacts with suspended microorganisms, bubbles, suspended particles in the ocean, etc., it will scatter. The amount of scattering will depend of the size of the scatterer and the wavelength of the sound. If the size of the object is comparable to, or bigger than, the wavelength of the sound, the amount of scattering will be significant.

The general goal of SONAR is to obtain signal-to-noise ratios above a certain threshold (i.e., where the signal features can be effectively distinguished from any noise). Noise, either ambient, due to reverberation or self-noise, inherently exists in the environment and interferes with the signal. Ambient noise is the noise that exists in the water whether SONAR transmitters are present or not. Its sources can be biological, thermal, resulting from surface processes such as wind, waves, and rain, or may be due to shipping. Ambient noise is highly dependent on frequency, location and depth, and is often highly directional; however, the simplest models of dealing with ambient noise are isotropic. Noise from reverberation is defined as the echo of a transmitted signal off the environment. The cause for the generation of such echoes can be boundaries like the water surface (which is a near-perfect acoustic reflector due to the great impedance difference between air and water), the seafloor, or the presence of scatterers in the water itself, in which case we speak of volume reverberation. Reverberation is directly proportional to signal energy and duration. Self-noise is the noise the SONAR and its transporting vehicle make in the environment, which is of important consideration when small signal amplitudes need to be evaluated. The sources can be manifold, such as from electrical, machinery, or the most commonly dominant source of self-noise, flow through water. The target itself can also emanate noise. While inconvenient for active detection, since the energy emanating from the SONAR and the target can interfere, it is very convenient for passive detection.

Finally, the target can emit an echo. This is particularly important in fisheries SONAR for example. The target often reflects a replica of the transmitted signal at an echo level that is proportional to the transmitted source level. The echo level itself depends on the target strength, which is a complex measure of the reflecting attributes of the target. For example, fish with swim bladders are very good targets, since air-bubbles reverberate very strongly and provide an excellent return signal. Fish without swim bladders are poor targets since it is primarily their impedance difference to the water that determines echo strength. The target echo of a swimming, or moving target, is Doppler-shifted relative to the transmitted signal.

8.2.3 Sending and Receiving the Signal

In acoustic remote sensing the sound sources used to generate acoustic waves are generally referred to as transducers (Fig. 8.4). Many different types of transducers exist, but in marine acoustics, so-called piezoelectric transducers are the most commonly encountered. Piezoelectric transducers most commonly consist of barium titanate ($BaTiO_3$) but also of lead zirconate titanate, lead titanate, or lead metaniobate. Such piezoelectric materials change in dimension when voltage is applied (in the case of a sound source) or emit a voltage when subjected to

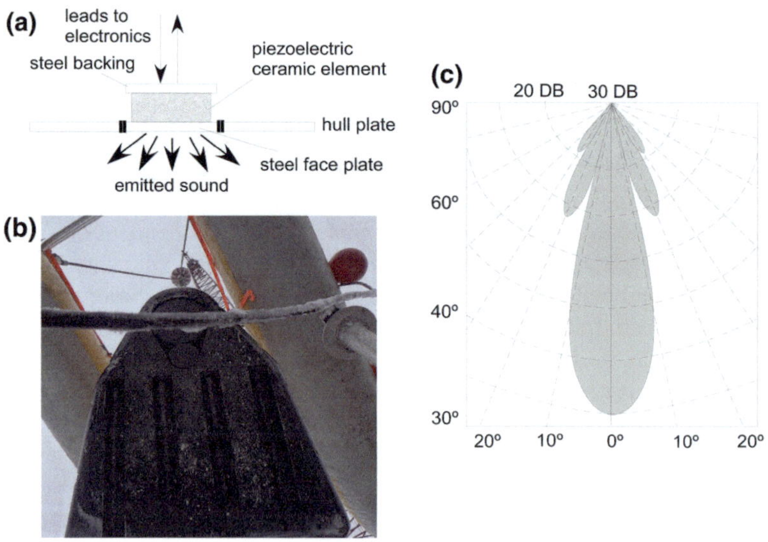

Fig. 8.4 **a** Schematic of a transducer. **b** The transmitter (*round*) and *long*, tubular receivers on a sub-bottom profiler. **c** The sound wave created by a transducer depends on its shape and configuration. In general a main lobe and several side lobes (shown in *grey*) are created. Most energy is transmitted along the nadir of the main lobe. The configuration illustrated in **c** is from a *circular* transducer with a diameter five times the wavelength of sound in water at the transmitted frequency (modified from Jones 1999, by permission of Wiley)

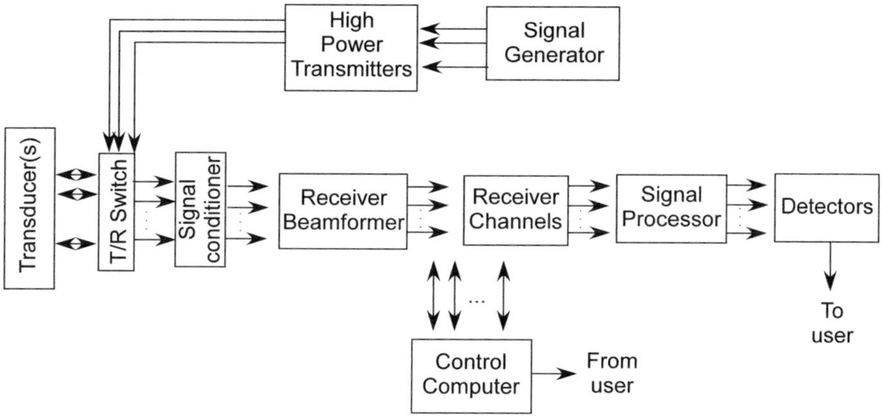

Fig. 8.5 Outline of components required for a SONAR system (modified from Mazur, personal communication)

pressure (in the case of a sound receiver). As an example, lead zirconate crystals will generate piezoelectricity when their static structure is deformed by ~0.1 % or, conversely, will change ~0.1 % of their static dimension when an external electric field is applied. To harness this physical property, the granular material is fused into a ceramic-like block that can be moulded into any shape, and when heat (~120 °C, the Curie temperature; Medwin and Clay 1998) and a DC polarizing voltage are applied the block becomes an assemblage of tetrahedron crystals with a preferred axis. Depending on the orientation of the electrodes, this causes the material to be piezoelectric for shear or compression, which thereby defines the type of transducer and its operating design.

To generate sound and harness its properties underwater, a cascade of processes needs to be set into motion (Fig. 8.5). The basic unit allowing the user to interface with the SONAR system is the computer, which maintains control of subsystems such as guidance, tracking and navigation. From the computer, the user determines what type of signal is to be generated by the signal generator, which in turn provides the drive signal to the high-power transmitters. The signal generator controls the amplitude and phase (frequency) of signal wave generation, multiple transmit sequencers, transmit beam shaping and steering, as well as own-Doppler nullification (ODN). The high power transmitters convert the low-level input signals from the signal generator to high-power signals that drive the transducer elements. A transmit/receive (T/R) switch connects the high-power transmitter to the transducer for active transmissions. And finally, the transducer(s) provide the actual interface between the ocean and the SONAR electronics, with the role of converting sound to electrical impulses and vice versa. Transducers can be single units or linear, planar, or volumetric arrays, and can either transmit, receive ("hydrophone"), or do both.

The returning signal is received by the transducer(s) and passed via the T/R switch to the signal conditioner, which limits peak signal levels to avoid damage to

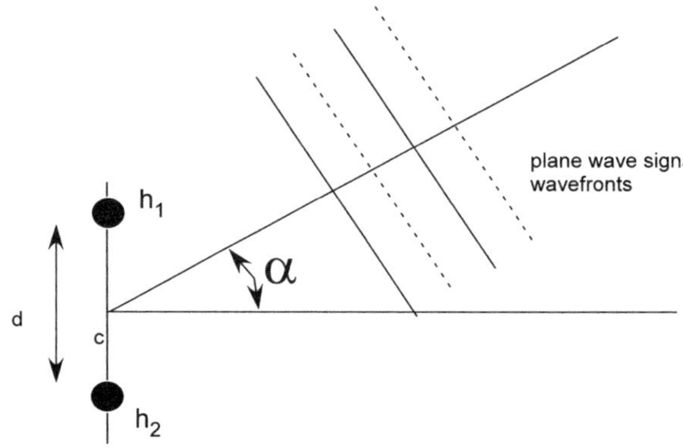

Fig. 8.6 Principle of a simple beam-former, where multiple receivers (h_1 and h_2) are sensitive to the signal that arrives offset at angle α to the plane of the array. Thus, h_1 will be excited first and h_2 second. This time-differential needs to be corrected. If considered as a beam-former of an outgoing signal, h_2 would transmit earlier than h_1. Hydrophones must add in phase according to the desired angle in order to maximize positive, rather than destructive interference (modified from Mazur, personal communication)

the receiver, matches impedance, and pre-amplifies if necessary. From there the signal passes to the receiver beam-former, which combines the conditioned transducer signals into beams, whereby each beam represents a spatial filter. Each receiver beam is then passed to its own receiver channel, which has the functions of band pass filtering and gain control. In the signal processor, the signal-to-noise ratio of the beams is enhanced, and other processes, such as matched filter processing and target angle calculations, are performed. The detectors finally apply a threshold, either fixed or variable, to the signal processor output signals, which are then recorded by the computer system.

In many advanced SONAR applications, such as swath mapping, the process of beam-forming is an important component. It is a method of special filtering with the goal of transmitting or receiving sound preferentially in some directions over others (Fig. 8.6). Beam-forming uses interference of signals to change directionality of an array. When transmitting, a beam-former controls phase and relative amplitude of a signal at each transmitter in order to create a pattern of constructive and destructive interference in the wave-front. When receiving, information from different sensors is combined in such a way that the expected pattern of radiation is preferentially observed. Beam-forming is analogous to the frequency domain Fourier analysis of time signals. While in time/frequency filtering the frequency content of a time-signal is investigated, in beam-forming the angular (directional) spectrum of a signal is analyzed. Beam-forming can be accomplished in one of three ways: (1) physically, by exciting different parts of the transducer (or different transducers in an array); (2) electrically, by analogue delay circuits; or (3) purely mathematical via digital signal processing. Beam-forming requires directivity (blocking the noise outside the direction of interest), side lobe control (not all acoustic energy produced by the transducer is exclusively transmitted in the main lobe; destructive and constructive interference at its edges create so-called side lobes, which need to be suppressed as much as possible; see Fig. 8.4), and beam-steering (the beam-former can be steered, which is the capability to modify the refracted angle of the beam). Beam-forming also requires a careful choice of frequency (the beam-former pattern is frequency dependent, where the higher the frequency, the narrower the lobes) and spacing of hydrophones if an array of discrete transducers are used (to avoid spatial aliasing).

Good positional accuracy of SONAR postings demands precise information on the vessel's location at all times during operation. Vessel location is commonly obtained using GPS. For more accurate locations, Differential GPS (DGPS) can be used for sub-meter positioning, or Real Time Kinematics (RTK) GPS can be used for centimetre level positioning, both of which utilize a base station in addition to the rover onboard the vessel. Vessel heading is measured with a gyrocompass located at the centreline of the vessel. Knowledge of the vessel attitude (heave, pitch and roll) also greatly improves the acoustic travel time measurement accuracy. Heave is the vessel going up or down, pitch is the movement of the bow going up or down and roll is the movement of the port and starboard side going up and down. These movements are measured with a motion sensor and reported to the data collection computer, along with the location and heading measurements, to make the appropriate corrections to the acoustic distance travelled. Any movement of the boat changes the orientation of the transducer to the substrate which increases or decreases distance.

In terms of acquisition, the geometry for the lines to be acquired is defined in such a manner as to completely cover the area, inclusive of the necessary overlap, at the desired spatial resolution. Line geometry is planned based on the charts, maps and environmental conditions of the area before the acoustic survey. The easiest geometry, known as a 'lawnmower' pattern, is to define a series of parallel lines crossed by either perpendicular or tie lines, which are used to validate the quality of the data and ensure consistency. During the survey, following these lines is called "mowing the lawn". While planning a survey with an acoustic system that uses a large footprint like side scan sonar or multi-beam, the amount of coverage overlap between the lines also has to be defined.

8.2.4 Processing Requirements

Once the data are collected, they have to be processed. The amount and sophistication of the signal processing depends on the system used and the intended objectives of the survey. Here, a signal is defined as a change or a disturbance in the normal "background" environment, where an acoustic signal is a disturbance in the background pressure of the seawater. The signal conveys information about both the nature of the disturbance and also of the environment. However, the recorded data include measurement of both signal and noise; hence, thresholds need to be set for the declaration of what is signal and what is noise (Fig. 8.7). This is usually accomplished based on models of signal and noise. Since noise creates the

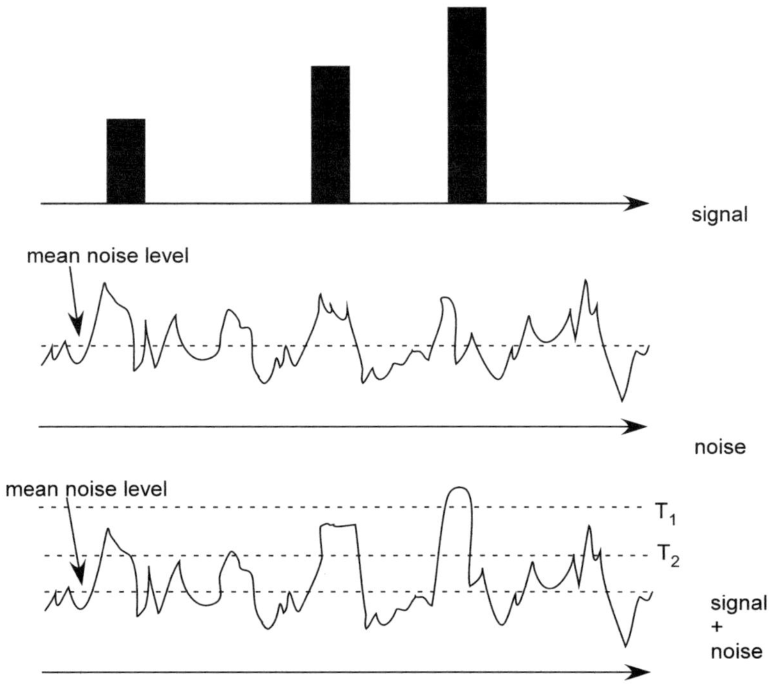

Fig. 8.7 The environment is full of noise that can easily obscure a signal. Only when the signal rises far enough above the mean noise level, can it reliably be detected. Line one represents the signal, line two the noise, and line three the signal added to the noise. In line 3 two thresholds are set (T_1 and T_2) which must be exceeded for a signal to be recognized. T_2 is much more likely to lead to false positives, since it is closer to the mean noise level; T_1 is, however, more likely to miss signals (modified from Mazur, personal communication)

possibility of false alarms, or false detections, the models are built to balance the probability of a false alarm versus that of detection. A delicate balance exists here: the more the detection threshold is lowered, the more signal will be detected but also the more false alarms due to noise will be incurred (resulting in, for example, erroneous depth measurements). Much work and experimental verification goes into the development of a useful receiver operating curve (ROC), which helps define these thresholds. Further processing is also done to improve the signal-to-noise ratio in order to enhance the probability of target detection or interpretation of signal shape. After this step, signal processing is typically performed using the temporal, spatial, or spectral variations in the recorded signals. For example, signal strength along a time axis can be evaluated for its shape or specific points of deflection, such as in the bottom identification algorithms of common depth sounders. Additional processing can incorporate cross-section graphs, surface models of survey area, seafloor mosaics, or volumetric representations.

Analysis and processing also requires the acoustic data to be tagged with the relevant navigation data (i.e., every data point is associated with the geographic information and vessel positioning attributes). This step is typically accomplished in real time while data are being acquired by means of synchronizing input from several sensors (e.g., DGPS and gyrocompass) in the data collection computer. These data are evaluated to remove any positional outliers observed while plotting the navigation lines to verify the geometry and confirm proper operation of the navigation instrument. Data also have to be adjusted for the variations of the sound speed profile using the data from a CTD, XBT, or other device used for measuring the sound speed in the water column. And the data have to be corrected for tide variations over the time of the survey. This tide correction can be done using harmonic tide predictions, like the ones supplied by the British Admiralty Manual, or achieved using a software program to generate tide corrections from manual observations, from high-low water times and heights, or from published tide data for the given area. A more accurate tide correction can be achieved by employing an actual tide gauge located at the survey site and/or by using an RTK GPS during data collection.

8.3 Applications of Acoustics

Active as well as passive acoustics are amongst the most commonly used applications in the marine sciences. In the marine geosciences, applications are found in mapping sediment transport, sediment classification, particle sizing, seafloor mapping, sub-bottom profiling and, in the very long frequencies, the detection of seismic events. In oceanography, the classical applications of acoustics are current measurements, wave measurements and the study of water properties using acoustic tomography (Medwin and Clay 1998). In marine biology, active acoustics are used for benthos detection (such as the mapping of submerged aquatic vegetation), and for the detection of plankton and fish in concert with stock assessments and population studies. Passive acoustics are becoming increasingly common for the tracking and identification of marine mammals and fish. In the commercial field, acoustics are staples for bathymetry and fish detection (a full gradient of sophistication exists from recreational depth-sounders and fish finders to survey-grade equipment), object detection, subsurface characterization for mineral resources, description of current and sediment/pollutant transport patterns, and many more.

The following examples are centered on acoustic applications related to the surrounding environment, while Chaps. 9 and 10 discuss acoustic systems used for benthic detection and classification in some detail.

8.3.1 Single Beam Bathymetry

The first and most fundamental use of sound in water is harnessing the time differential between the sent signal and its received echo to measure distance to the seafloor or any other object. This is the principle behind single beam echo-sounding, which calculates depth as the product of half the two-way transit time of a signal and the mean vertical sounding velocity. Measurements require a sharp leading edge in the signal, precise timing of the bottom reflection, and an accurate estimation of mean sound velocity in the measured water body. Frequency is also important, since sound attenuation becomes significant at greater distances. It is, for example, not possible to measure distances over 20 km (roughly the travel distance to and back from the greatest depth of the ocean, which is 10,008 m at the Mariana trench) with signal frequencies below 50 kHz (Jones 1999). Thus, different transducers are used for deep and shallow work.

After the sound pulse leaves the transducer, it travels towards the seafloor and is subject to conical beam spreading. The further away the reflecting object, or the smaller the object, the narrower the opening angle of the transducer should be in order to maintain a small footprint. This is a particularly important consideration if the seafloor is steeply sloping or highly rugose, since a large footprint does not adequately capture the details of the surface. Further, for especially steep slopes, or very rough surfaces, reflections may travel at an oblique, rather than perpendicular, angle back to the ship, causing a slight upward measurement error (i.e., the seafloor is perceived deeper than it is). This need for a smaller footprint, however, is balanced against the limitation that single beam systems only provide soundings along the survey track. Areas between the soundings must be interpolated if a closed surface between the soundings is desired. To avoid this, swath methods, which allow full and complete insonification of the seafloor, were developed. In the simplest configuration, there are several single beam transducers arranged along outriggers that obtain a mapped swath. More modern systems use beam-forming and beam-steering to generate a fan of sounding paths originating from one centralized transducer or transducer array.

8.3.2 Side Scan SONAR

The primary purpose of side scan SONAR is to produce accurate maps of seafloor topography. Side scan SONAR (SSS) can be ship-mounted or towed. In the latter case, the transducer arrays are packaged in a streamlined "towfish" or "fish" that is linked by a cable to the towing vessel (Fig. 8.8). The fish is used to decouple the sonar from ship movements and to limit the impacts of surface noise. Ideally, the fish is placed a few meters above the seafloor to obtain optimum coverage, where fish height is controlled by adjusting the amount of cable and the speed of the vessel. Many different configurations of SSS are available for different

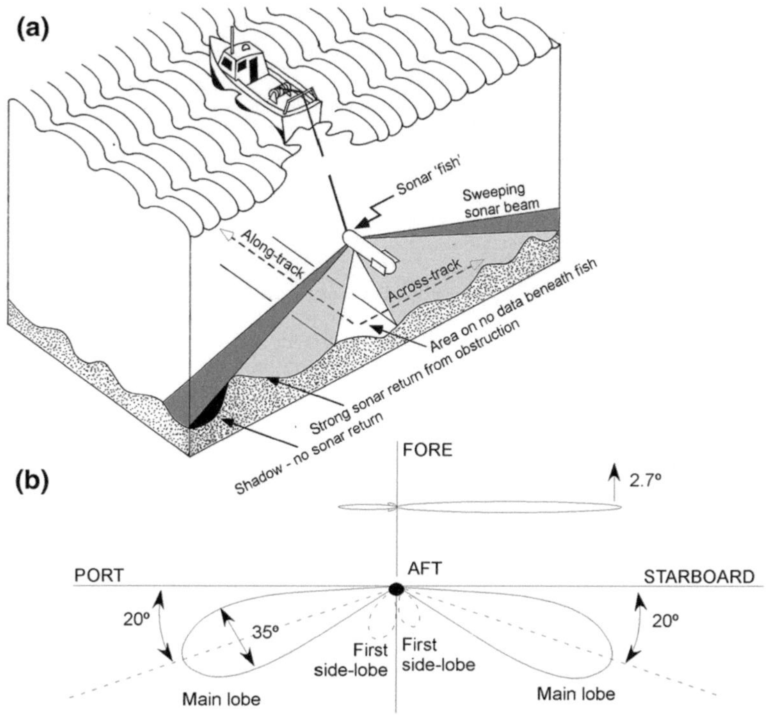

Fig. 8.8 Typical configuration of a side scan SONAR system (modified from Purkis and Klemas 2011 and Jones 1999, by permission of Wiley): **a** towfish and operational principle **b** sound configuration (beam pattern). Panel above is view from above. Negligible insonification in the fore-aft direction as opposed to the main energy directed to the sides. Only one beam is shown. Panel below is a view at the fish from behind (aft-fore view). The main lobes are perpendicularly oriented, and the first side lobes are used to collect data near nadir. A data-free zone at nadir remains

applications that can range from extremely wide beam and high frequency systems for work in shallow water to high-energy and low-frequency systems for deep-ocean work, such as Geological Long Range Inclined Asdic (GLORIA), Towed Ocean Bottom Instrument, (TOBI) and others.

SSS provide an 'aerial view' of the seafloor by utilizing two fan shaped beams generated by acoustic transducers mounted on the sides of a towfish (Fig. 8.9). In contrast to conventional echo sounders, where the transmission pattern is cone-shaped, SSS are configured with a narrow horizontal beam and a wide vertical beam. The smaller side-lobes are utilized for recording returns close to the ship. The transducers are mounted in linear arrays and tend to be operated at higher frequencies. Short pulses are emitted from the transducers and returned to the array directly by the seafloor and indirectly by backscatter and specular reflection from the seafloor to the side. The receiver has a time-variable gain to compensate for differences in range.

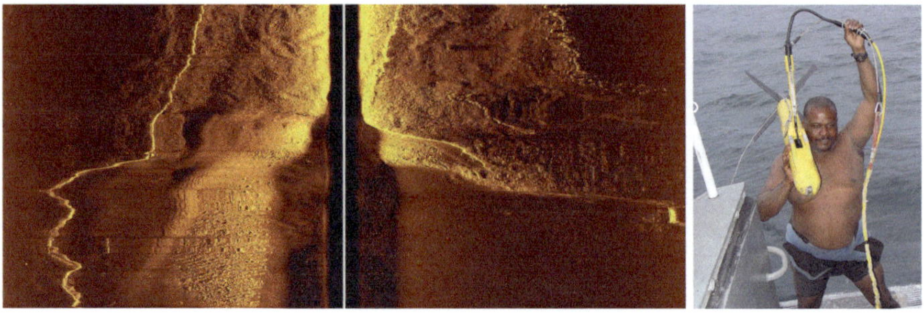

Fig. 8.9 (*Left*) view of a mud diapir on a backscatter intensity image (from Purkis and Klemas 2011, by permission of Wiley-Blackwell) (*Right*) the hardware, a side scan "towfish" or "fish"

Most SSS systems are used primarily as imaging devices to detect rough features on the seafloor that generate strong scatter and acoustic shadows. The intensity of backscattering from the seafloor is governed by the backscattering coefficient, which is a ratio of the intensity of sound scattered per unit area, and the intensity of the incident plane sound wave. Backscatter intensity is closely related to differences in seafloor roughness, sound velocity and density across the seafloor. Side scan data are therefore primarily represented as scaled (usually grey-scale) images of backscatter intensity. The capacity of SSS to produce accurate sea bottom topography is a function of the sonar operating frequency, the signal pulse length, transmitted power, the transmission method and the receiver bandwidth. The latest systems utilize image correcting techniques to compensate for variations in the slant range, ship speed and signal amplitude.

New improvements in SSS allow production of high quality, high resolution seafloor images through use of CHIRP technology, multi-ping technology, extra long arrays and simultaneous dual frequency systems. A limitation in conventional SSS is that the sonar has to wait until it receives the echo data from the farthest range before the cycle can be repeated, which imposes an inherent limitation in the towing speed. As a solution, CHIRP multi-ping systems use coded pulses. This coding eliminates cross pulse interference and allows keeping track of the independent pulses. Therefore, these systems can transmit several coded pulses per cycle, which makes it possible to tow the system at higher speeds without losing resolution. Or, alternatively, at lower speeds higher resolution is achieved by illuminating the seafloor with more pulses. A multi-ping system can be operated at twice the speed of a conventional system without missing data and also has twice the data density for equivalent tow speed. In such systems the operator can select between High Definition Mode (HDM) or High Speed Mode (HSM) with towing speeds up to 14 knots. Modern systems also have the ability to simultaneously use two frequencies: a low frequency (300 kHz) to enhance range, and a high frequency (900 kHz) to enhance resolution (as much as 1 cm resolution across track).

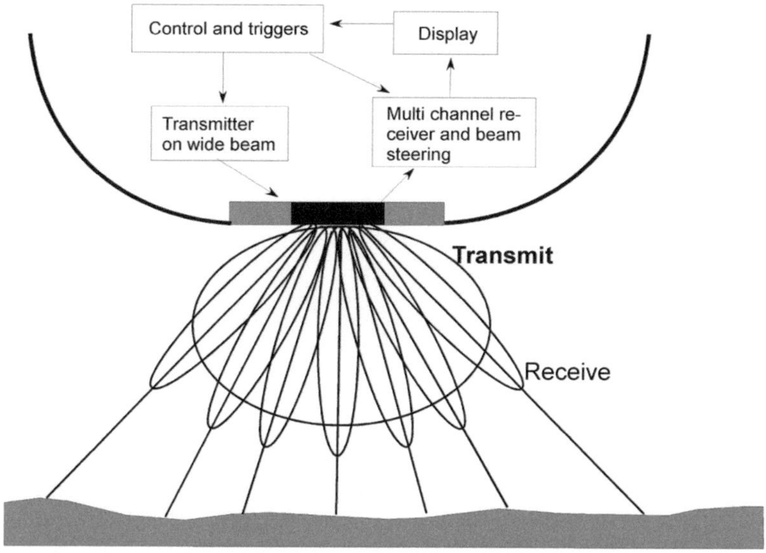

Fig. 8.10 Concept of a multi-beam system with preformed beams. Transmission occurs as one broad beam. Time delays in the receivers allow the receiving array to be preformed to a set of narrow beams that measure the depth at various positions (from Medwin and Clay 1998, by permission of Academic Press)

8.3.3 Multi-Beam SONAR

Multi-beam SONAR systems (MBS) are essentially an extension of single-beam systems that use an array of emitters (either a physical array or a virtual array created by beam-forming) to generate a multitude of raypaths from several or single sound pulses (Figs. 8.10 and 8.11). MBS can map more than one location on the ocean floor with a single ping and with higher resolution than those of conventional single-beam echo sounders. It thus has the effect of performing the duties of several narrow single-beam echo sounders, measuring multiple locations at once. Originally developed by the US Navy in the 1960s as SONAR Array Sounding System (SASS), the idea is to generate and receive a large number of sideways reflected signals where the differences in range and bearing are used to record the off-track depths. Such systems can rely on arrays of several transducers producing several discrete signals, or can rely on signal processing, performed via beam-forming and beam-steering, to produce the multiple signals.

Multi-beam measurements are arranged such that they map a contiguous area of the seabed, usually a strip of points in a direction perpendicular to the path of the survey vessel. This area is called a swath. The dimension of the swath in the across-track or athwartship (perpendicular to the path of the ship) direction is called the swath width, and can be measured either as a fixed angle or as a physical size that changes with depth. The sonar transducer emits acoustic pulses

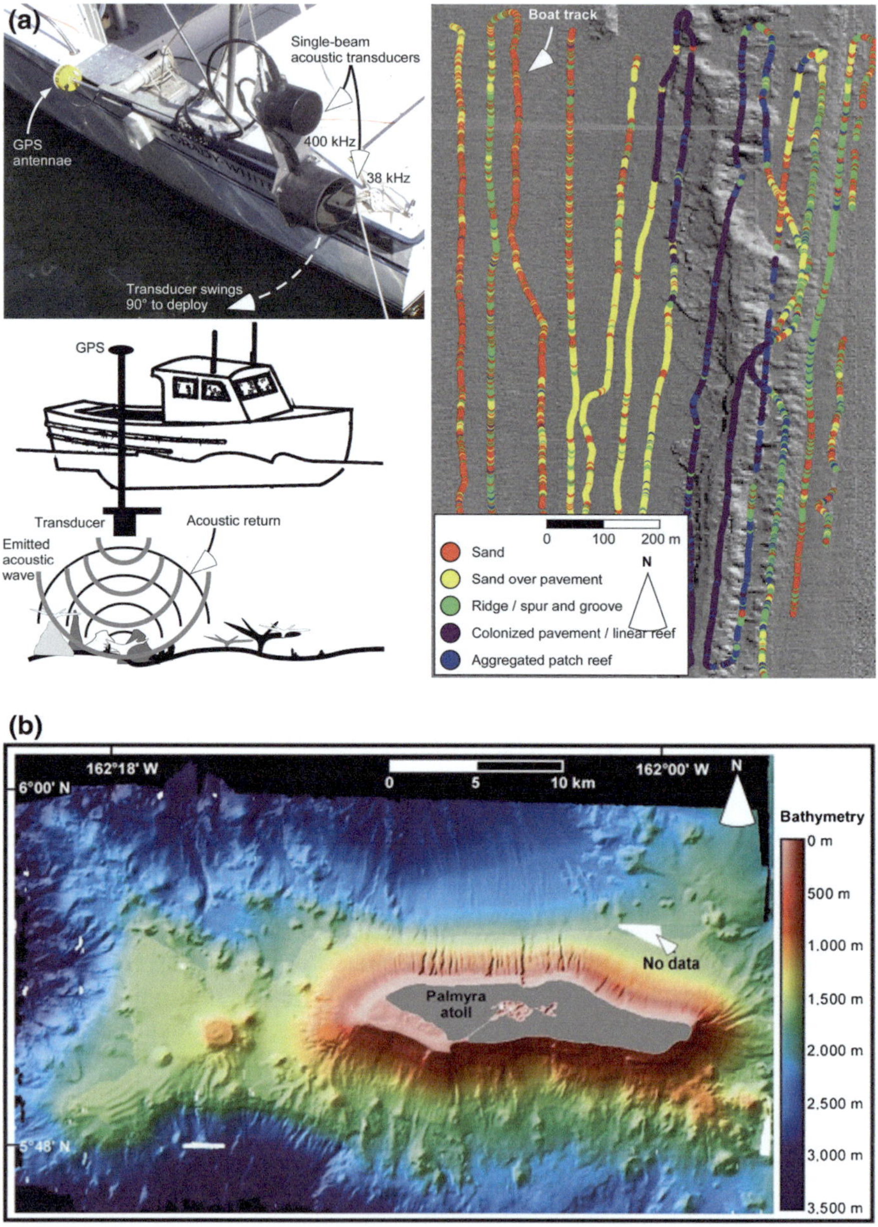

Fig. 8.11 **a** Single beam acoustics provides coverage only along discrete lines underneath the boat track. While single beam returns can be harvested to provide information about the nature of the seafloor (see Chap. 9), no continuous surface of measurements is available (from Purkis and Klemas 2011, by permission of Wiley-Blackwell) **b** Swath systems, such as multi-beam bathymetry provides complete surfaces of bathymetric information and allow imaging of the seafloor with more complete information (image courtesy NOAA)

propagated inside a wide across-track and narrow along-track sector, while the receiver array is usually directed perpendicularly to the transmit array. The receiver array is steered simultaneously across the different narrow across-track directions by a beam-forming process, and the system then performs spatial filtering of the acoustic signals backscattered from the different directions.

The resolution within the swath is referred to as across-track resolution (the greater number of individual SONAR beams formed, the finer the resolution the survey). Many multi-beam SONAR systems can counteract the degradation in survey resolution caused by conical beam-spreading inherent in single-beam systems (i.e., counteract the issue that when the insonified footprint gets larger, many more surface irregularities are included within a single sonar footprint and hence not resolved). The along-track resolution of multi-beam SONAR is the ratio of acoustic wavelength to the length of the array, which defines the resolution at a given depth. For typical multi-beam systems this ratio is on the order 1:60–1:400 (i.e., resolutions from 1 m at 60 m to 1 m at 400 m ranges, respectively). A longer array will increase this ratio, but fitting such a long array is not always possible or practical. Higher frequencies will also increase the ratio, but limit the achievable range due to higher absorption of the signal. Synthetic Aperture SONARs can be used to overcome these limitations by utilizing data from several consecutive pings to synthesize a longer sonar array.

Different frequencies are used to map different water depths. Higher frequencies (>100 kHz) are generally used in shallow water, while the lower frequencies (<30 kHz), are preferred for deep water. With different frequencies there is a trade-off in resolution, with higher frequency systems providing greater spatial resolution than lower frequency systems. Importantly, due to conical beam-spreading, swath width varies with survey depth (i.e., the deeper the water, the wider the swath) and the coverage area of these systems is thus a direct function of water depth. Most systems provide coverage ranging from approximately 2 to 7 times the water depth.

An early SONAR system, the Seabeam, is used here to illustrate the functioning of a multi-beam system. Several transducers are mounted in an array along the keel of a vessel and transmit pulses that insonify an area of seafloor normal to the ship's track. A hydrophone (receiving transducer) array on the ship's keel has its axis set in the fore and aft direction. Thus, two separate sonar arrays are oriented orthogonal to one another, one for transmitting and one for receiving. This arrangement is referred to as a Mills Cross Array. The arrays and the associated analogue electronics provide 90 × 1°-wide unstabilized beams. Roll and pitch compensation reduces this to 60 × 1°-wide stabilized beams, which permits mapping a 60° swath of the sea floor with each ping. This system allows survey vessels to produce high-resolution coverage of wide swaths of the ocean bottom in far less ship time than would have been required for a single-beam echo sounder, greatly reducing the costs of mapping.

In more modern systems, most of the signal processing, including beam-forming, has been moved from analogue signal processing into the digital (discrete) signal processing domain using digital signal microprocessor (DSPmP) chips. Their availability permits the implementation of sophisticated detection algorithms and increases the number of beams in the swath. Incoming signals are formed into several beams by vector summation, where each beam is related to returns within a defined angle normal to the ship's track in the fore-aft direction (Fig. 8.12). The processed acoustic signals come from those areas of seafloor where the transmitting and receiving beams overlap (Jones 1999). As with other SONAR systems, depth is calculated from travel time, and backscatter images are produced from the intensity of the returns. Although such multi-beam backscatter images can be used for the interpretation of seafloor material, most multi-beam systems are currently used only to produce measurements of water depth.

8.3.4 Acoustic Doppler Current Profiling

An Acoustic Doppler Current Profiler (ADCP) is used to measure water velocity, or the velocity of objects in the water. These systems make use of the phase-shift in frequency of reflected signals, known as the Doppler Effect. Between the time when the sound pulse is transmitted and received, it is shifted in frequency by the relative velocity of the water. The sound may also be shifted in frequency by scatterers in the water if there is a difference in relative velocity of water to the scatterer. Trigonometry, averaging, and some critical assumptions are used to calculate the velocity of the water, or the velocity of a group of echoing scatters in a volume of water. The return echo is repetitively sampled, and only a part of the returning sound wave is evaluated at each step (a process referred to as "gating the return data in time"). The ADCP can thus produce a "profile" of water currents

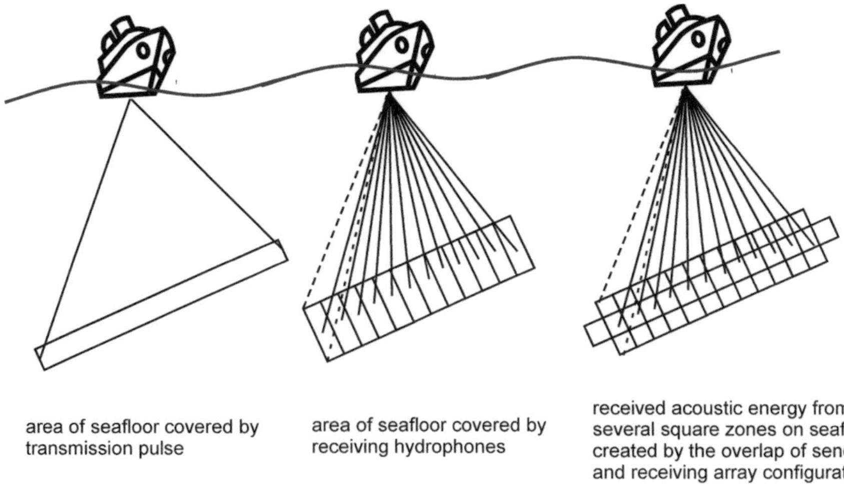

Fig. 8.12 Operational principle of a multi-beam system (modified from Jones 1999, by permission of Wiley)

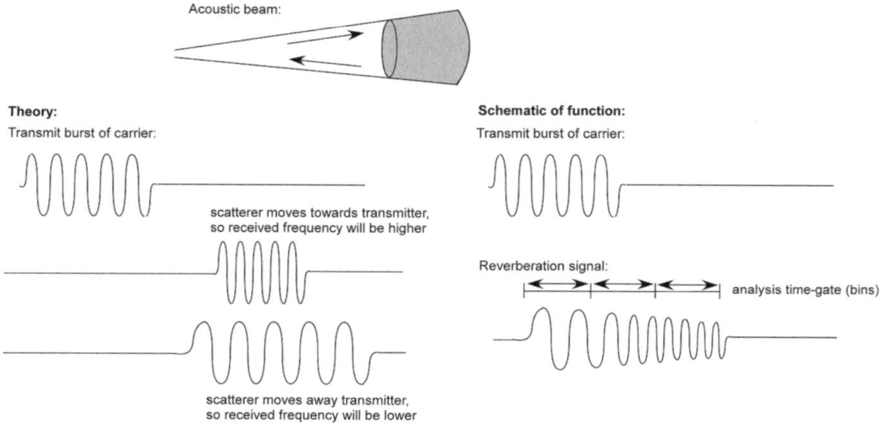

Fig. 8.13 Schematic of ADCP functioning. Three to four acoustic beams are emitted as a standard carrier signal. After emitting sound, the transducer listens to and receives a complex, multi-frequency signal from the entire insonified distance. The signal is time-gated into distinct bins, within which the signal is decomposed for frequency analysis. According to the Doppler principle, received frequency will be higher than the emitted carrier signal if scatterers move towards the sound source, lower otherwise. Thus, a directional component can be calculated for each bin

over a range of depths (Fig. 8.14). Phased array techniques are also used to aim the sound (acoustic) energy, allowing for economical production of smaller ADCPs to accommodate a range of frequencies from 38 kHz to several megahertz. In effect, these transducers are aimed such that the sound pulse travels through the water in different, but known directions.

In addition to the transducers, an ADCP typically has an electronic amplifier, receiver, mixer, oscillator, accurate clock, temperature sensor, compass, pitch and roll sensor, analog-to-digital converters, memory, and digital signal processor. The analog-to-digital converters (ADCs) and digital signal processor (DSP) are used to sample the returning signal, determine the Doppler shift, and sample the compass and other sensors in order to calculate the range and velocity relative to a known orientation (Fig. 8.13). Examples of ADCP applications are the measurement of currents and suspended sediment levels near reefs (Hoitink and Hoekstra 2005).

8.3.5 Fisheries Acoustics

Fisheries acoustics makes use of two important physical properties of fish: (1) the fact that many fish have swim-bladders (i.e., large bubbles enclosed within their tissues), and (2) that they form dense schools. Bubbles are an important aspect of general marine acoustics, and with the accompanying knowledge that has been developed in this field, swim-bladders are excellent objects for fish detection. Medwin and Clay (1998) developed a bio-acoustical pyramid that relates animal

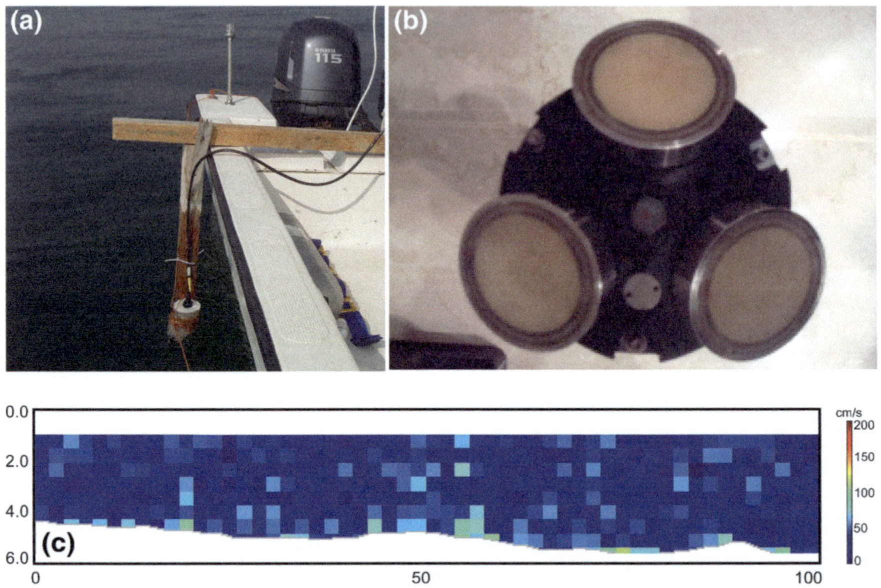

Fig. 8.14 **a** A downward-looking ADCP deployed from a small vessel. The instrument subtracts vessel speed from current speed; **b** typical sensor configuration on an ADCP; **c** profile of a downward-looking ADPC transect across a gently flowing water body. The vertical (y) axis is depth. Distance travelled by the instrument is shown on the x-axis, horizontally increasing from *left* to *right*. Distance is expressed by the number of adjacent, processed pings. Each ping is mathematically 'cut' into several sectors of pre-determined length within which current speed is calculated from the Doppler shift of sound bouncing off scatterers. Thus, each ping appears as a stack of differently colored cells (strength of Doppler shift gives current strength, which is coded along a color ramp visible to the *right* of the graph) that in sum make up a patchwork of cells along the entire transect that fully characterize current speed

length or equivalent spherical radius (a mathematical tool expressing the equivalent radius of a non-cylindrical object as that of a cylindrical object exhibiting identical acoustic properties) to the sound frequencies suitable for their detection (Fig. 8.15). Note that the smaller the body, the higher the frequency needed to detect the animals. In fish, the impedance mismatch between the body and the swim-bladder or other air bodies, has a very similar structure to the biological pyramid. The swim-bladder in fish is the dominant scattering organ. As illustration, the swim-bladder produces about 80 % of the scatter in perch (a fish that strongly utilizes the entire water column) and about 22 % in northern hog sucker (a fish that tends to stay near the bottom) (Sun et al. 1985). But even if no air-bodies are present in an animal, it can still be detected based on its other scattering characteristics. For acoustical models of fish, the scatter produced by the swim-bladder must be added to that produced by the fluid-filled body (i.e., the muscles, bones and blood vessels) and then solved for different aspects of the fish relative to the incident sound wave (Nakken and Olsen 1977; Clay and Horne 1994).

Hydroacoustic methods have been successfully used in many applications, for example the search for spawning aggregations conducted by Johnston et al. (2006).

To measure and track smaller particles (ranging from plankton to fish), estimates can be made if both the target size is known and the scattering lengths are known as a function of frequency (e.g., the smaller the animal, the smaller its scattered echo). Since small animals tend to live in dense schools (e.g., plankton), the small echo and their physical proximity to each other makes individual resolution of the animal often impossible, but they are nonetheless often detected by mass-scatter. A striking example is the detection of the deep-scattering layers (DSL) in the ocean, which represents the dial vertical migrations of plankton. Judicious choice of sampling frequency, or the use of broadband signals, can also allow for the identification of different plankton (Medwin and Clay 1998).

In a different fisheries application, the phrases 'passive acoustics' and 'acoustic daylight' were coined by Buckingham et al. (1992), referring to the use of the backscattered sound intensity reflected off objects in the environment, or created by them, as an analogue to light. Either the environment can be insonified and the reflected sound intensity is used, or an entirely passive device just records the emitted sound. By differentiation from the acoustic background, objects (e.g., fish) from which ambient sound is reflected can be detected since they modify the sound in characteristic ways. In these systems, a receiver picks up the modified noise

Swimbladder Low frequency maximum	Animal L or a_{es}		Fish body (flesh) Frequency for ka=1
?	Mammals and large fish	L>2m	<1.2 kHz
15-60 Hz 150-600 Hz	Large nekton: cods, tuna, etc. swimbladders?	200 cm L 20 cm	1.2 kHz 12 kHz
150-600 Hz 1500-6000 Hz	Small nekton and macrozooplankton: anchovy, shrimp, etc.	20 cm L 2 cm	12 kHz 120 kHz
1.5-6 kHz 15-60 kHz	Zooplankton, fish larvae amphipods, euphausiid etc.	20 mm L or a_{es} 2 mm	120 kHz 1200 kHz
?	Copepods, etc.	2 mm L or a_{es} 0.2 mm	1.2 MHz 12 MHz

Fig. 8.15 The marine bioacoustical pyramid from Medwin and Clay (1998). It shows the levels of animal lengths (L), or equivalent spherical radius, a_{es} (i.e., a mathematical tool expressing the equivalent radius of a non-spherical particle as that of a spherical particle exhibiting identical acoustic properties), and effective sound frequencies for their detection. Two frequency bands are given for each level. The *left* side shows the radial resonance of an equivalent spherical bubble at the surface, where the range of values represents a variety of swim bladders and different fractions of fish volume. The *right* side gives the frequency for ka = 1, where a is an equivalent cylindrical radius of the fish body (i.e., a mathematical tool expressing the equivalent radius of a non-cylindrical object as that of a cylindrical object exhibiting identical acoustic properties) or equivalent spherical radius (a_{es}) of small zooplankton

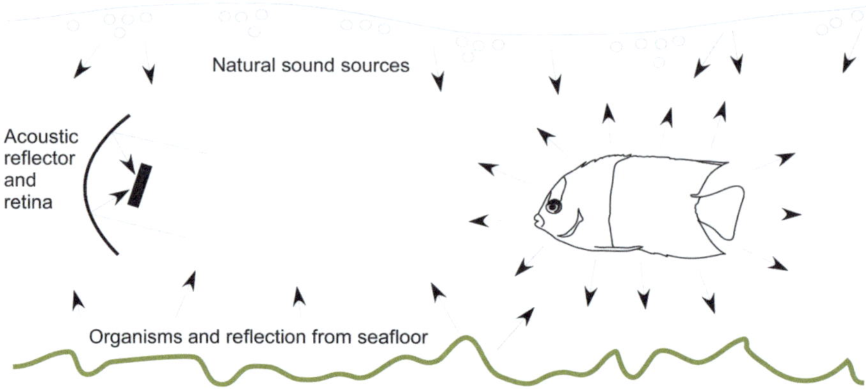

Fig. 8.16 Object (*a fish*) insonified by acoustic daylight. Natural sources generate sound that illuminate the object whose reflection can be imaged by an acoustic retina (modified from Medwin and Clay 1998, by permission of Academic Press)

signals and sends the information to a computer (Fig. 8.16). After image processing and enhancement, the 'acoustic daylight imaging' system can form false-color moving images (Buckingham et al. 1996). Listening devices are also useful to identify organisms inhabiting the sea since, by comparison with a library of typical frequencies, the identity of the sound-producing organisms can then be revealed.

8.4 Conclusion

Developed initially for the military's need to detect submerged objects, such as submarines and mines, marine acoustics has since developed into an expansive field and diversified into a plethora of civilian and scientific applications. The physical properties of sound in water are well understood and increases in computing power and advent of new polymer materials have facilitated dramatic advances in acoustic survey hardware and signal processing. Within marine resource management, acoustic methods are routinely used in the development of baseline information for marine spatial planning, mainly through the development of bathymetric maps and habitat maps based on seafloor backscattering properties. The use of advanced multi-beam or side scanning SONAR systems for seafloor imaging and bathymetric mapping is now standard in most survey environments. Upward, downward, and sideways looking acoustic current profilers are deployed around many of the world's coastline to routinely map currents. Fisheries acoustics, based on the detection of moving targets in the water column, play an increasingly important role in stock assessments and understanding movement patterns. While acoustics have found their firm place in the management of marine resources, their importance can be expected to expand as technological advances increasingly reduce instrument size and increase detection power.

Suggested Reading

Jackson DR, Richardson MD (2010) High-frequency seafloor acoustics. Springer, New York
Jones EJW (1999) Marine geophysics, 5th edn. Wiley, New York
Lurton X (2010) An introduction to underwater acoustics. Springer, Berlin
Medwin H, Clay CS (1998) Fundamentals of acoustical oceanography. Academic, London
Urick RJ (1983) Principles of underwater sound. McGraw Hill, New York
Wille PC (2005) Sound images of the ocean. Springer, Berlin

References

Benenson W, Harris JW, Stocker H, Lutz H (2002) Handbook of physics. Springer, Newyork, p 1181

Buckingham MJ, Berkhout BV, Glegg SAL (1992) Imaging the ocean with ambient noise. Nature 356:327–329

Buckingham MJ, Potter JR, Epifanio CL (1996) Seeing under water with background noise. Sci Am 274:40–44

Chen C-T, Millero FJ (1977) Speed of sound in seawater at high pressures. J Acoust Soc Am 62:1129–1135

Chilowsky C, Langevin P (1916) Procedes et appareils pour la production de signaux sous-marins diriges et pour la localization a distance d'obstacles sous-marins. Brevet francais 502913

Clay CS, Horne JK (1994) Acoustic models of fish: the Atlantic cod (*Gadus morhua*). J Acoust Soc Am 96:1661–1668

Colladon JD, Sturm JKF (1827) Speed of sound in liquids. Ann Chim Phys Ser 2, part IV

Hoitink AJF, Hoekstra P (2005) Observations of suspended sediment from ADCP and OBS measurements in a mud-dominated environment. Coast Eng 52(2):103–118

Johnston SV, Rivera JA, Rosario A, Timko MA, Nealson PA, Kumagai KK (2006) Hydroacoustic evaluation of spawning red hind (*Epinephelus guttatus*) aggregations along the coast of Puerto Rico in 2002 and 2003. NOAA Prof Pap NMFS (5). NOAA, Seattle, WA, pp 10–17

Jones EJW (1999) Marine geophysics, 5th edn. Wiley, New York

Medwin H, Clay CS (1998) Fundamentals of acoustical oceanography. Academic, London

Nakken O, Olsen K (1977) Target strength measurements of fish. Rapp P-V Reun Cons Int Expl Mers 170:52–69

Purkis SJ, Klemas V (2011) Remote sensing and global environmental change. Wiley-Blackwell, Oxford, p 368

Sun Y, Nash R, Clay CS (1985) Acoustic measurements of the anatomy of fish at 220 kHz. J Acoust Soc Am 78:1772–1776

Tolstoy I, Clay CS (1966) Ocean acoustics. Theory and experiment in underwater sound. McGraw Hill, NY, p 293

Wilson WD (1960) Speed of sound in sea water as a function of temperature, pressure, and salinity. J Acoust Soc Am 32(6):641–644

Chapter 9
Acoustic Applications

Greg Foster, Arthur Gleason, Bryan Costa, Tim Battista and Chris Taylor

Abstract In the past few decades aerial and satellite imagery have been the primary source for producing regional scale maps of coral reefs. Although these tools are well-suited for the clear and shallow waters of the typical coral reef setting, they are less useful for mapping in shallow turbid coastal waters or in mesophotic depths (30–75 m), where considerable coral reefs are known to exist. Acoustic remote sensing systems are relatively insensitive to water column transparency and turbidity, making them ideal tools for mapping reefs in settings inappropriate for optical mapping. Acoustic mapping of coral reefs was in its nascent stage in the 1990s at the time when optical aerial, satellite, and more recently LiDAR technologies were maturing into widely accepted methods for coral reef mapping.

G. Foster (✉)
National Coral Reef Institute, Oceanographic Center,
Nova Southeastern University, 8000 N. Ocean Drive, Dania, FL 33004, USA
e-mail: fjohn@nova.edu

A. Gleason
Physics Department, University of Miami, 1320 Campo Sano Ave,
Coral Gables 33146, USA
e-mail: art.gleason@miami.edu

B. Costa · T. Battista
National Oceanic and Atmospheric Administration, 1305 East West Highway,
Silver Spring, MD 20910, USA
e-mail: bryan.costa@noaa.gov

T. Battista
e-mail: tim.battista@noaa.gov

C. Taylor
National Oceanic and Atmospheric Administration,
Center for Coastal Fisheries and Habitat Research, 101 Pivers Island Road,
Beaufort, NC 28516, USA
e-mail: chris.taylor@noaa.gov

J. A. Goodman et al. (eds.), *Coral Reef Remote Sensing*,
DOI: 10.1007/978-90-481-9292-2_9,

Recent developments in hardware, software, and post-processing, however, have rapidly elevated acoustic remote sensing to a viable alternative for producing accurate high-resolution maps of coral reefs. In addition to delineating habitats, the information encoded in acoustic echoes can also provide information about the water column and substrate, including suspended solids, fish size and biomass, bathymetry, grain size, and epifaunal height and abundance.

9.1 Introduction

9.1.1 Relevance to Coral Reef Management

The broad applicability and underlying physics of the various acoustic remote sensing platforms was introduced in the preceding chapter (Chap. 8), along with application examples for assessing the bathymetry and environment around reefs. This chapter will focus specifically on the use of acoustic remote sensing tools for the production of benthic habitat maps of coral reef ecosystems. Current, accurate and consistent benthic habitat maps are a vital component of many aspects of resource preservation and management, including (1) inventorying coral reef resources, (2) monitoring health, cover, and species assemblages, (3) characterizing habitats for place-based conservation measures, such as marine protected areas (MPAs) and essential fish habitats (EFHs), (4) enabling scientific understanding of the large-scale oceanographic and ecological processes affecting reef health, and (5) acting as a proxy for the spatial distribution and abundance of marine flora and fauna. Other common applications of acoustic remote sensing in reef environments include fisheries stock assessment, nautical charting, coastal engineering, and environmental change detection.

9.1.2 Role of Acoustics in Benthic Habitat Mapping

Although aerial and satellite imagery have traditionally been the primary source of information for producing regional scale maps of benthic habitat in shallow tropical waters, acoustic remote sensing has several important roles to fill. The science of acoustic benthic habitat mapping is undergoing rapid developments in hardware, software, and post-processing methodologies, similar to the rapid development of optical remote sensing for coastal habitat mapping in the 1990s (Andrefouet and Riegl 2004). Nevertheless, it can be anticipated that the particular roles to be played by the various acoustic platforms will largely be a function of their inherent attributes, including spatial coverage, thematic resolution, and cost effectiveness.

The primary role of acoustics in benthic habitat mapping is for operating in water that is too deep or turbid for optical techniques. This limit typically occurs at approximately 30 m in the clearest tropical waters and much less for coastal or temperate waters. Areas that cannot be mapped with satellite or aerial imagery are both extensive and ecologically important. For example, over 55 % of the Florida Keys National Marine Sanctuary (about 1,540 square nautical miles) has not been mapped due to water depth or clarity limitations (FMRI 1998). The Tortugas Bank, Pulley Ridge, and Flower Garden Banks are three examples from the Gulf of Mexico that illustrate the potential for luxuriant communities of shallow-water (zooxanthellate) corals to exist at "mesophotic" depths of 30–75 m (Miller et al. 2001; Hickerson and Schmahl 2005; Jarrett et al. 2005). In addition, coral communities can exist below the photic zone, where deep-water (azooxanthellate) corals form mounds up to several hundred meters high (see Chap. 10). Recent ocean exploration initiatives indicate that such deep corals are much more extensive than previously thought (Roberts et al. 2006). Deep corals provide important habitat for fishes, and shallow coral species may potentially find refuge from warming surface waters at mesophotic depths (Riegl and Piller 2003). Acoustics can also be used to obtain information that is not readily accessible by optical techniques, including but not limited to grain size distribution, bedform patterns, abundance and canopy height of epibiota, bathymetry, and measures of topographic complexity. Acoustics can also be merged with optical datasets to enhance classification accuracy (Bejarano et al. 2010).

9.1.3 Acoustic Remote Sensing Platforms

Single-beam echo sounders (SBES): The simplest systems are vertical-incident single-beam echo sounders that measure depth only. Inexpensive bottom finders, which are categorized as SBES, are capable of producing reasonably accurate bathymetry. Advanced SBES can run from shallow to full ocean depth at very high accuracy. When coupled with a tide gauge and multiplexed with a GPS, SBES can reveal seafloor zonation patterns in areas where little is known. Heyman et al. (2007), for example, exploited an inexpensive, off the-shelf SBES system to map bathymetry at two reef fish spawning aggregation sites in Belize.

Acoustic seabed classification systems (ASC): ASC systems are more sophisticated scientific versions of single-beam echo sounders. ASC systems extract information from the returned echo waveforms that, in conjunction with ground validation, can be related to physical and biological properties of the seabed. As with other acoustic systems, ground validation can be accomplished by visual observations using divers, towed video, drop cameras, or by measurement of sediment physical properties.

Most commerically available ASC systems fall into one of two approaches for signal processing. The first approach, exemplified by the RoxAnn and ECHOplus systems, exploits the intensity of both the first and second seabed echoes (Chivers et al. 1990). When acquired over relatively flat and monotonous seabeds, E1 (the trailing edge of the first echo envelope) and E2 (the complete second echo envelope resulting from a portion of the first echo reflecting off the air–water interface) generally relate to seabed roughness and hardness, respectively. In one analysis approach, user-defined polygons can be drawn around clusters of points on a plot of E1 versus E2 for known bottom types, typically distinguished by grain size. Subsequent survey data are then classified according to which box each E1:E2 data pair falls within. A second approach, exemplified by QTC IMPACT (Quester Tangent Corporation 2002; Preston et al. 2004) and other non-commercial systems (e.g., van Walree et al. 2005) analyzes features derived from the first echo only. Using this approach, a large number of features describing the shape, duration, and for some systems the amplitude, of the first echo are computed, reduced to a smaller number of uncorrelated variables using principal components analysis and clustered. The clusters are assigned different bottom types and used to classify subsequent survey data.

While the capability of both approaches to infer grain size is well-established, the theoretical basis for using ASC to classify reef environments is more complicated than the analysis of sedimentary environments alone, due to factors such as high topographic complexity, large slopes, and variable assemblages of epibenthic biota. ASC systems that store digitized waveforms (e.g., Biosonics, Simrad, QTC) offer greater scope and utility compared to analog systems that only output parameters derived from the raw waveforms. Customized methods of waveform analysis, potentially at multiple frequencies, can be used to produce multivariate datasets that provide greater bottom type discrimination for a larger variety of applications.

Split-beam echo sounder systems: These are another specialization of single-beam echo sounders. In contrast to simple single-beam systems, the returning echo is received across four quadrants, and it is the phase difference across the four quadrants that determine the angular position of the target in the beam. The ability to determine accurate angular positioning in the beam allows for precise determination of target strength. Split-beam echo sounders are used primarily for scientific surveys of fishery resources, where accurate estimates of echo intensity (e.g., fish target strength) are necessary to infer fish sizes and for scaling to biomass and density. Split-beam echo sounders can also be used as bottom finders, as part of an ASC system, or added to the compliment of instrumentation (e.g., MBES, SSS) in a benthic habitat mapping system. The benefit of the split-beam echo sounders is the ability to detect fish and fauna in the water column simultaneous with bottom detection, and to make inferences on the distribution of biomass and habitat use by fish on coral reefs.

Sidescan sonar (SSS): SSS are essentially two echo sounders in a single housing that transmit beams port and starboard. Frequencies are typically in the range of 100–500 kHz. Higher frequencies produce higher resolutions (1–10s of cm) but at a reduced range (i.e., narrower swath width; see Chap. 8). SSS can be operated over a broad range of depths, from sub-meter to 100s of meters, but geo-location of the towed instrument can be problematic. Unlike the other acoustic systems, SSS does not directly produce bathymetry, but instead focuses on backscatter intensity. A related sensor, known alternatively as interferometric sonar (IS), or as phase differencing bathymetric sonar (PDBS), produces a backcatter intensity sonograph similar to SSS but also provides highly accurate co-located bathymetry. Under optimal conditions SSS is capable of producing full-coverage high-resolution two-dimensional (2D) imagery of the seabed. The analog grayscale image can be visually interpreted for characterizing sediment grain size and bedforms, delineating hardbottom features, or detecting objects on the order of 10s of cm in size. The analog grayscale image can also be digitized and quantitatively classified using first order (tonal) and second order (textural) statistics (Blondel and Gomez-Sichi 2009). Tonal statistics use a variety of indices (e.g., extrema or median values) to describe the amount of acoustic energy returned as backscatter. Textural statistics contain the most useful information for classification and relate to the relative spatial variation of backscatter intensities, providing quantitative measures of roughness, patchiness, randomness, etc.

Multi-beam echo sounders (MBES): MBES transmit between 100 and 500 narrow beams at varying angles to cover a fan-shaped swath of seabed 3–7× the water depth. The complicated beam geometry necessitates the integration of precise vessel attitude and positioning into data processing in order to produce an accurate measurement of bathymetry. This is accomplished using the output of a global positioning system (GPS) and an Inertial Motion Unit (IMU), which together records the location and orientation of the vessel (and hence the beam) at the time of echo generation. Early MBES transducers were exclusively hull-mounted, but recent reductions in the size of the various components have produced systems that can be pole-mounted (albeit with compromised signal quality) or mounted on Autonomous and Remotely Operated Underwater Vehicles. MBES can be operated over a very wide range of depths (0.5 m to full ocean depth), but are most efficient in deeper water where the spreading beams cover a wider swath. MBES simultaneously acquire accurate high density bathymetry and backscatter intensity information, which can be interpolated into high-resolution seafloor topography and morphometrics. The current generation of MBES also support the capture of water column information to elucidate abiotic (e.g., oil spill) or biotic (e.g., fish, plankton, diatoms, marine mammals) features. In a typical early MBES application, the topographic surface would be visually interpreted as classes of geomorphological structure. Recent applications have also utilized bathymetry-derived indices (e.g., slope, rugosity) and backscatter intensity (related to seabed roughness and hardness) either as additional layers for subjective classification, or for quantitative multi-dimensional clustering techniques.

9.1.4 Selecting an Acoustic System

Choosing between acoustic systems involves considerations such as budget (e.g., deployment costs, hardware, software), survey objectives (e.g., resolution, coverage), output (e.g., bathymetry, backscatter, thematic classification, object detection), survey area (e.g., extent, range of depth, heterogeneity), and post-processing (e.g., expertise, data storage, computing requirements).

When complete coverage of the seabed is required, SSS and MBES are the tools of choice. MBES are the most attractive system for mapping coral reef habitats in depths 30 m and greater, where the wide swath (3–7× depth) allows for cost-effective data acquisition. The dense narrow-beam point bathymetry of a MBES can be interpolated into high-resolution hill-shaded topographic maps that are visually intuitive, informative, and GIS-ready. MBES topographic maps can also be complemented by slope analysis and texture, as well as integrated with analyses of backscatter intensity. Such maps are ideal for expert-driven visual interpretation of geomorphological features, delineation of MPA boundaries, or identification of essential fish habitat. In shallow waters, PDBS produces bathymetric and backscatter intensity information similar to MBES, but at greater swath width (10–12× depth).

SSS is less expensive to own and operate than MBES and can be used from sub-meter to 100s of meters depth, but the quality of the 2D imagery is generally less consistent than MBES and lacks the bathymetric component. Also, the signal amplitude of SSS frequently varies significantly within a survey. The backscatter intensity of individual swaths must be balanced to achieve a consistent composite image, which is best achieved if there are distinct seabed features displaying consistent characteristic backscatter responses to be used as reference standards. Kenny et al. (2003) discussed SSS versus MBES trade-offs, including coverage, survey depth, and object detection limits. The imagery produced by SSS and MBES swath systems can both be used to deduce dynamic processes (e.g., sediment transport deduced from the directionality of seabed forms). Extracting textural properties (i.e., mean, standard deviation, and higher order moments, amplitude quantiles and histograms, power spectral ratio, grey-level co-occurrence features, fractal dimension) from SSS and MBES imagery also allows for statistical benthic habitat classification.

ASC systems are relatively inexpensive to purchase, operate, deploy, and process, but their along-track bathymetry does not make them suitable for producing full coverage maps of spatially complex habitats. For example, at 15 m depth and 50 m line spacing, a narrow (10°) beamwidth transducer would cover only 5 % of the area between lines. Interpolating such large information gaps results in a low resolution bathymetric surface and the potential for generating false features, particularly when surveying along parallel lines. However, the temporally resolved waveforms and fixed-geometry of single-beam systems allows for direct detection of benthic habitats and other environmental properties such as vegetative biomass, epibiotic canopy height, or suspended solids in the water column. In situations with a relatively homogeneous setting, such as some back reef lagoons, an ASC system could be used to map the abundance and distribution of seagrass, epibiota, infauna, or sediment grain size. An ASC system could also be used to complement habitat maps produced from MBES or LiDAR topography. For example, the ASC system could be used to map the abundance and distribution of epibiota, adding a biological component to the topographic map. ASC systems are also useful for identifying essential fish habitat, either by bottom classification or from measures derived from the along-track bathymetry (e.g., rugosity or slope analysis).

The closely related split-beam echo sounders can provide another layer for habitat interpretation by quantifying the size and density of fishes.

9.2 Applications

The principle of optical discrimination (i.e., photography, multispectral and hyperspectral) is generally intuitive at a fundamental level, relating directly to the spectral properties of the benthos. Although acoustic discrimination is not so straightforward, rapid progress is currently being made in creating benthic habitat maps from acoustic data. In the field of ASC mapping, recent advances in interpreting acoustic backscatter have led to benthic habitat maps with greater thematic accuracy and resolution. In the fields of SSS and MBES mapping, application of tonal and textural classifiers to backscatter imagery have greatly expanded the output of these systems to include a rapidly expanding suite of seafloor properties. Examples of such applications are presented below.

9.2.1 Single-Beam Acoustic Seabed Classification

Mumby and Harborne (1999) argued that a standardized approach to mapping tropical coastal habitats greatly increased the utility of the resulting maps. Greene et al. (1999) made the same argument for deep seabed habitats. Anderson et al. (2008) proposed a list of ten priorities for research that would advance the field of acoustic seabed classification. At least five of these priority research areas fall under the general topic of standardizing instruments and methods. There is widespread agreement, therefore, on the benefits of standardization, but almost no studies employing single-beam ASC in coral reef study sites have used the same classification scheme (Table 9.1).

Developing a standardized approach to mapping relies on an objective and systematic method for defining seabed classes; it does not require producing every habitat map at the same spatial scale, thematic resolution, or using the same data source (Mumby and Harborne 1999). Hierarchical classification schemes are one way to provide both commonality among sites and flexibility to identify detailed classes where needed. The span of thematic and spatial resolutions required for these tasks also dictates a hierarchical classification scheme that can be expanded or collapsed to the desired level of detail or resolution of available data (Mumby and Harborne 1999). Examples of hierarchical classification schemes include those developed by the Florida Marine Research Institute (Madley et al. 2002) and the NOAA Biogeography Branch (Costa et al. 2009a).

This section describes two recent developments that advance single-beam ASC toward the goal of employing a hierarchical classification scheme based on geomorphology and biotic cover that would be applicable to multiple sites and

Table 9.1 Classes for single-beam ASC surveys in coral reef environments

Source	# classes	Defined by biota	Defined by geomorphology	Defined by relief	Defined by substrate only	Defined by sediment properties	System (frequency)
1	5	Coral Seagrass on Sand	Patch Reef/Hardbottom			Sand Fine Sand	RoxAnn (not given)
2a	5					5 classes of grain size	RoxAnn (50 kHz)
2b	5					5 classes of grain size	QTC IV (38 kHz)
3	3	Coral Dominated				Sand Mud	RoxAnn (200 kHz)
4	4			High Rugosity-Hard Low Rugosity-Hard High Rugosity-Soft Low Rugosity-Soft			QTC V (50 and 200 kHz)
5	2				Sand Hard Substrate		QTC V (50 kHz)
6a	4		Rocky Ridges		Rock and Hardground	Less-Sorted Sand Well-Sorted Sand	QTC V (50 kHz)
6b	2				Hardground Sand		ECHOplus (50 kHz)
7	2				Hardbottom Sediment		QTC V (50 kHz)
8	5			>2m Relief and 5–50% Hardbottom >2m Relief and 50–100% Hardbottom <2m Relief and 5–50% Hardbottom <2m Relief and 50–100% Hardbottom 95–100% Sediment			QTC V (50 kHz)
9a	5		Colonized Pavement/Linear Reef Aggregated Patch Reef/Ridge Spur and Groove		Sand over Hardbottom	Sand/Sand-Deep	BioSonics DT-X (38 kHz)
9b	5		Colonized Pavement/Linear Reef Aggregated Patch Reef Ridge/Spur and Groove		Sand over Hardbottom	Sand/Sand-Deep	BioSonics DT-X (418 kHz)

(continued)

Table 9.1 (continued)

Source	# classes	Defined by biota	Defined by geomorphology	Defined by relief	Defined by substrate only	Defined by sediment properties	System (frequency)
10	5	Sparse SAV	Pavement	Rugose Hardbottom			BioSonics DT-X (418 kHz)
		Branching coral			Sand		
11	8	Colonized Pavement			Sand over Hardbottom	Sand	BioSonics DT-X
			Spur and Groove			Sand-Deep	
			Aggregated Patch Reef				
			Linear Reef				
			Ridge				(38 + 418 kHz)

The first column references the following sources: (1) Murphy et al. (1995), (2) Hamilton et al. (1999), (3) White et al. (2003), (4) Riegl and Purkis (2005), (5) Moyer et al. (2005), (6) Riegl et al. (2007), (7) Gleason et al. (2006, 2009, 2011) and this chapter (Sect. 9.2.1; supervised classification), (8) Miller et al. (2008), (9) Foster et al. (2009, (10) Foster et al. (2011), and (11) Foster (unpublished) this chapter (Sect. 9.2.1: supervised classification). Note that the only scheme used at more than one site is hardbottom/sediment (sources 6b, 7) have high accuracy even at a fine level of thematic detail. The first subsection reviews the supervised classification approach of Foster et al. (2011), which is a promising development towards increasing classification accuracy at a moderate level of thematic detail. The second is the unsupervised classification approach of Gleason et al. (2009, 2011), which provides consistent classifications at multiple sites with little to no ground truth required.

Supervised classification: The overall accuracy of ASC-derived maps of coral reef environments has been found to decrease rapidly as more acoustic classes are used (Fig. 9.1 *open symbols*). Methods are therefore needed to improve classification accuracy at higher levels of thematic detail. Foster et al. (2009) demonstrated a method to improve accuracy by selective filtering of ground-validated E1:E2 data pairs, where a simple 20–80 percentile filter applied to class-specific values of E1 and E2 increased overall accuracy from 52 to 80 % at 38 kHz and from 58 to 82 % at 418 kHz (Fig. 9.1 *open* and *closed circles*), but at the cost of discarding 40 % of the data. Foster et al. (2011) described another method to improve overall accuracy by using multiple iterations of discriminant analysis (DA) to refine training samples acquired on a Palauan coral reef into six "pure" endmember habitats.

In this section a small portion of a 2006 ASC survey of Palm Beach county, FL was used to demonstrate how the supervised classification methodology of Foster et al. (2011) can be used to: (1) map a coral reef using the same benthic habitat definitions used for optics-based mapping, and (2) provide supplementary geomorphological and biological information.

A dual-frequency (38 and 418 kHz) single-beam survey utilizing a BioSonics DT-X echo sounder was performed off the coast of Palm Beach along N–S lines spaced 75 m apart (Fig. 9.2). Depth in the survey area ranged from 5–40 m. Hydroacoustic data were processed with BioSonics Visual Bottom Typer (VBT) seabed classification software to obtain values of E0 (pre-bottom backscatter), E1′

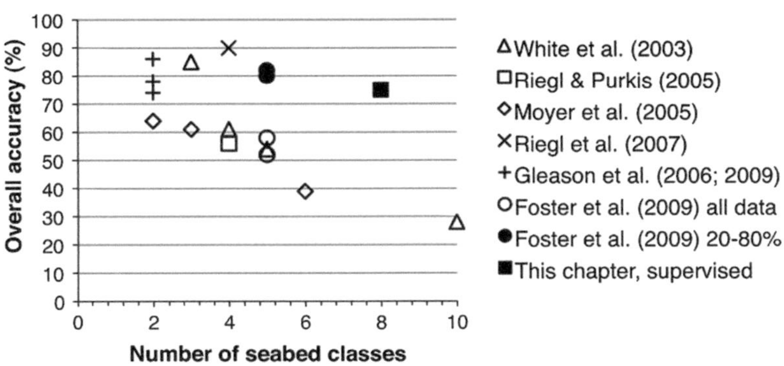

Fig. 9.1 Overall accuracy of single-beam ASC in coral reef surveys as a function of the number of acoustic classes mapped. Generally, accuracy reported in previous studies using all available data declines rapidly with increasing numbers of classes (*open symbols*). Improved processing techniques may lead to increased accuracy at higher levels of thematic detail (*closed symbols*)

(leading edge of the first echo), E1 (trailing edge of first echo), E2 (complete second echo), FD (fractal dimension of 1st echo), and depth. The raw sonar datasets were empirically normalized to depth using sand as a calibration standard. Approximately 10 % of the raw data were removed by a series of filters designed to detect aberrant waveforms, particularly those not obtained at near vertical incidence (e.g., those obtained during excessive pitch/roll of the survey vessel).

A subset of the survey data was used to construct an acoustic training dataset by pairing select acoustic data points with a spatially coincident geomorphological classification (Fig. 9.2) derived from visual interpretation of high-resolution LiDAR bathymetry (Walker et al. 2009). The LiDAR interpretation identified

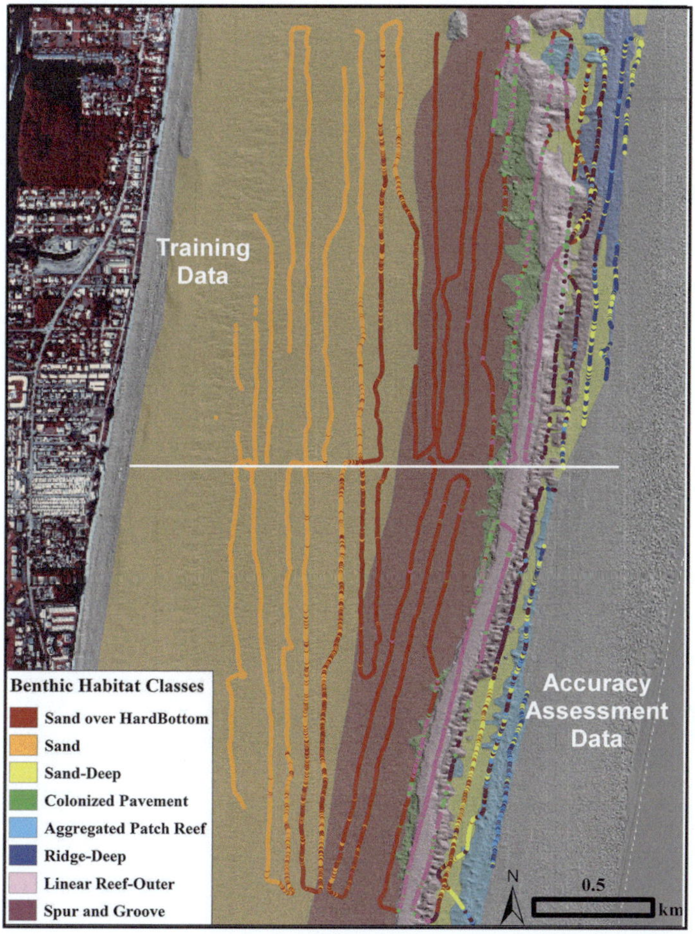

Fig. 9.2 Sub-set of a 2006 single-beam (ASC) survey of Palm Beach county, USA, displaying the classified acoustic track plot of training and accuracy assessment data using Linear Discriminant Functions from the 3rd-Pass Discriminant Analysis of a combined 38 and 418 kHz training dataset. Acoustic track plot is displayed over visual-interpretation of LiDAR bathymetry

seven benthic classes in this survey area: sand, sand-deep, colonized pavement, ridge-deep, aggregated patch reef, linear reef-outer, and spur and groove. The 38 kHz signal revealed an eighth class, a thin (5–10 cm) veneer of sand over hardbottom. The merged 38 and 418 kHz training dataset was then clustered into the eight classes by supervised classification, using a series of three discriminant analyses (DA). Eleven predictor variables were used, consisting of the 418 kHz depth and the 38 and 418 kHz E0, E1′, E1, E2, and FD acoustic parameters.

Only those records that (1) were correctly classed by the discriminant analysis and (2) exceeded a minimum probability of group membership were passed onto the next DA. Approximately 40 % of acoustic records were selectively removed from the training dataset, which had the effect of refining the continuous data cloud into relatively discrete clusters of geomorphology. This can be seen in the plots of the first two of seven (i.e., k−1) canonical discriminant functions before and after the refinement process (Fig. 9.3). The Fisher's linear discriminant coefficients obtained from the third DA were used to classify (1) the original training dataset (Fig. 9.2, *top*) and (2) the accuracy assessment data (Fig. 9.2, *bottom*). The classified acoustic track plots can be seen to agree closely with the LiDAR-derived classifications, as quantified by the high overall predictive accuracy ($P_o = 75.3$ %) in the confusion matrix populated by the accuracy assessment data (Table 9.2). The acoustic classifications also provided a measure of within habitat variability of the relatively large LiDAR polygons (one acre minimum mapping unit). Additionally, the acoustic interpretation of the LiDAR 'sand-deep' class was 75 % 'sand' and 25 % 'sand over hardbottom', thereby effectively quantifying the protrusion of the seaward escarpment.

To create a biological layer to accompany the LiDAR-derived geomorphological layer, 700+ records from 25 discrete acoustic samples acquired over short (<0.5 m) and tall (0.5–1.25 m) gorgonians were added to the training dataset and

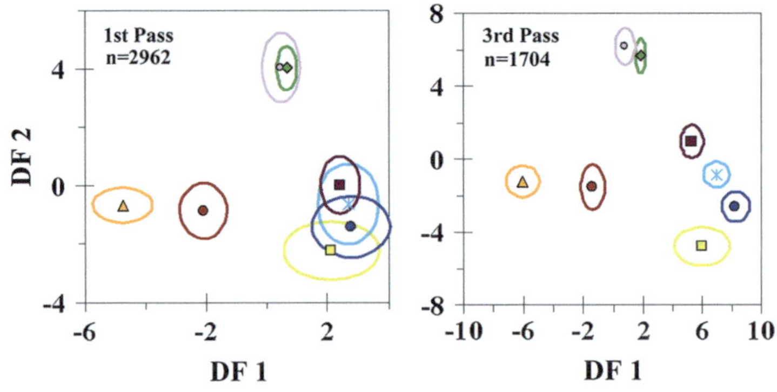

Fig. 9.3 Supervised clustering of acoustic training dataset (38 and 418 kHz; E0, E1′, E1, E2, FD, and depth) into eight geomorphological classes by multiple discriminant analysis (DA) passes. Plots of first 2 of 7 discriminant functions for (*left*) 1st and (*right*) 3rd DA Pass. Records that were classified correctly and exceeded a minimum probability of group membership were passed onto the next DA. Dispersion shown as 2 standard deviations about the mean

Table 9.2 Confusion matrix of acoustically classified accuracy assessment data

	Lider-delineated habitat class									
Discriminant analysis	Sand over HB	Sand	Sand-deep	Col pav	Agg patch	Ridge	Linear	S&G	n	User
Sand over HB	1675	1021	166	17	2	0	0	4	2885	58.1 %
Sand	45	2911	0	0	0	0	0	0	2956	98.5 %
Sand-deep	0	0	239	0	115	0	0	5	359	66.6 %
Col pav	14	0	0	69	0	0	46	14	143	48.3 %
Agg patch	0	0	1	0	200	0	0	9	210	95.2 %
Ridge	0	0	1	0	167	0	0	0	168	n/a
Linear	17	1	0	75	0	0	429	33	55	77.3 %
S&G	0	0	34	0	125	0	0	297	456	65.1 %
n	1751	3933	441	161	609	0	475	362	7732	
Producer	95.7 %	74.0 %	54.2 %	42.9 %	32.8 %	n/a	90.3 %	82.0 %	Po = 75.3 %	

Analysis performed using Linear Discriminant Functions from the 3rd-Pass Discriminant Analysis for 2006 ASC survey of Palm Beach county, USA clustered using the same multi-pass discriminant analysis methodology. The two classes of gorgonians, short and tall, clustered appropriately according to the substrate types that would be expected, emphasizing the point that acoustic discrimination is informed by a combination of substrate and epibiota. Gorgonian "hits" were tallied per polygon to produce maps of short and tall gorgonian abundance (Fig. 9.4). Note that an ASC could be similarly utilized to provide within and between habitat characterizations of other epifaunal and infaunal biota (e.g., seagrass, macroalgae, oyster beds).

Beyond demonstrating that ASC can be used to map coral reef environments at moderate-high thematic resolution at an acceptable accuracy, this case study also demonstrates how the output of different platforms can be merged in a GIS environment to create mapping products with integrated complimentary layers of geomorphological and biological information.

Unsupervised classification: Unsupervised classification exploits the advantages of statistical segmentation to find natural boundaries in a dataset. Numerous techniques for unsupervised classification are available, but all follow three steps. First, statistically segment a dataset into clusters. Second, label the clusters. Third, assess the thematic accuracy of the labeled clusters. Many methods exist for segmentation (e.g., Legendre and Legendre 1998), and accuracy assessment techniques are well established for remotely sensed maps (Congalton and Green 1999). The class-labeling step poses the greatest difficulties to productive

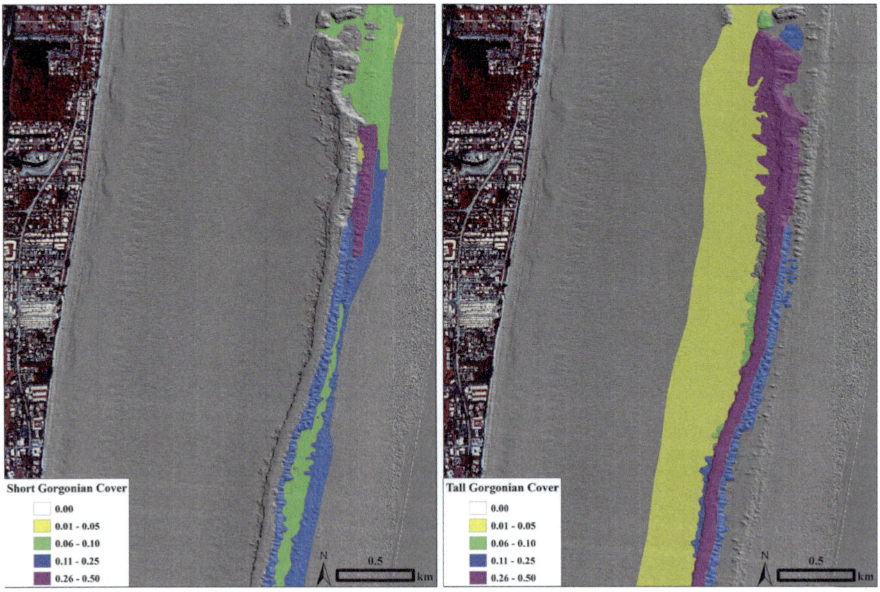

Fig. 9.4 Acoustic predictions of short (<0.5 m) and tall (0.5–1.25 m) gorgonian abundance obtained from ASC survey of Palm Beach county, USA, obtained from supervised classification of 38 and 418 kHz data from a BioSonics DT-X single-beam echo sounder

implementation of unsupervised classification. First, selecting the appropriate number of classes is not always straightforward. Second, the segments and classes found are not always useful, are not always possible to fit into a pre-determined hierarchical system, or are not easy to identify why they were segmented out. Third, available ground-truth data, which is often limited, must be divided into two subsets for labeling and accuracy assessment. Fourth, clusters are usually site-specific and sometimes sensitive to the size of the dataset. Nonetheless, unsupervised classification can still be an effective tool.

One method for labeling the clusters found in an acoustic dataset is to compare the shapes of the average echoes for each cluster identified in the segmentation step. Clusters comprised of echoes with long average duration are labeled hard-bottom, and those with short average echoes are labeled sediment. The advantages of this approach are, first, that class labeling can be performed with minimal to no ground truth data, and, second, that the same classes can be mapped in different areas or using different SONAR systems. The disadvantage of this approach is that the result is simply a two-class map (i.e., a map with low thematic detail). This section will illustrate the results of this technique applied to survey areas in the Florida Keys and the Bahamas, illustrating that even a map with low thematic detail can be effective and useful.

Gleason et al. (2009) describe four surveys using single-beam data from a 50 kHz QTCView Series V (QTCV): one from Lee Stocking Island (LSI), Bahamas; a second from Carysfort Reef, in the Florida Keys; a third from Fowey Rocks, also in the Florida Keys; and a fourth more recent survey from Andros Island, Bahamas. The Andros survey consists of approximately 73 km of track lines acquired in depths from 1 to 8 m. One portion of the bank was covered with a grid of 19 approximately 1,500 m long transects spaced 100 m apart. A second portion of the bank top was covered by nine widely spaced, cross-shelf transects that, in turn, were connected by segments of an along-shelf transect formed while transiting between the cross-shelf areas.

The processing steps for each dataset were threefold. First, cluster the data using the IMPACT software package (Quester Tangent Corporation 2002). Second, use ancillary datasets such as satellite imagery, snorkeler observations, mean echoes for each cluster, and previous seabed maps to label the largest clusters (those comprising over 90 % of the dataset) as either hardbottom or sediment. Third, assess the accuracy of the hardbottom/sediment classified map using independent measurements from divers or towed video. Overall accuracy for the hardbottom/sediment maps was 86, 78, 74, and 73 % for the Carysfort, Fowey, LSI, and Andros survey areas, respectively.

The mean echoes for all clusters show a steep rise in amplitude corresponding to the initial reflection from the seabed followed by a gradual decay (Fig. 9.5). The amplitude of the mean hardbottom echoes decayed more slowly with time (i.e., they were longer) than the sediment echoes in all four areas (Fig. 9.5), which was expected since rock has stronger off-nadir backscattering than sediment (APL-UW 1994). There was significant variation in the shapes of the hardbottom echoes even though all of the mean echoes from the hardbottom clusters have slower decay

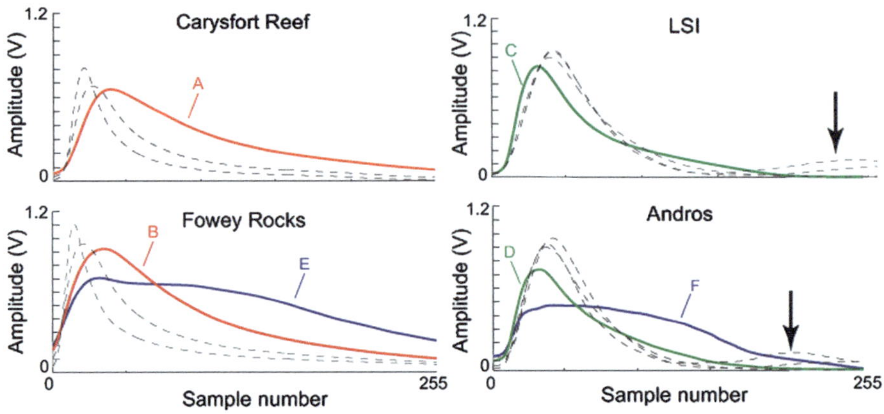

Fig. 9.5 Mean echoes for four acoustic classes grouped by survey area. Sample number is proportional to time (i.e., time increases to the right; Preston 2004). All of the sediment classes are plotted as *dashed lines*. The hardbottom classes are plotted with *solid lines*, colored according to their general shape, and labeled with letters corresponding to pictures in Fig. 9.6. *Black arrows* point to second echoes visible for some of the LSI and Andros classes. Note that, discounting the second echoes, the hardbottom echoes have longer duration than the sediment echoes

times than the mean echoes from the sediment clusters. Three basic shapes of hardbottom echoes were observed in the four survey areas. Hardbottom classes A and B, observed over hardbottom with approximately 0.5 m of relief (Fig. 9.6), had a slower rise time than echoes from sediment, a well-defined peak value, and exponential decay (Fig. 9.5). Classes C and D, observed over "pavement" hardbottom with extremely low relief (Fig. 9.6), had faster rise times and earlier amplitude peaks than the sediment classes at those sites. Finally, classes E and F, which correspond to areas with at least 1 m relief (Fig. 9.6), had slower rise times and delayed peaks relative to the sediment classes at those sites, like classes A and B. Unlike classes A and B, however, classes E and F had nearly linear decay with time, leading to the longest echoes from any survey.

The results of these supervised and unsupervised ASC classification studies show how the balance between thematic resolution, classification accuracy, and turnaround time is greatly influenced by the selection of a classification scheme, which in turn is guided by project objectives. When the objective is to coarsely reconnoiter an area, perhaps in advance of a more detailed study, rapid turnaround is the paramount factor. Using unsupervised classification at four different sites with little or no ground-truthing, hardbottom was accurately (73–86 %) distinguished from sediment. That this should be possible is hardly surprising; people have been doing this for decades by eye. What is new is to demonstrate that it is possible to do this in a systematic and objective way with minimal to no training data. The capability to interpret classes by their mean echo shape alone means that multiple sites can be mapped using a consistent classification scheme.

When the objective is to produce a detailed benthic habitat map, considerably more time and effort will need to be spent collecting and preparing a training

Fig. 9.6 Oblique underwater photographs from sites corresponding to the six hardbottom classes plotted in Fig. 9.5. **a** and **b** are low-relief hardbottom at Carysfort Reef and Fowey Rocks, respectively. **c** and **d** are nearly flat "pavement" at Andros and LSI, respectively. **e** and **f** are areas of relatively higher relief hardbottom at Fowey Rocks and Andros respectively. *White arrow* points to a 1 m-long, T-shaped scale bar

dataset. Using supervised classification it was possible to achieve high thematic detail (8 classes) at high classification accuracy (75 %). This is a significant first step towards the long-term goal of developing a hierarchical classification scheme based on geomorphology and biotic cover that would be applicable to multiple sites and have high accuracy, even at a fine level of thematic detail. Future directions should seek to merge these methods or otherwise build on them to achieve the same ends.

9.2.2 Multi-Beam Echo Sounder Application

High-resolution MBES bathymetric imagery is ideal for delineating the complex geomorphological structures of coral reef habitats, given that it is synoptic, highly resolved and positionally accurate. The correlation of acoustic signatures with specific coral reef habitat types, however, is difficult because the magnitude of the acoustic return varies widely within habitat types, and overlaps across different habitat types. In order to address this problem, new classification techniques are being developed to better mine and extract information on the geomorphological and biological attributes of the seafloor from these highly variable datasets. One of these techniques, developed by Costa et al. (2009a) and described below, uses principal components analysis (PCA), edge-based segmentation (Jin 2009), and Quick, Unbiased, Efficient Statistical Tree (QUEST) algorithms (Loh and Shih 1997) to create a benthic habitat map (Fig. 9.7).

The habitat map was created for the moderate-depth (30–60 m) areas in and around the Virgin Islands Coral Reef National Monument (VICRNM) in St. John, U.S. Virgin Islands (Fig. 9.8). It describes the location of habitat features (in relation to the shoreline), their physical composition (i.e., geomorphological structure) and the types of organisms that colonize them (i.e., biological cover and live coral cover). Bathymetry and backscatter were collected in this area using a Reson Seabat 8101 240 kHz Extended Range (ER) multi-beam echo sounder. A

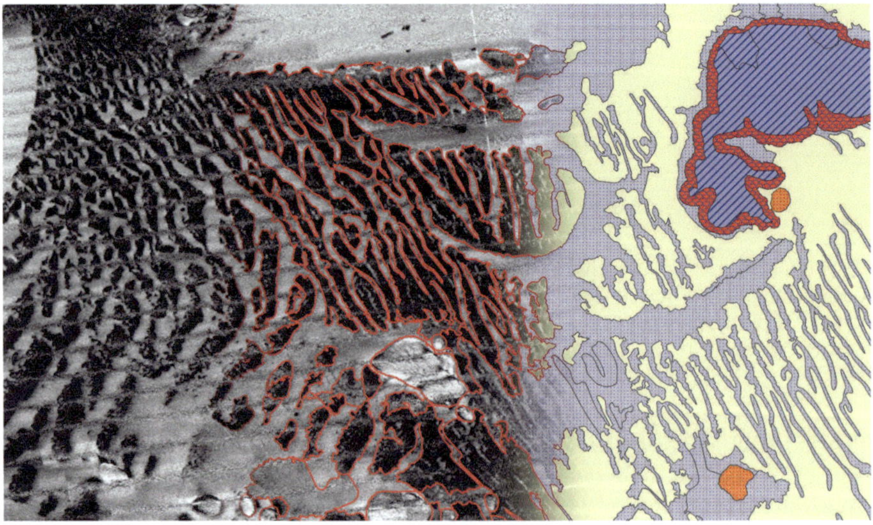

Fig. 9.7 The process used to create a benthic habitat map from acoustic imagery. The *left third* of the figure depicts the principal component surface derived from the MBES imagery. The *middle third* depicts the delineation and segmentation of seafloor features in the principal components surface using edge detection algorithms. The *right third* depicts the classification of seafloor features extracted by the edge detection algorithm by QUEST

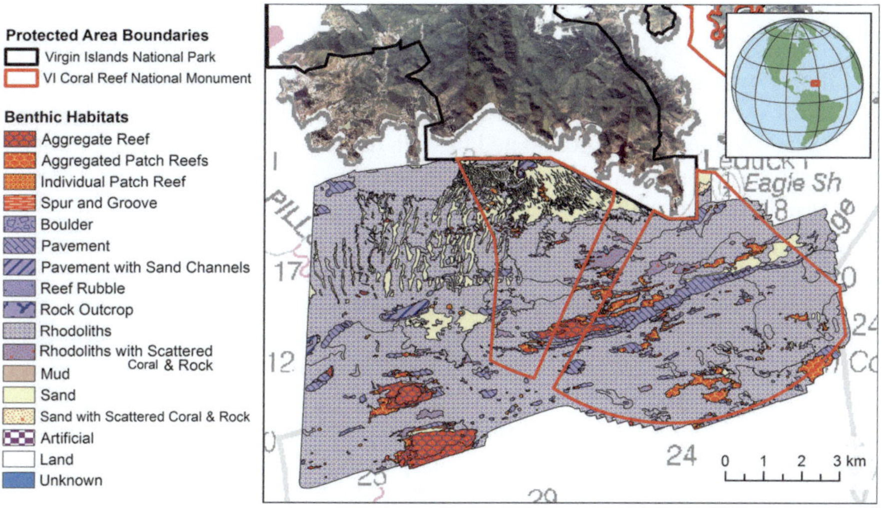

Fig. 9.8 Depicts the benthic habitat map created from MBES imagery for the moderate-depth (30–60 m) areas in and around the Virgin Islands Coral Reef National Monument in St. John, U.S. Virgin Islands. The map is symbolized to denote physical composition (i.e., geomorphological structure) of the seafloor

suite of complexity metrics was derived from the bathymetry surface in order to highlight the differences between distinct habitat structures on the seafloor. These metrics specifically included: (1) mean depth, (2) standard deviation of depth, (3) curvature, (4) plan curvature, (5) profile curvature, (6) rugosity, (7) slope, and (8) slope-of-slope. These metrics were collectively transformed into their principal components to remove redundant information and retain unique information that described the complexity and structure of the seafloor. Using this principal component image, discrete seafloor features were then segmented using edge detection algorithms. A variety of spatial, spectral and textual attributes were next calculated for each seafloor feature, describing its size, shape and color.

Using a supervised classification approach, the seafloor attributes, along with known points denoting habitat types at particular locations, were used to train the QUEST algorithm to classify all of the seafloor features extracted by the segmentation process. QUEST is a type of Classification and Regression Tree (CART) algorithm (Breiman et al. 1984) that efficiently splits an image into two parts recursively, until all seafloor features are classified or the algorithm is instructed to stop. QUEST partitioned the acoustic imagery into 35 unique combinations of geomorphological structure, detailed biological cover and live coral cover types. This classified habitat map was manually reviewed and edited before its thematic accuracy was assessed. The thematic accuracies (corrected for proportional bias) for the major and detailed structure, major and detailed biological cover, and live coral cover classes were: 95.7, 88.7, 95.0, 74.0 and 88.3 %, respectively.

In total, 90.2 km^2 of the seafloor in and around the VICRNM was characterized using MBES imagery. This area (both outside and inside the Monument's boundaries) was dominated by rhodoliths (i.e., calcareous algal nodules). Hard and soft corals were present in low densities ($0 \leq 10$ %), although one 0.25 km^2 area outside the Monument was dominated by high density live coral (i.e., $50 \geq 90$ %). In general, the habitat map denoted the presence of slightly more live coral outside (than inside) the current VICRNM boundaries. These quantitative results illustrate the utility of MBES imagery for moderate-depth habitat mapping and for ecosystem-based resource management. With that in mind, expediting the rate of MBES seafloor mapping will require the collection of datasets that concurrently address the needs of multiple users (Costa et al. 2009b), such as the "collect once, use many times" approach of the Interagency Working Group on Ocean and Coastal Mapping (IWG-OCM). For ecosystem-based management and marine spatial planning purposes, maps with high thematic accuracies and resolutions are important to have because predictions of species diversity, abundances and distributions may differ depending on the thematic properties of the input maps (Kendall and Miller 2008).

9.2.3 Phase Differencing Bathymetric Sonar

Benthic habitats in shallow (<30 m), perpetually turbid coral reef environments are exceptionally challenging to characterize. In particular, they are challenging because many conventional mapping technologies are unable to synoptically map these areas (e.g., passive and active optical sensors), or are inefficient and costly for mapping these areas (e.g., MBES systems). The use of interferometric sonars (IS), also known as phase differencing bathymetric SONAR (PDBS), may fill in this informational gap, where conditions are not optimal for the operation of other sensors. Like MBES systems, PDBS can collect co-located bathymetric and backscatter intensity information. These two pieces of information (along with underwater video and photos) can be used to develop maps of habitats on the seafloor. Unlike MBES systems, however, PDBS can collect these spatially coincident datasets over wide swaths in shallow waters (<30 m), typically 10–12× the depth versus 3–5× depth for MBES systems (Gosnell 2005). PDBS are able to collect wide swaths in shallow waters because they are not beam forming but rather they accurately measure depths by precisely measuring the phase offsets of acoustic returns (Gosnell 2005). These phase offsets are used to calculate the angle from which the return was received (e.g., Denbigh 1989). This angle is combined with measurements of range (based on two-way travel time) to calculate the position (and depth) of the seafloor.

Given that both depth and intensity surfaces are collected, PDBS systems can be used to develop benthic habitat maps of shallow, turbid coral reef ecosystems. Such habitat maps, describing the geographic location, geomorphological structure and biological cover of seafloor habitats, were developed from an interferometric

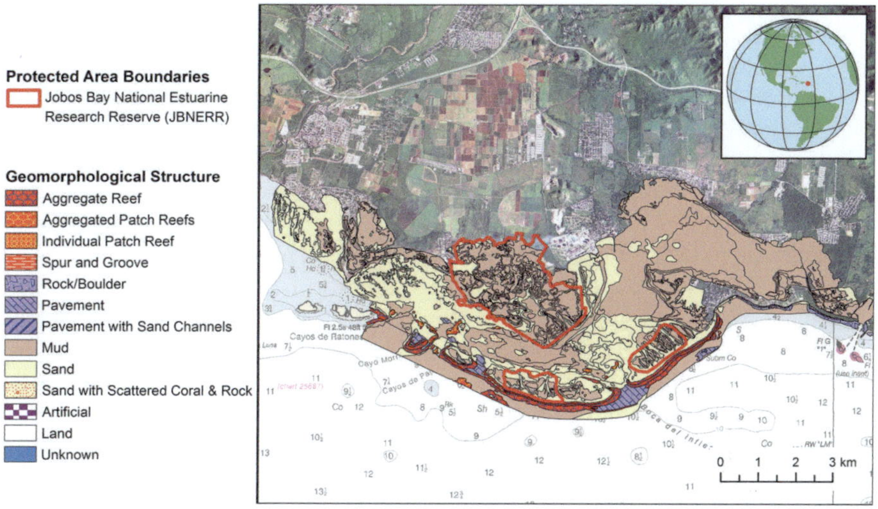

Fig. 9.9 Depicts the benthic habitat map created from aerial photographs and acoustic imagery for the shallow water (<30 m) areas in and around Jobos Bay National Estuarine Research Reserve in Puerto Rico. The map is symbolized to denote physical composition (i.e., geomorphological structure) of the seafloor

dataset for the Jobos Bay National Estuarine Research Reserve (JBNERR) in southeast Puerto Rico (Fig. 9.9). Specifically, 4 × 4 m bathymetry and 1 × 1 m intensity images were collected using a Teledyne Benthos C3D 200 kHz Lightweight Pole Mount (LPM) system covering an area ranging from about 1–25 m in depth. The bathymetric imagery was used to derive a suite of complexity surfaces to better describe the geomorphological structure of the seafloor, following the same process discussed above in the MBES application and outlined in Costa et al. (2009a). These complexity surfaces were transformed into their first three principal components, which were used, in concert with the intensity surface, to visually delineate and describe habitats on the seafloor in areas where the water was turbid. It is important to note, however, that the bathymetry (and consequently, principal components) collected by the C3D system was prohibitively noisy in some locations. This noise was the result of a combination of: (1) rough weather conditions, (2) limited accuracy of the motion sensor, and (3) vertical and horizontal uncertainty inherent in the system. In these noisy areas, and elsewhere within the study area where water visibility allowed, aerial photographs collected by the U.S. Army Corp of Engineers (USACE) at 1 × 1 ft resolution were used to visually identify, delineate and characterize habitats on the seafloor (Zitello et al. 2009). Together, the PDBS dataset and aerial photographs were used to create a seamless habitat map for the coral reef ecosystem in and around Jobos Bay from the shoreline to approximately 25 m in depth. This map will provide JBNERR with increased technical capacity for ocean exploration, management, and stewardship.

9.2.4 Split-Beam Application

Coral reef resource users and managers have need for highly resolved and highly detailed maps of the benthic habitats; however, management mandates also include the inhabitants (i.e., the fishes and invertebrates) that occupy the habitats. Surveys of inhabitants of reef communities have relied on direct visual observations at fine spatial scale using scuba divers, remotely operated vehicles (ROVs), drop cameras, or extractive methods such as traps and nets. These surveys provide highly detailed data on species composition, but can be costly and are limited in maximum depths, environmental conditions (e.g., sea conditions, light levels, visibility), and overall extent. Additionally, high spatial variation in fish densities or community assemblages can result in difficulties in capturing trends or spatial patterns from which to interpret impacts from natural or anthropogenic impacts. Adding split-beam echo sounder surveys to existing coral reef habitat survey platforms is a modest investment that will provide significant payoff in terms of adding value and interpretive power to habitat maps and related products from coral reef benthic habitat maps.

Similar to commercially available bottom- and fish-finders, scientific split-beam echo sounders can be used to detect fishes in the water column and near-bottom at high vertical and horizontal resolution. Unlike most commercial echo sounders, however, these scientific digital echo sounders have the capability of acquiring and storing digitized echo return data for later analysis. In moderate depths (<100 m), short pulse lengths (0.1–0.3 ms) can result in vertical resolutions of <20 cm, while high pulse repetition (ca. 5–10 Hz) typically results in numerous echo returns as fish pass through the acoustic beam.

Target tracking algorithms accumulate repeated echo returns from individual fish, from which attributes are calculated for each target (Fig. 9.10). Individual fish are identified and attributed a target strength based on the intensity of the returning echo, which can then be converted to length using a generalized relationship. Position of individual fish targets include the range from transducer, based on the time delay of the acoustic echo return, relative horizontal position within the acoustic beam, determined from phase differencing in the split-beam quadrants, and a geographic coordinate from the ship's positioning system.

When fish are in dense schools or aggregations, individual fish tracks are indiscernible. Instead, the total acoustic energy that is returned from the insonification of the school is assumed to represent the sum of the acoustic energies of the individual fish. In this way, the acoustic energy is integrated over the school and the density of fish is estimated following theories of echo-integration (Simmonds and MacLennan 2005). Additional metrics can be derived from fish schools and aggregations, such as size, spatial structure and average acoustic energy return. Fish density is calculated for a discrete segment by weighting individual fish by their position within the acoustic beam, accounting for the higher probability of detecting a fish at greater range from the transducer as the beam becomes

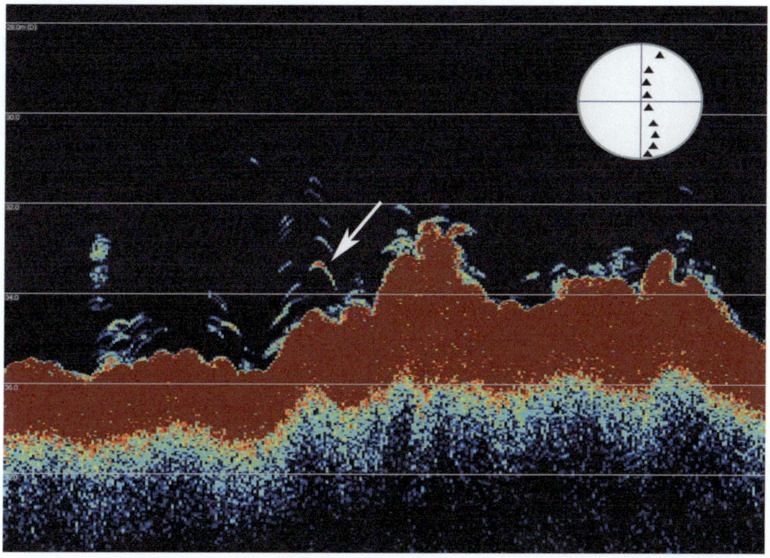

Fig. 9.10 Split-beam echogram showing numerous fish observed over high-relief coral reef habitat. *Horizontal lines* provide 15 and 20 m references and show fish can be resolved within 1 m of the bottom. *Arrow* in echogram corresponds to *inset*, which shows *top-view* of individual returning echoes (*triangles*) as the fish path through transducer beam from the survey vessel

wider. Fish weights are summed over a transect or segment and divided by the segment length.

As an example from the US Virgin Islands, split-beam echo sounders were integrated into a sensor package on a benthic habitat mapping survey vessel and used to map the distribution of fishes and mid-water invertebrates throughout the water column and near-bottom. Surveys were conducted in two areas identified by regional management partners as high priority areas for habitat mapping. The first location was near Virgin Islands Passage, about 16 km SW of St Thomas, US Virgin Islands and 6 km NE of Vieques Island, Puerto Rico. The second was south of St. Johns, US Virgin Islands. The survey design was optimized for MBES hydrographic surveys, where survey transects were conducted either parallel to one another, or following depth contours, spaced 50–100 m apart. The split-beam echo sounder was a Simrad EK60 operating at two frequencies, 120 and 38 kHz, though only the 120 kHz data are discussed here. Fish lengths were estimated from the average target strength (TS) using a generalized TS-length relationship (Love 1971). Results were then divided into three fish size classes: (1) less than 12 cm, which represent small, zooplanktivorous or reef associated species, (2) between 12 and 28 cm, representing adults of reef associated species and some juveniles of commercially important species of snapper and grouper, and (3) greater than 29 cm, representing larger pelagic and reef associated species including snappers and groupers. Densities were computed for each size class in 100 m segments along transects, producing units of fish per 100 m^2.

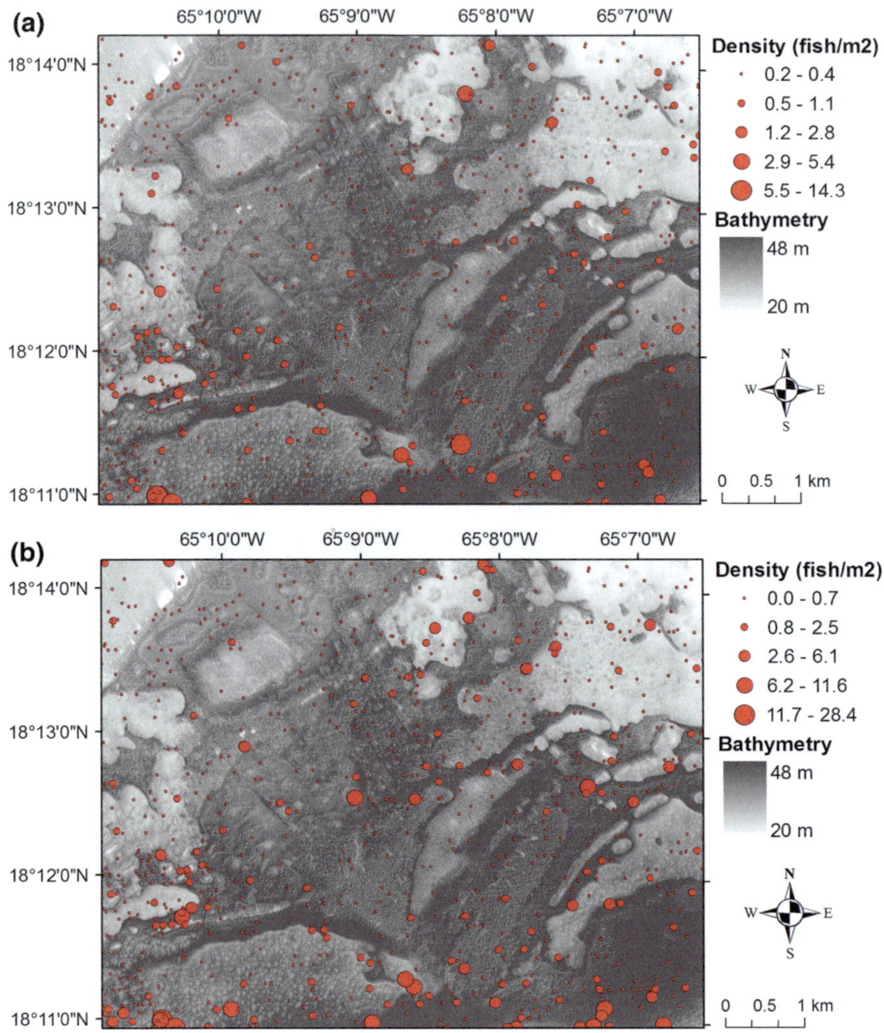

Fig. 9.11 Map of fish density near Virgin Passage, US Virgin Islands. Bathymetry is provided from a MBES hydrographic survey that included a split-beam echo sounder for fish detections. Fish densities were calculated in 100 m segments for two size classes: **a** fish between 12 and 28 cm, and **b** fish greater than 28 cm

The distribution of fish densities was determined to be highly variable throughout the sampling region of Virgin Island Passage (Fig. 9.11), though relatively higher densities were found at edges of areas that had high bottom relief. In contrast to Virgin Passage, fish distributions on the St. John Shelf were primarily restricted to the high relief regions along the shelf break (Fig. 9.12). Further analysis of these data, and a spatially-explicit comparison with habitat

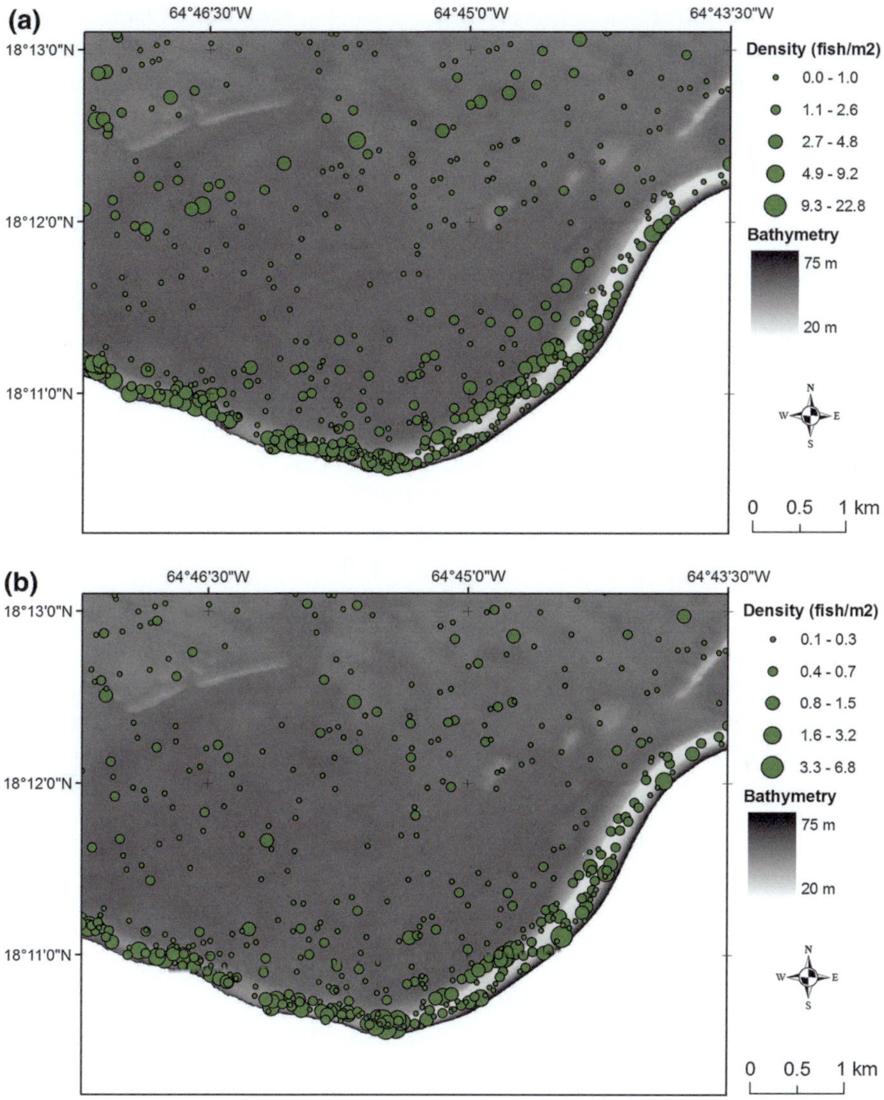

Fig. 9.12 Map of fish density on St. John Shelf, south of St. Johns, US Virgin Island. Bathymetry is provided from hydrographic survey that included a split-beam echo sounder for fish detections. Fish densities were calculated in 100 m segments along survey transects for two size classes: **a** fish between 12 and 28 cm, and **b** fish greater than 28 cm

classification maps derived from the MBES data, can determine the habitat and landscape features that help to explain the distribution patterns of fish in this region.

There are some limitations to the data available from split-beam echo sounder surveys of fish communities in coral reef ecosystems. While the split-beam technology can provide detections of fishes throughout the water column at high spatial resolution, and precise estimates of acoustic target strengths from which to infer sizes, it is not yet possible to infer species from the acoustic signatures alone. This may pose a problem in diverse systems such as coral reefs, and particularly when the objective for reef fish assessments is for the purpose of monitoring species-specific abundances or responses to management measures. In contrast, identifying and studying large, single-species aggregations of reef fish, such as reef fish spawning aggregations can benefit from using split-beam echo sounders. Large areas can be rapidly surveyed and analysis of split-beam data can provide accurate estimates of density and abundance for large groups of fish, a task that can be very difficult to accomplish using divers alone (Taylor et al. 2006).

The utility of these coral reef fish habitat maps to coral reef management are several-fold. First, these maps provide a broad depiction of fish biomass distribution over a coral reef ecosystem, which will help identify regions of high and low fish density and inform marine spatial planning and management. Second, in regions which have not yet been surveyed visually for fish communities, the maps can guide survey designs by identifying locations of relatively high or low fish densities. Third, the maps can be used to help interpret fine-scale, and limited-extent, direct visual observations in a larger spatial context. Fourth, the survey technique is rapid, repeatable and consistent, and can be conducted over a range of spatial and temporal scales. As such, split-beam echo sounder surveys can also add value to coral reef assessments by (1) tracking changes in fish biomass over time, or (2) inferring movements and migrations of fishes over daily or seasonal time scales. The value of these surveys could be particularly important when monitoring or assessing the efficacy of marine reserves or other spatial management measures prior to and following implementation. Additionally, extending fish habitat relationships through statistical and process-based modeling will guide further interpretation and utility of these products.

9.3 State of the Science and Future Directions

In response to the growing need by resource managers for current, accurate and consistent benthic habitat maps across a range of spatial scales, an international group of scientists met to review the state of acoustic remote sensing (Anderson et al. 2008). The group concluded that acoustic seabed classification using single-beam, multi-beam, and sidescan systems is in its nascence and prioritized the top ten issues in need of immediate and future attention by the international scientific community to advance the utility of acoustic remote sensing for mapping marine ecosystems. The main issues raised by Anderson et al. (2008) are recounted below, including progress that has been made in the intervening years.

- *Statistical versus interpretive classification*: In the interest of advancing repeatability of results, statistical classification is preferred over interpretive classification. The advancements here are twofold: statistical and other machine learning techniques allow for further automation and increased efficiency in the production of management-ready habitat maps of coral reef ecosystems as well as reduction in biases and process-based errors in interpretation. The applications presented in this chapter are predominantly departures from subjective interpretative methodologies and are advances towards the goal of repeatable classification techniques.

- *Spatial scales and sampling resolution*: Hierarchical benthic habitat classification schemes have recently been developed to meet the needs for resource management over large spatial scales (Madley et al. 2002; Costa et al. 2009a). Although acoustic bathymetry has long sufficed for the largest scales of seafloor classification (e.g., shelves and basins), more information and advanced methods are needed for classifying coral reef habitats. Acoustic studies are in the early stages of demonstrating whether and how such fine degrees of discrimination can be extracted from backscatter and textural properties.
- *Ground verification scale*: Consistently matching the scale of verification with that of insonfication within and between studies can be a challenge given the range of beam and swath widths. The acoustic footprint of an ASC operated in 10 m of water can range from 3 m^2 for a 6.4° beam operated at 5 Hz, versus 200 m^2 for a 42° beam operated at 1 Hz. In contrast, multi-beam and interferometric sonar offer beam footprints an order of magnitude higher resolution than ASC, while achieving swath coverage greater than 3× water depth. In both cases, ground verification needs to strike the balance between effort in the field and minimizing uncertainty.
- *Temporal variability*: The potential for temporal variability in topography, reflectivity, and biological attributes have rarely been accounted for in acoustic remote sensing studies. This could be especially problematic in ephemeral nearshore hardbottom habitats or seagrass beds, due to sediment transport by winter storms and annual periods of expansion and retreat, respectively. Remote sensing platforms such as acoustics can be applied to detect these changes when coupled with objective classification techniques. Indeed, the ability to conduct repeated and repeatable surveys is a notable strength of acoustic systems and remote sensing systems in general.
- *Reference areas*: While resampling a known area of seabed is often used for internal calibration within or between surveys, it is only truly useful if it is accompanied by proper ground verification to determine the extent to which the reference patch might have changed between samplings. Reference patches do not address the current lack of an universal reference standard, which impedes corroborative research efforts.
- *Calibration of acoustic systems*: The degree to which acoustic systems can be calibrated differs both within and between acoustic platforms. Calibration varies widely between ASC systems, varying from turn-key configurations to numerous manual and automatic gain adjustments. Moreover, commercial ASC vary widely in their approaches to removing depth dependency via time-varied gain and normalization of echo length to a reference depth. Calibration of fishery

split-beam echo sounders is a necessity to accurately calculate fish sizes from acoustic target strengths. Additionally, backscatter intensity of SSS swaths vary within surveys, necessitating distinct seabed features present throughout the survey area that can be used as a reference to calibrate grayscale images. Variable beam geometry also adds complexity to MBES systems. Given such wide latitude of calibration scenarios, inter-comparability of acoustic studies awaits the development of universal calibration standards.

- *Characterization of the acoustic signal*: Corroboration of acoustic classification requires specific knowledge of how the acoustic signals are processed. Within ASCs, the multi-echo classification scheme (i.e., E1 and E2) used by RoxAnn is relatively straightforward compared to the clustering of proprietary PCA-reduced acoustic parameters used by QTC. The different beam geometries of various MBES systems affect feature extraction and are likewise difficult to replicate. As with calibration and a need for reference areas, inter-comparability requires greater uniformity in signal processing.
- *Single versus multiple frequencies*: Integrating multiple frequencies into single datasets provides greater scope for classifying seabeds or biological characteristics, as both scattering and volume scattering vary with frequency. Anderson et al. (2008) suggest incorporating multi-frequency ASC with single frequency MBES during surveys may be a cost-effective way of improving seabed classification. For mapping habitat use by fishes, species identification will remain a challenge. Future development of multi-frequency and broadband signal processing may hold promise in classifying species groups or higher levels of biological organization; however, more work is needed in this area.
- *Survey design*: The current practice of conducting surveys along systematic line transects needs to be reconsidered in favor of a strategy that adds nested randomized lines. Conducting an SBES or ASC survey along parallel line transects introduces bias and error into continuous surfaces obtained from spatial interpolation. Similarly, conducting an MBES survey along bathymetric contours to minimize variability of backscatter intensity may fail to detect small-scale spatial variability of seabed features. Survey design is particularly important when conducting surveys of habitat use by fishes. Interpretation of spatial patterns of fish distributions need to account for a variety of fish behaviors, including feeding migrations and diurnal/nocturnal activity patterns.
- *Design in national habitat programs*: Advances described in this and preceding chapters may suggest that the field of ASC is well matured. On the contrary, significant improvements can still be made in survey design, efficiency and accuracy, as well as research and development in sensor technologies. For this reason, we echo the recommendation by Anderson et al. (2008) that formal mechanisms be established to integrate acoustic remote sensing research and development into national classification and mapping programs.
- *Defining fish habitats and habitat use*: Many SBES and some MBES can simultaneously acquire water column data for detecting and enumerating fishes. These synoptic fish and habitat data are elucidating important spatial and temporal variation in habitat use, and showing that not all habitats (even those

classified similarly) are created equal. Fine-scale maps are most likely to define areas of high site fidelity for a specific life-stage of a fish species; whereas, landscape-scale mapping would encompass distribution patterns for a wider range of life stages and species. Beyond maps depicting habitat classes and types, we are likely to learn a great deal about how the arrangement of habitats in a mosaic drive fish distribution and abundance patterns and how best to meet management needs by prioritizing habitats and regions that are important for sustaining both fishes and ecosystem services.

Suggested Reading

Hamilton LJ (2001) Acoustic seabed classification systems. Department of Defence, Defence Science and Technology Organisation Victoria (Australia) Aeronautical and Maritime Research Lab

Penrose JD, Siwabessy PJW, Gavrilov A, Parnum I, Hamilton LJ, Bickers A, Brooke B, Ryan DA, Kennedy P (2005) Acoustic techniques for seabed classification. Cooperative Research Centre for Coastal Zone Estuary and Waterway Management, Technical Report 32

International Council for the Exploration of the Sea (2007) Acoustic seabed classification of marine physical and biological landscapes, ICES Cooperative Research Report No. 286, pp 183

References

Anderson JT, Holliday DV, Kloser R, Reid DG, Simard Y (2008) Acoustic seabed classification: current practice and future directions. ICES J Mar Sci 65:1004–1011

Andrefouet S, Riegl B (2004) Remote sensing: a key tool for interdisciplinary assessment of coral reef processes. Coral Reefs 23:1–4

Applied Physics Laboratory: University of Washington (APL-UW) (1994) APL-UW high-frequency ocean environmental acoustic models handbook, Technical Report APL-UW TR9407 AEAS 9501. Applied Physics Laboratory, University of Washington, Seattle

Bejarano S, Mumby J, Hedley JD, Sotheran I (2010) Combining optical and acoustic data to enhance the detection of Caribbean forereef habitats. Remote Sens Environ 114:2768–2778

Blondel PH, Gomez-Sichi O (2009) Textural analyses of multibeam sonar imagery from Stanton Banks, Northern Ireland continental shelf. Appl Acoust 70:1288–1297

Breiman L, Friedman JH, Stone CJ, Olshen RA (1984) Classification and regression trees. Wadsworth and Brooks/Cole, Monterey

Chivers RC, Emerson N, Burns DR (1990) New acoustic processing for underway surveying. Hydrogr J 56:8–17

Congalton RG, Green K (1999) Assessing the accuracy of remotely sensed data: principles and practices. Lewis Publishers, Boca Raton

Costa BM, Bauer LJ, Battista TA, Mueller PW, Monaco ME (2009a) Moderate-depth benthic habitats of St. John, U.S. Virgin Islands. NOAA Technical Memorandum NOS NCCOS 105, Silver Spring

Costa BM, Battista TA, Pittman SJ (2009b) Comparative evaluation of airborne LiDAR and ship-based multibeam SoNAR bathymetry and intensity for mapping coral reef ecosystems. Remote Sens Environ 113:1082–1100

Denbigh PN (1989) Swath bathymetry: principles of operation and an analysis of errors. IEEE J Oceanic Eng 14:289–298

Florida Marine Research Institute (FMRI) (1998) Benthic habitats of the Florida Keys. FMRI Technical Report TR-4. Florida Marine Research Institute/Florida Department of Environmental Protection and the National Oceanic and Atmospheric Administration, St. Petersburg

Foster G, Walker BK, Riegl B (2009) Interpretation of single-beam acoustic backscatter using Lidar-derived topographic complexity and benthic habitat classifications in a coral reef environment. J Coast Res SI53, pp 16–26

Foster G, Ticzon VS, Riegl B, Mumby PJ (2011) Detecting end-member structural and biological elements of a coral reef using a single-beam acoustic ground discrimination system. Int J Remote Sens 32:7749–7776

Gleason ACR, Eklund A-M, Reid RP, Koch V (2006) Acoustic signatures of the seafloor: tools for predicting grouper habitat. In: Taylor JC (ed) Emerging technologies for reef fisheries research and management. NOAA Professional Papers NMFS #5

Gleason ACR, Reid RP, Kellison GT (2009) Single-beam acoustic remote sensing for coral reef mapping. In: Proceedings of 11th international coral reef symposium, Ft. Lauderdale, pp 611–615

Gleason ACR, Reid RP, Kellison GT (2011) Geomorphic characterization of reef fish aggregation sites in the upper Florida Keys, USA, using single-beam acoustics. Prof Geogr 63:443–455

Gosnell K (2005) Efficacy of an interferometric sonar for hydrographic surveying: do interferometers warrant and in-depth examination? Hydrogr J 118:17–24

Greene HG, Yoklavich MM, Starr RM, O'Connell VM, Wakefield WW, Sullivan DE, McRea Jr JE, Cailliet GM (1999) A classification scheme for deep seafloor habitats. Oceanol Acta 22(6):663–678

Hamilton LJ, Mulhearn PJ, Poechert R (1999) Comparison of RoxAnn and QTC-View acoustic bottom classification system performance for the Cairns area, Great Barrier Reef, Australia. Cont Shelf Res 19:1577–1597

Heyman WD, Ecochard JLB, Biasi FB (2007) Low-cost bathymetric mapping for tropical marine conservation: a focus on reef fish spawning aggregation sites. Mar Geod 30:37–50

Hickerson EL, Schmahl GP (2005) Flower garden National Marine Sanctuary: introduction. Gulf Mex Sci 23:2–4

Jarrett BD, Hine AC, Halley RB, Naar DF, Locker SD, Neumann AC, Twichell D, Hu C, Donahue BT, Jaap WC, Palandro D, Ciembronowicz K (2005) Strange bedfellows: a deep-water hermatypic coral reef superimposed on a drowned barrier island; Southern Pulley Ridge, SW Florida platform margin. Mar Geol 214:295–307

Jin X (2009) Segmentation-based image processing system. US Patent 20,090,123,070. Filed 14 Nov 2007. Issued 14 May 2009

Kendall MS, Miller T (2008) The influence of thematic and spatial resolution on maps of a coral reef ecosystem. Mar Geod 31:75–102

Kenny AJ, Cato I, Desprez M, Fader G, Schuttenhelm RTE, Side J (2003) An overview of seabed-mapping technologies in the context of marine habitat classification. ICES J Mar Sci 60:411–418

Legendre P, Legendre L (1998) Numerical ecology, 2nd edn. Elsevier, New York

Loh W-Y, Shih Y-S (1997) Split selection methods for classification trees. Stat Sinica 7:815–840

Love RH (1971) Measurements of fish target strength: a review. US NMFS Fish Bull 69:703–715

Madley KA, Sargent B, Sargent FJ (2002) Development of a system for classification of habitats in estuarine and marine environments (SCHEME) for Florida. Unpublished report to the U.S. Environmental Protection Agency, Gulf of Mexico Program (Grant Assistance Agreement MX-97408100). Florida Marine Research Institute, Florida Fish and Wildlife Conservation Commission, St. Petersburg

Miller SL, Chiappone M, Swanson DW, Ault J, Smith S, Meester G, Luo J, Franklin E, Bohnsack J, Harper D, McClellan DB (2001) An extensive deep reef terrace on the Tortugas Bank, Florida Keys National Marine Sanctuary. Coral Reefs 20:299–300

Miller MW, Halley RB, Gleason ACR (2008) Reef geology and biology on Navassa Island. In: Riegl B, Deodge RE (eds) Coral reefs of the USA. Springer, Berlin

Moyer RP, Riegl B, Banks K, Dodge RE (2005) Assessing the accuracy of acoustic seabed classification for mapping coral reef environments in South Florida (Broward County, USA). Revta Biologia Trop 53(1):175–184

Mumby PJ, Harborne AR (1999) Development of a systematic classification scheme of marine habitats to facilitate regional management and mapping of Caribbean coral reefs. Biol Conserv 88:155–163

Murphy L, Leary L, Williamson A (1995) Standardizing seabed classification techniques. Sea Tech 36:15–19

Preston JM (2004) Resampling sonar echo time series primarily for seabed sediment classification. United States Patent and Trademark Office, Patent Number US 6,801,474 B2

Preston JM, Christney AC, Beran LS, Collins WT (2004) Statistical seabed segmentation—from images and echoes to objective clustering. In: Proceedings of 7th European conference on underwater acoustics, vol 813, p 818

Quester Tangent Corporation (2002) QTC IMPACT acoustic seabed classification, user guide version 3.00. Integrated mapping, processing and classification Toolkit. Sidney, Canada

Riegl B, Piller WE (2003) Possible refugia for reefs in times of environmental stress. Int J Earth Sci 92:520–531

Riegl BM, Purkis SJ (2005) Detection of shallow subtidal corals from IKONOS satellite and QTC View (50, 200 kHz) single-beam sonar data (Arabian Gulf; Dubai, UAE). Remote Sens Environ 95:96–114

Riegl BM, Halfar J, Purkis SJ, Godinez-Orta L (2007) Sedimentary facies of the Eastern Pacific's northernmost reef-like setting (Cabo Pulmo, Mexico). Mar Geol 236:61–77

Roberts JM, Wheeler AJ, Freiwald A (2006) Reefs of the deep: the biology and geology of cold-water coral ecosystems. Science 312:543–547

Simmonds J, MacLennan D (2005) Fisheries acoustics: theory and practice, 2nd edn., Fish and Aquatic Resources SeriesWiley-Blackwell, New York

Taylor JC, Rand PS, Eggleston DB (2006) Nassau grouper (*Epinephelus striatus*) spawning aggregations: hydroacoustic surveys and geostatistical analysis. In: Taylor JC (ed) Emerging technologies for reef fisheries research and management. NOAA Professional Paper NMFS 5

van Walree PA, Tegowski J, Laban C, Simons DG (2005) Acoustic seafloor discrimination with echo shape parameters: a comparison with the ground truth. Cont Shelf Res 25:2273–2293

Walker BK, Riegl BM, Dodge RE (2009) Mapping coral reef habitats in Southeast Florida using a combined technique approach. J Coast Res SI 53:16–26

White WH, Harborne AR, Sotheran IS, Walton R, Foster-Smith RL (2003) Using an acoustic ground discrimination system to map coral reef benthic classes. Int J Remote Sens 24:2641–2660

Zitello AG, Bauer LJ, Battista TA, Mueller PW, Kendall MS, Monaco ME (2009) Shallow-water benthic habitats of St. John, U.S. Virgin Islands. NOAA Technical Memorandum NOS NCCOS 96, Silver Spring

Chapter 10
Deep Acoustic Applications

Thiago B. S. Correa, Mark Grasmueck, Gregor P. Eberli, Klaas Verwer and Samuel J. Purkis

Abstract Because cold-water coral ecosystems exist at relatively inaccessible depths of 500–1,000 m, only a limited number of accurate maps have been produced for this resource. This chapter describes a combined acoustic survey approach used to acquire high-spatial resolution (up to 0.5 m) maps from two cold-water coral sites in the Straits of Florida. The approach consists of reconnaissance surveys using hull-mounted multi-beam systems, followed by deployment of multi-beam and side-scan sonar systems on an autonomous underwater vehicle (AUV). The wide swath of the reconnaissance survey tools permitted coverage of large areas, producing coarse (20 and 50 m) resolution maps that identified coral-building mounds larger than 2,600 m^2. Areas of interest identified

T. B. S. Correa (✉) · M. Grasmueck · G. P. Eberli · K. Verwer
Rosenstiel School of Marine and Atmospheric Science, University of Miami, 4600 Rickenbacker Cswy, Miami FL 33149, USA
e-mail: tcorrea@rsmas.miami.edu and thiago.b.s.correa@conocophillips.com

M. Grasmueck
e-mail: mgrasmueck@rsmas.miami.edu

G. P. Eberli
e-mail: geberli@rsmas.miami.edu

K. Verwer
e-mail: klver@statoil.com

T. B. S. Correa
ConocoPhillips Company, 600 North Dairy, Ashford, Houston TX 77029, USA

K. Verwer
Statoil, Sandsliveien 90 5254 Bergen, Norway

S. J. Purkis
National Coral Reef Institute, Nova Southeastern University, Oceanographic Center, 8000 North Ocean Drive, Dania Beach FL 33004, USA
e-mail: purkis@nova.edu

J. A. Goodman et al. (eds.), *Coral Reef Remote Sensing*,
DOI: 10.1007/978-90-481-9292-2_10,

using the reconnaissance tools were then surveyed with the AUV platform, which resolved cold-water coral fields at 0.5–3 m resolution. The AUV maps detected mounds as small as 81 m^2 and revealed fine-scale coral ridges up to 20 m high that were not resolved by the reconnaissance maps. The AUV maps, and other remotely acquired data, were ground-truthed with submersible dives to produce an integrated, geo-referenced dataset. Spatial and quantitative analyses were applied to this dataset in order to characterize the morphology and distribution of coral-building features in each surveyed site. In the Miami Terrace site, where corals build low-relief ridges, a habitat classification map and spatial analyses show that coral patches preferentially grow on and along the northern sides of the ridges. A southward flowing bottom current, measured by the AUV, dictates the observed asymmetrical coral distribution. In the site on the lower slope of Great Bahama Bank, where corals form individual mounds, morphometric analyses show a lack of correlation between bottom current regime and mound morphology. Results from these analyses indicate that the two cold-water coral sites in the Straits of Florida are highly variable in terms of coral distribution, spatial parameters, and current regime. Given its high-resolution, the approach presented here is ideal for determining the biophysical processes that underlie these and other remote, fragile ecosystems. Assessment and monitoring of coral distribution and mound abundance based on geophysical data is crucial for managing cold-water coral habitats and is an important research priority.

10.1 Introduction

Scleractinian (stony) cold-water corals are branching, colonial organisms (Fig. 10.1) that are distributed throughout aphotic water depths of 50–3,000 m (Freiwald et al. 1997; Roberts et al. 2006). These corals can baffle and trap mobile sediments to form large mounds, which in addition to corals contain numerous organisms such as sponges, hydroids and anemones (Mullins et al. 1981; Roger 1999; Reed et al. 2006; Roberts et al. 2006). Mounds are also habitats for mobile species including economically important populations of fishes (Fosså et al. 2002; Reed 2002; Costello et al. 2005). Cold-water coral mound fields differ among geographic locations in terms of their size-frequency distribution, morphology, spatial pattern, and the relative exposure or burial of the mounds they contain (Neumann et al. 1977; Del Mol et al. 2002; Huvenne et al. 2003; Wheeler et al. 2007; Correa et al. 2011). Within a given field, individual mounds can also vary significantly in height from 1 to 300 m, and in shape from individual cone-like forms to elongated features (Van Weering et al. 2003; Wheeler et al. 2005a; Grasmueck et al. 2006). This variability is the result of local hydrodynamics, antecedent topography, and/or other factors, such as sedimentation rate (White et al. 2005; Mienis et al. 2007; Dorschel et al. 2007; Correa et al. 2011).

Fig. 10.1 **a** Branching thickets of cold-water scleractinian corals (mostly *Lophelia pertusa* and *Enallopsammia* spp.) at the Miami Terrace, Straits of Florida. **b** Close-up of live (bright white) corals growing atop a dense framework of dead coral skeletons. **c** Individual mound on the slope of Great Bahama Bank. *White line* represents the submersible track and the *yellow star* the location of image **d**, which shows dense cold-water coral framework on the mound flanks. Where visible, two green laser dots are 0.25 m apart

Few detailed maps of these ecosystems are available since most coral fields are at relatively inaccessible depths of 500–1,000 m. Conventional optical remote sensing tools, which have been useful for mapping shallow-water coral reefs, cannot typically map environments at depths >30 m, because of high light absorption through the water column. Until recently, the only mapping tools that could be applied to deep environments were submersibles and single-beam acoustic sounders, which both produce spatially limited data. The development of new acoustic survey techniques, however, has allowed data acquisition for large areas covered with cold-water corals. For example, the wide swath corridors of side-scan and multi-beam sonars have allowed scientists to explore the extent and variability of these ecosystems, as well as to observe anthropogenic impacts to these environments (e.g., destruction from trawling; Wheeler et al. 2005b; Roberts et al. 2006). These ship-based acoustic tools have produced maps of greater area and increased quality, but still have low positioning accuracy (~50 m; Wheeler et al. 2005a) and/or reasonably coarse resolution (~30 m; Guinan et al. 2009).

It is important to be aware of such limitations, the situations in which they are most likely to occur, and the extent of our ability to correct associated errors.

Given that each acoustic mapping tool has different strengths and weaknesses, integrated surveys combining multiple acoustic mapping tools along with geo-referenced environmental parameters and 'ground-truth' data represent an optimal approach for studying cold-water coral ecosystems. This chapter describes a survey approach using both an autonomous underwater vehicle (AUV) and submersible dives. The AUV acquires geo-referenced parameters (i.e., bathymetry, backscatter, current regime data, sub-bottom profiles, and measurements of chemical-physical water properties) that are unprecedented in quality and resolution. These parameters are then ground-truthed via submersible video transects and bottom samples. Two case studies from the Straits of Florida are described here as examples of quantitative analyses that can be performed with such datasets, and to illustrate the level of resolution (e.g., sub-meter or decameter scales) necessary for studying cold-water coral ecosystems.

10.2 History of Mapping Cold-Water Coral Habitats

Samples of cold-water coral species have been retrieved from the seafloor via dredging for over a century (Pourtales 1868; Cairns 1979). However, it was not until the 1960s that single-beam echosounding surveys revealed that cold-water corals form high relief mounds, similar to the reefs generated by shallow water corals in the tropics and sub-tropics (Teichert 1958; Stetson et al. 1962). Following this discovery, submersible dives were used to investigate the distribution of cold-water corals and associated fauna across mounds (Neumann and Ball 1970; Neumann et al. 1977; Reed 1980; Messing et al. 1990). Hypotheses regarding mound morphology and its relation to the local bottom current were important outcomes of these surveys. The Neumann et al. (1977) description of streamlined mounds aligned parallel to the northward flowing Florida Current on the Bahama Bank slope became a model for deep-water mounds, and strongly influenced subsequent studies in the field. Despite this progress in characterizing cold-water coral fields, the precise locations of coral mounds, their sizes, and abundances remained poorly understood due to the limited spatial coverage of early surveys.

Concurrent advances in the application of industrial seismic data to deep-water environments revealed additional sites covered by cold-water coral mounds, and also stimulated new hypotheses regarding the processes controlling mound distribution and development (e.g., Hovland et al. 1994; Del Mol et al. 2002). In some areas, tectonic faults were identified and mapped beneath mound structures. These faults were interpreted to be conduits of hydrocarbons (mainly methane) that serve as a food basis for cold-water corals and associated fauna (Hovland et al. 1990, 1994). High-resolution seismic data often showed that mounds were rooted on truncated reflectors formed during erosive events, indicating that bottom current influences the initiation and distribution of coral mounds (Del Mol et al. 2002; van Weering et al. 2003; van Rooij et al. 2003).

The finding of large mound structures along the European margin (e.g., Kenyon et al. 2003) as well as the improvement of acoustic mapping tools triggered a new phase in the study of cold-water coral ecosystems in the last decade. Deep-towed side-scan sonar and more recently multi-beam systems are now the most common acoustic sensors used to map these habitats (Paul et al. 2000; Huvenne et al. 2002; Foubert et al. 2005; Wheeler et al. 2005a; Roberts et al. 2005; Mienis et al. 2006; Dolan et al. 2008; Guinan et al. 2009; Dorschel et al. 2009). Both sensors can cover a reasonable subset (e.g., tens of km^2) of the spatial area of a typical cold-water coral field. However, the datasets that these sensors generate are often limited by a poor underwater positioning system and/or coarse resolution.

Side-scan sonar (SSS) produces images of the seabed by transmitting acoustic waves through side antennas that intercept the seafloor at grazing incidence (Blondel 2009). SSS is frequently used to map cold-water coral ecosystems because it can readily differentiate coral habitats from the surrounding seabed (Fosså et al. 2005). Coral habitats produce high acoustic amplitudes and thus are particularly visible in SSS images against a soft and smooth seabed, which are characterized by relatively low amplitudes. Most SSS sensors are deep-towed near the seafloor which allows the use of higher frequencies. As frequency increases, the acoustic wavelength shortens and so does the time interval between two successive measurements of acoustic energy (i.e., ping rate). Thus, towing the sensor near the seabed makes it possible to generate maps at meter-resolution (e.g., Mienis et al. 2006). The resulting layback (i.e., distance) between the sensor and the mother ship, however, creates operational challenges that can affect survey effectiveness and data quality (Northcutt et al. 2000).

Pulling the 'towfish' on which the SSS is mounted requires an extensive amount of tow cable (up to 10,000 m), which substantially increases drag on the mother ship (Northcutt et al. 2000). This limits survey speed to $\sim$2.5 knots, restricting the area that can be covered in a given cruise, and thus reducing the cost effectiveness of deep-towed SSS surveys (Northcutt et al. 2000). Traditionally, the towfish position is calculated using trigonometric relationships between cable length, towfish depth, and ship speed. However, sinuosity of the ship's track and drifting of the sensor due to strong ocean bottom currents, which are often observed in cold-water coral fields, can result in significant positioning errors (e.g., $\pm$50 m; Wheeler et al. 2005a). One strategy for reducing these errors is to place an additional acoustic beacon on the cable or the towfish. This beacon relays the position of the towfish to the mother ship; however, this extra beacon can add noise to the collected data due to acoustic interference and/or can destabilize the cable when directly attached to it (Fosså et al. 2005). Alternatively, the towfish position can be acquired by placing an encompassing grid of acoustic transponders on the seafloor, but this requires significant time and finances (Blondel and Murton 1997).

Another major limitation of acoustic images produced from SSS sensors is that they do not acquire topographic data (see Chap. 8), although the height of objects such as mounds can be roughly estimated from trigonometry (Blondel 2009).

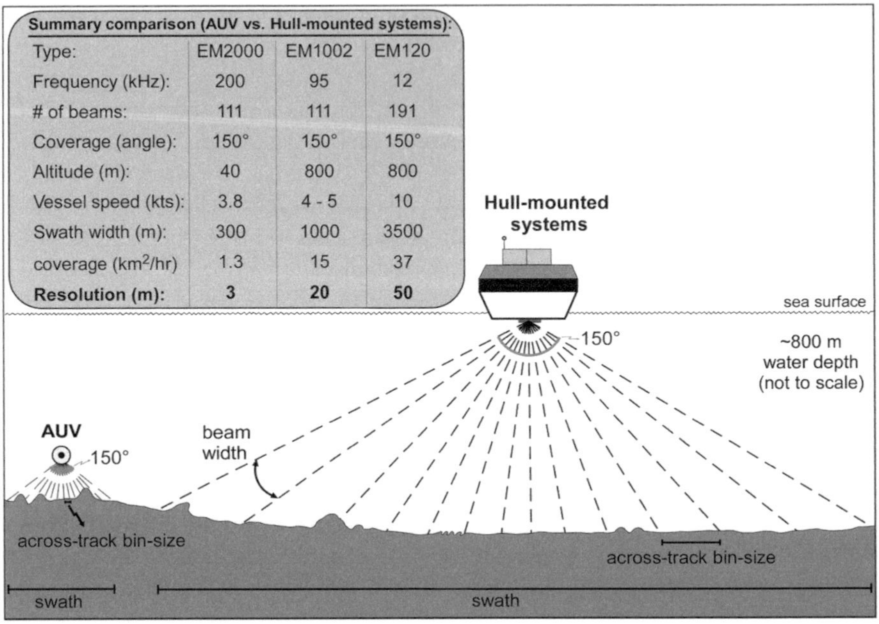

Fig. 10.2 Schematic diagram illustrating the differences in geometries between the hull-mounted (EM120 and EM1002) and the AUV-mounted (EM2000) multi-beam systems. Inset *grey* box provides comparative information for these multi-Beam systems (figure adapted from Courtney and Shaw 2000)

Without accurate topographic information, however, it can be difficult to reconcile variations observed between video transects and seabed acoustic response in a cold-water coral habitat (Dolan et al. 2008). In summary, despite the large coverage and the high resolution of the maps acquired using deep-towed SSS, the use of this sensor alone is not an efficient approach for mapping cold-water corals due to poor positioning accuracy, lack of topographic data, and operational challenges related to this type of survey.

Multi-beam sonar is a more recent system used for mapping cold-water coral habitats. Multi-beam systems have traditionally been hull-mounted on the mother ship (Roberts et al. 2005; Dolan et al. 2008; Guinan et al. 2009) and more recently on ROV (i.e., remotely operated vehicle; Foubert et al. 2011). Multi-beam systems transmit acoustic waves in a fan-like configuration that is perpendicular to the vessel track (Fig. 10.2). Depth soundings are recorded within each beam, creating a 3D digital elevation model (DEM) with an area much larger and more contiguous than results from single-beam surveys. In addition to bathymetry, multi-beam systems can record the amplitude of backscattered signals. Like SSS images, the backscatter amplitudes can be used to infer superficial sediment type, and to distinguish soft sediment seabed from coral habitats (e.g., Fosså et al. 2005).

The number and spacing of beams in multi-beam sonar varies among models and manufacturers. One of the most commonly used hull-mounted multi-beams, the 12 kHz Simrad EM 120 system, contains 191 beams over a range of 150°, producing a beam width of $\sim 1°$ in cross-track direction (Kongsberg 2005). The bin-size resolution (i.e., the seafloor area over which sounding occurs and backscatter is measured) and the swath width of the DEM depend on the width of the beams when they encounter the seabed (Fig. 10.2). Beam width expands as distance increases between the sensor and the seafloor. On the hull-mounted multi-beam systems, the sensor is far from the cold-water coral habitats they measure. Thus hull-mounted systems have generated maps with relatively large swath but low resolution (Fig. 10.2).

Advantages of mounting the system on the mother vessel are that the ship's motion, heading, roll, pitch, heave, and position can be accurately measured by ancillary sensors and GPS receivers (Courtney and Shaw 2000). These ship-related measurements can be used to correct the multi-beam data and increase DEM quality. Even with these corrections, however, hull-mounted multi-beam sonar systems produce cold-water coral habitat maps that are approximately 10-fold coarser in resolution than the maps acquired by deep-towed SSS platforms. In addition, the acoustic image recorded by the hull-mounted multi-beam system is usually of lower quality in terms of backscatter intensity contrast than the data produced by deep-towed SSS systems. This is mainly because the incident angle in hull-mounted multi-beam systems is more variable in multi-beam systems relative to those from deep-towed SSS, and the measured reflectivity is averaged within each beam across its entire swath (Lurton 2002; Fosså et al. 2005).

Taken together, deep-towed SSS and hull-mounted multi-beam sonars have different advantages and limitations: SSS is ideal for imaging the seabed, whereas multi-beam provides accurate topography. Together these data are critical for assessing the distribution of cold-water coral habitats and the environmental forces that influence them. Therefore, the most powerful survey approach to produce both high quality and fine-scale resolution topographic and sonar maps is to combine both sensors in a single vehicle that can: (1) travel autonomously near the seafloor, and (2) carry internal sensors that precisely measure the motion and position of the platform. AUVs and ROVs have no umbilical cable to the mother vessel. AUVs are typically torpedo-shaped platforms and therefore can maneuver on a more exact route and at higher speeds relative to ROVs and deep-towed SSS (George et al. 2003). For example, AUVs can travel up to 4 knots, approximately twice as fast as most deep-towed SSS surveys and ROVs (Northcut et al. 2000). The autonomy of an AUV also increases survey efficiency relative to deep-towed platforms. For example, in a gridded survey, a vehicle must turn 180° when it reaches the end of each line, and begin a new line adjacent to the previous line (i.e., line turn). The line turn duration of an AUV is about 5 min; the same procedure can take up to six hours in a deep-towed survey (Northcut et al. 2000). Furthermore, AUVs can employ multiple navigation and mapping tools in order to acquire simultaneously bathymetry, backscatter data, and sub-surface profiles as well as environmental parameters such as current data and temperature. Thus, AUVs can produce highly informative and accurate integrated datasets.

10.3 Cold-Water Coral Mapping Example

10.3.1 Sonar and AUV Configuration

AUVs vary in their mapping capabilities. Here we describe one possible configuration of the C-Surveyor-II AUV from C&C Technologies Inc. (Lafayette, Louisiana, USA). A crucial component of the C-Surveyor-II AUV is its inertial navigation system, which allows a post-processing positioning accuracy of 3 m, given a survey conducted at 800 m water depth (Jalving et al. 2003). A Kalman filter combines the AUV orientation, speed over ground, and depth outputs obtained from a gyrocompass, a 300 kHz acoustic doppler current profiler (ADCP), and a high precision pressure sensor, respectively. Positioning drift is minimized with measurements from an ultra short baseline (USBL) acoustic system and a differential global positioning system installed on the deployment vessel (Chance and Northcutt 2001; George 2006). The AUV is also equipped with obstacle-avoidance sonar that enables the vehicle to detect and respond to abrupt changes in relief. Mission endurance is up to 55 h due to the aluminium oxygen fuel-cell batteries carried in the AUV (George 2006). Considering the AUV's travelling speed of ~3.8 kts, approximately 50–60 km^2 can be mapped per mission, with parallel line spacing of 200 m.

The seabed mapping sensors on the C-Surveyor-II AUV used in this study consisted of a 200 kHz Simrad EM2000 multi-beam sonar and a 120 kHz Edgetech SSS system. The multi-beam system contains 111 beams across a 300 m swath, with 100 m of overlap between swaths. Overlapping the swaths is done to increase sounding accuracy at the swath edges, since the outermost beams are usually of lower quality (Lurton 2002). To further increase multi-beam data quality, raw echo sounding data are corrected for heave, pitch, and roll of the vehicle using the measurements recorded by the inertial navigation sensors in the AUV. Although the multi-beam system itself is capable of producing a DEM of 1 m bin size, positioning accuracy limits gridding to a 3 m bin size (George 2006). The SSS system transmits acoustic waves at about three times per second, which results in an along-track ping distance of ~60 cm (AUV speed of 3.8 kts). The associated SSS swath is about 400 m, with a 200 m overlap for each line.

The AUV also acquires sub-bottom profiles and environmental parameters throughout each mission. A 3 kHz Edgetech Chirp system collects sub-bottom profiles at 200 m spacing with a time penetration of up to 40 m (Correa et al. 2011, 2012). Temperature and salinity are also measured as seawater runs continuously through a SEB FastCAT CTD sensor. In addition to acquiring the AUV ground velocity, the ADCP data (combined with gyrocompass measurements) are also used to determine the velocity vector of the 40 m water column between the vehicle and the seafloor at one second intervals.

The cruising altitude (40 m) of the AUV is too high for video ground-truthing of the seafloor; these data must be collected separately. In the surveys described in subsequent sections of this chapter, ground-truthing was performed using the

Johnson-Sea-Link II submersible. The submersible's acrylic sphere provides observers with a field of view >180°, while an external pan and tilt video camera can record images along the track. In addition, the submersible is equipped with a manipulator arm adapted to grab, scoop, or suck up target samples such as rock slabs, sediments and organisms. The submersible navigates with an USBL positioning system that records the submersible's real-time position every 4 s (Reed et al. 2006). Analysis of USBL tracking accuracy for a worst-case tracking scenario estimated a maximum statistical positioning error of 9.6 m at a depth of 500 m (Opderbecke 1997). This positioning error is most easily accounted for and corrected in sites that contain variable topographic relief. During data analysis, regular changes in relief allow the observer to better connect actual submersible position with the high-resolution DEM. Overall, video ground-truthing is performed with greatest confidence on habitat patches that have radii larger than the statistical positioning error of the submersible (i.e., <9.6 m). Despite these positioning limitations of the submersible, the combination of the video ground-truthing with the AUV data permits cold-water coral habitats to be characterized in terms of terrain, benthic fauna, sedimentary features, and environmental parameters at unprecedented resolution.

10.3.2 Survey Design and Data Analysis

The workflow for the surveyed sites described in this chapter consists of reconnaissance mapping, AUV survey, submersible ground-truthing, data processing, and spatial and quantitative analyses of the integrated dataset. Reconnaissance mapping was performed with hull-mounted multi-beam sonar systems, EM120 and EM1002, which operate at 12 and 95 kHz frequencies, respectively (Table 10.1). There is an inherent trade-off between operating the EM120 versus the EM1002. The EM120 system covers a larger area due to its wider swath and faster survey speeds, but produces a DEM map of lower resolution than the EM1002 (Fig. 10.2; Table 10.1). Given this trade-off, the EM120 sensor was employed to explore areas in which coral colonies and/or mounds had not been documented previously; whereas in areas known to contain coral colonies and mounds, the EM1002 was utilized to conduct more detailed reconnaissance surveys. Based on the reconnaissance maps, two sites in the Straits of Florida were selected for the C-Surveyor-II AUV deployment (Fig. 10.3). The sites are located at the base of the Miami Terrace and on the slope of Great Bahama Bank (GBB), ranging in water depths from 580 to 870 m (Fig. 10.3). Together these sites comprise 75 km^2 of high-resolution mapped area, which were subsequently ground-truthed by a total of six submersible transects ($\sim$8 km long).

The multi-beam datasets from the hull-mounted multi-beam systems were rendered to produce DEMs of each study area. The EM120 and EM1002 multibeam systems generated maps with bin-size resolution of 50 and 20 m, respectively (Table 10.1). In contrast, the AUV-mounted EM2000 system produced a

Table 10.1 Summary of the survey platforms used for surveying the Grand Bahama Bank and Miami Terrace study areas

Survey platform	Vessel speed	Sensor (frequency)	Swath width/resolution/sampling	Output	Objective
Hull-mounted Multi-beam (Depth ~800 m)	10 kts	Simrad EM 120 Multi-beam (12 kHz)	3500 m width; 50 m bin-size	DEM	Reconnaissance mapping
Hull-mounted multi-beam (depth ~800 m)	4–5 kts	Simrad EM 1002 multi-beam (95 kHz)	1000 m width; 20 m bin-size	DEM and acoustic image	Reconnaissance mapping
C-Surveyor-II AUV (40 m above seabed)	3.8 kts	Simrad EM 2000 multi-beam (200 kHz)	300 m width (100 m overlap); 3 m bin-size	DEM and acoustic image	High-resolution mapping and spatial analyses
		Edgetech side-scan sonar (120 kHz)	400 m width (200 m overlap); 0.5 m bin-size	Acoustic image	Habitat classification map
		Edgetech chirp profiler (3 kHz)	200 m line spacing	Sub-bottom profiles	Sedimentation rates and echo-character
		ADCP	Sampling every second	Current velocity, direction	AUV positioning; bottom current
		FastCAT Conductivity-Temperature-Salinity (CTD)	Sampling every second	Temperature, Salinity	Water mass property
Johnson-Sea-Link II Submersible	<1 kts	Video camera, manipulator-arm	Maximum positioning error of 9.6 m (at 500 m depth)*	Video, bottom sampling	Ground-truthing

[a] As in Reed et al. 2006

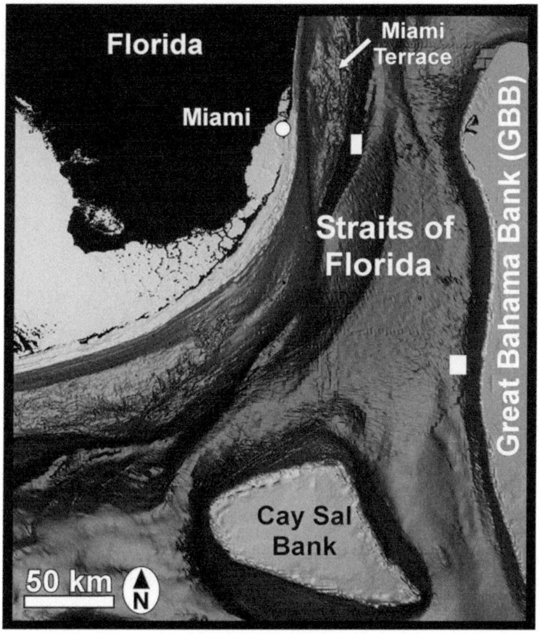

Fig. 10.3 Bathymetric map of the Straits of Florida with the location of the two surveyed sites (*white rectangles*, not to scale) at the base of the Miami Terrace and the toe-of-slope of Great Bahama Bank

DEM of 3 m bin-size resolution, as well as a 1 m resolution backscatter image of the seafloor. These acoustic images, together with the images produced from the SSS, were draped onto the DEMs. The backscatter images and DEMs were then loaded in a GIS project and 'overlaid' by the submersible track coordinates, where the timed submersible navigation tracks were fused with the timestamps of the video footage. The bottom facies types discriminated from the submersible video could then be correlated with the acoustic backscatter patterns observed on the AUV maps. These maps were used to distinguish muddy sediments from coral rubble and standing coral thickets for the mapped areas. Finally, the time and coordinates of each bottom current measurement collected by the AUV were plotted in the GIS project for spatial and temporal analyses.

The terrain, acoustic reflectivity pattern, sediment features, current regime, and coral distribution recorded in these surveys varied significantly within and between the two sites. In the site located on the GBB slope, for example, coral distribution is associated with isolated, individual mounds (Fig. 10.1c). Here, mound morphometrics (i.e., quantitative analyses used to describe mound parameters) were useful for identifying correlations between mound orientation and the prevailing current direction. In contrast, corals of the slope of the Miami Terrace site are related to low-relief ridges that cover a large part of the mapped area, and habitat classification followed by terrain spatial analyses was applied to assess relationships between coral distribution and bottom currents. Although some of the differences between sites are visually apparent on the AUV maps, others are evident only after systematic quantification of site attributes (e.g., mound footprint area and coral distribution). The variables selected for quantification were determined on a site-by-site basis as a function of the overall characteristics of each site. Both of these approaches are described in detail in the sections below, as well as an assessment of the resolution necessary to properly map cold-water coral habitats in the Straits of Florida.

10.3.3 Cold-Water Coral Mound Characterization

Cold-water coral fields in the Straits of Florida have been mainly reported as individual mound build-ups that can reach up to 50 m in relief and 1,000 m in footprint area (Neumann et al. 1977; Mullins et al. 1981; Messing et al. 1990; Paul et al. 2000; Reed et al. 2006). The distribution and spatial characteristics (e.g., size, morphology, complexity) of the mounds in these fields are poorly described due to the coarse resolution of available maps. To determine the minimum level of resolution required for accurately assessing cold-water coral mounds in the Straits of Florida, this study analyzed mound size-frequency distribution of three different DEMs (50, 20, and 3 m grid-size resolution; Table 10.1) collected over the 47 km^2-surveyed area of the GBB site.

Mounds are visible on all three DEM maps of the GBB site (Fig. 10.4). Mound perimeters are, however, complex and difficult to consistently define. To systematically assess mound size-frequency distributions for each DEM, an automated mound extraction approach was developed. This approach relies on the change of slope angle between mound and surrounding area. First, slope angle maps are generated from each DEM (Fig. 10.5a), and closed polygons are created along the contour line where the slope angle exceeds 8° (Fig. 10.5b). This 8° cutoff was determined following an attempt to delineate mound perimeters manually. This manual delineation indicated that the majority of the mounds rise out of the surrounding seabed with a cutoff plane of ~8°. Because the slope angle can vary within a mound feature by more than 8°, the algorithm can also create new polygons within a given mound (Fig. 10.5b). Therefore, all polygons within another polygon were filtered out, so that only the outermost polygons would represent mound perimeters (i.e., mound footprint; Fig. 10.5c). Data from each original DEM is then removed within the areas enclosed by mound perimeter (Fig. 10.5d). The DEM is re-gridded to generate new bathymetric maps without the mounds themselves, where the vertical relief within each removed mound was interpolated from the mound perimeters (Fig. 10.5e). The newly gridded surfaces are then subtracted from the original DEMs to produce maps in which only the vertical relief within mound perimeters is displayed (Fig. 10.5f). Finally, a Matlab routine calculates the maximum thickness (i.e., height) within each mound polygon. In this study, a mound feature in any given DEM is defined as every closed polygon that is >1 m in height and has a footprint area >81 m^2. This minimum mound area for calculating morphometrics is based on the fact that a 3 × 3 pixel matrix is 81 m^2 (given that each pixel = 3 m). Smaller matrices (e.g., 2 × 2) do not contain sufficient pixels to represent free-form mound footprints.

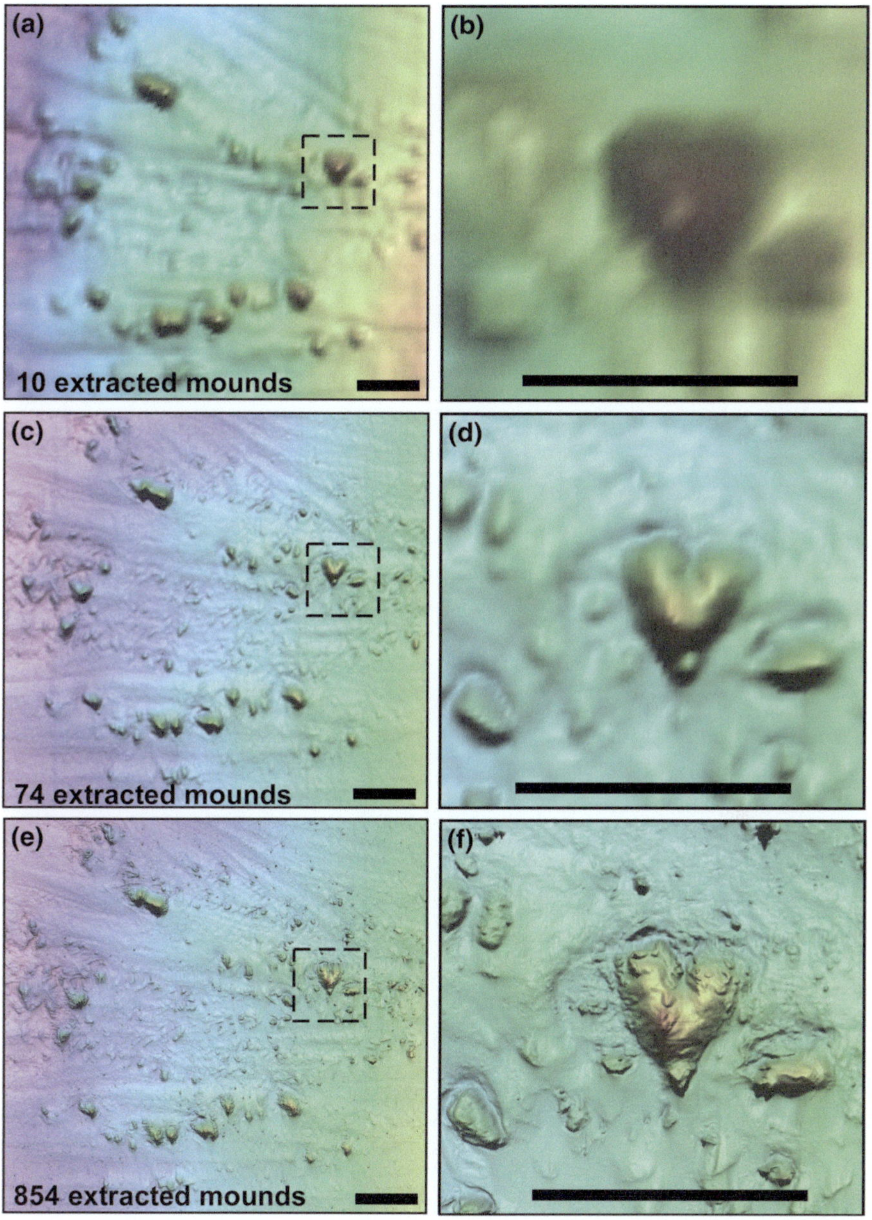

Fig. 10.4 Digital elevation models (DEMs) from the Great Bahama Bank study area (DEM overview—*left* column; DEM enlargement—*right* column), produced by multi-beam systems of different resolutions. EM120 (**a**–**b**), EM1002 (**c**–**d**), and EM2000 (**e**–**f**) contain grid resolutions of 50, 20, and 3 m, respectively. The ability to detect complex mound morphologies increases as DEM resolution increases. *Black* scale bars are 1 km

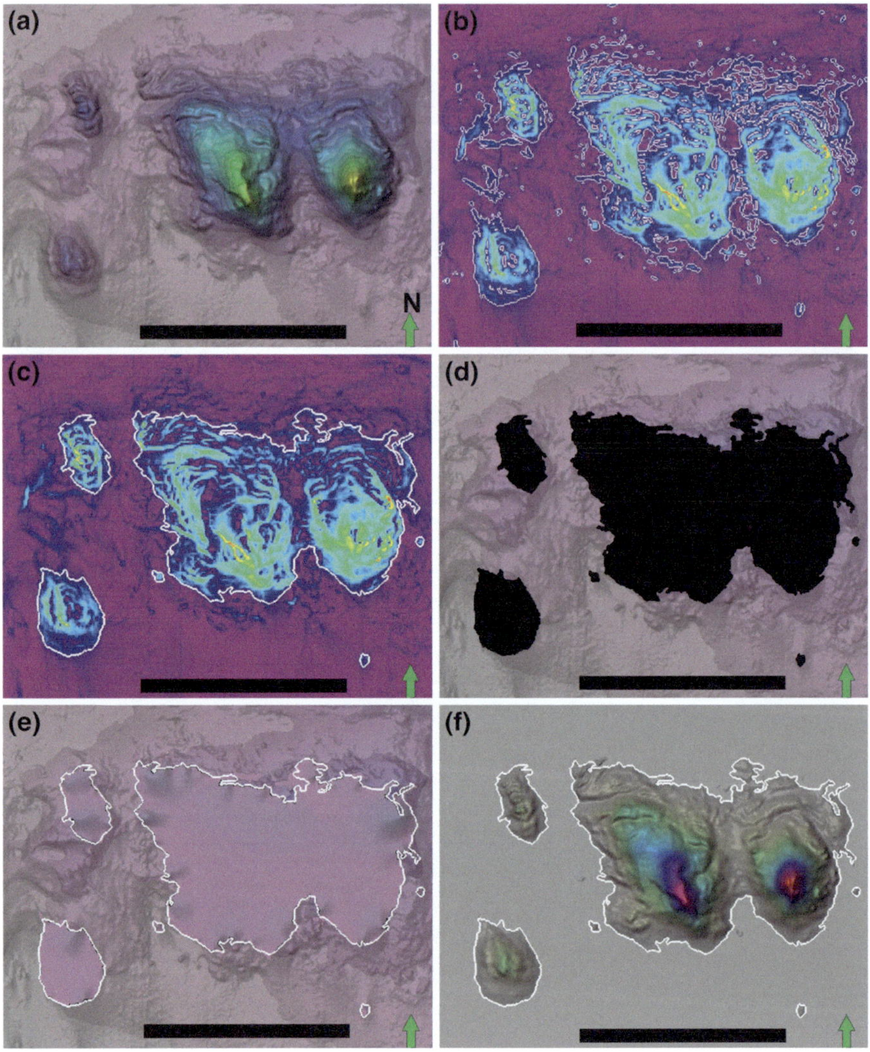

Fig. 10.5 Automated workflow to extract and delineate mound perimeters from a digital elevation model (DEM). **a** High-resolution DEM showing a mound feature in plan-view of the Great Bahama Bank study area. **b** A slope angle map created based on the DEM, where closed polygons (*white lines*) are generated along the contour line where the slope angle exceeds 8°. **c** All polygons contained within another polygon are filtered out, and the remaining polygon represents the mound perimeter (mound footprint). **d** Data from the original DEM is removed within the areas enclosed by mound perimeter. **e** The DEM is re-interpolated with mound data removed so that vertical relief at the mound perimeter is extended to fill the entire mound area. **f** The re-interpolated surface is then subtracted from the original DEM to produce a map in which vertical relief within mound perimeter is displayed in meters. *Black* scale bars are 500 m

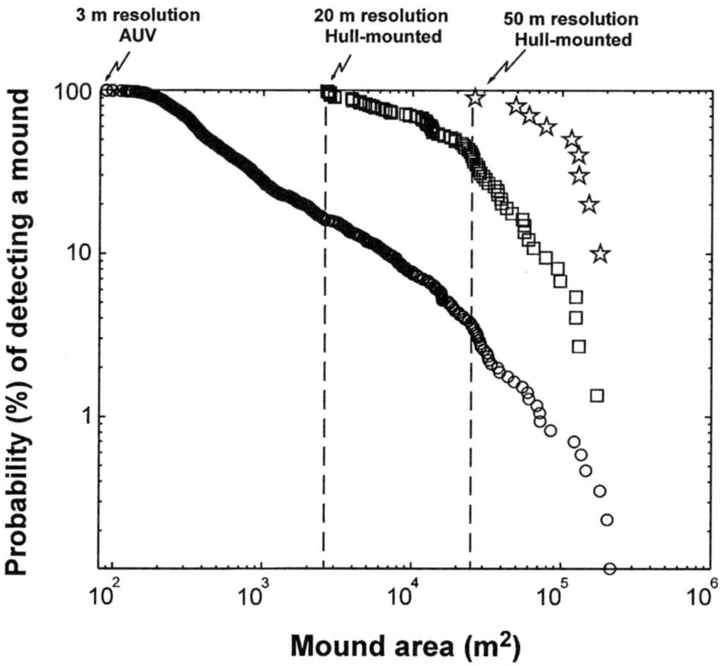

Fig. 10.6 Bi-logarithmic plot of mound footprint area versus exceedance probability for cold-water coral mounds on the Great Bahama Bank extracted from DEMs derived using three different multi-beam systems (3, 20 and 50 m resolution). The exceedance probability (y-axis) represents the likelihood that a given mound will be equal to or greater than a given area (x-axis). The three resolutions contain similar mound size-frequency distributions, with higher probability of observing small mounds than large ones. The 3 m resolution probability curve, collected by the AUV, covers greater mound size range and frequency than the maps collected using the hull-mounted systems (20 m and 50 m resolutions). The *dashed lines* show the threshold for minimum mound detection for each system

The DEMs from the three analyzed multi-beam datasets show overall similar mound size-frequency distributions for the GBB site (Figs. 10.4 and 10.6). The size-frequency distribution is determined by plotting exceedance probability versus mound footprint area (Fig. 10.6), where exceedance probability represents the likelihood (y-axis) that a given mound will be equal to or greater than a given area (x-axis). For example, in the study set, there is a 50 % probability of encountering a mound 440 m^2 or larger, but only a 1 % chance of observing a mound equal to or larger than 60,000 m^2 in area. This indicates that a high number of small mounds and a limited number of large structures characterize the mound size-frequency distribution of the GBB site in all three DEM maps. The reconnaissance maps are, however, skewed towards the large mounds (Fig. 10.6). On the 50 m resolution DEM produced from the EM120 multi-beam, for example, 10 mounds are detected in total (Figs. 10.4 and 10.6). These mounds are all larger than 26,000 m^2. In

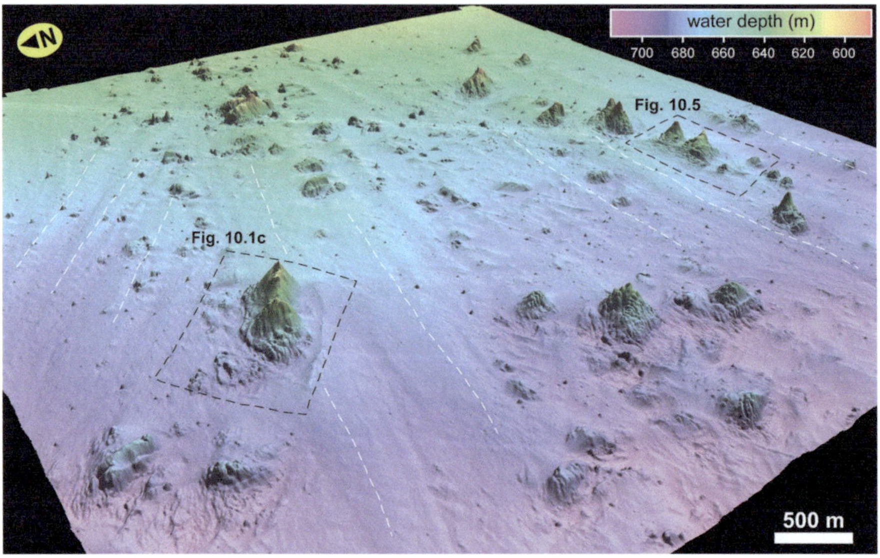

Fig. 10.7 Oblique view of a DEM near the Great Bahama Bank (GBB). Numerous mounds are visible. Mounds vary significantly in size and shape. *Dashed white lines* highlight the topographic highs that are up to 5 m in relief and diverge slightly westward (basinward)

contrast, on the 20 m resolution DEM derived from the EM120, only mounds larger than 2,600 m^2 could be detected (Fig. 10.6). As an additional comparison, on the EM120 survey the GBB site contains a total of just 74 mounds (Fig. 10.6). On the 3 m resolution AUV-derived DEM, however, 854 mounds are identified, with mounds as small as 81 m^2 in footprint area (Figs. 10.6 and 10.7).

The ecological role(s) of the small mounds have yet to be fully understood. However, Correa et al. (2011) assess the influence of sedimentary regime on mound distribution based on the same high-resolution AUV-derived maps. Small mounds are absent in areas experiencing high sedimentary rates, but present where sedimentation is moderate to low (Correa et al. 2011). This finding was not possible using data from just the 50 or 20 m resolution DEMs, since the small mounds were not extracted from the areas experiencing moderate to low sedimentation. Figure 10.4 illustrates that changes in resolution further affect the ability to accurately observe mound morphology and complexity. As DEM resolution increases, so does the ability to detect complex mound morphologies and mounds with small footprint areas (Figs. 10.4 and 10.6). Thus, these findings show that the 3 m resolution AUV-derived maps are necessary to study cold-water coral mound distribution and morphology on the Straits of Florida. This allows morphometrics to be calculated on the high resolution AUV dataset to assess correlations between mound morphology (i.e., shape and orientation) and the prevailing bottom current direction on the GBB site.

10.3.4 Mound Morphometrics

Morphometrics (e.g., height, shape, orientation, etc.) of the 854 mounds analyzed from the high-resolution map of the GBB site (47 km^2 total area) show that mounds range from 81 to ~268,000 m^2 in footprint area and from 1 to 83 m in height. Mounds in this site are mostly located on a series of topographic highs that are up to 5 m in relief, 500–1500 m wide and extend along a divergent pattern in an east–west direction (Fig. 10.8). The nature of these highs are unknown but are commonly observed on the slope of carbonate platforms (e.g., Mullins et al. 1984). The highest mounds at the GBB site are located on these topographic highs (Figs. 10.1c and 10.7), but large mounds up to 60 m in relief are also present within lows (Fig. 10.7). The largest mound was ground-truthed by submersible and was found to contain dense coral colonies of predominantly *Lophelia pertusa* and *Enallopsammia profunda* (Fig. 10.1c–d).

The footprint shape of each mound was quantified in terms of the ratio between its longest and shortest diameters (i.e., principle axes ratio). For this parameter, a mound with a ratio of 1 is circular, while an elongated ellipsoidal mound has a ratio approaching 0.5, and a long linear mound has a ratio close to 0.1 (e.g., Purkis et al. 2007; Correa et al. 2011). For the mounds that have a ratio <1, the azimuth of their longest axis is calculated (i.e., footprint orientation). Analyses of the mound footprint shape and orientation show that no correlation exists between area and shape (Fig. 10.8a). This indicates that mounds do not become streamlined as their area increases. A predominant mound orientation is also not documented (Fig. 10.8b).

The AUV measured bottom current data depict a north–south (N–S) flowing current that reverses approximately every 6 h over 45 h (Fig. 10.8c). Grasmueck et al. (2006), using the same current data, show that the changes in bottom current direction correlate with the modeled curve for the nearby North Bimini tidal gauge. This indicates that a diurnal tide is the dominant current regime along on the GBB slope. The current data therefore show that there is no preferred alignment of the mounds along the dominant N–S bottom current directions (Fig. 10.8). The mound footprint shapes also show no correlation with the N–S current regime. These findings therefore contradict the previous studies that interpreted mounds in the Straits of Florida as teardrop-shaped and aligned parallel to the northward current (Neumann et al. 1977; Messing et al. 1990).

10.3.5 Habitat Classification Map

The Miami Terrace site lies on the eastern side of the Florida Peninsula at water depths from ~630 to 870 m (Fig. 10.9a). The AUV high-resolution DEM reveals that the slope of the lower Miami Terrace is covered with a series of linear ridges that extend downslope for up to 2,000 m and are as much as 20 m in relief (Fig. 10.9b). The ridges are oriented perpendicularly to the terrace break and have

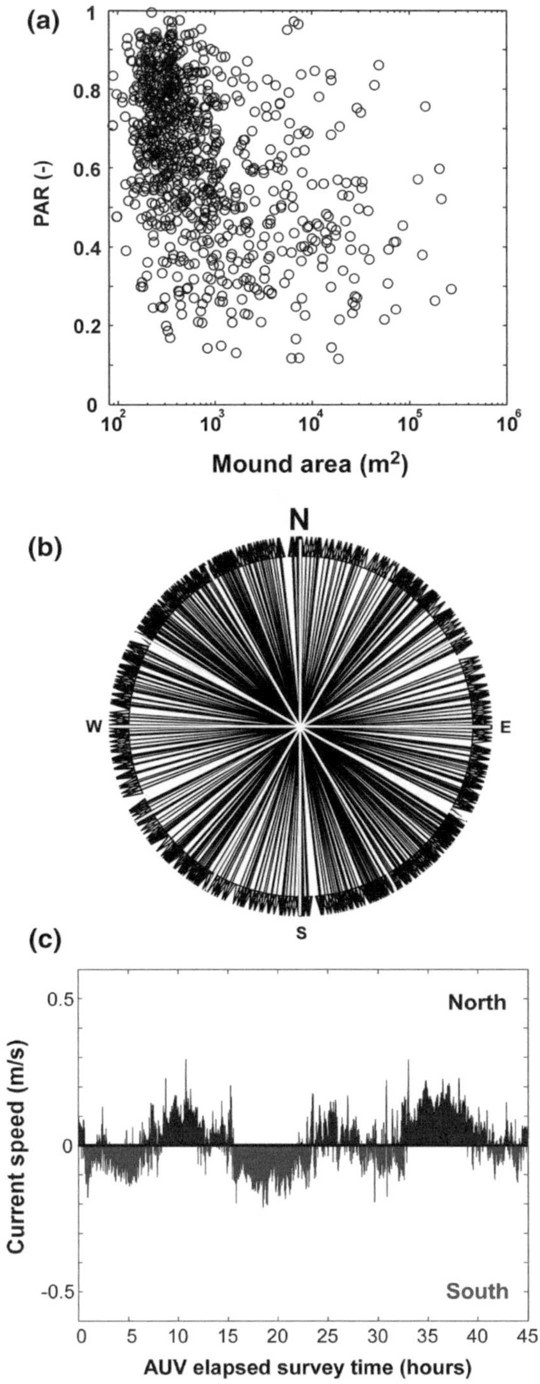

Fig. 10.8 Plots showing the morphometric analyses and current data from the Grand Bahama Bank study area. **a** Mound footprint area versus mound footprint shape. Mound shape is based on the principal axis ratio (PAR, y-axis): values approaching one indicate circular shapes, whereas values approaching zero represent elongated forms. Graph shows that mounds do not become elongated as their size increases. **b** Mound footprint orientation based on the azimuth of the principal axis (*black arrow*) of every mound analyzed. Graph shows that mounds do not have a preferential footprint orientation. **c** North–south current velocity component versus elapsed AUV survey time (x-axis). The current reverses approximately every six hours, indicating tidal bottom current regime

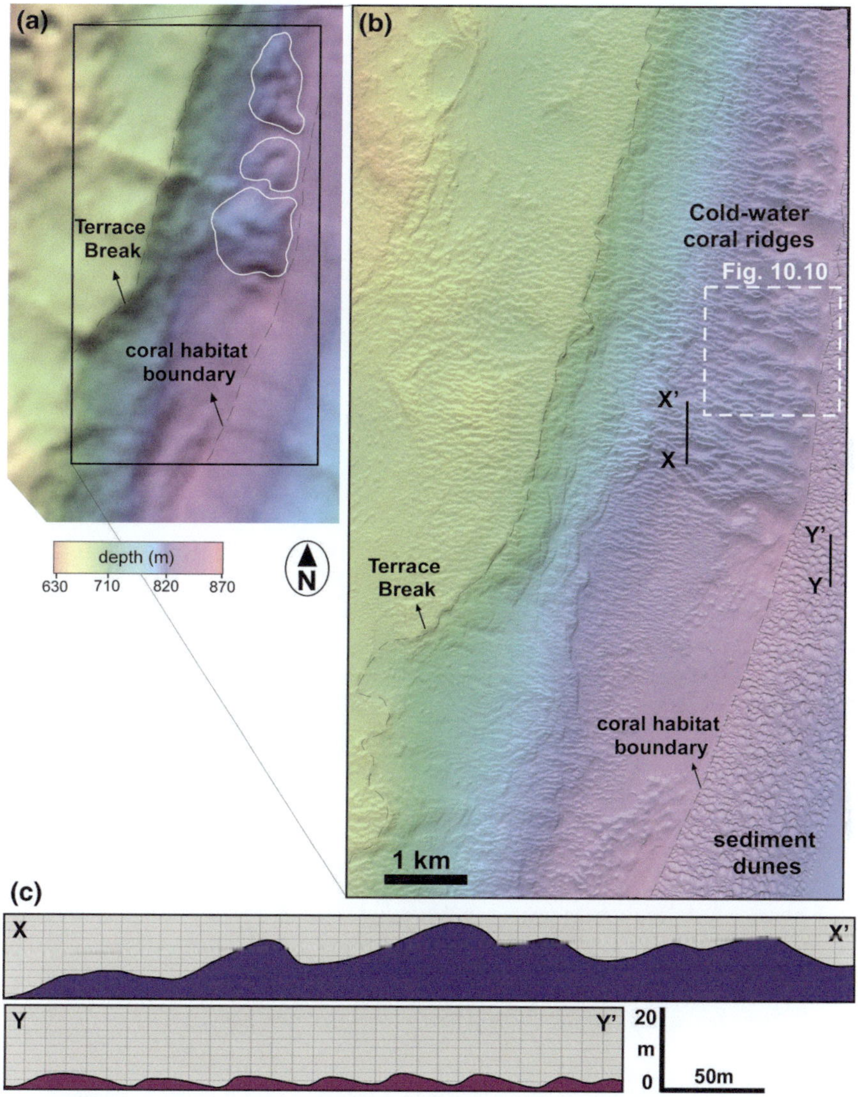

Fig. 10.9 Digital elevation models (DEMs) of different resolutions of the Miami Terrace study area. **a** The reconnaissance 50 m resolution DEM image (collected from the EM120 system) shows three large mound features (outlined in *white*) up to 1.5 km^2 in area at the base of the Miami Terrace. **b** The 3 m resolution DEM from data collected using the C-Surveyor-II AUV shows that what appeared to be mounds (based on the 50 m resolution DEM) are in fact a regular set of ridges that extend perpendicularly from the terrace break into the Straits of Florida. *Black dashed lines* indicate the terrace break and transition from coral field to sediment dunes in both **a** and **b**. The dunes appear featureless in the 50 m resolution map, indicating that only the 3 m DEM has sufficient resolution for mapping fine-scale geomorphological features at these deep sites. **c** (*bottom*) Representative profiles illustrate differences in morphology between the coral ridges (X-X') and the sand dune (Y-Y') fields

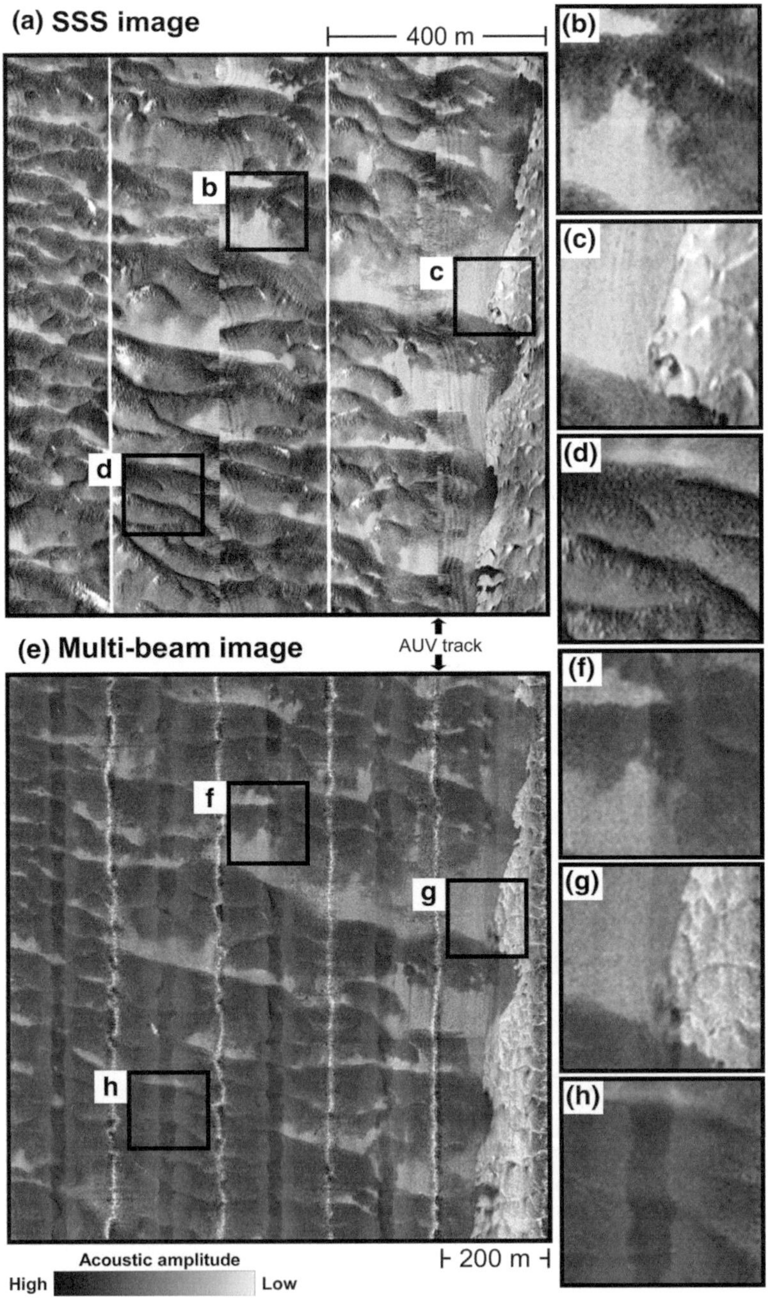
(a) SSS image
400 m
b
c
d
(b)
(c)
(d)
AUV track
(e) Multi-beam image
f
g
h
(f)
(g)
(h)
Acoustic amplitude
High
Low
200 m

◀ **Fig. 10.10** Comparison of the acoustic variability depicted in sidescan sonar (SSS) and multi-beam images captured from the same area using the C-Surveyor-II AUV at the Miami Terrace study area. **a** Acoustic SSS image and related close-up views **b–d**. **e** Acoustic image from the multi-beam system and related close-up views (**f–h**). Sharp changes in backscatter across-ridge features (at transitions from ridges to troughs) are observed in both the SSS **b**, and the multi-beam acoustic image **f**. Sharp changes in backscatter where the ridges meet the dunes are also observed in both the SSS **c**, and the multi-beam acoustic image **g**. The ridges are characterized by relatively high amplitude values, whereas the troughs between ridges and the dune field contain low amplitudes. **d** The SSS image depicts acoustic variability across individual ridges, but multi-beam acoustic image **h** does not resolve this variability. In the multi-beam images, N–S gray stripes are artefacts produced from the outmost beams while white parallel lines are the blanked nadir beams below the AUV track lines

a slightly asymmetric profile, with their steeper sides facing north (Fig. 10.9c). These fine-scale ridges, however, can only be detected in the 3 m resolution AUV map (Fig. 10.9). In the 50 m resolution reconnaissance map, ridges are resolved as three large mound features up to 1.5 km^2 in footprint area (Fig. 10.9a). The contrasting results from these two maps indicate that other areas previously surveyed solely with coarse-resolution, hull-mounted multi-beam systems might also contain cold-water coral ridges, which have previously been interpreted as individual mounds.

The coral ridges terminate abruptly against an eastward field of sediment dunes that are up to 5 m high. These dunes have their steeper sides facing south, in opposite direction to the ridges' profiles (Fig. 10.9c). Where the ridges meet the dunes, a sharp change in backscatter is observed on both the SSS and multi-beam acoustic images. A sharp backscatter change is also observed between the ridges and the troughs on both images (Fig. 10.10). These acoustic changes are visible because the ridges are characterized by high amplitude values, relative to the low reflectivity values of the troughs and dunes (Fig. 10.10). On the SSS image, a gradual acoustic variability is observed across individual ridges. This contrasts with the multi-beam image that can only depict acoustic variability in areas where seabed characteristics change abruptly (i.e., ridge vs. trough and dunes; Fig. 10.10). The acoustic variability across the ridges on the SSS is subtle and therefore difficult to assess using visual interpretation alone. Automated image analyses can be applied to extract small-scale (<1 m) textural variations. The main constraint in automated image analyses of SSS data is that across-track changes in insonification angles noticeably affect image quality, especially at the nadir zones of the SSS swaths (Fig. 10.11). In this survey, SSS acoustic variability across individual ridges occurs mostly along-track (Fig. 10.11), so insonification angle error is relatively small. To reduce this error further, data were removed from the nadir area based on a 30 m buffer zone (Fig. 10.11). Although this decreased the final analyzed swath width from 400 to 370 m, it ensured that the reflectivity patterns documented within ridges represent physical differences in the seabed rather than sensor limitations.

Ground-truthing of the study area by five submersible transects revealed that cold-water coral thickets cover the ridges at the Miami Terrace site (Fig. 10.1a).

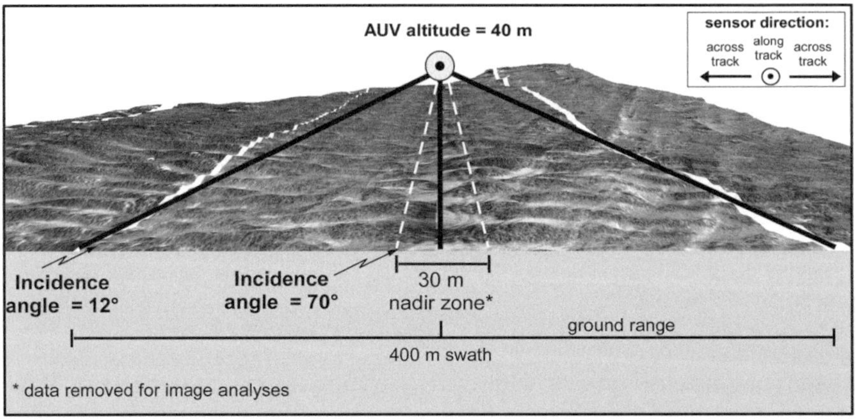

Fig. 10.11 Geometric parameters of the SSS data collected using the C-Surveyor-II AUV with background image from the Miami Terrace study area. Note that the acoustic reflectivity is distorted along the nadir zone (shown in the background image as *white lines* where the distorted data has been removed from analysis)

Most of the live and dead coral colonies are *E. profunda* and *L. pertusa* (Fig. 10.1b). The coral thickets are mainly located on the ridge crests, whereas the flanks are dominated by coral rubble that gradually dissipates into the troughs. Based on coral cover and particle grain size, six habitat classes were discriminated along the submersible transect: (1) live dense coral thickets (25–100 % of the seafloor covered by live and dead coral colonies in growth position); (2) dead dense coral thickets (25–100 % of the seafloor covered by dead coral colonies only); (3) isolated coral thickets (<25 % of the seafloor covered by dead coral colonies in growth position); (4) coral rubble (debris on sediment bottom); (5) soft mud-sized sediment (devoid of coral); and (6) coarse bioclastic sand (mostly pteropods and planktonic foraminifera). Habitat classes were assigned different color codes, and for every class, a dot of the appropriate color was plotted onto the SSS image, with a polygon then centered on each dot for every class (Fig. 10.12). From these polygons, the acoustic signature for each habitat class was extracted using ENVI image analysis software (Exelis Visual Information Solutions).

Based on this process, it was determined that five habitat classes could be distinguished acoustically. The 'live' and 'dead' dense coral thicket classes could not be acoustically differentiated from one another and were thus combined into a single 'dense coral thickets' class. The acoustic values of the five habitat classes were further used to classify the entire SSS image using a supervised classification algorithm (ENVI; Mahalanobis distance classifier). This classification approach segments the SSS image according to a pixel-by-pixel classification, whereby each classified pixel represents a small homogeneous area characterized by unique acoustic properties that are distinguishable from other classes. To convert this pixel classification into a vector-based classification (i.e., polygons), a 3 × 3 pixel

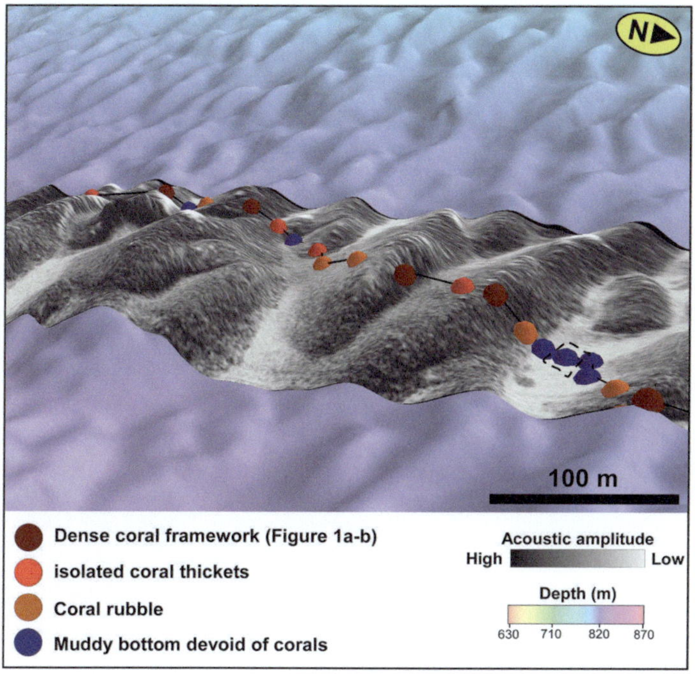

Fig. 10.12 Integrated map of sidescan sonar (SSS) image draped onto digital elevation model of the ridges from the Miami Terrace study area. The *colored dots* represent habitat classes discriminated along the submersible transect. The dense coral class is correlated with higher acoustic amplitudes on the ridge crests, whereas lower acoustic amplitudes characterize soft mud-sized sediment bottom class at the ridge topographic lows. The *black dashed* polygon (*lower right*) illustrates a representative patch selected to extract the acoustic signature of the soft mud-sized sediment bottom class

median filter was first passed over the image. This filter reduces noise in the classification by eliminating pixels without similar neighboring pixels. It also diminishes the map's original resolution from 0.5 to 1.5 m, whereby the minimum threshold for patch size analyses was finally 2.25 m^2. The resulting classified SSS image was then converted to vector format by generating polygons around groups of similar pixels. The final polygons from each class were next recorded as a single shapefile (i.e., ArcGIS vector format), and all habitat class shapefiles were draped over the DEM to yield a high-resolution, 3-D habitat classification map (e.g., dense coral thicket patches in Fig. 10.13).

The workflow described above indicates that the mapped portion of the Miami Terrace is a major cold-water coral field: the coral habitat classes (i.e., dense coral thickets, isolated coral thickets, and coral rubble) cover approximately 76 % ($\sim$13 km^2) of the site (Fig. 10.8). Coral rubble is the most abundant class (48 %), followed by dense coral thicket (16 %), and isolated coral thicket (12 %). The bioclastic sand dune field comprises 14 % of the mapped area, and mud-sized sediments cover only $\sim$8 % of the site. Analyzing only ridge features, coral rubble

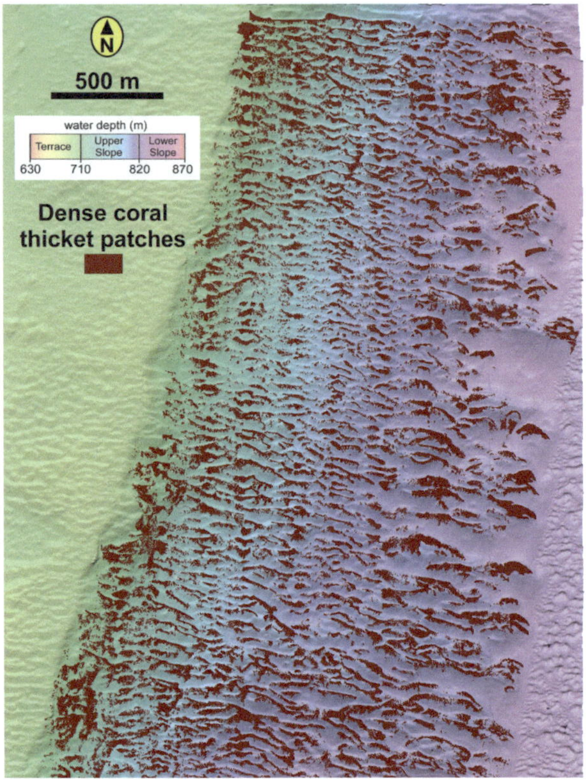

Fig. 10.13 Habitat classification map of the dense coral thicket class of the Miami Terrace study area. Maroon colored patches represent the dense coral class polygons extracted via supervised classification. These patches cover ~16 % of the analyzed area and are preferentially located on ridges' crest

is even more dominant (62 % of total ridge area), and is still followed by the dense coral class (22 %), and the isolated coral class (16 %). Visual analysis suggests that the different coral habitat classes are non-randomly distributed throughout the site. For example, the dense coral class patches are mostly on the ridge crests, whereas the troughs between ridges contain mostly patches of coral rubble or mud-sized sediment bottom, both of which are devoid of standing coral thickets.

A subset of the dense coral habitat class was further analyzed using bathymetric parameters in order to infer the controlling processes on the distribution of cold-water corals at the Miami Terrace. Bathymetric parameters such as slope angle and azimuth were calculated from the DEM using a Matlab routine. The slope angle indicates the slope of a patch at a given location, and therefore its position with respect to topographic relief (i.e., crest, slope, or trough). The azimuth shows the cardinal orientation of a patch (i.e., north, south, and so on). Together with the habitat class information these two variables can indicate whether a given habitat class exhibits a non-random distribution in a remotely sensed area.

Of the 1,086 patches of dense coral analyzed using the bathymetric parameters, the slope angle varies from 0 to 25°, with most values at 5° (Fig. 10.14a). Azimuth values are mostly within the first quadrant with a dominant orientation towards the north-northeast (Fig. 10.14b). Together, these values indicate that dense coral

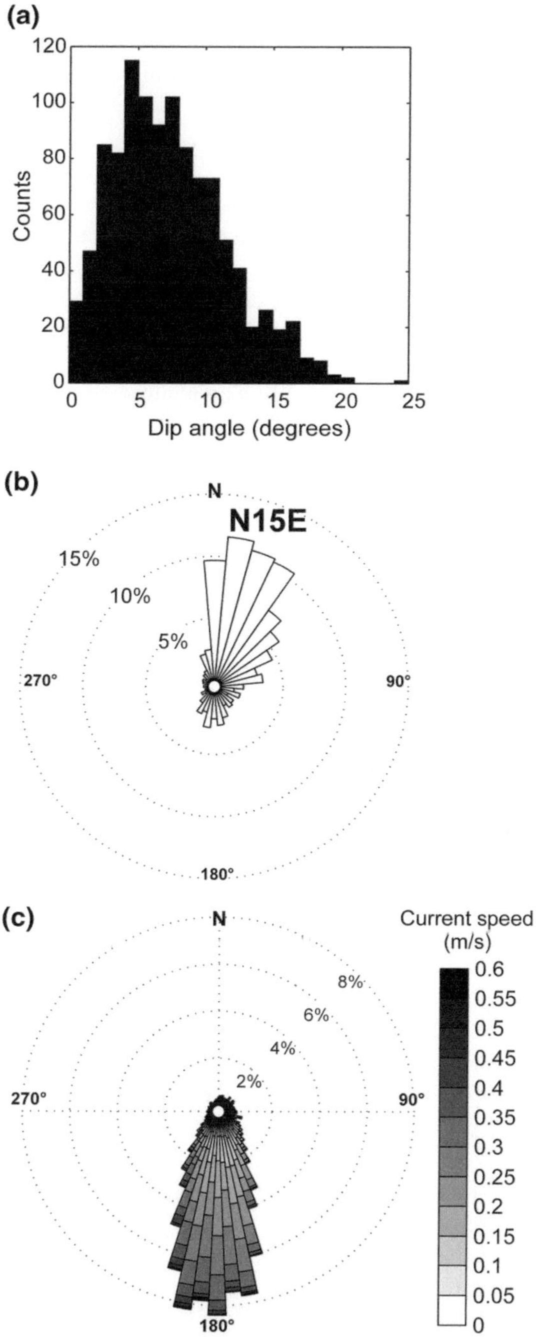

Fig. 10.14 Plots showing the spatial bathymetric analysis and current data from a subset of the Miami Terrace study area. **a** Histogram of the slope angles of the dense coral thicket patches shows that they slope between 0 and 25° with a peak at 5°. **b** Azimuth rose diagram displays the predominant slope orientation of the patches towards north-northeast. **c** Rose diagram of current directions collected every second for over continuous 24.5 h by the AUV. Predominant current direction is due south with average velocity of 18 cm/s, which is in the approximately opposite direction to the main coral patch orientation

thicket patches are mainly located on the crest of the ridges and are facing generally northwards (Fig. 10.13). The observations made from the submersible videos corroborate the remotely sensed acoustic data, and thus indicate that the asymmetric distribution of dense coral thicket patches are a defining characteristic of the ridges at the Miami Terrace.

Along the Miami Terrace site, the non-random distribution of dense corals is driven by local hydrodynamics. Bottom current data, recorded continuously by the AUV for over 24.5 h, indicates a constant southward flowing current, with an average velocity of 18 cm/s and maximum velocity of 60 cm/s (Fig. 10.14c). Thus the AUV current data corroborate previous findings of a regional deep counter-current at the base of the Miami Terrace, flowing in the opposite direction to the main northward Florida Current (Hurley and Fink 1963; Düing and Johnson 1971; Neumann and Ball 1970). The southward current flows into the northern sides of the ridges where most dense coral patches are located. The correlation between current direction and coral spatial distribution suggests that coral survivorship is highest on the current-facing sides of the Miami Terrace ridges. We hypothesize that the current-facing coral colonies intercept a greater number of food particles than the colonies positioned in the lee of the current (e.g., Messing et al. 1990; Dorschel et al. 2007). Thus the integrated dataset and spatial analyses presented here not only provide quantitative information on the distribution of cold-water corals, but also contribute to hypotheses regarding the environmental parameters that underlie coral distribution at an ecological scale. Such information is crucial for predicting where additional cold-water coral habitats most likely exist, and thus, prioritizing regions for future exploration.

10.4 Conclusions and Recommendations

This chapter described an integrated survey approach for remote sensing cold-water coral fields at regional and local scales. Based on the comparison of different mapping tools, it is clear that high-resolution (0.5–3 m) AUV-generated maps are crucial for characterizing the detailed distribution of coral ridges and mounds in the Straits of Florida. However, AUVs are costly to operate and have relatively narrow swaths. Thus, to maximize survey cost effectiveness, reconnaissance mapping at regional scales followed by deployment of the AUV in identified areas of interest is highly recommended. Hull-mounted multi-beam systems are ideal for reconnaissance surveys given their wide swaths and relatively low operating costs. In unexplored, potential cold-water coral habitats, the EM120 is recommended given that its swath width, survey speed, and thus, total area covered per survey are greatest. In areas where cold-water coral features have previously been documented, the EM1002 system is recommend for reconnaissance because it provides a moderate level of resolution while still covering a relatively large area.

In identified areas of interest, the combination of AUV-generated maps with ground-truthing data enables the quantitative assessment of cold-water coral abundance over significant areas (measuring 10 s of square kilometres). For example, in the Miami Terrace, this approach shows that coral habitats cover up to 76 % of the observed ridges ($\sim$13 km^2). Submersible observations indicate, however, that many of the 'coral thickets' detected using backscatter at this site are dead colonies. In contrast, submersible ground-truthing reveals a higher proportion of live corals at the GBB site, but that their distribution is restricted to isolated mounds. These examples demonstrate that submersible ground-truthing is critical for validating the outputs of remote sensing tools. Further, 'coral thickets' identified by the presented remote sensing workflow should be interpreted as standing coral thickets (live and dead), and not as a measure of live coral cover.

The integrated survey approach and quantitative data analyses described here have significant applications for characterizing cold-water coral ecosystems. We recommend the selection of representative sites throughout regions known to contain cold-water coral ecosystems for the acquisition of high-resolution baseline datasets, followed by long-term monitoring. Each recommended survey effort should include preliminary large-scale mapping using hull-mounted multi-beam systems to detect and/or confirm the extent and distribution of coral habitats at regional scales. Sites selected for monitoring should then be mapped using an AUV for morphological characterization of coral features. These AUV maps can also be used for accurately planning ground-truthing surveys using submersible or ROV platforms. All geo-referenced data should then be integrated in a GIS system in order to produce comprehensive coral habitat maps. Quantitative analyses can then be performed on these maps to systematically calculate coral abundance and distribution. These baseline assessments of cold-water coral habitats can critically inform long-term monitoring programs and policy efforts, such as the designation of Habitat Areas of Particular Concern (HAPCs) and Essential Fish Habitats (EFHs). Additional surveys of each designated site can then be performed and compared to baseline datasets in order to track site condition, particularly following exposure to acute (e.g., bottom trawling) and/or chronic (e.g., ocean acidification) disturbances. The application of high-resolution remote sensing and ground-truthing tools to the management of cold-water coral ecosystems represents a critical step forward in our ability to identify and protect these valuable ecosystems.

Acknowledgments The authors thank NOAA Oceans Explorer program for AUV ship time, as well as *C&C Technologies*, the crew of the R/V Northern Resolution, R/V Seward Johnson, and the JSL-II submersible. A grant provided by the State of Florida, Medicines from Florida's Oceans Project (HBOI Project # S2156, S2168) provided funding for the submersible cruises. We also thank David Viggiano for current data processing, and John K. Reed for his input on the manuscript. Post-cruise analyses are supported by the American Chemical Society Petroleum Research Fund (#49017ND8) and by the Industrial Associates of the Comparative Sedimentology Laboratory (CSL) at the University of Miami.

Suggested Reading

Freiwald A, Murray JR (2005) Cold water corals and ecosystems. Springer, Berlin, p 1243

Lurton X (2002b) An introduction to underwater acoustics: principle and application. Springer, Chichester

Roberts MJ, Wheeler AJ, Freiwald A, Cairns SD (2009) Cold-water corals: the biology and geology of deep-sea coral habitats. Cambridge University Press, New York

References

Blondel P (2009) The handbook of sidescan sonar. Springer, Chichester

Blondel P, Murton BJ (1997) Handbook of seafloor sonar imagery. Wiley-Praxis, Chichester

Cairns S (1979) The deep-water scleractinia of the Caribbean Sea and adjacent waters. Stud Fauna Curacao Other Caribbean Isl 57:1–341

Correa TBS, Grasmueck M, Eberli GP, Reed JK, Verwer K, Purkis SJ (2012) Variability of cold-water coral mounds in a high sediment input and tidal current regime, straits of Florida. Sedimentology 59:1278–1304

Correa TBS, Eberli GP, Grasmueck M, Reed KJ (2012) Genesis and morphology of cold-water coral ridges in a unidirectional current regime. Marine Geol 326–328:14–27

Courtney R, Shaw J (2000) Environmental marine geoscience 2. Multibeam bathymetry and backscatter imaging of the Canadian continental shelf. Geosci Can 27:31–42

De Mol B, Van Rensbergen P, Pillen S, Van Herreweghe K, Van Rooij D, McDonnell A, Huvenne V, Ivanov M, Swennen R, Henriet JP (2002) Large deep-water coral banks in the porcupine basin, Southwest of Ireland. Mar Geol 188:193–231

Dolan MFJ, Grehan AJ, Guinan JC, Brown C (2008) Modeling the local distribution of cold-water corals in relation to bathymetric variables: adding spatial context to deep-sea video data. Deep-sea research part I-oceanographic research papers, 55:1564–1579

Dorschel B, Hebbeln D, Foubert A, White M, Wheeler AJ (2007) Hydrodynamics and cold-water coral facies distribution related to recent sedimentary processes at Galway Mound West of Ireland. Mar Geol 244:184–195

Dorschel B, Wheeler AJ, Huvenne VAI, de Haas H (2009) Cold-water coral mounds in an erosive environmental setting: TOBI side-scan sonar data and ROV video footage from the Northwest Porcupine Bank, NE Atlantic. Mar Geol 264:218–229

Duing W, Johnson D (1971) Southward flow under the Florida current. Science 173:428–430

Fosså JH, Lindberg B, Christensen O, Lundalv T, Svellingen I, Mortensen PB, Alvsvag J (2005) Mapping of Lophelia reefs in Norway: experiences and survey methods. In: Freiwald A, Murray JR (eds) Cold water corals and ecosystems. Springer, Berlin, pp 359–391

Fosså JH, Mortensen PB, Furevik DM (2002) The deep-water coral *Lophelia Pertusa* in Norwegian waters: distribution and fishery impacts. Hydrobiologia 471:1–12

Foubert A, Beck T, Wheeler AJ, Opderbecke J, Grehan A, Klages M, Thiede J, Henriet JP (2005) New view of the Belgica mounds, Porcupine Seabight, NE Atlantic: preliminary results from the Polarstern ARK-XIX/3a ROV cruise. In: Freiwald A, Murray JR (eds) Cold water corals and ecosystems. Springer, Berlin, pp 403–415

Foubert A, Huvenne VAI, Wheeler A, Kozachenko M, Opderbecke J, Henriet JP (2011) The Moira mounds, small cold-water coral mounds in the Porcupine Seabight, NE Atlantic: part B: evaluating the impact of sediment dynamics through high-resolution ROV-borne bathymetric mapping. Mar Geol 282:65–78

Freiwald A, Henrich R, Patzold J (1997) Anatomy of deep-water coral reef mound from Stjernsund, West Finnmark. In: James NP, Clarke JAD (eds) Cool-water carbonates, vol 56. SEPM Special Publication, Tulsa, pp 741–762

George RA, Shuy JP, Cauquil E (2003) Deep-water AUV logs 25,000 km under the sea—technology provides high-quality remote sensing data for deep-water seabed engineering projects in half the time. Sea Technol 44:10–15

George RA (2006) Advances in AUV remote-sensing technology for imaging deep-water geohazards. Lead Edge 25:1478–1483

Grasmueck M, Eberli GP, Viggiano DA, Correa TBS, Rathwell G, Luo JG (2006) Autonomous Underwater Vehicle (AUV) mapping reveals coral mound distribution, morphology, and oceanography in deep water of the straits of Florida. Geophy Res Lett, vol 33. doi:10.1029/2006GL027734

Guinan J, Grehan AJ, Dolan MFJ, Brown C (2009) Quantifying relationships between video observations of cold-water coral cover and seafloor features in Rockall Trough, west of Ireland. Mar Ecol Prog Ser 375:125–138

Hovland M, Croker PF, Martin M (1994) Fault-associated seabed mounds (carbonate knolls?) off Western Ireland and Northwest Australia. Mar Pet Geol 11:232–246

Hovland M (1990) Do carbonate reefs form due to fluid seepage? Terra 2:8–18

Hurley RJ, Fink LK (1963) Ripple marks show that countercurrent exists in Florida straits. Science 139:603–605

Huvenne VAI, Blondel P, Henriet JP (2002) Textural analyses of sidescan sonar imagery from two mound provinces in the Porcupine Seabight. Mar Geol 189:323–341

Huvenne VAI, De Mol B, Henriet JP (2003) A 3D seismic study of the morphology and spatial distribution of buried coral banks in the porcupine basin, SW of Ireland. Marine Geol 198:5–25

Jalving B, Gade K, Hagen OK, Vestgård K (2003) A Toolbox of aiding techniques for the HUGIN AUV integrated inertial navigation system. In: Proceedings from Oceans, San Diego, pp 1146–1153

Lurton X (2002) An introduction to underwater acoustics: principle and application. Springer, Chichester, p 347

Kenyon NH, Akhmetzhanov AM, Wheeler AJ, van Weering TCE, de Haas H, Ivanov MK (2003) Giant carbonate mud mounds in the Southern Rockall Trough. Mar Geol 195:5–30

Kongsberg (2005) 12 KHz multi-beam echo sounder—seabed mapping to full Ocean depth, http://www.kongsberg-simrad.de, Kongsberg Newsletter. Accessed November 2010

Messing CG, Neumann AC, Lang JC (1990) Biozonation of deep-water lithoherms and associated hardgrounds in the Northeastern Straits of Florida. Palaios 5:15–33

Mienis F, van Weering T, de Haas H, de Stigter H, Huvenne VAI, Wheeler AJ (2006) Carbonate 699 mound development at the SW Rockall Trough margin based on high resolution TOBI and 700 seismic recording. Mar Geol 233:1–19

Mienis F, de Stigter HC, White M, Dulneveldc G, de Haas H, van Weering TCE (2007) Hydrodynamic controls on cold-water coral growth and carbonate-mound development at the SW and SE Rockall Trough margin, NE Atlantic Ocean. Deep-Sea Research Part I-Oceanographic Research Papers 54:1655–1674

Mullins HT, Newton CR, Heath K, Vanburen HM (1981) Modern deep-water coral mounds North of Little Bahama Bank—criteria for recognition of deep-water coral bioherms in the rock record. J Sediment Petrol 51:999–1013

Mullins HT, Heath KC, Van Buren HM, Newton CR (1984) Anatomy of a modern open ocean carbonate slope: Northern Little Bahama Bank. Sedimentology 31:141–168

Neumann AC, Ball MM (1970) Submersible observations in straits of Florida—geology and bottom currents. Geol Soc Am Bull 81:2861–2873

Neumann AC, Kofoed JW, Keller GH (1977) Lithoherms in straits of Florida. Geology 5:4–10

Northcutt JG, Kleiner AA, Chance TS, Lee J (2000) Cable route surveys utilizing autonomous underwater vehicles (AUVs). Mar Technol Soc J 34:11–16

Opderbecke J (1997) At-sea calibration of a USBL underwater vehicle positioning system, Oceans 1997 MTS/IEEE 1:721–726

Paull CK, Neumann AC, Ende BAA, Ussler W, Rodriguez NM (2000) Lithoherms on the Florida-Hatteras slope. Mar Geol 166:83–101

Pourtales LF (1868) Contributions to the fauna of the gulf stream at great depths. Bull Mus Comp Zool Harvard 1:121–142

Purkis SJ, Kohler K, Riegl BM, Rohmann SO (2007) The statistics of natural shapes in modern coral reef landscapes. J Geol 115:493–508

Reed JK (1980) Distribution and structure of deep-water *Oculina-Varicosa* coral reefs off Central Eastern Florida. Bull Mar Sci 30:667–677

Reed JK (2002) Comparison of deep-water coral reefs and lithoherms off Southeastern USA. Hydrobiologia 471:57–69

Reed JK, Weaver D, Pomponi SA (2006) Habitat and Fauna of deep-water *Lophelia pertusa* coral reefs off the Southeastern USA: blake plateau, straits of Florida, and Gulf of Mexico. Bull Mar Sci 78:343–375

Roberts JM, Brown CJ, Long D, Bates CR (2005) Acoustic mapping using a Multibeam Echosounder reveals cold-water coral reefs and surrounding habitats. Coral Reefs 24:654–669

Roberts JM, Wheeler AJ, Freiwald A (2006) Reefs of the deep: the biology and geology of cold-water coral ecosystems. Science 312:543–547

Rogers AD (1999) The Biology of *Lophelia pertusa* (LINNAEUS 1758) and other deep-water reef-forming corals and impacts from human activities. Int Rev Hydrobiol 84:315–406

Stetson TR, Squires DF, Pratt RM (1962) Coral Banks occurring in deep-water on the Blake Plateau. Am Museum Novitates 2114:1–39

Teichert C (1958) Cold- and deep-water coral banks. Am Assoc Pet Geol Bull 42:1064–1082

Van Rooij D, De Mol B, Huvenne VAI, Ivanov M, Henriet JP (2003) Seismic evidence of current-controlled sedimentation in the Belgica Mound Province, upper porcupine slope, Southwest of Ireland. Mar Geol 195:31–53

Van Weering TCE, de Haas H, de Stigter HC, Lykke-Andersen H, Kouvaev I (2003) Structure and development of giant carbonate mounds at the SW and SE Rockall Trough margins, NE Atlantic Ocean. Mar Geol 198:67–81

Wheeler AJ, Beyer A, Freiwald A, de Haas H, Huvenne VAI, Kozachenko M, Roy KOL, Opderbecke J (2007) Morphology and environment of cold-water coral carbonate mounds on the NW European margin. Int J Earth Sci 96:37–56

Wheeler AJ, Kozachenko M, Beyer A, Foubert A, Huvenne VA, Klages M, Masson DG, Olu-LeRoy K, Thiede J (2005a) Sedimentary processes and carbonate mounds in the Belgica mound province, Porcupine Seabight, NE Atlantic In: Freiwald, Murray JR (eds) Cold water corals and ecosystems, Springer, Berlin, pp 571–603

Wheeler AJ, Bett BJ, Billett DSM, Masson DG, Mayor D (2005b) The Impact of demersal trawling on NE Atlantic deep-water coral habitats: the case of the Darwin mounds, UK. In: Barnes PW, Thomas JP (eds) Benthic habitats and the effects of fishing. American Fisheries Society, Maryland, pp 807–817

White M, Mohn C, de Stigter H, Mottram G (2005) Deep-water coral development as a function of hydrodynamics and surface productivity around the submarine banks of the Rockall Trough, NE Atlantic. In: Murray JR, Freiwald A (eds) Cold water corals and ecosystems. Springer, Berlin, pp 503–514

Section IV
Thermal and Radar

Chapter 11
Thermal and Radar Overview

Scott F. Heron, Malcolm L. Heron and William G. Pichel

Abstract This chapter presents an overview of the remote sensing technologies that measure sea surface temperature (SST) based on emitted radiation (passive sensors), as well as the suite of properties that are measured using transmitting radar systems (active sensors). Remote sensing of these parameters is undertaken using the infrared, microwave and radio-wave regions of the electromagnetic spectrum. The overview describes the underlying physics for each remote sensing technique, presents issues that influence the acquisition of various parameters, and describes the environmental variables that can be monitored. Chapters 12 and 13 then describe the applications for coral reefs of thermal and radar remote sensing, respectively.

S. F. Heron (✉)
Coral Reef Watch, National Oceanic and Atmospheric Administration,
675 Ross River Rd, Townsville, Queensland 4817, Australia
e-mail: scott.heron@noaa.gov

S. F. Heron
Marine Geophysical Laboratory, Physics Department, School of Engineering and Physical Sciences, James Cook University, Townsville, Queensland 4811, Australia

M. L. Heron
Marine Geophysical Laboratory, School of Environmental and Earth Sciences,
James Cook University, Townsville, Queensland 4811, Australia
e-mail: mal.heron@ieee.org

M. L. Heron
Australian Institute of Marine Science, Townsville, Queensland 4810, Australia

W. G. Pichel
Center for Satellite Applications and Research, National Oceanic and Atmospheric Administration, NCWCP, E/RA3, Room 3229, 5830 University Research Ct,
College Park, MD 30740-3818, USA
e-mail: william.g.pichel@noaa.gov

J. A. Goodman et al. (eds.), *Coral Reef Remote Sensing*,
DOI: 10.1007/978-90-481-9292-2_11,

11.1 Introduction

The discovery of "calorific rays", now known as infrared radiation, by William Herschel in 1800 was the first step in the development of the field of thermal remote sensing. Following the landmark work of James Clerk Maxwell in electromagnetism during the 1860s and 1870s, passive sensing of emitted or reflected radiation was augmented with the ability to undertake active sensing, wherein radiation is produced and the backscatter measured. The development of Radio Detection and Ranging (RADAR) in the lead-up to World War II was significant, not only in determining the outcome of many battles, but in monitoring atmospheric conditions. From these and many other stepping stones, environmental remote sensing has developed into an area upon which our society relies every day.

11.2 Thermal Overview

11.2.1 Thermal Physical Principles

Remote sensing of temperature is undertaken by passive detection of radiation emitted by the source object. All objects with temperature above absolute zero (i.e., 0 K = −273.15 °C) emit radiation. The efficiency of emission, or emissivity (ε), is defined as the ratio of energy radiated by a body to that of a perfect emitter (blackbody) at the same temperature and is, therefore, unitless. By this definition, the emissivity of a blackbody is $\varepsilon = 1$. Emissivity can be a function of wavelength, radiation angle and temperature; however, it is often assumed that the value is constant for a particular material. It is of note that absorptivity, the efficiency of radiation absorption, is equal to emissivity. By corollary, a blackbody is an object that absorbs all radiation incident upon it (i.e., it reflects none), and a whitebody is an object with $\varepsilon = 0$ (i.e., a perfect reflector). Objects with emissivity between these extremes are termed greybodies (e.g., the emissivity of water is around 0.96).

The energy density of radiation of a blackbody, ρ, varies with wavelength and temperature according to Planck's Law (Atkins 1994),

$$\rho\,(\lambda,\ T) = \frac{8\pi hc}{\lambda^5}\frac{1}{e^{hc\,/\,\lambda kT} - 1}, \tag{11.1}$$

where λ is wavelength, T is temperature in kelvin (K), h is Planck's constant (6.63×10^{-34} J s), c is the speed of light (3.00×10^{8} m s^{-1}) and k is Boltzmann's constant (1.38×10^{-23} J K^{-1}). This relationship between energy density and wavelength is illustrated for various temperatures in Fig. 11.1. It is of note that, for a given wavelength and temperature, there is a unique value for the radiation energy density. The importance of this for remote sensing of temperature is that the level of radiation measured at a particular wavelength provides the

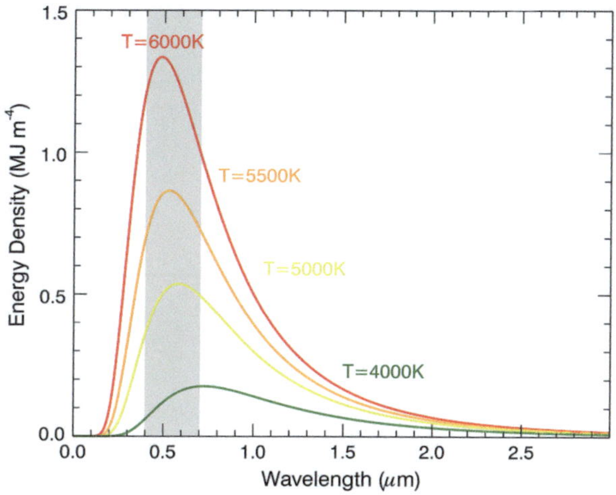

Fig. 11.1 Spectral energy density of blackbody radiation for various temperatures. Note that the wavelength corresponding to the peak increases with decreasing temperature. The *shaded area* indicates the visible wavelengths (0.4–0.7 μm)

corresponding temperature of the radiating object. For greybodies the emissivity will scale the energy density given in Eq. 11.1.

There are two notable relationships that are derived from Planck's Law. The first of these produces the wavelength at which the peak energy density occurs, determined by setting the derivative of Eq. 11.1 to zero. For high temperatures, this results in Wien's Displacement Law (Atkins 1994):

$$\lambda_{\max} = \frac{b}{T} \tag{11.2}$$

where $b = 2.898 \times 10^{-3}$ m K. This relationship indicates that a radiating body with temperature of 5,700 K has peak energy at around 500 nm, characteristic of our "yellow" Sun.

The second relationship derived from Planck's Law provides the total energy of radiation. This is obtained by integrating the energy density of Eq. 11.1 over all wavelengths; the result is dependent only upon temperature (Atkins 1994):

$$E = \sigma T^4. \tag{11.3}$$

This relationship is commonly known as the Stefan-Boltzmann Law, where $\sigma = 5.67 \times 10^{-8}$ W m^{-2} K^{-4} is the Stefan-Boltzmann constant.

For the purpose of thermal remote sensing, two regions of the electromagnetic spectrum are employed (it should be noted that the exact boundaries between these regions are somewhat arbitrary and vary between different applications). The first region, the infrared, is immediately adjacent to visible light in the electromagnetic spectrum (the name infrared literally refers to frequencies "below red"; Fig. 11.2). The entirety of this region covers wavelengths from approximately 700 nm to 1 mm (frequencies 400 THz to 300 GHz). Closest to the visible wavelengths, the sub-region of 'near infrared' radiation (700 nm to ~2.5 μm) shows similar application in remote sensing as visible light in that it is reflected by target bodies.

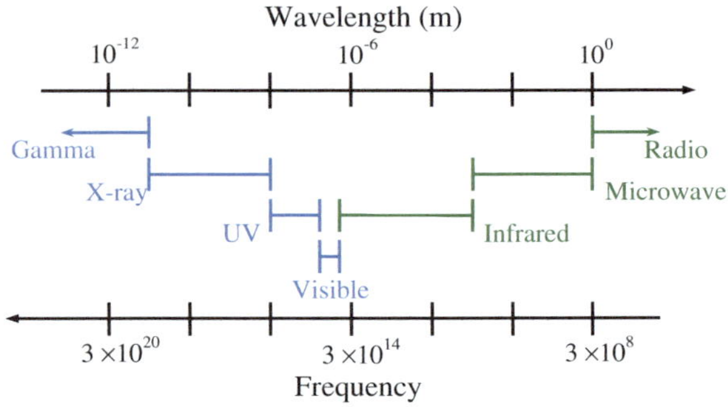

Fig. 11.2 Schematic of the electromagnetic spectrum. Note that boundaries between spectral bands can vary with the application. The spectral bands discussed in this chapter are highlighted in *green* (after Sabins 1997)

In contrast, the remaining infrared wavelengths (~2.5 μm to 1 mm) are emitted by target bodies providing their own thermal signature. The second spectral region for thermal remote sensing, the microwave, encompasses wavelengths from 1 mm to 1 m (300 to 0.3 GHz). The advantages and disadvantages for measuring temperature remotely using each of these regions are discussed here.

Satellite sensors are used to remotely sense ocean temperature (and related environmental parameters) around coral reefs, and typically observe bands of electromagnetic radiation (channels) in the wavelength range 400 nm to 15.0 μm, covering the visible and parts of the infrared spectrum. The specific spectral bands within this range depend upon the instrument design, which in turn is prescribed by the parameters desired for study (see Table 11.1 for examples). The effect of the atmosphere on surface emissions is pertinent to the channel design and is discussed in Sect. 11.2.2. Monitoring of temperature and other parameters (e.g., rainfall, wind speed, water vapor, cloud vapor, snow, ice and soil moisture) is also conducted using combinations of microwave channels. For example, the Advanced Microwave Scanning Radiometer for EOS (AMSR-E) instrument passively measures horizontally- and vertically-polarized signals in the frequency bands 6.925, 10.65, 18.7, 23.8, 36.5 and 89.0 GHz (wavelengths ranging from 43 mm down to 3 mm, respectively).

The satellite orbit characteristics impact the spatial and temporal resolution of satellite data. Modern environmental satellites generally orbit in one of two regimes: sun-synchronous and geosynchronous. Sun-synchronous satellites orbit in an approximately North–South plane at relatively low altitude (~800 km), passing approximately over Earth's poles multiple times each day to provide coverage over most ocean locations twice-daily. These satellites cross the equator at the same local time for each overpass. In contrast, geo-synchronous satellites are designed to sit at high altitude (~36,000 km) above a single location near the Earth's equator, orbiting in an approximately East–West plane at the same rotational speed as the Earth. As such, there is a nearly constant but only partial coverage of the Earth's surface that a single geo-synchronous satellite can measure—but with the advantage of increased temporal resolution. It should be noted that current and historical satellites have also had orbit regimes operating between these North–South and East–West extremes (e.g., the International Space Station

Table 11.1 Examples of satellite instrument applications and relevant infrared channels. Instrument channel number is given in parentheses

Application	AVHRR (μm) (NOAA POES, ESA MetOp)	Imager (μm) (NOAA GOES)	MODIS (μm) (NASA aqua/terra)
Cloud and surface mapping	0.580–0.680 (1)	0.55–0.75 (1)	0.620–0.670 (1)
	0.725–1.10 (2)		0.841–0.876 (2)
Land/cloud properties	1.580–1.640 (3a)		0.459–0.479 (3)
			0.545–0.565 (4)
			1.230–1.250 (5)
			1.628–1.652 (6)
			2.105–2.155 (7)
Water vapor	0.725–1.10 (2)		0.890–0.920 (17)
			0.931–0.941 (18)
			0.915–0.965 (19)
Surface/cloud temperature	3.55–3.93 (3)	3.8–4.0 (2)	3.660–3.840 (20)
			3.929–3.989 (21)
			3.929–3.989 (22)
			4.020–4.080 (23)
Atmospheric temperature			4.433–4.498 (24)
			4.482–4.549 (25)
Water vapor		6.5–7.5 (3)	1.360–1.390 (26)
			6.535–6.895 (27)
			7.175–7.475 (28)
			8.400–8.700 (29)
Surface/cloud temperature	10.3–11.3 (4)	10.2–11.2 (4)	10.780–11.280 (31)
	11.5–12.5 (5)	11.5–12.5 (5)	11.770–12.270 (32)

orbit is inclined such that the latitude range is 51.6°S–51.6°N; and the Tropical Rainfall Measuring Mission satellite orbits with inclination 35°, with coverage 38°S–38°N).

The geometry of satellite orbits is an important consideration for properly understanding the resolution of satellite observations and thereby effectively interpreting and applying the information. Key to this understanding is the difference between the Earth frame-of-reference, where measurements are actually interpreted and used, and the orbital (satellite) frame-of-reference, in which measurements are acquired. The first such consideration arises with the field-of-view and look-angle of the satellite scan. Though often described by a distance or area on the ground, the resolution of a satellite pixel is better described by the angle of the satellite instrument field-of-view, since actual ground-distance changes depending upon the look-angle (Fig. 11.3a). For example, the location

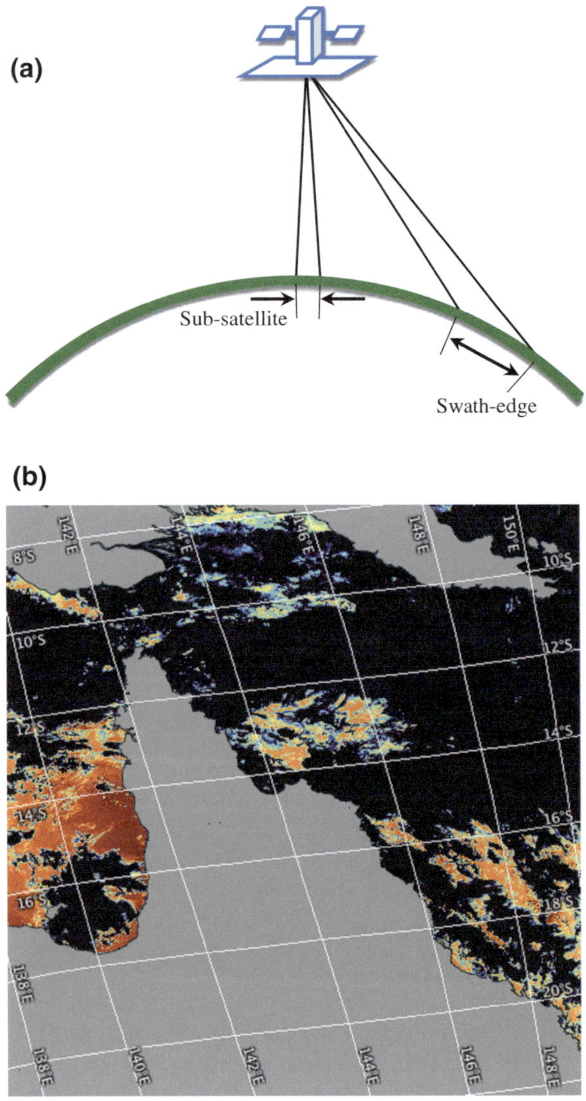

Fig. 11.3 **a** Schematic of ground resolution differences for different look-angles of satellite swath. **b** SST swath in orbital frame-of-reference acquired at 23:19 12 Jan 2010 (UTC) over NE Australia. Note the variation in the size, angle and curvature of *grid lines* (and therein pixels). Image produced using the NOAA CoastWatch data analysis tool

immediately below the satellite location (i.e., nadir view) will have the smallest areal footprint, whereas locations at the edge of the swath will have a significantly greater footprint in the cross-track dimension. Additionally, since the sensor measures across a specific angle, the satellite altitude, and variations in relative satellite altitude with respect to topographic differences, will also affect the footprint dimension. As a result, beyond the simplicity of the schematic in Fig. 11.3a, the three-dimensional nature of the dynamic satellite-Earth system is such that, if the footprint of the satellite were inscribed on the Earth's surface, the footprint would not necessarily be of the same shape or of equal dimension from location to location.

For sun-synchronous satellites, the rotation of the Earth beneath the satellite means that the satellite track is not directly oriented North–South along the Earth's surface, and because the orbiting satellite moves as the cross-track scan (e.g., for AVHRR; Table 11.1) is acquired, neither is the scan East–West oriented (Fig. 11.3b). In combination with variations in footprint size, this results in raw data being acquired in irregular grids. However, data users often prefer products aligned into regular grids with known, and fixed, longitude and latitude bounds, requiring that orbital data be spatially transformed, which may involve averaging and/or sub-sampling. Many users may not consider such impacts when using the data. An understanding of the geometry of satellite retrieval and the potential impact on interpretation and data resolution is an important consideration for users.

As discussed previously, satellite data resolution is best described by the instrument's field-of-view and, as such, the corresponding areal footprint at the Earth's surface depends upon the satellite altitude. Orbital period (i.e., the time to orbit the Earth) increases with orbit radius (i.e., satellite altitude) according to the gravitational laws, as does the ground-footprint for a given field-of-view angle. For example, a geo-synchronous satellite (with orbital period 24 h) is further from the Earth's surface and has lower spatial resolution than a polar-orbiter (of period approximately 100 min) with the same field-of-view angle. Thus there is an application-dependent compromise between frequency of coverage (and sensor field-of-view characteristics) and spatial resolution. One must therefore take into consideration how instrument capability (i.e., field-of-view angle), the resulting spatial resolution, and the return period impact each remote sensing application.

11.2.2 Acquisition Logistics

The most significant factor influencing satellite remote sensing of the ocean surface is the effect of the atmosphere. Of the radiation emitted by the ocean surface, aerosols in the atmosphere (e.g., water vapor, H_2O; ozone, O_3; oxygen, O_2; carbon dioxide, CO_2; nitrous oxide, N_2O; methane, CH_4; nitrogen dioxide, NO_2; nitrogen, N_2; dust; and particulates) attenuate particular wavelengths via absorption and scattering (see Fig. 11.4). Note that the atmosphere is fairly transparent at visible wavelengths (400–700 nm), which is beneficial for visible remote sensing (e.g., ocean color), and also transparent at other specific "windows", which are employed in instrument channel design (see Table 11.1). For measurement of SST, the general technique of using information from multiple channels within these optically transparent windows includes "dual window", "split window" and "triple window" algorithms (see Li et al. 2001 and references therein). These algorithms rely upon parameters derived from regression of satellite data with in situ measurements of temperature, and, in some cases, an a priori (reference) estimate from an external SST dataset.

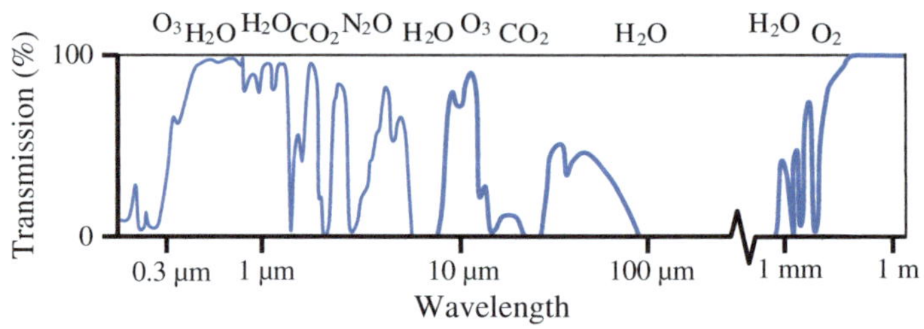

Fig. 11.4 Atmospheric transmission spectrum showing absorption lines due to various aerosols (figure developed from: Canadian Centre for Remote Sensing 2007; Gibson 2000; Sabins 1997; Woodhouse 2006)

Attenuation by the atmosphere also provides the potential to define certain satellite channels for measuring atmospheric components (e.g., water vapor; Table 11.1). The information gained from these channels can be useful to refine knowledge of the state of the atmosphere and can be incorporated into models of atmospheric radiative transfer (e.g., for atmospheric correction). This allows the potential for improving SST measurements using "physical retrieval" methods (Nalli and Smith 1998). These methods use "first guess" SST and atmospheric profile information, along with satellite radiances, radiative transfer models, and knowledge of uncertainties in the measurements and procedures, to thereby determine a refined SST measurement. Improvements in accuracy using physical retrieval are also possible using local information rather than employing globally-averaged regression coefficients; however, these potential improvements can be hard to achieve in practice.

Clouds provide a significant issue for measurement of SST at infrared wavelengths, as in remote sensing using visible wavelengths, since the sea surface radiation at these wavelengths is scattered and thus does not reach the satellite sensor. The sensor instead detects the radiation from the cloud-top and therefore results in a "gap" in the SST measurement. However, longer-wavelength microwave radiation passes through clouds and other masses of particulates (e.g., dust and haze), allowing surface data to always be collected, with the only exception being during heavy rainfall. Microwave sensors detect reflected, surface-emitted and atmosphere-emitted radiation. Similar to techniques used for infrared wavelengths, comparison of acquisitions from different microwave channels can provide atmospheric information (e.g., water and ozone) that can be used to refine SST measurement.

It is important to note that the different wavelength bands acquire different temperature measurements within the sea surface layer. This is because (a) the penetration depth of microwaves is an order of magnitude greater than that of infrared radiation; and (b) a strong temperature gradient generally exists at the surface of the ocean (referred to as the skin layer) due to radiation and heat fluxes (Donlon et al. 2002; Fig. 11.5). Infrared remote sensing measures the temperature of the upper ~10 μm of the ocean (SST_{skin}), within the skin layer. In contrast, microwave measurements originate from the upper ~1 mm of ocean, representative of the temperature at the lower boundary of the skin layer ($SST_{subskin}$).

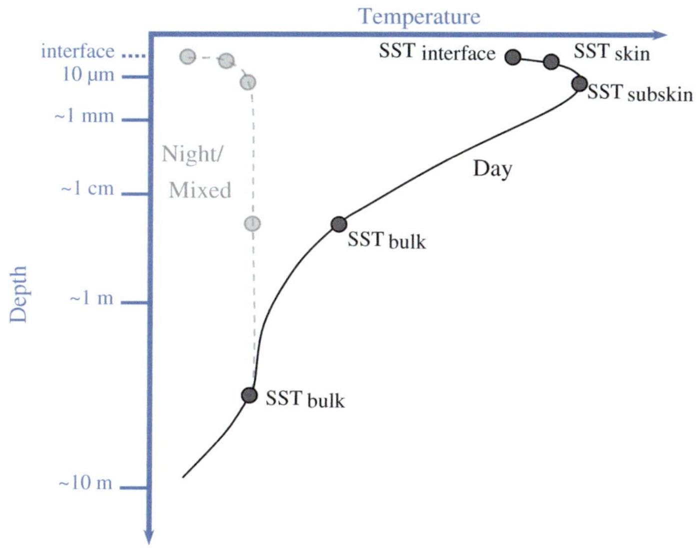

Fig. 11.5 Schematic profile of near-surface ocean temperature with depth, showing the skin layer and locations of measurements relevant to satellite SST monitoring. The *solid black profile* shows the surface daytime variation and stratification of the water column during high insolation and low-mixing conditions that are typical during coral bleaching events. The *grey dashed line* shows temperature consistent with depth near the surface, characteristically occurring at night and when the water column is well-mixed. The difference in SST_{skin} between *black* and *grey lines* is typically 0.6 K and as high as 2.8 K (after Donlon et al. 2002)

Below this, the bulk temperature (SST_{bulk}) varies with depth within the water column. As a result, in situ temperature measurements are of bulk temperature and should be described with the associated depth of acquisition.

Coral bleaching events have been linked to conditions of clear sunny skies and low mixing when the ocean surface becomes stratified, enhancing near-surface heating during the daytime (Skirving et al. 2006). The daily variation (i.e., diel fluctuation) in temperature is typically around 0.6 K but has been reported as high as 2.8 K (Zeng et al. 1999) and, as such, is an important consideration for satellite monitoring of conditions on coral reefs. While there may be differences between the absolute temperature measured by satellite and that at the depth of corals, monitoring applications often utilize temperature anomalies to describe the level of thermal stress experienced at specific locations (see Chap. 12). Inherent within this approach is an assumption that the anomaly at the skin is consistent with that experienced by the corals, which leads some monitoring efforts (e.g., Coral Reef Watch) to use night-only data so as to avoid the issue of diel warming.

A final issue for satellite monitoring is sunglint, where under certain conditions, particularly during periods of low wind, sunlight is reflected off the ocean surface into the view of the sensor. Sunglint "contaminates" the measured satellite retrievals and results in false temperatures due to this reflected thermal radiation from the sun. In SST analysis, areas of sunglint are identified during processing, based on knowledge of sun-position and time, and the corresponding areas are removed from further processing, resulting in gaps in the data.

11.2.3 History of Thermal Monitoring

In 1960, the U.S. government launched the first weather satellite, Television Infrared Observation Satellite (TIROS-1), providing the first repeat images of Earth from space. Infrared sensors were developed and installed on subsequent satellite missions from 1965, culminating in 1972 with the launch of the two-channel Scanning Radiometer (SR) and the higher-resolution two-channel Very High Resolution Radiometer (VHRR) onboard NOAA-2. Further refinements led to the four-channel Advanced Very High Resolution Radiometer (AVHRR), first launched in 1978 onboard TIROS-N, the five-channel AVHRR/2 in 1981 onboard NOAA-7 and the six-channel AVHRR/3 in 1998 onboard NOAA-15 (see Table 11.2). To date there are 3 decades of continuous data from AVHRR on polar-orbiting satellites. Augmenting these NOAA satellites, an AVHRR was also installed on MetOp-A (launched in 2006), the first of three MetOp satellites in a program jointly established by the European Space Agency (ESA) and EUMETSAT. The second satellite in this series, MetOp-B was launched in September 2012. In addition, the Visible/Infrared Imager/Radiometer Suite (VIIRS), a 22-band radiometer, was carried on the first satellite of the National Polar-orbiting Operational Environmental Satellite System (NPOESS) Preparatory Project (NPP), launched in October 2011.

The Moderate Resolution Imaging Spectroradiometer (MODIS) is deployed on two polar-orbiting satellites of NASA's Earth Observing System (EOS): Terra, launched in 2000, and Aqua, launched in 2002. The MODIS sensor has 36 channels designed to cover a suite of applications, including ocean color and ozone monitoring (see Table 11.1 for select channel information). In addition to MODIS, the Aqua satellite also carries the Advanced Microwave Scanning Radiometer for EOS (AMSR-E). A precursor to this instrument, AMSR, was onboard the unfortunately brief ADEOS-II satellite mission of 2002–2003, a joint effort of the Japan Aerospace Exploration Agency (JAXA), NASA, and the Centre National d'Études Spatiales (CNES, France). Microwave remote sensing of temperature has also been undertaken, since 1978, using the Scanning Multichannel Microwave Radiometer (SMMR; 1978–1987, onboard Seasat and Nimbus 7); Special Sensor Microwave Imager (SSM/I; 1987–present, onboard U.S. Defense Meteorological Satellite Program satellites, now administered by NOAA); and Tropical Rainfall Measuring Mission (TRMM) Microwave Imager (TMI; 1997–present). The TRMM satellite monitors only the tropical regions due to a 35° inclination and also carries the Visible and Infrared Scanner (VIRS), making it the forerunner of direct comparisons between microwave and infrared measurements of temperature.

The first geostationary satellite for environmental observation, SMS-1, was launched in 1974 to monitor conditions over the central Atlantic Ocean (45°W) and SMS-2 was launched and positioned over the Pacific Ocean (135°W) the following year.

Table 11.2 Select historical, present and future (**a**) polar-orbiting and other low-altitude satellite series, and (**b**) geostationary satellite series with instruments for temperature monitoring

Satellite series *Administering organization/s*	Date range	Milestone satellites (*planned*)	Thermal instrument/s	Spatial resolution
(a) Polar-orbiting and other low-altitude satellites				
NOAA Polar (TIROS, ESSA) *NOAA* (*U.S.*)	1960–present	NOAA-2 TIROS-N NOAA-7 NOAA-15 NOAA-19 (latest)	SR and VHRR (1972) AVHRR (1978) AVHRR/2 (1981) AVHRR/3 (1998)	~4.4 km global area coverage ~1.1 km local area coverage
DMSP *DoD/NOAA* (*U.S.*)	1987–present	F17 (latest)	SMMR (1987) SSM/I (1987)	~25 km
ADEOS *JAXA* (*Japan*) *NASA* (*U.S.*) *CNES* (*France*)	1996–1997 2002–2003	ADEOS (1996) ADEOS-II (2002)	AVNIR (ADEOS) AMSR (ADEOS-II)	8–16 m 5–50 km (channel specific)
TRMM[a] *JAXA* *NASA*	1997–present	TRMM	TMI VIRS	~25 km ~12.5 km (binned from 2 km)
EOS *NASA*	2000–present	Terra (2000) Aqua (2002)	MODIS AMSR-E (Aqua)	0.25–1.1 km (channel specific) 5.4–56 km (channel specific)
MetOp *ESA* (*Europe*) *EUMETSAT* (*Europe*)	2006–present	MetOp-A (2006) (*MetOp-B*, *mid* 2012) (*MetOp-C*, 2016/17)	AVHRR/3	~4.4 km global area coverage ~1.1 km local area coverage

(continued)

Table 11.2 (continued)

Satellite series *Administering organization/s*	Date range	Milestone satellites *(planned)*	Thermal instrument/s	Spatial resolution
(b) Geostationary satellites				
GOES (SMS) *NOAA*	1974–present	SMS-1 GOES-11 (West, 135°W, 2000) GOES-13 (East, 75°W, 2006) GOES-15 (on standby, 2010)	Imager	1–8 km
METEOSAT *EUMETSAT*	1977–present	Meteosat-1 (0°, 1977) Meteosat-7 (57.5°E, 1997) Meteosat-8 (9.5°E, 2002, MSG) Meteosat-9 (0°, 2005, MSG)	MVIRI SEVIRI	2.5–5 km 1–3 km
GMS/MTSAT *JAXA*	1978–present	GMS-1 (140°E, 1978) MTSAT-2 (145°E, 2006)	VISSR Imager	1.25–5 km 1–4 km
Electro *RFSA (Russia)*	1994–1998 2011–present	Electro-1 (76°50'E) Electro-L (76°E)	STR MSU-GS	1.25–6.25 km 1–4 km
Kalpana *ISRO (India)*	2002–present	Kalpana-1 (74°E)	VHRR	2–8 km

[a] TRMM orbits with 35° inclination (coverage 38°S–38°N); Davis (2007); Harris (1987)

The late-1970s saw the continuation of SMS into the GOES satellite series (U.S.), and also the launch of other geostationary satellite series that have continued to the present: GMS (Japan; now MTSAT) and METEOSAT (Europe; now Meteosat Second Generation, MSG). With the more recent additions of the Electro series (Russia, 1994–1998, 2011) and Kalpana-1 (India, 2002) there is the potential, through data sharing arrangements, for overlapping coverage of SST and other environmental parameters from geostationary satellites orbiting at longitudes 135°W, 75°W, 60°W, 0°, 57.5°E, 74°E and 76.8°E. It is of note that the Feng Yun 2 GEO satellite of the China Meteorological Administration briefly operated at 105°E in 1997–1998.

11.2.4 Thermal Processing Requirements

Satellite-borne instruments measure the radiation from the target in specified wavelength channels, which can then be converted to an apparent temperature by way of Planck's law (Eq. 11.1). This assumes perfect emissivity (unity), which is inaccurate for Earth emissions and must be corrected. As such, the physical temperature can be extracted through empirical relationships developed during sensor calibration using temperature-brightness relationships derived from multiple bands (split-window algorithm). Most algorithms rely upon a reference temperature, often based on lower-resolution data, to provide an initial estimate of the temperature. With improvements in measurements and modeling of the atmosphere and related impacts on absorption and scattering of emissions, so-called "physical retrievals" of temperature can now be determined using local conditions rather than global calibration parameters.

Processing necessarily also includes geo-registration to reference data to the Earth's surface through modeling of the satellite location and correlation of image features (e.g., coastlines). Limitations of onboard storage capacity may require sub-sampling of data prior to download (e.g., the AVHRR Global Area Coverage stores the average of measured values from four of every five cross-track pixels and every third scan line, resulting in a 4 × 4 km pixel value derived from an approximately 1.1 × 4 km area). For the coral reef user, it is important to recognize the constraints (and associated uncertainties) that impact the accuracy of measured values and reported location.

It is of note that microwave emissions also include information on sea surface salinity, whereas infrared emissions do not. Because of this, comparison of observations from these different bands leads to salinity measurement. Changes in salinity can result in stress to corals. However, satellite observations of sea surface salinity are relatively nascent and do not have spatial resolution that is currently applicable to coral reef management.

Sea surface temperature derived from satellite is widely available for use by coral reef stakeholders. Significant research into ecosystem impacts related to thermal variation has been undertaken and has resulted in management tools that are distributed via the internet. Chapter 12 outlines these efforts and provides examples of their applicability for coral reef management.

11.2.5 Thermal Validation

Validation of SST can be undertaken by comparison with in situ measurements. As previously discussed, care must be taken when comparing measured temperatures based upon their measurement technique (i.e., SST_{skin} and $SST_{subskin}$ versus SST_{bulk}; Fig. 11.5). Satellite SST products are often validated for broad patterns of agreement with near-surface temperatures measured on deep-water buoys (e.g., Kilpatrick et al. 2001). However, to provide a like-for-like comparison, satellite-derived temperature can be compared with temperature measured by a ship-mounted radiometer (e.g., Minnett et al. 2001). An important element of sensor calibration is to understand how the instrument responds at the specific channels and how these responses change with time. As such, it is useful to maintain regular validation of derived parameters (e.g., temperature). Chapter 12 provides a thorough discussion of temperature validation, particularly related to the monitoring of coral reefs using currently available technology.

11.3 Radar Overview

11.3.1 Radar Physical Principles

Remote sensing of the ocean surface and the air-sea boundary region can be carried out by measuring backscatter of transmitted (active) radar signals across a variety of frequency bands (Table 11.3). Backscatter parameters that are used to indicate the state of the ocean include travel time, frequency shift, phase difference and polarity change. Here we describe principles of radar remote sensing in the microwave and radio portions of the electromagnetic spectrum (Fig. 11.2).

Table 11.3 Band designations for the communications bands (3–3,000 MHz) and the IEEE standard 521 for radar bands at higher frequencies

Designation	Frequency band	Wavelength band
HF	3–30 MHz	100–10 m
VHF	30–300 MHz	10–1 m
UHF	300–3,000 MHz	1–0.1 m
L-band	1–2 GHz	30–15 cm
S-band	2–4 GHz	15–7.5 cm
C-band	4–8 GHz	7.5–3.75 cm
X-band	8–12 GHz	3.75–2.50 cm
Ku band	12–18 GHz	2.50–1.67 cm
K band	18–27 GHz	1.67–1.11 cm
Ka band	27–40 GHz	1.11–0.75 cm

Ground-wave radars typically operate in the HF and VHF, and SAR instruments generally operate between L-band and X-band

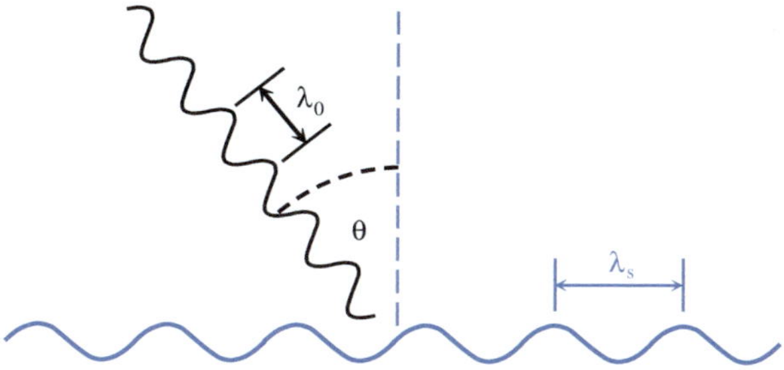

Fig. 11.6 Schematic of Bragg relationship between a radar signal of wavelength λ_0, incident at angle θ to the normal, and a sea surface wave of wavelength λ_S

The backscatter process is an interaction between incident radar signals and ocean surface gravity waves (Fig. 11.6). In all cases except near-vertical incidence, the dominant radar echo from the sea surface is produced by a Bragg scatter from waves on the sea surface. For a given angle of incidence to the normal, θ, there is a precise relationship between the radar wavelength, λ_0, and the wavelength of the sea surface wave, λ_S:

$$\lambda_s = \frac{\lambda_0}{2\sin\theta}. \quad (11.4)$$

There are two limiting cases of incidence angle for the Bragg mechanism. The first is normal incidence ($\theta = 0°$) for which the ocean wave has infinite wavelength according to Eq. 11.4. This is the case for downward looking radar (e.g., from satellite), where ocean information must be gained from other factors (e.g., travel time for altimetry; relative amplitude for scatterometry). A second limiting case of incidence angle is grazing incidence ($\theta = 90°$), known as ground-wave radar, which senses ocean wavelengths half that of the transmitted signal. Spaceborne radars generally operate between these extremes and sense ocean targets at a specific angle either alongside the satellite track (broadside mode), or forward- or backward-looking along the satellite track (squint mode).

The returned spectra (i.e., echo spectra) from High Frequency (HF) radar systems, with wavelengths 10–100 m (30–3 MHz), typically have well-defined Bragg peaks, which are due to scattering from approaching and receding surface gravity waves, as well as clear second-order structure. This allows information on surface currents, wave characteristics and wind direction to be extracted. A typical backscatter spectrum from ground-wave HF radar is shown in Fig. 11.7a. The two backscatter energy peaks are formed due to Bragg resonance (i.e., the scattering of radio waves of wavelength λ_0 by ocean surface waves of wavelength $\lambda_0/2$ that travel directly toward, or away from, the radar source). The scattering invokes a positive (negative) Doppler frequency shift for ocean waves moving toward (away from) the radar source, which is dependent upon the gravity wave speed. With the assumption of deep-water waves (i.e., the water depth is much greater than the wavelength of sea-surface waves), which is reasonable for the great majority of ocean monitoring, the Bragg frequency shift is easily calculated as a function of the transmitted radar

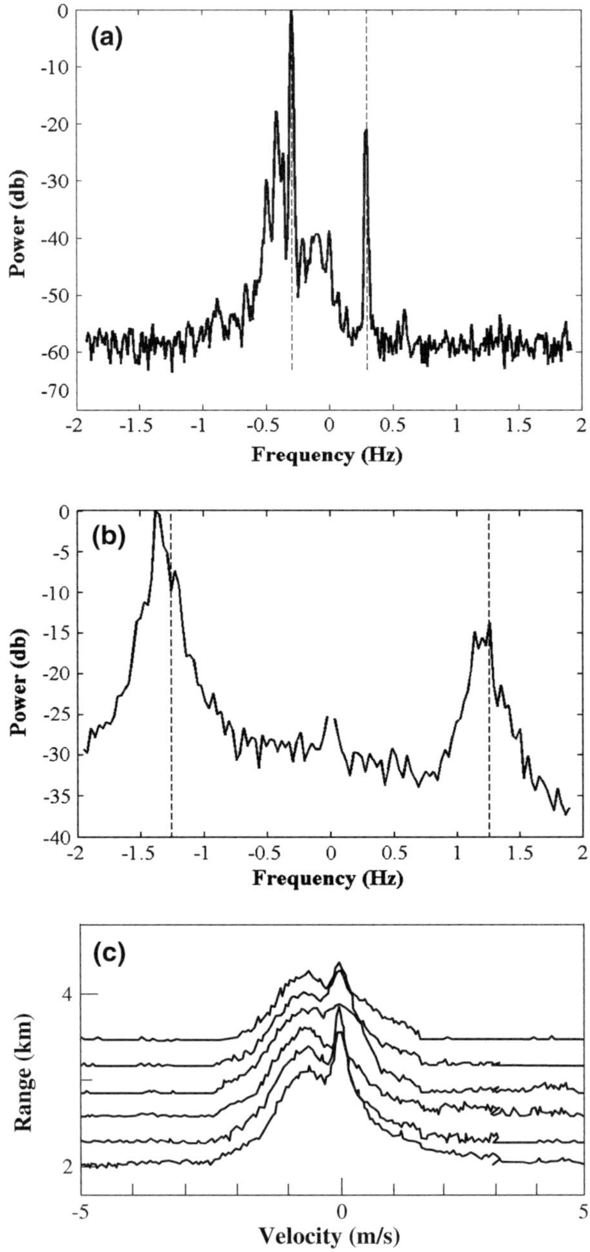

Fig. 11.7 Typical backscatter power spectra from: **a** High Frequency radar (HF; 8.348 MHz) showing the observed Bragg peaks, with positions of Bragg lines at ±0.295 Hz, representing zero surface current (*dash line*), and with distinct bands of second-order energy. **b** Very High Frequency radar (VHF; 152.2 MHz) showing the first- and second-order energy merged into broadened Bragg peaks, with the positions of Bragg lines at ±1.259 Hz for zero surface current (*dash line*) and **c** X-band radar where scatter from capillary waves swamps the separation of Bragg lines (modified from Heron et al. 1996). In each case the energy at zero corresponds to echoes from stationary objects

frequency. Any additional frequency offset of the Bragg peaks, compared with their theoretical offset, indicates the component of the surface current in the radial direction of the radar. In general, if a particular ocean region is monitored from multiple locations (multiple and/or moving radars), the surface current can be fully resolved. The wind direction can also be determined from the relative energy magnitude of the two Bragg peaks.

Second-order processes (including double scattering and non-linearities in the ocean waves) provide spectral energy at other frequency shifts from the incident radar frequency. For example, Barrick (1977) derived a relationship between the ratio of second-order to first-order energy, *R*, and the root-mean-square (rms) wave height, h_{rms}. Empirical testing of this relationship by Maresca and Georges (1980) and Heron and Heron (1998) has verified the capability of monitoring of ocean waves using radar. Further work by Wyatt (1991, 2011) has involved inverting the radar backscatter spectrum to produce full wave spectra.

Useful observation of the ocean around and between reefs out to a range of 100–200 km is a unique feature of ground-wave HF technology (spatial resolution of these observations is generally within 3–50 km). Radar systems with higher frequencies also detect Bragg scatter; however, there are necessarily different ways to interpret the echo spectrum, as the separation of first-order and second-order spectral information is more difficult (Fig. 11.7b, c).

Very High Frequency (VHF) radar, with wavelengths 1–10 m (300–30 MHz), exhibit a merging of second-order energy, due mainly to underlying swell, with first-order backscatter from surface gravity waves, thereby broadening identifiable Bragg peaks. The broadened Bragg peaks may nonetheless still be used to determine surface currents and significant wave heights. VHF ground-wave radars have spatial resolution down to about 25 m, producing detailed maps of surface currents at a resolution unique to VHF radars, and for an operating range of 3–20 km. For coral reef applications, VHF ground-wave radar stations need to be placed on a nearby atoll or island.

A significant consideration for the use of radars in coral reef regions is the assumption of deep-water gravity waves in the theoretical formulation of Bragg scatter. Gravity waves may be considered "deep" when water depth is on the order of one-sixth of the wavelength of the Bragg waves. For example, a ground-wave HF radar (15 MHz, wavelength 20 m) will have Bragg scatter from gravity waves of wavelength 10 m, which require the water depth to be over about 2 m for the deep-water assumption to hold. As such, retrievals over reef flats may be compromised. However, VHF radar does not have this issue due to the order-of-magnitude shorter wavelength and is therefore effective at monitoring shallower environs.

At radar frequencies in the C (4–8 GHz) and X (8–12 GHz) bands (Table 11.3), Bragg scatter is from capillary waves, and modulation by underlying gravity waves generally causes so much line broadening that the approaching and receding Bragg waves are indistinguishable. In these bands, amplitude is used to determine sea surface parameters, and it is possible to form very narrow beams from large antennas to achieve high spatial resolutions. Satellite-borne scatterometers measure echo energy from different directions in order to determine wind speed and direction at spatial resolutions of 12.5–50 km and, with special processing, experimentally down to 2.5 km (Plagge 2009). One application of X-band

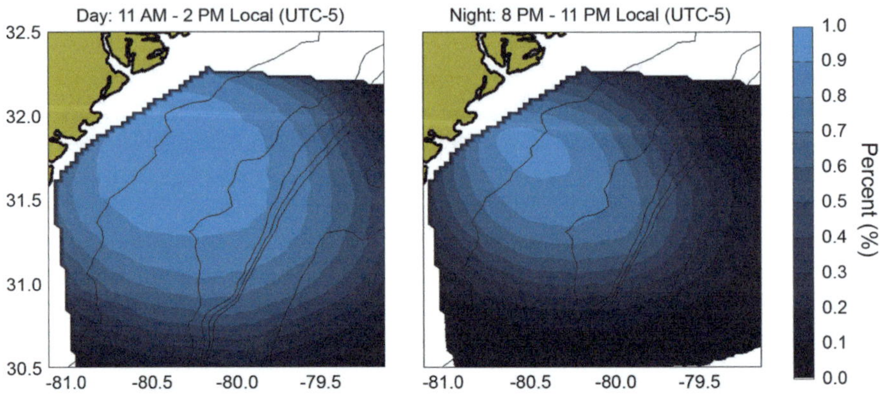

Fig. 11.8 Maps of fractional coverage of surface currents by a WERA HF radar system operated on the southeastern United States coastline, April 2006–May 2007, showing the variation in range with time-of-day: (*left*) Daytime 16:00–19:00 UTC; (*right*) Nighttime 01:00–04:00 UTC. Bathymetry contours at 20, 40, 60, 80, 100 and 500 m (courtesy of D. Savidge)

horizontally propagating radar is to observe echo amplitudes from wave crests and hence deduce 2D wave spectra; these are usually called X-band wave radars.

Radar resolution should be considered separately as radial and angular components. The angular resolution is dependent upon the transmitted wavelength (linearly) and antenna length (inversely), so that resolution is improved for shorter wavelengths and longer antennae. It is of note that a radar pixel dimension corresponding to this angle increases with the distance from the antenna, such that more-distant targets have larger cell sizes. In the radial direction, the range resolution is linearly dependent upon the duration of the transmitted pulse. However, shorter pulses imply lesser transmit energy, which degrades the range performance of the system. One technique to overcome these limitations is to impose a modulation on a longer pulse, called pulse coding (e.g., chirp). In such systems the backscatter signal is correlated with the transmitted coding during processing.

Measurement using radar can be complicated by noise from other undesirable sources. Detection of this noise is an important aspect of signal processing. The most dramatic impact occurs in relation to changes in the ionosphere between day and night, which can alter the background noise and significantly reduce the quality and range of signal. Solar radiation causes separation of electrons (ionization) that reflect radar signals, whereas at night, neutralization of free electrons and ions reduces this reflection capacity. Savidge et al. (2011) examined radar returns at day and night through a 13 month deployment of HF radar on the Southeastern United States coastline and showed a significant reduction in coverage at night (Fig. 11.8). Other sources of noise in radar backscatter include external radio signals and interference, and echoes from bodies moving across the radar domain (e.g., ships).

11.3.2 Radar Systems

Here we discuss in more detail the two types of radar systems that are presently used for monitoring environmental conditions around coral reefs: ground-wave radar; and air- or space-borne Synthetic Aperture Radar (SAR).

Ground-wave radar systems transmit vertically-polarized signals that graze the ocean surface. Existing systems, either HF or VHF, are located at near-shore terrestrial locations. There are two genres of ground-wave radar systems for monitoring conditions around coral reefs: phased-array and direction-finding.

Phased-array radar systems use separate transmit (usually omni-directional) and receive antenna arrays to extract ocean surface parameters across the domain of interest. Setting the range to specific ocean target regions is achieved by considering the time delay between transmission and reception (range), and through steering the receiver beam through prescribing the relative phase at each different individual receive antenna (direction). In modern systems the time delay used for ranging is coded in a frequency modulated chirp arrangement. Coverage of a broad ocean region is undertaken by scanning across the region (i.e., modifying the antenna phases) and acquiring echo spectra for each radial cell in the beam direction. The Doppler shift of the Bragg peaks provides the radial-component of the surface current at the target. Wave spectrum information is derived for each ocean target using the second-order backscatter spectrum; however, the lessened magnitude due to double-reflection reduces the range to which wave data can be extracted to about half the range for currents. The radial resolution of the radar is inversely proportional to the bandwidth of the transmitted radiation, while the azimuthal resolution depends upon the length of the antenna array, which is generally of the order of 6–8 wavelengths.

Direction-finding radars operate on the principle of transmitting the radar signal in all directions and separating the backscatter after it is returned to three independent, orthogonal (x-y-z) receive antennas to determine the oceanic conditions. Monitoring of a 360° ocean region is thus simultaneous, with ocean state parameters extracted during post-analysis. As with the phased-array, the time delay between transmission and reception provides the range to the ocean target (coded as frequency chirp) and the surface current is determined from the Doppler shift of the Bragg echoes. The bearing of the ocean target is determined using the relative levels of backscatter energy acquired simultaneously at the three orthogonal receive antennas. As with phased-array radar, the radial resolution is dependent upon the transmission bandwidth; however, the azimuthal resolution depends upon the amplitude resolution of the orthogonal receive antennas. A significant benefit of the direction-finding systems is their relatively small antenna footprint (normally 2 poles with guy-ropes).

HF and VHF radar systems have been developed since the 1970s to the point where commercial systems are now readily available (Table 11.4). Off-the-shelf systems currently deployed over coral reef regions include the WERA (phased-

Table 11.4 Select historical and present High Frequency (HF) and Very High Frequency (VHF) ground-wave ocean radar systems with specifications

Radar system *Institution*	Date range	Radar type	Frequency range (MHz)
CODAR *NOAA, CODAR Ocean Sensing*	From late 1960s	Phased-array, direction-finding	25–35
SeaSonde *CODAR Ocean Sensing*	1990–present	Direction-finding	4.3–5.4, 11.5–14 24–27, 40–45
COSRAD *James Cook University*	1983–2001	Phased-array	30
OSCR *Rutherford Appleton Laboratory, Marex, Marconi*	1982–2000	Phased-array	15–27
PISCES *University of Birmingham, Neptune Radar*	1981–present	Phased-array	3–38
WERA *University of Hamburg, Helzel Messtechnik GmbH*	1995–present	Phased-array	5–50
PortMap *Portmap Remote Ocean Sensing Pty Ltd*	2006–present	Phased-array	60–180

The approximate date range indicates system prototype/production; existing units may have continued operation following end date

array) and CODAR SeaSonde (direction-finding). A discussion of considerations between these systems for deployment is provided in Chap. 13.

Synthetic Aperture Radar (SAR) instruments, which are active microwave imaging radars, provide unique environmental remote sensing capability. These instruments, whether flown on an aircraft or carried on a satellite, send out pulses of electromagnetic radiation at a wavelength of a few centimetres and measure the backscattered radiation from the Earth's surface. By recording both the phase and amplitude of the returning radiation, undertaking coherent addition of the returned signals over periods of a few seconds, and Doppler processing with sophisticated signal processing techniques, it is possible to create a high-resolution image, essentially synthesizing the resolution that would have been achieved if the antenna were many kilometers in length (thus the term synthetic aperture; Fig. 11.9). The synthetic aperture procedure creates enhanced resolution in the along-track direction (i.e., in the direction of satellite or aircraft motion). High-resolution in the cross-track direction (i.e., perpendicular to the direction of

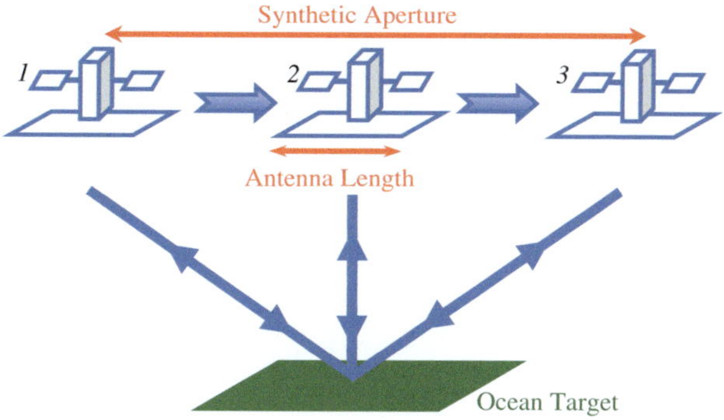

Fig. 11.9 Schematic representation of the synthetic aperture produced due to the motion of the radar (i.e., synthesizing the resolution that would have been achieved if the antenna were many kilometers in length)

motion), is achieved by frequency modulation (i.e., chirping) of the radar pulses and frequency processing of the returns.

The motion of the SAR allows it to collect information for each target location multiple times, with the accumulated signal providing enhanced resolution (down to 1 m) in the along-track direction. Over the ocean, backscattering is predominantly caused by Bragg scattering from capillary or small gravity waves of the same scale as the radar wavelength, but also by direct reflection off wave facets oriented perpendicular to the incident radar beam. Backscatter from stationary targets ahead of the SAR will have a positive Doppler shift due to the relative velocity toward the SAR, while the reverse is true of targets behind the SAR; these are resolved in range calculations since the motion of the instrument is known. SAR measurement is also complicated by Doppler shifts due to motion within the ocean target itself, which, when superposed on the pulse coding, can influence the accuracy of the ranging. This can result in: image shift (where an object moving toward or away from the satellite ground track is displaced sideways in the processed image because it has a Doppler feature identical to a nearby target); range walk (blurring of the object because it spans multiple range cells); and amplitude reduction (resulting from each of these degradations). Amplitude of backscatter is also influenced by the roughness of the ocean, which represents further information that can be obtained concerning sea state.

The constructive interference of many small waves with wavelength similar to that of the radar results in a strong signal from the ocean's surface, which increases with wave height and is well correlated with wind speed. Therefore, ocean phenomena that affect the small-scale roughness of the ocean's surface at a scale of a few centimeters can produce a detectable signal in a SAR image. These include the following phenomena: wind; waves (swell, wind and internal); oil and other surface surfactants; upwelling; frontal and current boundaries; and shallow water bathymetry. With this suite of parameters and very high spatial resolution, SAR technology is well conditioned for monitoring the environment around coral reefs and on reef flats. There is also the potential of SAR to characterize surface slicks produced by coral spawn (Jones et al. 2006).

Table 11.5 Select historical, present and future satellite-borne synthetic aperture radar systems with specifications. Resolution and swath width given for multiple modes of SAR operation

Program *Institution/Country*	Date range	Satellite (*planned*)	Band	Wavelength (cm)	Average ground resolution (m)	Swath width (km)
Seasat, *NASA*	1978	SAR	L	23.5	25	100
Shuttle imaging radar, *NASA*	1981	SIR-A	L	23.5	40	50
	1984	SIR-B	L	23.5	20–50	25–58
	1994 (2)	SIR-C/X-SAR	L	23.5	10–200	15–90
			C	5.8		
			X	3.1	25	15–40
	2000	SRTM	C	5.6	30	225
			X	3.1	25	45
ALMAZ, *Russia*	1987	Cosmos 1870	S	10.0	15	40
	1991–1992	ALMAZ				
ERS, *ESA*	1991–2000	ERS-1	C	5.6	25	100
	1995–2011	ERS-2			30/150/1000	100/400/400
	2002–2012	ENVISAT				
	2013–2015	(*Sentinel*-1*A*)			5/20/40	60/250/400
		(*Sentinel*-1*B*)				
JERS, *Japan*	1992–1998	JERS-1	L	23.5	18	75
	2006–2011	ALOS			10/20/30/100	70/70/30/250-350
	2013	[*ALOS*-2]			3/6/10/100	50/50/70/350

(continued)

Table 11.5 (continued)

Program *Institution/Country*	Date range	Satellite (*planned*)	Band	Wavelength (cm)	Average ground resolution (m)	Swath width (km)
RADARSAT, *Canada*	1995–present	RADARSAT-1	C	5.6	8/25/30/50/100	50/100/150/300/500
	2007–present	RADARSAT-2			3/8/28/28/50/100	20/50/100/150/300/500
	2016–2017	(*RADARSAT Constellation Mission, 3 satellites*)			3/5/30/50/100	20/30/125/350/500
TerraSAR, *Germany*	2007–present	TerraSAR-X	X	3.1	1/3/16	10/30/100
	2010–present	TANDEM-X				
	2013	*[TerraSAR-X2]*				
COSMO-SkyMed *Italy*	2007–present	COSMO-1	X	3.1	1/3/15/30/100	10/40/30/100/200
		COSMO-2				
		COSMO-3				
		COSMO-4				

SAR instruments generally operate between L-band (23.5 cm, 1.28 GHz) and X-band (3 cm, 10 GHz; Tables 11.3 and 11.5). L-Band is the limiting case for capillary waves, prior to the transition to wind waves, so there is linearity between wave height and wind speed (see further discussion in Sect. 13.5.1). X-band is affected by moderate precipitation and thus exhibits more atmospheric effects than the longer wavelength bands. Modern SAR instruments can also be operated in one of a number of "modes" of differing resolution and swath width. These range from a resolution of 1 m in "spotlight" mode, covering a 10×10 km region, to a resolution of 1 km with a swath width of 1,000 km. The most typical modes are "standard" mode (resolution 25 m, swath width 100 km) and "ScanSAR" mode (resolution 50–100 m, swath width 300–500 km). The standard mode is preferred for wave measurements, oil spill mapping and coastal ship detection, while ScanSAR mode is most useful for wind monitoring, and open ocean oil spill mapping and ship detection.

The first satellite synthetic aperture radar was deployed in 1978 on Seasat, a mission that was prematurely terminated after only 106 days due to an electrical failure. The data did, however, demonstrate the use of radar from space and paved the way for future missions. The next phase was the Shuttle Imaging Radar (SIR) series, the first of which, SIR-A, was comprised of spare parts from Seasat and operated for eight days aboard the Space Shuttle Columbia, collecting data over 10 million km^2 of the Earth's surface (Ford et al. 1982). Various improvements through the SIR program included mechanical tilting to allow multiple view-angles of terrain, multiple radar frequencies, capability for both horizontal and vertical polarization, and electronic antenna steering. The program culminated with the Shuttle Radar Topography Mission (SRTM) onboard Space Shuttle Endeavour, which mapped 80 % of the Earth's land area to provide an accurate high-resolution, topographic map. At present, multiple satellites from various agencies (Table 11.5) provide the potential for global coverage with SAR instruments and make available a wide variety of modes with a diversity of resolution, coverage, polarization and SAR frequency (McCandless and Jackson 2004).

11.3.3 Radar Processing Requirements

Processing radar backscatter spectra requires identification of the reflected signal from within the background noise in order to extract parameters on the ocean state. For ground-wave radar, this is often accomplished by determining the level of the power spectrum at frequencies away from that of the transmitted signal (i.e., the Bragg peaks and second-order scatter). However, this may be unduly influenced by bursts of energy other than "background noise" (e.g., from a moving ship). Heron and Heron (2001) provide a theory-based assessment of the noise level by rank-ordering spectral responses to differentiate signal from noise. This approach allows improved characterization of noise, ultimately leading to a more accurate description of the signal.

As described previously, the offset of first-order (Bragg) peaks provides information on the radial components of currents; thus two or more radar systems (or multiple views from a single moving radar) can resolve ocean currents. This gives information on connectivity between reefs and kinetic energy (mixing) at/near reefs. Measurement of wind (first-order energy) and waves (second-order energy) can also indicate levels of mixing around reefs. Examples of these applications are provided in Chap. 13.

Studies using early SAR data from Seasat were initially hindered by the vast quantity of data produced by the Seasat SAR instrument and the need to develop digital SAR signal processing techniques. Also, observations were only available for analysis when transmitted in real-time to ground stations, primarily due to the lack of storage capacity onboard the satellite. While this onboard storage issue no longer remains and digital SAR processing techniques are now very mature, the large volume of SAR data generated by modern SAR satellite instruments still leads to challenges in data acquisition, communication, and signal/product processing, particularly for near real-time applications. Some applications, such as wind, also require accurate SAR calibration, necessitating use of active or passive ground calibration targets and precise determination of antenna-pattern corrections using distributed targets, such as the Amazon rain forest. Nonetheless, the high-resolution information that can be extracted from these data is a worthy goal in spite of the constraints.

11.3.4 Radar Validation

Validation of radar-measured currents has been undertaken using ocean drifter position and in situ instrumentation (either direct measurement by a current meter or remotely sensed by acoustic sounders, which employ the same Doppler mechanism as radar measurement but through sound waves, see Sect. 8.3.4). Direct comparisons of current measurements from radar versus Acoustic Doppler Current Profilers (ADCPs) have shown general agreement, with root-mean-square (RMS) differences of only 4–20 cm s^{-1} (Graber et al. 1997; Shay et al. 2007), but are often limited by velocity shears between the surface radar measuring point and the uppermost usable ADCP bin (Kohut et al. 2006). Currents inferred from ocean drifter position (i.e., rate and direction of movement between locations) have shown similarly strong correlation with radar-measured currents (Paduan 2006). The technique of integrating radar currents to emulate path tracking has shown promise (Ullman et al. 2006) and more recent advances in processing HF radar data have shown improved comparisons with drifter tracks (Mantovanelli et al. 2010; refer also to Sect. 13.2.2). This is significant for the application of particle tracking, whether environmental (e.g., coral spawning) or for safety reasons (e.g., man overboard). Radar wind measurements have been validated through comparison with co-located meteorological buoys (Monaldo et al. 2001), dropwindsondes and passive microwave wind measurements from aircraft (i.e., using the Stepped

Frequency Microwave Radiometer; Fernandez et al. 2006), and comparisons with scatterometer/model winds (Monaldo et al. 2004; Horstmann et al. 2003). To determine the accuracy of radar wave measurements, wave measurements from buoys have been employed as well as comparisons with wave model predictions (Monaldo and Beal 1998; Kerbaol et al. 2004; Collard et al. 2005, 2009). Comparison of ENVISAT Advanced SAR swell height measurements with co-located buoy data (within 200 km and 1 h) found RMS error of 0.38 m (Collard et al. 2009). The comparison data set was limited to SAR images with waves that were well imaged, having peak wave periods in the 12–18 s range, and with wind speeds of 3–9 m s^{-1}. These restrictions limited contamination by wind sea spectra and minimized difficulties with measuring short azimuth-traveling waves.

11.4 Conclusion

An understanding of the thermal and radar technologies is paramount for ensuring appropriate application of the resulting data. Chapters 12 and 13 examine the use of these systems in the coral reef environment. Chapter 12 provides examples of available temperature data sets, as well as the use of temperature data in identifying recent trends, and in monitoring and managing coral reef ecosystems. Chapter 13 illustrates applications of ground- and satellite-based radar systems in studying and monitoring ocean and atmosphere conditions around coral reefs, including identification of ocean features, and tracking of biological- and human-sourced buoyant matter.

Acknowledgments The authors would like to thank William Skirving for advice in the initial layout for this chapter, and Al Strong for helpful comments during the review process. The manuscript contents are solely the opinions of the authors and do not constitute a statement of policy, decision, or position on behalf of NOAA or the US Government.

Suggested Reading

Barrett EC (1992) Introduction to environmental remote sensing, 3rd edn. Chapman and Hall, London

Campbell JB (2007) Introduction to remote sensing, 4th edn. The Guildford Press, New York

Jackson CR, Apel JR (eds) (2004) Synthetic aperture radar marine user's manual. U.S. National Oceanic and Atmospheric Administration, Washington

Robinson IS (1985) Satellite oceanography: an introduction for oceanographers and remote-sensing scientists. Ellis Horwood, Chichester

References

Atkins PW (1994) Physical chemistry, 5th edn. Oxford University Press, Oxford

Barrick DE (1977) The ocean wave height non-directional spectrum from inversion of the HF sea-echo Doppler spectrum. Remote Sens Environ 6:201–227

Canadian Centre for Remote Sensing (2007) Tutorial: fundamentals of remote sensing. http://www.ccrs.nrcan.gc.ca/resource/tutor/fundam/index_e.php. Accessed 17 Oct 2011

Collard F, Ardhuin F, Chapron B (2005) Extraction of coastal ocean wave fields from SAR images. IEEE J Ocean Eng 30(3):526–533

Collard F, Ardhuin F, Chapron B (2009) Monitoring and analysis of ocean swell fields from space: New methods for routine observations. J Geophys Res. doi:10.1029/2008JC005215

Davis G (2007) History of the NOAA satellite program. J Appl Remote Sens. doi:10.1117/1.2642347

Donlon CJ, Minnett PJ, Gentemann CL, Nightingale TJ, Barton IJ, Ward B, Murry MJ (2002) Toward improved validation of satellite sea surface skin temperature measurements for climate research. J Clim 15(4):353–369

Fernandez DE, Carswell JR, Frasier S, Chang PS, Black PG, Marks FD (2006) Dual-polarized C- and Ku-band ocean backscatter response to hurricane-force winds. J Geophys Res. doi:10.1029/2005JC003048

Ford JP, Cimino JB, Elachi C (1982) Space Shuttle Columbia view the world with imaging radar: the SIR-A experiment. JPL publication 82–95, Jet Propulsion Laboratory, Pasadena

Gibson PJ (2000) Introductory remote sensing: principles and concepts. Routledge, London

Graber HC, Haus BK, Shay LK, Chapman RD (1997) HF radar comparisons with moored estimates of current speed and direction: expected differences and implications. J Geophys Res 102:18749–18766

Harris R (1987) Satellite remote sensing: an introduction. Routledge & Keegan Paul, London

Heron ML, Heron SF (2001) Cumulative probability noise analysis in geophysical spectral records. Int J Remote Sens 22(13):2537–2544

Heron ML, Nadai A, Masuda Y (1996) An estimate of Doppler frequency shift and broadening for grazing incidence C-band ocean surface backscatter. In: Proceedings Pacific Ocean remote sensing conference 12–16 Aug 1996 Victoria, pp. 151–159

Heron SF, Heron ML (1998) A comparison of algorithms for extracting significant wave height from HF ocean backscatter spectra. J Atmos Ocean Technol 15(5):1157–1163

Horstmann J, Schiller H, Schulz-Stellenfleth J, Lehner S (2003) Global wind speed retrieval from SAR. IEEE Trans Geosci Remote Sens 41:2277–2286

Jones AT, Thankappan M, Logan GA, Kennard JM, Smith CJ, Williams AK, Lawrence GM (2006) Coral spawn and bathymetric slicks in synthetic aperture radar (SAR) data from the Timor Sea, north west Australia. Int J Remote Sens 27:2063–2069

Kerbaol V, Johnsen H, Chapron B, Rosich B (2004) Quality assessment of ENVISAT ASAR wave mode products based on regional and seasonal comparisons with WAM model output. In: Proceedings 2004 envisat and ERS symposium (ESA SP-572), Salzburg, CDROM #2.1, 6–10 Sep 2004

Kilpatrick KA, Podestá GP, Evans R (2001) Overview of the NOAA/NASA advanced very high resolution radiometer Pathfinder algorithm for sea surface temperature and associated matchup database. J Geophys Res 106:9179–9197

Kohut JT, Roarty HJ, Glenn SM (2006) Characterizing observed environmental variability with HF Doppler radar surface currents mappers and acoustic Doppler current profilers. IEEE J Ocean Eng 31:876–884

Li X, Pichel W, Maturi E, Clemente-Colón P, Sapper J (2001) Deriving the operational nonlinear multichannel sea surface temperature algorithm coefficients for NOAA-15 AVHRR/3. Int J Remote Sens 22:699–704

Mantovanelli A, Heron ML, Prytz A (2010) The use of HF radar surface currents for computing Lagrangian trajectories: benefits and issues. In: Proceedings oceans 2010 IEEE, Sydney, 24–27 May 2010

Maresca JW Jr, Georges TM (1980) Measuring RMS wave height and the scalar ocean wave spectrum with HF skywave radar. J Geophys Res 85:2759–2771

McCandless SW Jr, Jackson CR (2004) Principles of synthetic aperture radar. In: Jackson CR, Apel JR (eds) Synthetic aperture radar marine user's manual. U.S. National Oceanic and Atmospheric Administration, Washington

Minnett PJ, Knuteson RO, Best FA, Osborne BJ, Hanafin JA, Brown OB (2001) The Marine-Atmosphere Emitted Radiance Interferometer (M-AERI), a high-accuracy, sea-going infrared spectroradiometer. J Atmos Ocean Technol 18:994–1013

Monaldo FM, Beal RC (1998) Comparison of SIR-C SAR wavenumber spectra with WAM model predictions. J Geophys Res 103(C9):18815–18825

Monaldo FM, Thompson DR, Beal RC, Pichel WG, Clemente-Colón P (2001) Comparisons of SAR-derived wind speed with model predictions and ocean buoy measurements. IEEE Trans Geosci Remote Sens 3(12):2587–2600

Monaldo FM, Thompson DR, Pichel WG, Clemente-Colón P (2004) A systematic comparison of QuikSCAT and SAR ocean surface wind speeds. IEEE Trans Geosci Remote Sens 42(2):283–291

Nalli NR, Smith WL (1998) Improved remote sensing of sea surface skin temperature using a physical retrieval method. J Geophys Res 103(C5):10527–10542

Paduan JD, Kim KC, Cook MS, Chavez FP (2006) Calibration and validation of direction-finding high-frequency radar ocean surface current observations. IEEE J Ocean Eng 31(4):862–875

Plagge AM, Vandemark DC, Long DG (2009) Coastal validation of ultra-high resolution wind vector retrieval from QuikSCAT in the Gulf of Maine. IEEE Geosci Remote Sens Lett 6(3): 413–417

Sabins FF (1997) Remote sensing: principles and interpretation, 3rd edn. W.H. Freeman & Co., New York

Savidge D, Amft J, Gargett A, Voulgaris G, Archer M, Conley D, Wyatt L (2011) Assessment of WERA long-range HF-radar performance from the user's perspective. In: Proceedings current, waves and turbulence measurements IEEE/OES/CWTM, Monterey, 20–23 Mar 2011, pp 31–38

Shay LK, Martinez-Pedraja J, Cook TM, Haus BK, Weisberg RH (2007) High-frequency radar mapping of surface currents using WERA. J Atmos Ocean Technol 24:484–503

Skirving WJ, Heron ML, Heron SF (2006) The hydrodynamics of a bleaching event: implications for management and monitoring. In: Phinney JT et al (eds) Coral reefs and climate change: science and management. American Geophysical Union, Washington

Ullman DS, O'Donnell J, Kohut J, Fake T, Allen A (2006) Trajectory prediction using HF radar surface currents: Monte Carlo simulations of prediction uncertainties. J Geophys Res 111:C12005

Woodhouse IH (2006) Introduction to microwave remote sensing. Taylor & Francis, Boca Raton

Wyatt LR (1991) HF radar measurements of the ocean wave directional spectrum. IEEE J Ocean Eng 16:163–169

Wyatt LR (2011) Wave mapping with HF radar. In: Proceedings current, waves and turbulence measurements IEEE/OES/CWTM, Monterey, 20–23 Mar 2011, pp 25–30

Zeng X, Zhao M, Dickinson RE, He Y (1999) A multiyear hourly sea surface skin temperature data set derived from the TOGA TAO bulk temperature and wind speed over the tropical Pacific. J Geophys Res 104:1525–1536

Chapter 12
Thermal Applications

Scarla J. Weeks, Ray Berkelmans and Scott F. Heron

Abstract Of all the remote sensing products, sea surface temperature (SST) is arguably the most widely used data stream, both by scientists and especially by coral reef managers and stakeholders. The technology is mature and considerable effort has gone into creating high-quality time-series, climatologies, and issue-specific products to help managers and lay-people assess the risk of thermal stress in marine ecosystems. The effort that has gone into developing and improving global thermal stress products such as *HotSpots*, *Degree Heating Weeks*, and regional products such as *ReefTemp*, is in large part a reflection of how useful they have become to reef managers, reef users and scientists across a wide range of disciplines. In recent years, further development has seen SST 'now-casting' products extended to seasonal forecasting products with lead times of up to 6 months, albeit with reduced accuracy on long time horizons. This has been achieved by incorporating real-time SST data into sophisticated statistical, as well

S. J. Weeks (✉)
Biophysical Oceanography Group, School of Geography, Planning and Environmental Management, University of Queensland, Brisbane, QLD 4072, Australia
e-mail: s.weeks@uq.edu.au

R. Berkelmans
Australian Institute of Marine Science, PMB 3, Townsville, QLD 4810, Australia
e-mail: r.berkelmans@aims.gov.au

S. F. Heron
Coral Reef Watch, National Oceanic and Atmospheric Administration, 675 Ross River Rd, Townsville, QLD 4817, Australia
e-mail: scott.heron@noaa.gov

S. F. Heron
Marine Geophysical Laboratory, Physics Department, School of Engineering and Physical Sciences, James Cook University, Townsville, QLD 4811, Australia

J. A. Goodman et al. (eds.), *Coral Reef Remote Sensing*,
DOI: 10.1007/978-90-481-9292-2_12,

as dynamical, coupled ocean–atmosphere models. Increasing use of real-time and historic SST data in models, in conjunction with other data sources, is helping address more complex ecological and management problems.

12.1 Introduction

12.1.1 Infrared and Microwave Sensors

Sea surface temperature (SST) is a key measurement routinely made by satellite thermal infrared (TIR) and passive microwave (PMW) radiometers, in situ moored and drifting buoys, and ships of opportunity. These measurements are used to create daily spatially-complete global maps of SST that are used for ocean forecasts, weather prediction, coastal applications such as coral reef management and fisheries, and also for research by oceanography, meteorology, and climate scientists.

TIR sensors have been deployed on earth observing satellites for over 30 years and TIR-derived SST measurements are available in near real-time from a wide variety of satellite missions (Table 11.2). Of these, the most commonly used are the Advanced Very High Resolution Radiometer (AVHRR) on the NOAA Polar-orbiting Operational Environmental Satellites and the Moderate Resolution Imaging Spectroradiometer (MODIS) aboard the NASA Earth Observing System Terra and Aqua Satellites. TIR sensors on polar-orbiting satellites have a characteristic spatial resolution of 1.1 km (at nadir) with quasi-global coverage on a daily basis. TIR satellite retrievals are derived from radiometric observations in two distinct wavebands located at ~4 and 10–13 μm. The 4 μm channel is more sensitive to SST but is used only for night-time measurements because of the relatively strong reflection of solar irradiation at this wavelength. SST retrievals at 10–13 μm are used for day and night-time measurements. Both bands are sensitive to the presence of clouds and scattering by aerosols and atmospheric water vapor, hence TIR-derived SST measurements require atmospheric correction of the retrieved signal and can only be made for cloud-free pixels. The on-board calibration of TIR sensors (e.g., Corlett et al. 2006) is therefore supplemented with vicarious calibration in SST retrieval algorithms that compensate for the atmospheric attenuation of water leaving radiances using in situ SST measurements (e.g., Kilpatrick et al. 2001; Zhang et al. 2009). Geostationary satellites carry radiometers with TIR channels similar to the AVHRR. While their spatial resolution is coarser (~5 km), geostationary satellites sample at a higher temporal resolution (repeat acquisitions ~30 min), enabling the measurement of SST over ocean regions experiencing ephemeral cloud.

Prior to 1998, SST retrievals were only available globally from TIR sensors, but the launch of the TRMM (Tropical Rainfall Measuring Mission) Microwave

Imager (TMI) in December 1997 made SST from PMW retrievals possible (Gentemann et al. 2010). Global coverage of PMW-derived SST was achieved in 2002 with the launch of the Advanced Microwave Scanning Radiometer (AMSR) instrument. The advantage gained with PMW is that radiation at these longer wavelengths is largely unaffected by clouds and it is generally easier to correct for atmospheric effects. The PMW signal is, however, affected by sea-surface roughness and precipitation. Fortunately the surface-roughness and atmosphere signatures are quite distinct from the SST signature, and their influence can be readily removed (Gentemann et al. 2010). SST retrieval by PMW is prevented only in regions with sunglint, rain and near land, providing almost complete global coverage every 2 days, although at a coarser spatial resolution than TIR-derived SST measurements (25 km compared to 1 km).

12.1.2 Measurement Accuracies

The radiance measured by TIR and PMW radiometers originates in the surface 'skin' ($\sim$0.1 mm) and 'sub-skin' ($\sim$1 mm) layers of the ocean respectively, and not in the body of the water below, as represented by the 'bulk' temperature, which is measured by in situ thermometers. Near-surface temperature gradients can result from various processes, such as the absorption of solar radiation, heat exchange with the atmosphere, and subsurface turbulent mixing (Minnett 2010). The surface skin-layer of the ocean is generally cooler by a few tenths of a degree than the underlying bulk temperature just below the surface because the heat flux is nearly always from the ocean to the atmosphere (Hanafin 2002; Hanafin and Minnett 2001). This relationship is fairly well defined and remains the same on average for temperatures during the night at depths of a few meters, as well as during the daytime for wind speed conditions >6 m/s, but is variable under low winds (Donlon et al. 2002; Minnett 2010). The magnitude of the vertical temperature gradient is influenced by the diurnal cycle, cloud cover, and wind speed, which influences the turbulent mixing (Price et al. 1986; Fairall et al. 1996; Gentemann and Minnett 2008). Under low wind speed conditions, the heat generated in the upper ocean by insolation is not well mixed through the surface layer and results in thermal stratification, with temperature differences between the uppermost layer and the water below. Hence, the difference between the skin or sub-skin temperature and that measured by an in situ thermometer under low winds is variable and highly dependent on the depth of the bulk measurement (Kearns et al. 2000; Minnett 2003; Ward 2006).

SST measurements derived from TIR sensors are corrupted by the presence of clouds, with the temperature of cloud-contaminated pixels generally colder than the actual SST (Donlon 2010). For applications that require *absolute* accuracy of the retrieved SST values, any pixel that is slightly cloud contaminated should be excluded. For applications that make use of the *relative* SST values of adjacent pixels, such as the location of an oceanographic feature, cloud contamination is less critical. Cloud screening algorithms focus on applications for which the absolute accuracy of the SST value is paramount and rely on differences in emissivity, reflectivity, temperature and spatial structure between the ocean surface and clouds (Donlon 2010; Cayula and Cornillon 1996).

Degradation of the accuracy of SST can occur when the atmosphere contains high concentrations of aerosols, such as Saharan or volcanic dust, which generally introduce a negative bias error (retrieved SST values are too cold) of up to a few degrees Celsius. For example, the eruption of Mt Pinatubo in June 1991 drastically affected the ability to monitor temperature globally (Reynolds 1993). Hence, the affected time periods must either be excluded from any data analysis, or a specific set of coefficients estimated for the atmospheric correction algorithm corresponding to the particular conditions must be utilized. This has been mitigated to a certain degree by upgrading algorithms to include a dust index scheme (Merchant et al. 2006).

Uncertainties in satellite-derived SST can occur as a result of the above mentioned skin effect and diurnal heating, the effects of clouds and aerosols, and imperfections in the atmospheric correction algorithm (Minnett and Barton 2010). The standard approach to determine the uncertainties in SST is to compare them with coincident measurements from independent sources. Several approaches to validating satellite-derived SST that use different instruments are reviewed in Minnett (2010). The optimal method to determine the error characteristics of the satellite SST retrievals is with a ship-based radiometer, such as the Marine-Atmospheric Emitted Radiance Interferometer (M-AERI; Minnett et al. 2001), which can mimic the satellite measurement most closely. For AVHRR, the coefficients used in the atmospheric correction algorithms are derived by robust regression between the AVHRR-derived temperatures and in situ measurements from drifting buoys (Kilpatrick et al. 2001). The measured SSTs are bulk temperature readings that are related to the skin SST through the mean temperature differences in the matchups between satellite and in situ data. Similarly for MODIS, SST retrievals are of skin temperature, and the mean biases that result from comparison with the buoy bulk temperature measured at a depth of a meter or more are a manifestation of the thermal skin effect (Donlon et al. 2002). A summary of the accuracies of SST derived from various satellite sensors determined by comparison with independent measurements is provided in Minnett (2010). For example, uncertainties in SST from cloud-free comparisons between Pathfinder AVHRR retrievals from 1985 to 1998 and drifting buoys provide a mean bias of 0.02 °C (SD = 0.53 °C). Uncertainties in SST retrievals from all MODIS 11 μm retrievals compared to bulk buoy temperatures provide a mean bias of −0.16 °C (SD = 0.55 °C). These are derived from >600,000 matchups with drifting buoys measuring the bulk temperature. However, when compared with the radiometric SST from M-AERI, the Pathfinder AVHRR retrievals provide a mean bias of 0.14 °C (SD = 0.36 °C), and the MODIS retrievals a mean bias of 0.02 °C (SD = 0.55 °C). This implies differences still exist between derived and in situ temperatures. In summary, all other things being equal, MODIS data will be cooler on average by ~0.16 °C than AVHRR data since MODIS is calibrated with respect to skin temperature while AVHRR uses a buoy bulk temperature (Minnett personal communication).

For scientific applications, TIR SST retrievals are generally used at ~1 km spatial resolution, based on heritage instruments such as the NOAA AVHRR and MODIS. For many purposes (e.g., global SST mapping), subsamples, or binned data, are used at lower resolution, such as 4 or 9 km pixels. Some applications of satellite-derived SST, such as monitoring the positions and evolution of surface expressions of thermal fronts, or meso-scale features in the ocean, require measurement precision or relative accuracy, with absolute accuracy of lesser importance. However, for many applications, the absolute accuracy of the satellite-derived SST is of prime importance (Minnett 2010). The application with the most demanding accuracy requirement is climate research, where a multi-decadal time series of global SST is required to detect small changes against a background of short-term variations (NRC 2000). Analysis of SST time-series data for signatures of change over long periods requires measurement uncertainties and errors to be less than the anticipated signal or magnitude of variation, which is likely to be <0.2 °C per decade and thus requires consistent and accurate SST retrievals (Donlon 2010; Minnett 2010). Given that data records of SST may span several satellite missions, it is crucial to ensure that the validating measurements are themselves accurate over the climate data record period. It is also critical to remove artifacts caused by orbital-decay effects. All of the NOAA polar-orbiting platforms lose altitude after launch, which can result in an artificial cooling trend in the satellite-derived temperature series if not corrected. The decay correction varies from year to year, being greatest during periods of maximum solar activity when higher levels of solar ultraviolet radiation heat the upper atmosphere and increase the drag on the spacecraft (Wentz and Schabel 1998). Removing the effect of orbital decay and the derivation of trends from multiple satellites is a complex process that requires accurate and robust climate-monitoring systems. The accuracy that can be obtained for SST derived from TIR data is now at the limit of its capabilities with respect to the currently available operational in situ infrastructure, with biases achieved in global SST compared to drifting-buoy measurements typically <0.15 °C (Donlon 2010; O'Carroll et al. 2008).

12.1.3 Quality Control

SST measurements are generally provided with a separate "quality" field that is often derived from the cloud-screening portion of the retrieval algorithm. The overall quality flag associated with each SST value is a relative assignment of SST quality based on a hierarchical suite of tests, which allows users to mask SST values based on the quality threshold that meets their specific needs. However, data providers derive quality fields differently, requiring special attention by the user to the different meanings listed in the relevant documentation (Donlon 2010). For example, MODIS SST quality levels are set from 0 to 4, where 0 indicates best quality and

4 is complete failure (oceancolor.gsfc.nasa.gov/DOCS/modis_sst). In contrast, AVHRR Pathfinder quality flags vary from 0 to 7, with 0 being the lowest and 7 the highest (www.nodc.noaa.gov/sog/pathfinder4km/userguide.html). For most applications, using SST observations with quality levels of 4–7 for AVHRR Pathfinder, or 0–2 for MODIS, is typical. For applications requiring the most accurate observations (at the expense of the number of observations), only the highest quality levels should be used (Kilpatrick et al. 2001) and remaining values masked out of the analysis. Significant progress on improving cloud-screening, atmospheric aerosol detection and flagging algorithms to provide increased sensitivity and performance is expected in the next decade as climate-quality SST data are derived from a variety of sensors (Casey et al. 2010; Donlon 2010).

12.2 Thermal Data Products and Analysis

12.2.1 AVHRR Pathfinder Series

Of the many oceanic parameters capable of being observed from space, only SST from the NOAA AVHRR series enjoys an unbroken, nearly 30 year long history of observation from the same class of instrument. The AVHRR observations date back to 1981, and are expected to continue for perhaps another 10 years (Casey et al. 2010). The Pathfinder Program was conceived in 1990 as a NOAA and NASA Cooperative Agreement to support the needs of the US Global Change Research Program (King and Greenstone 1999) and to provide unprecedented access to large volumes of consistently processed satellite datasets. The official goals of the AVHRR Pathfinder project, as presented in 1992, were to produce: (a) long-term global datasets for research, modeling, and trend analysis; (b) datasets consistently processed with the best available community consensus algorithms; (c) multiple geophysical products from a common input stream using integrated processing concepts; and (d) a consistent, low maintenance, readily accessible archive for AVHRR data with browse and online access capabilities (Casey et al. 2010). Thus, the Pathfinder objective was to "find the path" forward in preparation for the ambitious Earth Observing System series of instruments and missions that followed (Casey et al. 2010).

Over its 20 year history, the Pathfinder SST Program has produced five distinct versions of SST products, and is currently developing its sixth. The Pathfinder SST algorithm is based on the Non-Linear SST (NLSST) algorithm (Walton et al. 1998; Kilpatrick et al. 2001), with numerous modifications over time to improve algorithm performance (Casey and Cornillon 1999; Kilpatrick et al. 2001; Casey et al. 2010). In 2002, the term "Climate Data Record" had emerged and the Pathfinder SST Program entered a new phase to deliver Pathfinder SST Version 5, a more accurate, consistent, and finer resolution SST data set than was available previously. Improvements in spatial resolution (4 km), the land mask, and the quality level determinations were implemented in Version 5, amongst others (Casey et al. 2010). Access to the Pathfinder Version 5.2 collection, including the entire record from 1981 to 2010 (www.nodc.noaa.gov/SatelliteData/pathfinder4km), is available online with full descriptions of the products and parameters provided in the User Guide. In addition to the SST values, Version 5.2 also provides wind speed, sea ice and aerosol content, as well as several other useful parameters that can be used to calculate higher-order statistics, individually determine what constitutes a "good" SST value, and examine the quality test output.

Pathfinder Version 6.0 is in preparation (Casey personal communication) and, significantly, will conform to the newest Group for High-Resolution SST (GHRSST) data format, data content, and metadata requirements. These improvements in Version 6 will enhance the compatibility of Pathfinder data with numerous GHRSST-compliant data streams, and will result in a more accurate, consistent, and useful climate data record for SST that will enable an even greater range of science and applications (Casey et al. 2010).

12.2.2 Group for High-Resolution SST

To address an emerging need for accurate, near real-time, moderate resolution SST products, the Global Ocean Data Assimilation Experiment (GODAE) High Resolution SST Pilot Project (GHRSST-PP) was initiated as an international collaboration in 2002, and replaced in 2009 by the Group for High-Resolution SST (GHRSST). The aim of the GHRSST is to provide the best quality SST data for applications in short, medium and decadal/climate time scales in the most cost effective and efficient manner through international collaboration and scientific innovation (www.ghrsst.org). Over the last decade, the GHRSST Project has provided a framework for the study and application of global SST at <10 km resolution using TIR and PMW satellite and in situ data, with applications demonstrated in ocean and atmospheric forecasting systems (Donlon et al. 2007). Key developments in the success of the GHRSST Project (Donlon et al. 2009) include, amongst others: (a) international agreement on the definitions of different SST parameters in the upper layer of the ocean that distinguish between measurements made by TIR radiometers, PMW radiometers, in situ sub-surface observations and merged SST outputs; (b) extensive research to ensure that SST diurnal variability is properly flagged within observational data; (c) methods to correct for bias in different satellite data sets; (d) methods to convert between radiometric skin SST and the bulk temperature measured by ships and buoys now used by operational SST analysis systems; (e) satellite SST data product formats and product content to include measurement uncertainty estimates for each derived SST value; and (f) operational implementation of new SST analysis products using in situ data merged with PMW and TIR satellite data. In essence, GHRSST has built a new generation of data products and services providing wide and open access in near real-time to many SST data products. Details of GHRSST SST products are available on the GHRSST website, including the spatial resolution and available time periods for each dataset (ghrsst.jpl.nasa.gov/GHRSST_product_table.html). A GHRSST international user community is currently testing and applying GHRSST data products and services within scientific projects and operational systems in real-time. The challenge for the GHRSST team is to deliver sustained production of stable, high-quality SST data products and services, and for the user community to demonstrate a corresponding requirement for sustained operations.

Long-term gridded datasets are also available that blend satellite and in situ data to provide greater context for long-term trends in SST. For example, the NOAA Extended Reconstruction Sea Surface Temperature (ERSST) analysis is generated using in situ SST data and statistical methods that allow stable reconstruction using sparse data. The monthly ERSST analysis extends from 1854 to present at 2° spatial resolution (Smith and Reynolds 2004). Although local and short-term variations have been smoothed, ERSST data are suitable for long-term global and basin-wide studies. As another example, the NOAA Optimum Interpolation (OI) SST analysis is a weekly and monthly product available from 1981 to present at 1° spatial resolution (Reynolds et al. 2002) that integrates in situ measurements, satellite SST, and SST simulated by sea ice cover. The use of re-analysis or reconstruction techniques, however, must be met with caution as differences in data density and acquisition characteristics can lead to different results within the derived SST output. Nonetheless, these datasets are useful for comparison with other variables and to simulate long-term trends in SST. In essence, in any application of thermal data, it is important to consider the characteristics, quality, limitations, caveats and time domain of the SST datasets in order to wisely select the product(s) most appropriate to the specific application.

12.2.3 Quantifying Trends and Changes

Variations in SST, as indicated through the satellite data record, show that most of the ocean has warmed. Example of studies undertaken to quantify this and other temporal trends in SST are presented below.

Using the retrospective AVHRR Pathfinder SST dataset for the period 1985–2006, Strong et al. (2009) evaluated trends in night-time SST of the tropical oceans (35°S–35°N) to examine implications for tropical ecosystems. Trends were calculated from monthly mean temperatures and showed that tropical oceans had generally experienced warming, or only slight cooling, with the exception of a small region of significant cooling in the north-eastern Pacific Ocean along 20°N (Fig. 12.1). Observing a shift in Pacific Ocean SST anomalies in the mid-1990s, Strong et al. (2009) also calculated trends for two eleven-year periods (1985–1995 and 1996–2006). This analysis showed significant regions of the global ocean, predominantly in the Southern Hemisphere, cooled in the earlier period (western Pacific, southern Indian and southern Atlantic); however, these same regions exhibited substantial warming trends during the second period. The observed shift may have coincided with a reversal of the Pacific Decadal Oscillation (PDO; Mantua et al. 1997), which moved from a so-called warm phase to a cool phase in the mid-1990s. Typical conditions during the warm phase of the PDO are positive SST anomalies in the equatorial Pacific, with negative anomalies in the polar regions of the ocean (the pattern is reversed during the cool phase). These findings suggest that warming trends in the tropical Pacific calculated for the overall 22 year period from 1985 to 2006 might have been lessened due to the PDO shift from warm to cool. This illustrates the significant point that the time domain through which trend analyses are undertaken must be carefully considered during interpretation of the outcomes.

Eakin et al. (2009) examined SST anomalies at 50 coral reef locations that were observed to have bleached during the 1997–1998 global bleaching event (Wilkinson 1998). Annual average SST anomalies were calculated for each of four regions as well as the long-term regional and global trends for 1985–2006 (Table 12.1). Each region showed significant warming through the 22 year period, with the Pacific Ocean warming at a slightly lower rate (though not statistically different). The authors noted that the calculated global trend was already within the range of predicted warming for the 21st century (2–4 °C/century; IPCC 2007) in the absence of a reduction in greenhouse gas emissions. Analysis of accumulated thermal stress through the satellite era showed an increasing trend for all regions. The greatest level of thermal stress was experienced both globally and regionally in 1998, with the exception of the Caribbean/Atlantic region for which stress levels were most severe during the bleaching event of 2005 (Eakin et al. 2010).

Eakin et al. (2009) also examined accumulated thermal stress dating from the late 19th century using historical SST datasets (HadISST, ERSST) for the same 50 reef locations. These analyses confirmed that the thermal stress levels observed in the Caribbean/Atlantic during 2005, and regionally and globally during 1998, were unprecedented in the data record. Heron et al. (2008) similarly calculated regional trends in SST anomalies of reef-containing pixels for the period 1880–2007 (ERSST data). Temperature increases were observed in all regions, ranging from 0.024 °C per decade in the eastern Pacific Ocean to more than 0.050 °C per decade in the Indian Ocean and south-western Pacific Ocean. Comparison of these values with those from the satellite era (Table 12.1) suggests that warming has accelerated through the last century.

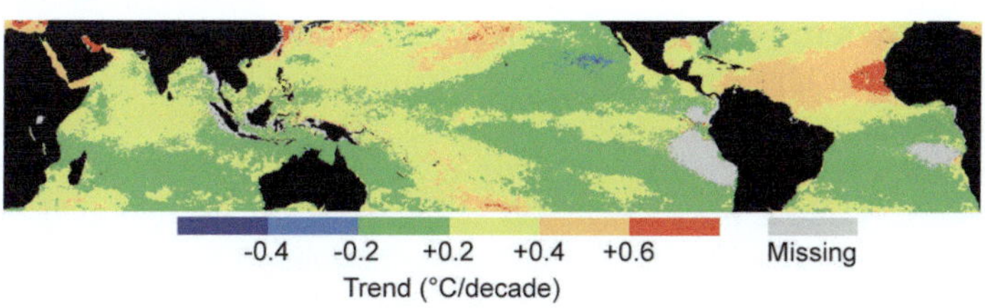

Fig. 12.1 SST trends (1985–2006) showing that the tropical ocean has generally warmed or slightly cooled during recent decades (after Strong et al. 2009)

Table 12.1 Trends in SST anomaly at previously-bleached reefs for five ocean regions, calculated from Pathfinder AVHRR SST data for 1985–2006

Region	No. of reef-containing pixels averaged	Trend in SST anomaly (°C/decade)
Indian Ocean and Middle East	18	0.261
Southeast Asia	9	0.232
Pacific Ocean	11	0.181
Caribbean and Atlantic Ocean	12	0.257
Global	**50**	**0.237**

Regional anomaly values were averaged across specific reef locations that were observed to bleach in 1997–1998 (after Eakin et al. 2009)

Heron et al. (2009) analyzed the Pathfinder data from 1985 to 2006 to determine trends in SST of the climatologically warmest month (which is specific to each location) to examine changes in the level of summertime thermal stress as an indicator of coral bleaching risk. This study showed warming trends over much of the globe, particularly in the Northern Hemisphere (Fig. 12.2a). Comparison of trend values with their standard error (Fig. 12.2b) showed only small oceanic areas cooling at a rate significantly different from zero, primarily in the southern ocean. However, much of the western and northern Pacific Ocean and the northern Atlantic Ocean experienced SST increases of more than double the standard error of the trend.

A major source of short-term climate variation is the well-known El Niño-Southern Oscillation (ENSO) phenomenon in which the extreme phases of El Niño and La Niña are linked to characteristic oceanic and atmospheric conditions that typically evolve over 12–18 months (McPhaden 2004). ENSO events do not, of themselves, cause coral bleaching; however, they can modulate the likelihood of bleaching occurring in different ocean basins by modifying general SST patterns. Eakin et al. (2009) examined how SST patterns compared across 20 El Niño, 20 La Niña and 20 ENSO-neutral events, considering average patterns in both the year of onset and the following year. The study indicated that much of the tropical ocean experiences significantly warmer maximum SST during El Niño events as compared with neutral years, initially in just the eastern Pacific Ocean and during the following year also in the Indian Ocean and greater Caribbean region. It is of note that a 'horseshoe' of cooler conditions 'arcs' from the north-central Pacific to the western Pacific, continuing to the south-central Pacific during El Niño years. However, this pattern reverses during the years of La Niña conditions, such that the locations in the horseshoe experience warmer conditions while other regions experience generally cooler conditions (compared with ENSO-neutral years). ENSO conditions can be monitored using ocean temperature indices such as the Niño 3.4 index, which is the average SST anomaly across the region [170–120°W, 5°S–5°N], and atmospheric indices such as the Southern Oscillation Index, which is the atmospheric pressure difference between Tahiti, French Polynesia and Darwin, Australia. It is of note that there has been recent evidence of anomalous ocean warming in the central Pacific Ocean that has not followed the

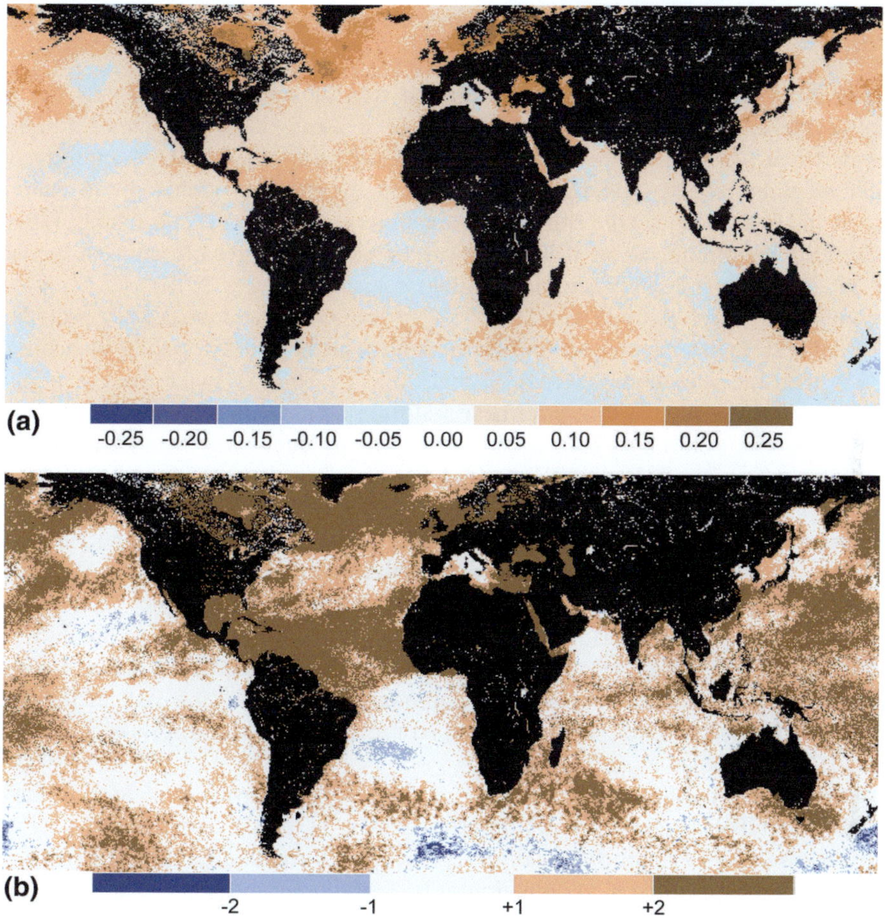

Fig. 12.2 **a** Trend in mean SST (°C/year) of the climatologically warmest month for the period 1985–2006. **b** Ratio of this trend to its standard error, showing where warming/cooling has been significantly different from zero (after Heron et al. 2009)

characteristics of traditional El Niño events (Ashok et al. 2007). Referred to as El Niño/La Niña Modoki, it has been suggested that these Modoki events are becoming more common with global warming and may be more typical of future conditions (Yeh et al. 2009; Lee and McPhaden 2010). Further, a recent study by Redondo-Rodriguez et al. (2011) has found that the influence of ENSO and Modoki events has a distinct spatial structure on the Great Barrier Reef (GBR). Classical ENSO events show no significant relationship with SSTs on the northern GBR, but have a strong signature on the southern GBR. Conversely, El Niño/La Niña Modoki is significantly related to summer SSTs on the northern GBR, but not for the southern GBR.

12.2.4 Application to Reef Management

Coral bleaching results from the stress and loss of symbiotic algae, known as zooxanthellae, from coral tissues during times of stress, often due to temperatures higher than the tolerance level of the coral colony (Glynn 1993). Monitoring of SST at global scales provides researchers and stakeholders with tools to understand and better manage the complex interactions leading to coral bleaching. When bleaching conditions occur, these tools can be used to trigger bleaching response plans and support appropriate management decisions (Beggs 2010).

NOAA's Coral Reef Watch (CRW; www.coralreefwatch.noaa.gov) provides current reef environmental conditions to identify areas at risk for coral bleaching. CRW uses the operational NOAA/NESDIS 0.5° (~50 km), twice weekly, night-time, AVHRR SST composites for near real-time monitoring of thermal stress that can cause coral bleaching (Liu et al. 2009). This product suite includes both data and global maps of SST and SST anomalies, as well as the coral-specific products HotSpot (positive anomaly from the warmest month climatology) and Degree Heating Weeks (DHW; accumulated thermal stress as given by HotSpot). A full description of these products is given by Skirving et al. (2006). More recently, additional management products have been added. The Bleaching Area Alert Product consolidates the HotSpot and DHW data to provide maps of the current status of thermal stress as it pertains to coral bleaching. Automated emails, through the Satellite Bleaching Alert system, summarize the bleaching risk at a user-nominated subset of (currently) 227 coral reef locations and inform stakeholders of status level changes at selected reefs.

In addition to NOAA's products, Australia's Great Barrier Reef Marine Park Authority (GBRMPA) uses ReefTemp (Maynard et al. 2008), a regionally-specific, higher resolution mapping product that provides information on bleaching risk in near real-time based on a combination of SST and other measures of temperature stress (www.cmar.csiro.au/remotesensing/reeftemp). ReefTemp builds upon the work by the NOAA CRW in that it 'nowcasts' bleaching risk daily at 0.018° (~2 km) resolution. Images of SST and associated measures of temperature stress produced by ReefTemp are based on comparing data from a 10 day composite of AVHRR SST to monthly average temperatures calculated from 1992 to 2002 (Griffin et al. 2004). Where the SST cannot be calculated due to cloud cover, the most recent SST is maintained for that location up to 10 days (for most locations SST values are typically updated every 1–4 days). The legend used to describe temperature stress is calibrated to in situ observations of bleaching stress from around Australia. ReefTemp is an important component of the GBRMPA's Coral Bleaching Response Plan (www.gbrmpa.gov.au/corp_site/key_issues/climate_change/management_responses) as it forms part of the early warning system, targeting assessment and monitoring plan for sites that experience stress levels conducive to bleaching.

Seasonal forecasting of potential thermal conditions leading to mass coral bleaching is performed by NOAA CRW, using output from SST forecast models, which can be used to predict anomalous SST several months in advance. CRW's current Bleaching Outlook products are based on the predictions of two forecast systems: (1) the statistical Linear Inverse Modeling (LIM) system provides deterministic weekly forecasts of SST at 2° resolution (Liu et al. 2009) and, (2) the more-recently employed dynamical Climate Forecast System (CFS) uses a 28-member ensemble to provide a range of SST predictions for each location and at each forecast time-step. The Bleaching Outlook is determined from these models by incorporating forecasts out to 18 weeks, with the potential risk level for stress updated each week. The LIM-based outlook indicates the forecast level of stress, while the CFS-based outlook provides the probability that identified stress thresholds will be exceeded. Interpretation of the Bleaching Outlook is provided as necessary on an approximately monthly basis during bleaching risk periods.

The Australian Bureau of Meteorology has implemented operational dynamical seasonal forecasts of SST anomalies to aid in the management of the GBR Marine Park using the Predictive Ocean Atmosphere Model for Australia (POAMA; Spillman and Alves 2009). POAMA is a global coupled ocean–atmosphere ensemble seasonal forecast system, where the resolution of the ocean model is 2° in the zonal direction and 0.5–1.5° in the meridional direction. This is the first application of a dynamical seasonal model for coral bleaching, providing valuable forecast tools for reef managers at useful time scales (Spillman and Alves 2009). Real-time forecasts of SST anomaly out to six months are produced daily based on the ensemble mean of the last 30 daily forecasts. The system also calculates a GBR index, the areal mean of monthly SST anomalies within the GBR study region, as an indicator of the average regional conditions. In addition, thermal stress referenced to the warmest month SST climatology and accumulated for the subsequent 3 months of predictions is presented as Degree Heating Months (Spillman et al. 2010). The use of a forecast ensemble allows the assessment of the likelihood of a particular SST anomaly or thermal stress threshold being exceeded (Spillman and Alves 2009). The model has shown good predictive skill in reproducing past monthly SST anomalies and indices with 0–2 months lead time in the GBR. However, uncertainty in the onset of the monsoon currently limits accuracy of summer predictions in the region at longer lead times (Spillman et al. 2010; Spillman 2011). Although the model has the capability to predict the Madden Julian Oscillation (MJO), and hence the onset of the monsoon, its effectiveness is ultimately limited by overall predictability of the system (Rashid et al. 2010). Future improvements in model resolution, initialization and ensemble creation, together with improved understanding of the predictability of these primary drivers, will lead to more skillful forecasts at longer lead-times.

Reef forecasting systems require highly accurate instrumentation to provide climate quality SST over shallow near-shore waters for long periods (Gramer et al. 2009; Liu et al. 2009). Higher spatial and temporal resolution is needed than the SST products currently used by NOAA's CRW in order to effectively model the coral bleaching response in distinct sub-regions of a reef system (Beggs 2010). SST observations at 1 km resolution from AVHRR and MODIS sensors are particularly useful for extreme shallow water coverage and for measuring SST close to coasts. These capabilities are currently used by a number of research groups for coral reef applications, but are not yet available on an operational basis.

Observations of diurnal variation of SST are necessary to accurately monitor thermal stress on coral (Maturi et al. 2008). Geostationary satellites, with repeat acquisitions every $\sim$30 min, provide a measure of the diurnal variation more effectively than SST products derived from polar-orbiters, which only observe a location, at most, twice per day. For example, the current methodology applied globally by NOAA CRW uses night-time only AVHRR (polar-orbiting) retrievals to determine daily SST, and then calculates SST anomaly values to represent the level of anomaly at the depth of corals during bleaching events (Skirving et al. 2006). Given that corals are sensitive to temperature extremes as well as thermal exposure over time, a bleaching prediction technique based on full diurnal observations of SST (Leichter et al. 2006) would be superior to techniques that are updated just once daily (Maturi et al. 2008). The GHRSST is currently conducting extensive research to ensure that SST diurnal variability is properly flagged within observational data and developing methods to correct for bias in different satellite datasets. NOAA is investing in blended SST products, which combine measurements from polar and geostationary satellites. It is of note that sub-surface variation of temperature needs be included in the interpretation and application of SST products derived from full diurnal coverage.

There are also several developing products, based on satellite SST, that describe various coral ecosystem parameters, currently being evaluated. For example, NOAA CRW experimental products include descriptions of the return period for bleaching-level stress, the duration of historical stress events, and recent short-term trends in SST. A regional product describing the risk of coral disease outbreak based on recent winter and summer temperatures is also available (Heron et al. 2010). ReefTemp produces experimental risk maps for research based on temperature conditions that are conducive to coral disease outbreaks (Maynard et al. 2010; www.cmar.csiro.au/remotesensing/reeftemp/web/ReefTemp_Disease.htm).
Another NOAA CRW experimental product combines the effects of satellite-derived temperature and solar radiation to provide an alternative predictor for coral bleaching. The Light Stress Damage (LSD) algorithm has been developed by specifically considering the photosystem of the coral symbiont (zooxanthellae) and examining the physiological responses of the coral-algal symbiosis under varying light and temperature conditions. While the existing CRW algorithms were developed on observed coral response to temperature alone, the LSD algorithm has been constructed using specially-designed physiological experiments to determine the influence of temperature on light thresholds (Nim and Skirving 2010). Preliminary results have led to improved bleaching prediction using the LSD algorithm over temperature-only products for the given test sites.

12.2.5 Limitations

The current limitations and concerns when using satellite-derived SST and model-derived products stem from a number of sources:

- First, and possibly most importantly, the accuracy of the output is entirely dependent on how well the ecology and physical processes of reefs are understood. Invariably, the reality is that we have at best only a limited understanding of these processes. Testing and validation procedures, which are a vital component of any model or product, should ensure that model outputs correlate strongly with observed patterns. However, correlation, no matter how strong, does not imply causation (Aldrich 1995). Interpretation of results, therefore, needs to be cautious with respect to the science, and tailored with respect to the application.
- The complexity of most models is often well beyond the understanding of the non-specialist; therefore applying products and interpreting outputs in the management environment must be done with specialists involved at every step.
- Most modeling applications encompass a number of diverse disciplines, each requiring their own specialist skills to ensure that the source data and the output products are appropriately used and applied. The key to successful modeling studies is close collaboration with experts in appropriate fields of science and management.
- Satellite thermal radiometers inherently map the surface properties of the water, not that of the subsurface, which is of predominant interest in most science and management applications. While considerable work has gone into calibrating thermal SST data to 'match' the bulk water temperature, bias errors do occur, especially during periods of extreme stratification of the water column and high atmospheric water vapor content. These conditions inevitably occur in the lead-up to coral bleaching events when accuracy is most important. Caution in interpretation and prudent validation of bleaching nowcasts are therefore in order.
- The spatial and temporal mismatch between different thermal products can lead to significant problems in the application and interpretation of data. A common source of error in the literature is the application of relatively coarse satellite data to ecological data and processes at specific locations. Spatial heterogeneity on the micro-scale may not be detected using coarse spatial resolution, and conversely, patterns on the macro-scale may not be detected using fine spatial resolution. While the application of remotely-sensed thermal data has greatly advanced our understanding of coral reef ecosystems, caution should be applied in selecting the optimal resolution for the application and in the interpretation of results.

12.3 Example Thermal Applications

While the direct application of remotely sensed temperature products has been highly beneficial to reef management and reef studies in general, the largest potential gain is the integration of these empirical data with other datasets, statistical correlations, expert opinion and model outputs. Integrating these disparate information sources into a modeling framework allows for much more complex problems to be objectively examined, thereby shedding new light on the physical and ecological dynamics of coral reefs. It also enables new tools to be developed for reef managers, allowing them to evaluate management strategies on the basis of sound science, taking into account competing ecological, social and economic demands. An example of such a tool is *Marxan*, an integrative software system used in terrestrial and marine protected area design. It was used in the rezoning of the GBR in 2003–2004 and has since become the most widely used conservation planning software (Fernandes et al. 2005; Watts et al. 2009). The following case studies are examples of recent inroads that remotely sensed SST data have made through integration and modeling.

12.3.1 Marine Protected Area Design

The concept of building resilience principles into Marine Protected Area (MPA) design and management is relatively new and incorporates four main principles: (1) spread the overall risk from catastrophic events, like bleaching, to increase likelihood of reef survival; (2) identify and protect key sites; (3) incorporate patterns of connectivity and replenishment; and (4) inform effective and adaptive management (Green et al. 2007). These principles are being put into practice in the design of new MPAs in Kimbe Bay, Papua New Guinea (www.nature.org) and the Bahamas (www.livingoceansfoundation.org). One of the critical new concepts in these programs is explicitly incorporating the effects of climate change into the planning of the marine reserves. Spreading the risk from bleaching is achieved by exploring and utilizing the complex mosaic of thermal environments that often exist around coral reefs. Identifying and mapping these areas allows a range of risk profiles to be incorporated into reserves, which reduces the likelihood of wholesale impact.

Mumby et al. (2011) mapped the chronic and acute thermal stress environments in the Bahamas and, in conjunction with connectivity models, used these to develop and test reserve designs with a view to maximize resilience of reefs in the future. Results show that scales of larval dispersal are commensurate with desirable connectivity outcomes, and that 'adaptation' (in the vernacular sense, i.e., either genotypic or phenotypic) scenarios can dramatically influence the optimal reserve network design. The study describes the following scenarios and prioritization strategies (Fig. 12.3):

- *'Adaptation' adequate*. If adaptation rates are adequate, it is best to prioritize connections among and between reefs with low chronic and acute stress, and reefs with low acute but high chronic stress. Reefs in high stress (chronic and acute) environments might serve as an 'evolutionary pump' where adaptation could be faster and more directional.

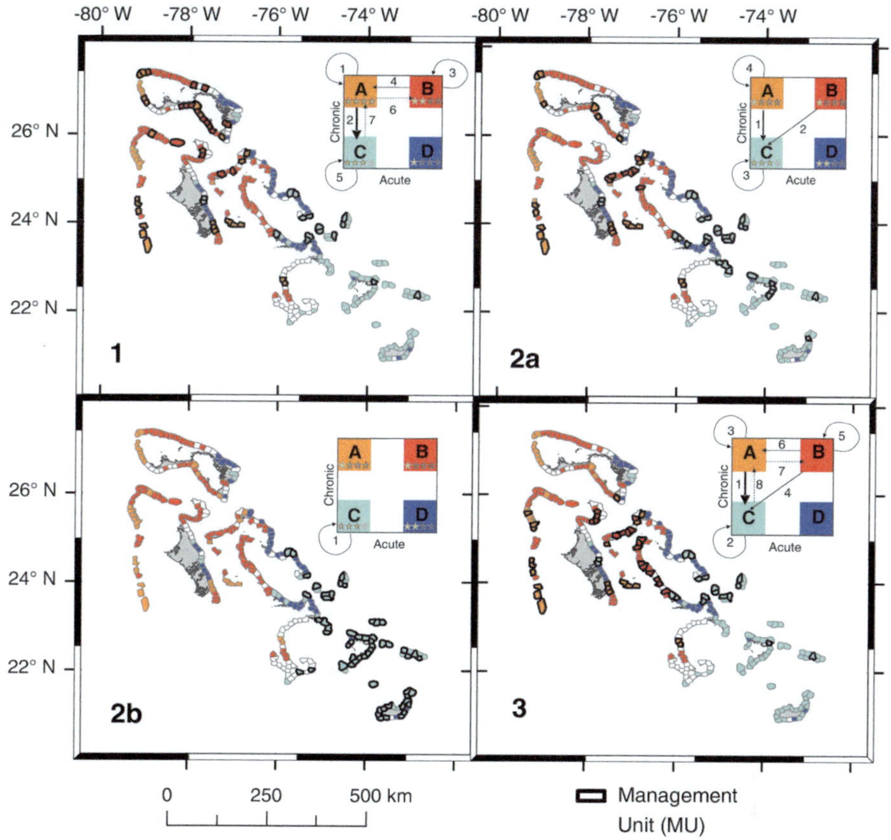

Fig. 12.3 Optimal reserve networks in the Bahamas for scenarios *1* ('adaptation' adequate), *2a* ('adaptation' inadequate but genetic), *2b* ('adaptation' inadequate but phenotypic), and *3* (the bet-hedging strategy). Figures show the location of individual reef planning units, where those with bold edges are possible reserves. Inset diagrams reveal the prioritization of connections for each scenario (after Mumby et al. 2011)

- *'Adaptation' inadequate, but genetic.* If adaptation rates are inadequate but existing patterns of phenotypic acclimation have a genetic basis, then it would be best to prioritize for the migration of thermally-resistant genotypes from reefs with high chronic stress today to those experiencing relatively low chronic and low acute stress.
- *'Adaptation' inadequate, but phenotypic.* If present patterns of local acclimation are purely phenotypic, reserve design would not require migration of stress-tolerant offspring. In this scenario it would be best to prioritize connections among those reefs that experience the lowest chronic and acute stress as these may have the capacity to acclimatize while stressed reefs might eventually experience lethal conditions.

- *Bet-hedging.* In the face of uncertainty on adaptation rates and whether any adaptation will be genetically or phenotypically driven, a 'bet-hedging' strategy may be more appropriate based on a combination of the number of connections prioritized in the above scenarios and the ranking of the most important connections.

In the Bahamian example, Mumby et al. (2011) found that the bet-hedging scenario produced a reserve design that was most similar to the 'Adaptation inadequate, but genetic' scenario, where a genetic basis for adaptation is present in existing communities. They conclude that although it was difficult to create a compromise strategy that hedged against all scenarios, there was a subset of 15 % of sites that were selected under all scenarios, making them ideal candidates for early protection and inclusion in a reserve network.

12.3.2 Water Quality and Coral Bleaching

Wooldridge and Done (2009) used a Bayesian approach for modeling and data integration to shed new light on ecological processes associated with a possible interaction between warming ocean temperatures and water quality. The mechanistic basis for this link is that elevated dissolved inorganic nitrogen (DIN) stimulates enlarged (suboptimal) coral symbiont populations with higher metabolic demands during periods of elevated irradiance and temperature. More energy consumed by the symbionts themselves in turn causes reduced translocation of energy to the host coral, dampening host productivity and CO_2 production. This in turn causes less CO_2 to diffuse to the symbionts. A limited CO_2 substrate around the Rubisco enzyme in the symbionts is postulated to disrupt their "dark" photosynthetic reactions, thereby leading to bleaching (Wooldridge 2009a). Thus, high levels of DIN make the coral-algal symbiosis less stable so that bleaching effectively happens at lower temperatures than would be the case if DIN levels were low.

Wooldridge and Done (2009) tested the DIN link by showing that observed bleaching patterns on the GBR in 1998 and 2002 (Berkelmans et al. 2004) correlated better when coastal DIN and thermal history were included as additional explanatory variables over temperature stress alone (predictive accuracy of 84 vs 73 % respectively). The onset of bleaching occurred $\sim$1–1.5 °C higher in reefs with good water quality and lower thermal history (Fig. 12.4). The implications for reef management are that managing nutrient runoff from rivers can have a tangible benefit to coral reefs and improve their resilience while they cope with the impacts of warming waters as a result of climate change (Wooldridge 2009b). These studies also provide opportunities for further integration by including socio-economic modeling to assess and optimize the tradeoffs between the economic costs of reducing fertilizer use and the benefits to the ecology of the reef and industries which are dependent on a healthy reef, such as tourism (Thomas et al. 2009).

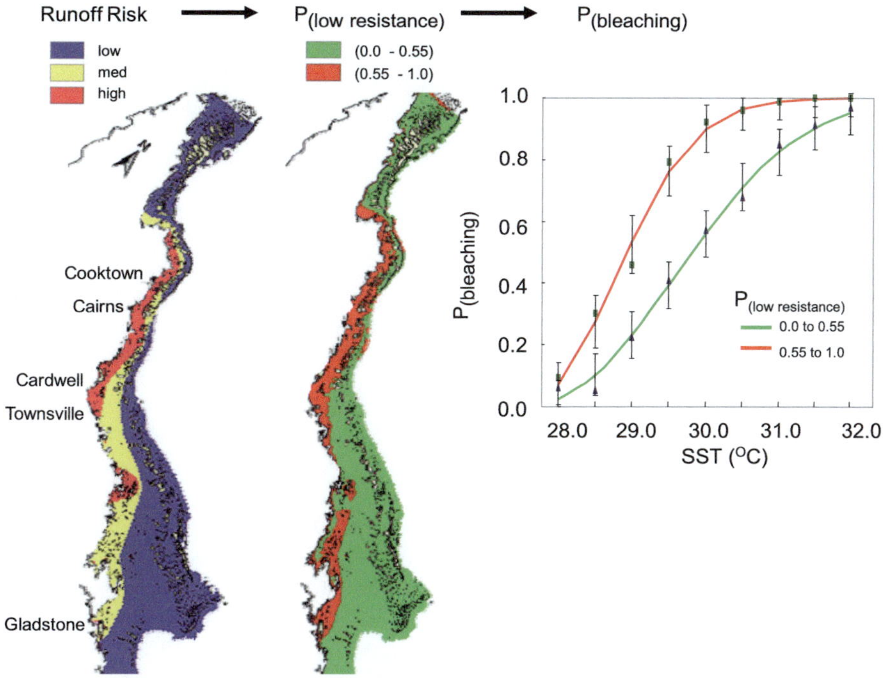

Fig. 12.4 Spatial distribution of the risk of compromised water quality and modeled resistance to bleaching in the Great Barrier Reef (GBR) based on thermal history, water quality and heat stress during the 1998 bleaching event. The probability of bleaching per unit of SST is reduced by 1–1.5 °C under good water quality conditions (after Wooldridge 2009b)

12.3.3 Coastal and Oceanic Upwelling

Understanding and predicting the impacts of global climate change is dependent on understanding the dynamics of ocean processes from global to local scales. While large-scale oceanographic processes are often well understood, smaller-scale processes, such as upwelling in tropical reef environments, are often not. For example, Berkelmans et al. (2010) found upwelling activity in the central GBR correlated strongly with local and regional SST, with the strongest upwelling occurring during the bleaching years of 1998 and 2002 (Fig. 12.5). In these situations, warm water still encompasses much of the GBR and the cold upwelled water remains mostly subsurface due to a heavily stratified water column. A link was also identified between upwelling activities and anomalously warm summers, when the East Australian Current speeds up during doldrum periods, leading to enhanced stratification of the water column. Enhanced upwelling signals are thus correlated with bleaching by a period of several weeks, and may also be used as a seasonal forecasting tool for a pending warm summer.

In another example, Weeks et al. (2010) use MODIS time series of SST and chlorophyll-*a* data in conjunction with in situ measurements and model

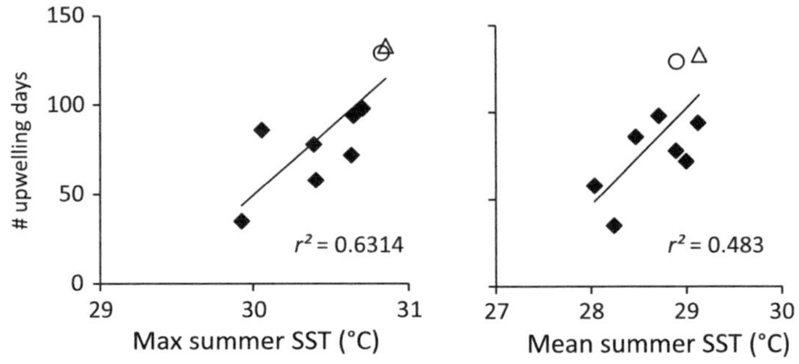

Fig. 12.5 Relationship between AVHRR Pathfinder summer (December, January, February) mean (*right*) and maximum (*left*) SST for the whole of the Great Barrier Reef (GBR) versus upwelling activity at Myrmidon Reef (central GBR). Non-bleaching summers—*closed diamond*; bleaching summer of 1997/98—*open circle*; bleaching summer of 2001/02—*open triangle* (after Berkelmans et al. 2010)

observations to identify and assess a meso-scale eddy feature off the east Australian coast (the Capricorn Eddy). This eddy ($\sim$150 km diameter) forms in the lee of the southern GBR as the East Australian Current continues its southward flow, entraining water behind the 'cape-like' coastal feature of the Great Barrier Reef (Fig. 12.6). Frictional drag resulting from the shallowing sea floor and nearby land boundary, together with the swiftly flowing offshore current, produces lateral stress on the water mass in this lee zone and exerts a cyclonic (clockwise) rotation on the water. Weeks et al. (2010) used an operational 3D ocean forecast model (OceanMAPS) that was able to replicate eddy patterns observed in the satellite imagery. In this model, wind stress and in situ sea temperature data (at 5 depths from 13 to 44 m) revealed the dynamic thermal structure of the eddy. The cyclonic rotation causes upwelling in the centre of the eddy as well as at the 'shear zone' of the East Australian Current (Fig. 12.7). The system is inherently dynamic, waxing and waning with the strength of the East Australian Current, local and regional wind stress, and internal perturbations of the thermocline.

This meso-scale oceanographic feature has a marked effect on the ecology of the reef. Weeks et al. (in review) show that the foraging ecology of the wedge-tailed shearwater (*Puffinus pacificus*) is inextricably linked to the Capricorn Eddy. This bird nests on Heron Island near the southwest edge of the eddy and feeds on planktivorous fish in the eddy zone. During periods of strong summer East Australian Current flow and eddy rotation, enhanced upwelling causes stratification of the water column, which is exacerbated during periods of low wind and high insolation in summer. In early 2006 these conditions caused mild bleaching at Heron Island, but more significantly had a direct effect on sea birds as their prey appeared to remain in the cooler water below the thermocline, unavailable to the surface-feeding birds. Meal sizes and the proportion of nests fed halved during the period of intense stratification, similar to conditions which led to a dramatic failure in the reproductive success of *P. pacificus* coinciding with the coral bleaching event of 2002.

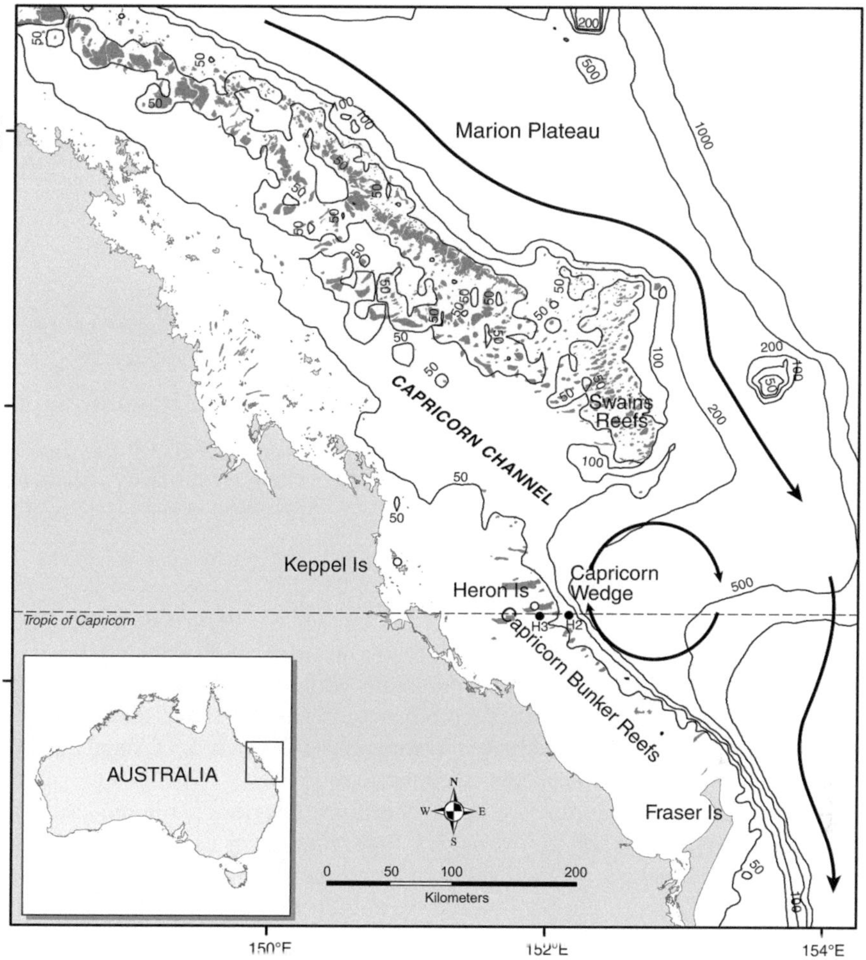

Fig. 12.6 The Capricorn Eddy forms in the lee of the southern end of the Great Barrier Reef (after Weeks et al. 2010)

12.4 Future Directions

The future of remotely sensed thermal data, both for management and for understanding physical/ecological processes, will undoubtedly be built on the successes achieved so far. In the 1980s and early 1990s, remote sensing was seen as the "tool without application" (Andrefouet and Riegl 2004). This stigma had largely dissipated by the late 1990s with the development of SST products that directly benefitted reef managers and researchers. The NOAA global HotSpot and DHW products described above have played a significant role in cementing the role of remote sensing in reef management as well as for the broader reef research

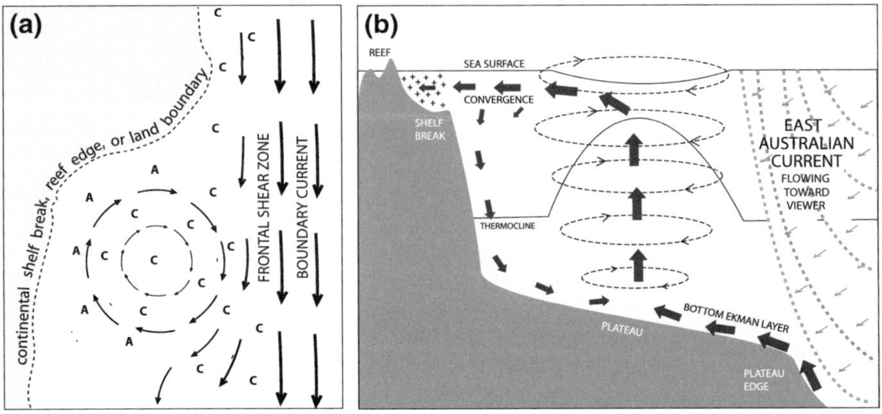

Fig. 12.7 **a** The rotational force imposed on the eddy as a result of the frontal shear zone from the East Australian Current and the frictional drag from the shallowing bathymetry adjacent to the coast causes upwelling in areas marked with a "C" and downwelling in areas marked with an "A". **b** Upwelled water is pushed onto the shelf break but mostly remains subsurface, causing stratification of the water column, especially during hot, doldrum weather conditions (after Weeks et al. 2010)

community. Integrating remotely sensed thermal data with other datasets and modeling will allow questions and applications with ever increasing complexity to be addressed. The experimental LSD product and the case studies described above are good examples of the increased complexity in which SST data are being applied, and this trend is expected to continue.

Aside from addressing more complex questions, integration and modeling have two distinct advantages for reef managers. First, they allow users and stakeholders to "see" where and how the data are used, exposing the linkages and assumptions in the models. Outputs can be adjusted or re-cast with alternate views, data and management scenarios. Such transparency is a keystone of modern marine reserve design and is especially important when decisions are made with incomplete or imperfect data, as is frequently the case (Fernandes et al. 2005; Causey 2010). Secondly, the education and outreach potential of models to stakeholders and the wider community cannot be underestimated. Reef managers are often unpopular with the community because management decisions inevitably do not suit all interests. Model outputs that bring stakeholders together, or successfully predict future events (e.g., bleaching, algal blooms, disease), allowing stakeholders to adapt, gives reef managers credibility and helps them build trust and partnerships that might otherwise be impossible (Keller and Causey 2005).

At a workshop on satellite monitoring of coral reefs in a changing climate (Nim and Skirving 2010), a focus group of key scientists and reef managers identified a number of new integrated remote sensing products that would benefit future science and management. Those products incorporating a significant link to thermal applications included the ability to: track and predict changes in climate oscillations (e.g., Pacific Decadal Oscillations); identify and monitor thermal fronts for aid in management of coral reef fisheries; detect and quantify ecosystem changes, such as coral mortality following bleaching and seagrass dieback; predict disease outbreaks in coral and sponge communities (see also Heron et al. 2010; Maynard et al. 2010); predict coral spawning; predict doldrum conditions ahead of bleaching and upwelling perturbations (Berkelmans et al. 2010; Weeks et al. 2010); and predict cold-water bleaching (Hoegh-Guldberg and Fine 2004). Such products will provide valuable information for both coral reef management and science applications.

Acknowledgments The authors would like to thank Peter Minnett, Ken Casey, Helen Beggs, Claire Spillman, Jeff Maynard and William Skirving for helpful comments on components of this chapter. Thanks to Gang Liu and Al Strong for assistance with Fig. 12.1. We also acknowledge and thank Peter Mumby, Scott Woolridge, and Brad Congdon for allowing us to use their work as case studies to solving more complex questions. The manuscript contents are solely the opinions of the authors and do not constitute a statement of policy, decision, or position on behalf of NOAA or the U.S. Government.

Suggested Reading

Barale V, Gower JFR, Alberotanza L (eds) (2010) Oceanography from space revisited. Springer, Dordrecht. doi:10.1007/978-90-481-9292-2_2

Martin S (2004) An Introduction to ocean remote sensing. Cambridge University Press, Cambridge, UK. ISBN 0521802806

Robinson IS (1985) Satellite oceanography: an introduction for oceanographers and remote-sensing scientists. Ellis Horwood, Chichester 455

Robinson IS (2010) Discovering the oceans from space: the unique applications of satellite oceanography. Springer, Berlin, p 638

Selig ER, Casey KS, Bruno JF (2010) New insights into global patterns of ocean temperature anomalies: implications for coral reef health and management. Glob Ecol Biogeogr. doi:10.1111/j.1466-8238.2009.00522.x

Zhang ZM, Tsai BK, Machin G (eds) Radiometric temperature measurements and applications. Academic/Elsevier, New York, pp 333–391

References

Aldrich J (1995) Correlations genuine and spurious in Pearson and Yule. Stat Sci 10:364–376

Andrefouet S, Riegl B (2004) Remote sensing: a key tool for interdisciplinary assessment of coral reef processes. Coral Reefs 23:1–4

Ashok K, Behera SK, Rao SA, Weng H, Yamagata T (2007) El Niño Modoki and its possible teleconnection. J Geophys Res 112:C11007. doi:10.1029/2006JC003798

Beggs HM (2010) Use of TIR from space in operational systems. In: Barale V, Gower JFR, Alberotanza L (eds) Oceanography from space revisited. Springer, Dordrecht. doi:10.1007/978-90-481-9292-2_2

Berkelmans R, De'ath G, Kininmonth S, Skirving WJ (2004) A comparison of the 1998 and 2002 coral bleaching events on the Great Barrier Reef: spatial correlation, patterns and predictions. Coral Reefs 23:74–83

Berkelmans R, Weeks S, Steinberg CR (2010) Upwelling linked to warm summers and bleaching on the Great Barrier Reef. Limnol Oceanogr 55(6):2634–2644

Casey KS, Cornillon P (1999) A comparison of satellite and in situ based sea surface temperature climatologies. J Clim 12:1848–1863

Casey KS, Brandon TB, Cornillon P, Evans R (2010) The past, present, and future of the AVHRR pathfinder SST program. In: Barale V, Gower JFR, Alberotanza L (eds) Oceanography from space revisited. Springer, Dordrecht. doi:10.1007/978-90-481-9292-2_2

Causey B (2010) Managing coral reefs in a changing climate. In: Nim CJ, Skirving W (eds) Satellite monitoring of reef vulnerability in a changing climate. NOAA Technical Report CRCP 1. NOAA Coral Reef Conservation Program. Silver Spring, p 53–56

Cayula JF, Cornillon P (1996) Cloud detection from a sequence of SST images. Remote Sens Environ 55:80–88

Corlett GK, Barton IJ, Donlon CJ, Edwards MC, Good SA, Horrocks LA, Llewellyn-Jone DT, Merchant CJ, Minnet PJ, Nightingale TJ, Noyes EJ, O'Carroll AG, Remedios JJ, Robinson IS, Saunders RW, Watts JG (2006) The accuracy of SST retrievals from AATSR: an initial assessment through geophysical validation against in situ radiometers, buoys and other SST data sets. Adv Sp Res 37(4):764–769

Donlon CJ (2010) Sea surface temperature measurements from thermal infrared satellite instruments: status and outlook 211–227. In: Barale V, Gower JFR, Alberotanza L (eds) Oceanography from space revisited. Springer, Dordrecht. doi:10.1007/978-90-481-9292-2_2

Donlon CJ, Minnett PJ, Gentemann C, Nightingale TJ, Barton IJ, Ward B, Murray MJ (2002) Toward improved validation of satellite sea surface skin temperature measurements for climate research. J Clim 15:353–369

Donlon CJ, Robinson I, Casey KS, Vazquez-Cuervo J, Armstrong E, Arino O, Gentemann C, May D, LeBorgne P, Piolle J, Barton I, Beggs H, Poulter DJS, Merchant CJ, Bingham A, Heinz S, Harris A, Wick G, Emery B, Minnett P, Evans R, Lewellyn-Jones D, Mutlow C, Reynolds RW, Kawamura H, Rayner N (2007) The global ocean data assimilation experiment (GODAE) high resolution sea surface temperature pilot project (GHRSST-PP). Bull Am Met Soc 88(8):1197–1213. doi:10.1175/BAMS-88-8-1197

Donlon CJ, Casey KS, Robinson IS, Gentemann CL, Reynolds RW, Barton I, Arino O, Stark J, Rayner N, LeBorgne P, Poulter D, Vazquez-Cuervo J, Armstrong E, Beggs H, Llewellyn Jones D, Minnett PJ, Merchant CJ, Evans R (2009) The GODAE high resolution sea surface temperature pilot project (GHRSST-PP). Oceanography 22(3):34–45

Eakin CM, Lough JM, Heron SF (2009) Climate variability and change: monitoring data and evidence for increased coral bleaching stress.I. In: Lough J, Van Oppen M (eds) Coral bleaching: patterns, processes, causes and consequences. Springer, Berlin, pp 41–67

Eakin CM, Morgan JA, Heron SF, Smith TB, Liu G, Alvarez-Filip L, Baca B, Bartels E, Bastidas C, Bouchon C, Brandt M, Bruckner AW, Bunkley-Williams L, Cameron A, Causey BD, Chiappone M, Christensen TRL, Crabbe MJC, Day O, de la Guardia E, Diaz-Pulido G, DiResta D, Gil-Agudelo DL, Gilliam DS, Ginsburg RN, Gore S, Guzman HM, Hendee JC, Hernandez-Delgado EA, Husain E, Jeffrey CFG, Jones RJ, Jordan-Dahlgren E, Kaufman LS, Kline DI, Kramer PA, Lang JC, Lirman D, Mallela J, Manfrino C, Marechal JP, Marks K, Mihaly J, Miller WJ, Mueller EM, Muller EM, Toro CAO, Oxenford HA, Ponce-Taylor D, Quinn N, Ritchie KB, Rodriguez S, Ramirez AR, Romano S, Samhouri JF, Sanchez JA, Schmahl GP, Shank BV, Skirving WJ, Steiner SCC, Villamizar E, Walsh SM, Walter C, Weil E, Williams EH, Roberson KW, Yusuf Y (2010) Caribbean corals in crisis: record thermal stress, bleaching, and mortality in 2005. PLoS ONE 5(11):e13969. doi:10.1371/journal.pone.0013969

Fairall C, Bradley E, Godfrey J, Wick G, Edson J, Young G (1996) Cool-skin and warm-layer effects on sea surface temperature. J Geophys Res 101:1295–1308

Fernandes L, Day J, Lewis A, Slegers S, Kerrigan B, Breen D, Cameron D, Jago B, Hall J, Lowe D, Innes J, Tanzer J, Chadwick V, Thompson L, Gorman K, Simmons M, Barnett B, Sampson K, De'ath G, Mapstone B, Marsh H, Possingham H, Ball I, Ward T, Dobbs K, Aumend J, Slater D, Stapleton K (2005) Establishing representative no-take areas in the Great Barrier

Reef: large-scale implementation of theory on marine protected areas. Conserv Biol 19:1733–1744

Gentemann CL, Minnett PJ (2008) Radiometric measurements of ocean surface thermal variability. J Geophys Res 113:C08017

Gentemann CL, Wentz FJ, Brewer M, Hilburn K, Smith D (2010) Passive microwave remote sensing of the Ocean: an overview 13–33. In: Barale V, Gower JFR, Alberotanza L (eds) Oceanography from space revisited. Springer, Dordrecht. doi:10.1007/978-90-481-9292-2_2

Glynn PW (1993) Coral reef bleaching: ecological perspectives. Coral Reefs 12:1–7

Gramer LJ, Hendee JC, Hu CM (2009) Integration of SST and other data for ecological forecasting on coral reefs. In: GHRSST 2009 International Users Symposium Conference Proceedings, Santa Rosa, USA, 29–30 May 2009, pp 110–113

Green AL, Lokani P, Sheppard S, Almany J, Keu S, Aitsi J, Karvon JW, Hamilton R, Lipsett-Moore G (2007) Scientific design of a resilient network of marine protected areas. Kimbe Bay, Papua New Guinea: the nature conservancy. Pacific Island Countries, report no 2/07

Griffin DA, Rathbone CE, Smith GP, Suber KD, Turner PJ (2004) A decade of SST satellite data. Final report for the National Oceans Office, contract NOOC2003/020, pp 1–8

Hanafin JA (2002) On sea surface properties and characteristics in the infrared. Ph.D. thesis, University of Miami, Miami, p 111

Hanafin JA, Minnett PJ (2001) Profiling temperature in the sea surface skin layer using FTIR measurements. In: Donelan MA, Drennan WM, Saltzmann ES, Wanninkhof R (eds) Gas transfer at water surfaces. American Geophysical Union Monograph, pp 161–166

Heron SF, Skirving WJ, Eakin CM (2008) Global climate change and coral reefs: reef temperature perspectives covering the last century. In: Wilkinson C (ed) Status of coral reefs of the world. Global Coral Reef Monitoring Network and Reef and Rainforest Research Centre, Townsville. ISSN 1447-6185, pp 35–40

Heron SF, Skirving WJ, Christensen TR, Eakin CM, Gledhill DK, Liu G, Morgan JA, Parker BA, Strong AE (2009) Satellite SST trends and climatologies—how many years is enough? Eos Trans AGU 90(22) Jt Assem Suppl, Abstract OS23B-03

Heron SF, Willis BL, Skirving WJ, Eakin CM, Page CA, Miller IR (2010) Summer hot snaps and winter conditions: modelling white syndrome outbreaks on Great Barrier Reef corals. PLoS ONE 5:e12210

Hoegh-Guldberg O, Fine M (2004) Low temperatures cause coral bleaching. Coral Reefs 23:444

IPCC (2007) Climate change 2007: summary for policymakers of the synthesis report of the IPCC fourth assessment report

Kearns EJ, Hanafin JA, Evans RH, Minnett PJ, Brown OB (2000) An independent assessment of pathfinder AVHRR sea surface temperature accuracy using the marine-atmosphere emitted radiance interferometer (M-AERI). Bull Am Meteorol Soc 81:1525–1536

Keller BD, Causey BD (2005) Linkages between the Florida Keys National Marine Sanctuary and the South Florida Ecosystem restoration initiative. Ocean Coast Manag 48:869–900

Kilpatrick KA, Podesta GP, Evans R (2001) Overview of the NOAA/NASA advanced very high resolution radiometer Pathfinder algorithm for sea surface temperature and associated matchup database. J Geophys Res 106(C5):9179–9197

King MD, Greenstone R (eds) (1999) Earth observing system (EOS) reference handbook. National Aeronautics and Space Administration, EOS Project Science Office

Lee T, McPhaden MJ (2010) Increasing intensity of El Niño in the central-equatorial Pacific. Geophys Res Lett 37:L14603

Leichter JJ, Helmuth B, Fischer AM (2006) Variation beneath the surface: quantifying complex thermal environments on coral reefs in the Caribbean, Bahamas and Florida. J Mar Res 64:563–588

Liu G, Christensen TRL, Eakin CM, Heron SF, Morgan JA, Parker BA, Skirving WJ, Strong AE (2009) NOAA coral reef watch's application of satellite sea surface temperature data in operational near real-time global monitoring and experimental outlook of coral health and potential application of GHRSST. GHRSST 2009 international users symposium conference proceedings, Santa Rosa, USA, 29–30 May 2009, pp 106–109

Mantua NJ, Hare SR, Zhang Y, Wallace JM, Francis RC (1997) A Pacific interdecadal climate oscillation with impacts on salmon production. Bull Am Meteorol Soc 78(6):1069–1079

Maturi E, Harris A, Merchant C, Mittaz J, Potash B, Meng W, Sapper J (2008) NOAA's sea surface temperature products from operational geostationary satellites. Bull Am Meteorol Soc 89(12):1877–1888

Maynard JA, Turner PJ, Anthony KRN, Baird AH, Berkelmans R, Eakin CM, Johnson J, Marshall PA, Packer GR, Rea A, Willis BL (2008) ReefTemp: an interactive monitoring system for coral bleaching using high-resolution SST and improved stress predictors. Geophys Res Lett 35:L05603

Maynard JA, Anthony KRN, Harvell CD, Burgman MA, Beeden R, Sweatman H, Heron SF, Lamb JB, Willis BL (2010) Predicting outbreaks of a climate-driven coral disease in the Great Barrier Reef. Coral Reefs 30:485–495. doi:10.1007/s00338-010-0708-0

McPhaden MJ (2004) Evolution of the 2002/03 El Niño. Bull Am Meteorol Soc 85(5):677–695

Merchant CJ, Embury O, Le Borgne P, Bellec B (2006) Saharan dust in nighttime thermal imagery: detection and reduction of related biases in retrieved sea surface temperature. Remote Sens Environ 104(1):15–30

Minnett PJ (2003) Radiometric measurements of the sea-surface skin temperature—the competing roles of the diurnal thermocline and the cool skin. Int J Rem Sens 24:5033–5047

Minnett PJ (2010) The validation of sea surface temperature retrievals from spaceborne infrared radiometers, 229-247. In: Barale V, Gower JFR, Alberotanza L (eds) Oceanography from space revisited. Springer, Dordrecht. doi:10.1007/978-90-481-9292-2_2

Minnett PJ, Barton IJ (2010) Remote sensing of the earth's surface temperature. In: Zhang ZM, Tsai BK, Machin G (eds) Radiometric temperature measurements and applications. Academic/Elsevier, New York

Minnett PJ, Knuteson RO, Best FA, Osborne BJ, Hanafin JA, Brown OB (2001) The marine-atmosphere emitted radiance interferometer (M-AERI), a high-accuracy, sea-going infrared spectroradiometer. J Atmos Oceanic Tech 18:994–1013

Mumby PJ, Eakin CM, Skirving WJ, Elliott IA, Paris CB, Edwards HJ, Enriquez S, Iglesias-Prieto R, Cherubin LM, Stevens JR (2011) Reserve design for uncertain responses of coral reefs to climate change. Ecol Lett 14:132–140

Nim CJ, Skirving W (eds) (2010) Satellite monitoring of reef vulnerability in a changing climate. NOAA Technical Report CRCP 1. NOAA Coral Reef Conservation Program, Silver Spring

NRC (2000) Issues in the integration of research and operational satellite systems for climate research: II. Implementation. National Academy of Sciences, Washington

O'Carroll AG, Eyre JR, Saunders RW (2008) Three-way error analysis between AATSR, AMSR-E and in situ sea surface temperature observations. J Atmos Ocean Technol 25:1197–1207

Price JF, Weller RA, Pinkel R (1986) Diurnal cycling: observations and models of the upper ocean response to diurnal heating, cooling and wind mixing. J Geophys Res 91:8411–8427

Rashid HA, Hendon HH, Wheeler MC, Alves O (2010) Prediction of the Madden-Julian Oscillation with the POAMA dynamical prediction system. Clim Dyn. doi:10.1007/s00382-010-0754-x

Redondo-Rodriguez A, Weeks SJ, Berkelmans RB, Hoegh-Guldberg O, Lough JM (2011) Climate variability of the Great Barrier Reef in relation to the tropical Pacific and El Niño-Southern oscillation. Mar Freshw Res 63:1–14. http://dx.doi.org/10.1071/MF11151

Reynolds RW (1993) Impact of Mount Pinatubo aerosols on satellite-derived sea surface temperatures. J Clim 6:768–774

Reynolds RW, Rayner NA, Smith TM, Stokes DC, Wang W (2002) An improved in situ and satellite SST analysis for climate. J Clim 15:1609–1625

Skirving WJ, Strong AE, Liu G, Liu C, Arzayus F, Sapper J (2006) Extreme events and perturbations of coastal ecosystems: Sea surface temperature change and coral bleaching. In: Richardson LL, LeDrew EF (eds) Remote sensing of aquatic coastal ecosystem processes. Springer, Dordrecht

Smith TM, Reynolds RW (2004) Improved extended reconstruction of SST (1854–1997). J Clim 17:2466–2477

Spillman CM (2011) Operational real-time seasonal forecasts for coral reef management. J Oper Oceanogr 4:13–22

Spillman CM, Alves O (2009) Dynamical seasonal prediction of summer sea surface temperatures in the Great Barrier Reef. Coral Reefs 28:197–206

Spillman CM, Alves O, Hudson DA (2010) Seasonal prediction of thermal stress accumulation for coral bleaching in the tropical oceans. Mon Weather Rev. doi:10.1175/2010MWR3526.1

Strong AE, Liu G, Eakin CM, Christensen TRL, Skirving WJ, Gledhill DK, Heron SF, Morgan JA (2009) Implications for our coral reefs in a changing climate over the next few decades—hints from the past 22 years. In: Proceedings of the 11th International Coral Reef Symposium, Ft. Lauderdale, Florida, pp 1324–1328, July 2008

Thomas CR, Gordon IJ, Wooldridge S, van Grieken M, Marshall P (2009) The development of an integrated systems model for balancing coral reef health, land management and tourism risks on the Great Barrier Reef. 18th world IMACS/MODSIM Congr, Cairns, Australia, pp 4346–4352, 13–17 July 2009

Walton CC, Pichel WG, Sapper JF (1998) The development and operational application of non-linear algorithms for the measurement of sea surface temperatures with the NOAA polar-orbiting environmental satellites. J Geophys Res 103:27999–28012

Ward B (2006) Near-surface ocean temperature. J Geophys Res 111:C02005

Watts ME, Ball IR, Stewart RS, Klein CJ, Wilson K, Steinback C, Lourival R, Kircher L, Possingham HP (2009) Marxan with Zones: software for optimal conservation based land- and sea-use zoning. Env Model Softw 24:1513–1521

Weeks S, Bakun A, Steinberg C, Brinkman R, Hoegh-Guldberg O (2010) The Capricorn Eddy: a prominent driver of the ecology and future of the Southern Great Barrier Reef. Coral Reefs 29:975–985. doi:0.1007.0.1007/s00338-010-0644-z

Weeks SJ, Steinberg C, Congdon BC (in review) Oceanography and seabird foraging: within-season impacts of increasing sea surface temperatures on the Great Barrier Reef. Mar Ecol Prog Ser

Wentz FJ, Schabel M (1998) Effects of orbital decay on satellite-derived lower-tropospheric temperature trends. Nature 394:661–666

Wilkinson CR (ed) (1998) Status of coral reefs of the World: 1998. Global Coral Reef Monitoring Network. Australian Institute of Marine Science, Townsville

Wooldridge SA (2009a) A new conceptual model for the warm-water breakdown of the coral-algae endosymbiosis. Mar Freshw Res 60:483–496

Wooldridge SA (2009b) Water quality and coral bleaching thresholds: formalising the linkage for the inshore reefs of the Great Barrier Reef, Australia. Mar Pollut Bull 58:45–751

Wooldridge SA, Done TJ (2009) Improved water quality can ameliorate effects of climate change on corals. Ecol Appl 19:1492–1499

Yeh SW, Ku JS, Dewitte B, Kwon MH, Kirtman BP, Jin FF (2009) El Niño in a changing climate. Nature 461:511–U70

Zhang HM, Reynolds RW, Lumpkin R, Molinari R, Arzayus K, Johnson M, Smith TM (2009) An integrated global observing system for sea surface temperature using satellites and in situ data: research to operations. Bull Am Meteorol Soc 90:31–38

Chapter 13
Radar Applications

Malcolm L. Heron, William G. Pichel and Scott F. Heron

Abstract Active remote sensing by radar relies dominantly on the interaction between electromagnetic waves and sea surface waves, being capillary waves for microwave radars and surface gravity waves for radars operating below about 1 GHz. High Frequency and Very High Frequency radars provide high quality mapping of surface currents and some systems map wave heights on grid scales 50 m to 20 km. Scatterometers are microwave radar systems that map surface winds and sea roughness on large scales of 1,000 km^2. Synthetic aperture radar makes observations at very fine resolution scales of a few meters, and is used to measure wave height and detect changes in surface roughness. Radar remote

M. L. Heron (✉)
Marine Geophysical Laboratory, School of Environmental and Earth Sciences, James Cook University, Townsville, QLD 4811, Australia
e-mail: mal.heron@ieee.org

M. L. Heron
Australian Institute of Marine Science, Townsville, QLD 4810, Australia

W. G. Pichel
Center for Satellite Applications and Research, National Oceanic and Atmospheric Administration, NCWCP, E/RA3, Room 3229, 5030 University Research Ct, College Park, MD 30740-3818, USA
e-mail: william.g.pichel@noaa.gov

S. F. Heron
Coral Reef Watch, National Oceanic and Atmospheric Administration, 675 Ross River Rd, Townville, QLD 4817, Australia
e-mail: scott.heron@noaa.gov

S. F. Heron
Marine Geophysical Laboratory, Physics Department, School of Engineering and Physical Sciences, James Cook University, Townsville, QLD 4811, Australia

J. A. Goodman et al. (eds.), *Coral Reef Remote Sensing*,
DOI: 10.1007/978-90-481-9292-2_13,

sensing from satellites, aircraft and shore platforms is a rapidly developing field that has the capacity to carry out monitoring observations on scales from reef lagoons to regional and global oceans.

13.1 Introduction

A wide range of frequencies are available in the electromagnetic spectrum for use in remote sensing of coral reefs (Fig. 11.2), which, in addition to the other remote sensing disciplines discussed in this book, include the capability at radar wavelengths for measuring a rich set of sea surface characteristics. The discovery of resonant peaks within radar scatter from the ocean was made by Crombie (1955) by investigating sea clutter echoes using an ionospheric sounder. However, only theoretical development of the concept of High Frequency (HF; 3–30 MHz) ocean radar occurred over the next 15 years, and practical progress was deferred until the mid-1970s when mini-computers could be taken into the field. For over a decade, work predominantly involved only discovery research and development, until the mid 1990s when commercial systems emerged and the focus shifted to applications using radar as an operational tool for measuring coastal currents. HF ocean radars are now being used to provide current and wave fields around coral reefs with specific applications related to: connectivity between reefs and other ecosystems; pollution management; and assessing physical stresses leading to coral bleaching.

Satellite/space-based radar technology followed a similar trajectory in the same era, where initial development was reliant upon the advent of reliable digital processing. NASA's Seasat satellite, launched in 1978, carried the first Synthetic Aperture Radar (SAR) and scatterometer instruments into space following a lengthy development period using aircraft. Initially, SAR data were processed optically; however, some digitally processed Seasat SAR data became available within a year after launch. Unfortunately, the Seasat satellite experienced a fatal power system failure, limiting its operating life to just over 100 days. This premature mission termination proved to be a major setback in satellite SAR availability for many years. NASA's SAR efforts for the next two decades shifted to the Shuttle Imaging Radar program, concentrating on a number of short-duration Shuttle missions, culminating with the Shuttle Radar Topography Mission (SRTM) in 2000. After Seasat, another free-flying satellite SAR was not available until the European Space Agency (ESA) launched the ERS-1 satellite in 1991. By that time, digital SAR processing systems were sufficiently developed to enable near real-time SAR image formation, whereupon ocean, land, cryospheric, and atmospheric SAR applications began developing rapidly.

This chapter begins with a description of HF radar technology, illustrating some applications in coral reef management, followed by the natural extension into the Very High Frequency (VHF; 30–300 MHz) high-resolution system for mapping ocean surface currents in smaller domains (e.g., lagoons, ports and harbors). Applications of SAR systems are then presented with case studies in mapping wind, oil spills and other ocean features. Following this, scatterometers are described for wind measurement, including a case study of doldrums off the west coast of the central Americas. Finally, the use of X-band radar for two-dimensional wave fields is described.

13.2 HF Ocean Radar

13.2.1 Analysis and Classification Techniques

The starting point for analysis of HF ocean radar echoes to extract sea surface parameters is the Doppler shift spectrum. Examples are shown in Fig. 13.1 for the two most common types of HF ocean radar systems: phased-array (Fig. 13.1a) and crossed-loop direction-finding (Fig. 13.1b). In these graphs the dominant peaks are due to first-order Bragg scatter, and the rest of the energy above the noise background is due to double-scatter mechanisms and higher order non-linearities in the waves (see Chap. 11). In both systems the radial component of the current towards the radar station is determined, and a second radar looking at a near-orthogonal angle is required to produce the surface current vector. The phased-array approach gives fine spatial and temporal resolution because measurements are made in individual pixels. In contrast, the direction-finding approach has more compact antennas but generally gives a broader view of the surface current field.

For long-range HF ocean radars operating up to 200 km, the errors in measuring the radial components of the surface currents vary on the order of 5–20 cm/s. Chapman and Graber (1997) compared radar surface currents with those from the top usable bin of a moored acoustic profiler and found rms differences of 15 cm/s. Kohut et al. (2006), however, pointed out that some of this difference is due to shears in the flow between the effective depth of the radar measurements and the uppermost usable bin on the acoustic profiler, concluding that the radar error was just 5 cm/s. Cook et al. (2007) estimated the error in crossed-loop systems using a simulation method and found that the errors were 6 cm/s at short ranges but increased to over 20 cm/s at long ranges where the signal-to-noise ratios are reduced. In comparison, the phased-array systems have errors of typically 6–12 cm/s and are invariant with range.

The spatial resolution of HF radar varies from about 1 km at short ranges, where it is limited by the operating bandwidth of the radar, to about 8–20 km at long ranges, where it is limited by the accuracy of the radar azimuth. Temporal resolutions vary from about 10 min to 3 h depending on the type of radar and the configuration.

Raw HF radar data are susceptible to inclusion of outliers due to a wide range of phenomena, including radio interference, spurious echoes from ships, and local effects like animals or vehicles affecting the performance of individual antennas.

There is a trend towards the adoption of rigorous quality control processing on HF radar data (Heron and Prytz 2011) by assessing the quality of the Bragg peaks as well as the properties of the data points themselves. In many cases, however, the quality control processing must rely solely on the consistency of the data points.

Observations of wind directions, one of the primary outputs of HF radar analysis, are useful in coastal and reef management and research because they bridge a gap between the terrestrial network of meteorological stations (which are often not representative of the marine environment) and the offshore global wind wave models (which are often not representative of the nearshore environment). Wind direction observations are derived from the ratio of energies in the two first-order Bragg lines in the spectrum. The principle is that if the wind has a component towards the radar station, then the positively shifted Bragg peak is more energetic than the negatively shifted Bragg peak; if the two peaks were of equal energy, for example, then we would deduce that the wind must be blowing orthogonally to the radar look direction. If we adopt a model shape for the directional spreading of the wind waves with the Bragg wavelength, and use the observed Bragg ratios from two radar stations, then the wind direction may be determined to about ± 10 degree (Heron and Prytz 2002). However, Heron and Marrone (2010) point out that the assumed model shapes for wave direction usually represent steady state conditions. When there is significant meso-scale meteorological structure (e.g., cold fronts), it may be better to use the Bragg energy ratios directly to gain insights into the spatial structure of the wind. Wyatt and Holden (1994), for example, applied an inversion method for directional wave spectra that does not assume steady state conditions and is therefore a more robust algorithm when the wind is not steady.

Wave height observations by HF ocean radar provide spatial data in the coastal zone where both wave-current interactions and wave shoaling occur. Wave heights are derived from the second-order part of the Doppler shift spectrum on phased-array systems as seen in Fig. 13.1a. The second-order energy originates from

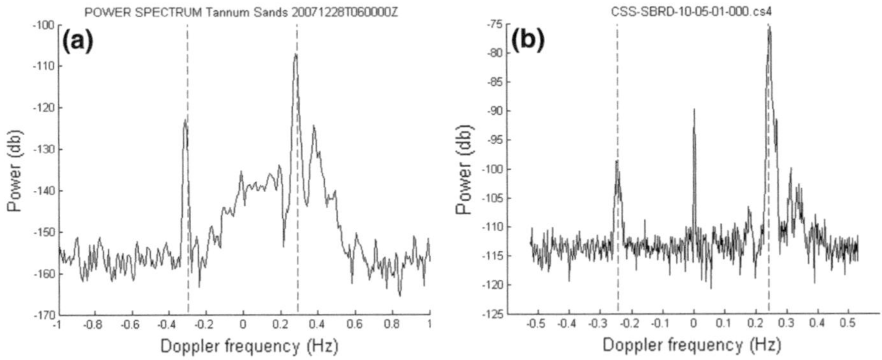

Fig. 13.1 Echo spectra for: **a** one 4 × 4 km pixel for the phased-array system in the Great Barrier Reef; and **b** one annular range cell 3 km × 180° for the crossed-loop system in Western Australia. The larger target area in (**b**) results in spreading of the Bragg lines

double-scatter and non-linearities in the gravity waves on the sea surface, and the first-order Bragg energy is used to normalize the second-order measurements (Barrick 1977). Because this analysis uses the second-order energy, which is typically >10 db below the first-order signals, it is generally limited by signal-to-noise ratios, and the spatial coverage is much reduced from that observed for surface currents and wind directions, which use first-order echoes (Heron and Heron 1998).

13.2.2 System Comparison

As mentioned, the two dominant HF ocean radar systems currently in use are the phased-array and the crossed-loop direction-finding system. An example of a phased-array system is presented here (Fig. 13.2) by a 12-element array of receive antennas in the Capricorn-Bunker Group of reefs and islands in the southern Great Barrier Reef World Heritage Area. This system operates at 8.34 MHz with a bandwidth of 33 kHz and maximum transmit power of 30 W. An example of a

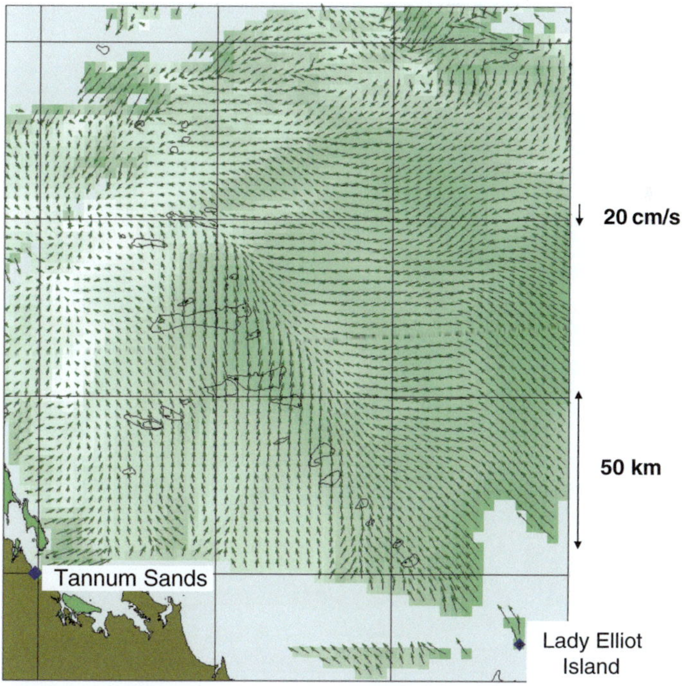

Fig. 13.2 A typical 10 min record of surface currents from the phased-array Integrated Marine Observing System (www.imos.org.au) radar in the southern GBR from the HF ocean radar stations at Tannum Sands and Lady Elliot Island. The length and direction of the *arrows* depict surface current speed and direction, and the *green shading* is darker where the speeds are greater

crossed-loop system (Fig. 13.3) is presented from the Turquoise Coast north of Perth in Western Australia, which operates at 5.2 MHz, a bandwidth of 50 kHz and maximum transmit power of 80 W. The bandwidth of 33 kHz transforms to a spatial resolution in the range dimension of about 4 km, while the 50 kHz bandwidth transforms to about 3 km. The two genres of HF ocean radar both use the strong Bragg energy in the Doppler shift spectra of the ocean echoes to derive environmental parameters. Both systems also use a ranging method based on the time delay that is coded into the linear frequency sweep.

One of the main differences between the two genres of HF radar is the manner in which they determine the azimuthal direction of echo energy and hence the spatial resolution of the resulting measurements. In both systems the fundamental pixels are in polar coordinates and standard processing techniques are used to resample the raw pixels in order to present the results on a regular rectangular grid. The phased-array uses the classical approach of forming a narrow beam. Along with the range resolution, this beam identifies a pixel area on the sea surface, centered on a (polar) grid point, from which the echoes are received. For the

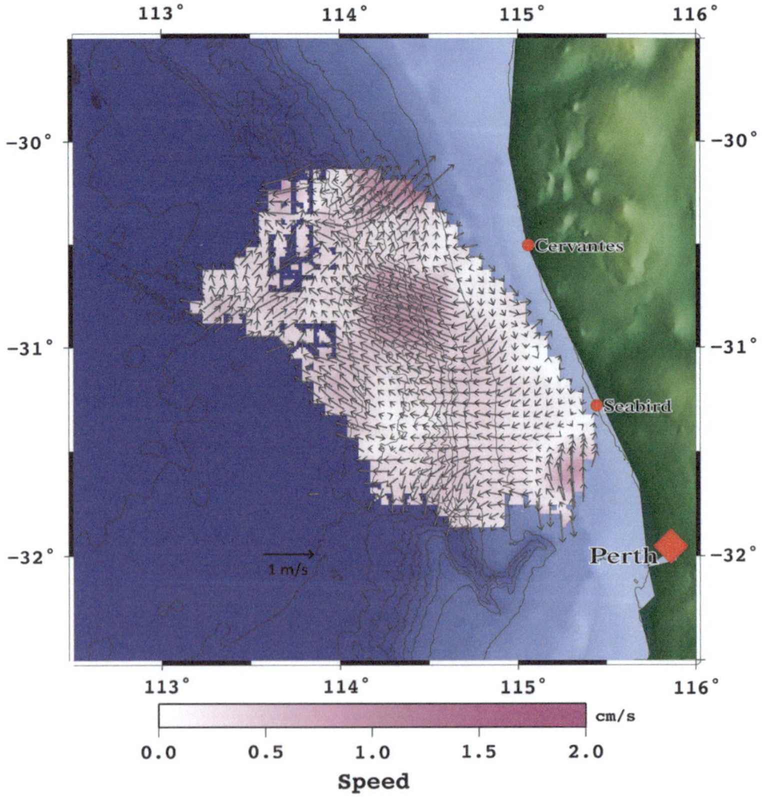

Fig. 13.3 A typical 80 min record of surface currents from the crossed-loop Integrated Marine Observing System (www.imos.org.au) radar in Western Australia (courtesy of D. Atwater)

12-element phased-array in the Great Barrier Reef, the width of the beam is 9.47°, which means the native polar pixels are 4 × 4 km at a range of 24 km. Closer to the radar the pixels retain the 4 km range resolution but are narrower in the azimuthal direction; further away the pixels still retain the 4 km range resolution but become extended in azimuth so that at a range of 100 km the pixels become 4 × 16 km. The phased-array data shown in Fig. 13.2 are not smoothed or interpolated and are presented on a fixed rectangular grid with spacing 4 km. This means the presentation does not capture the full spatial resolution at close ranges, but over-samples at long ranges. The time resolution is 10 min between independent surface current maps.

The crossed-loop radar systems have fewer antennas but the trade-off is in poorer resolution in both time and space. For the crossed-loop system on the Turquoise Coast in Western Australia, the uncertainty in the azimuths is of the order of 18°, which means the native polar pixels are 3 × 3 km at a range of 10 km. Closer to the radar the pixels are 3 km in range resolution but narrower in the azimuthal direction; further away the pixels retain the 3 km in range resolution but become extended in azimuth so that at a range of 100 km they are 3 × 31 km. The crossed-loop data shown in Fig. 13.3 are smoothed as they are transformed to a rectangular grid with spacing 3 km. Due to this smoothing process, the effective spatial resolution becomes 20 km near to shore, increasing to over 50 km at the most distant grid points. The time resolution is 80 min with output products issued every 60 min.

The crossed-loop system is useful for measuring large scale oceanic currents, such as in Western Australia where it is appropriate for following the meanders of the Leeuwin current. The phased-array system performs well when assessing the detailed flow near reefs and along the edge of the continental shelf. A typical phased-array system has receiver antennas extending over a distance of ~200 m, while crossed-loop systems require only three (larger) antennas spaced about 60 m apart. If space for the antennas is a limiting issue then the crossed-loop solution is required, but, as shown in the case studies below, the phased-array system often provides better performance.

13.2.3 Example Applications

Flow around ribbon reefs An early example of the use of HF ocean radar on coral reefs is given by Young et al. (1994), where a radar station deployed on Lizard Island, Australia, observed flow around the ribbon reefs about 18 km to the east. The single radar station observed only radial currents, which included the flow through the narrow channels and into the leeward side of the reefs on a flood tide (Fig. 13.4; dashed lines). The solid lines show the modeled circulation, indicating that oceanic water was brought directly on to the sheltered side of the reef, which affects the distribution of nutrients on the reef and is expected to have significant biological consequences.

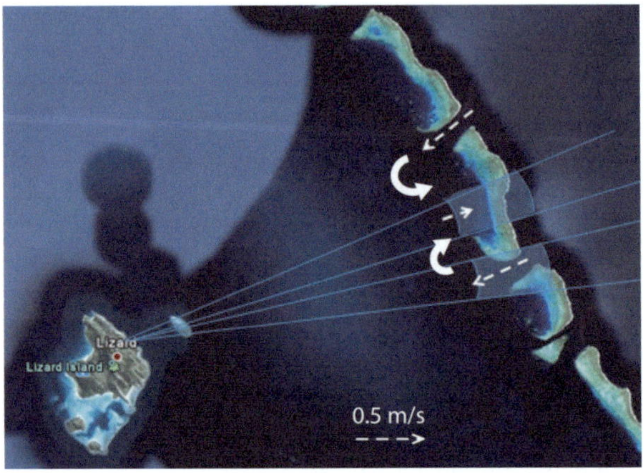

Fig. 13.4 Schematic of current flow around a ribbon reef in the northern section of the Great Barrier Reef. The single radar station deployed on Lizard Island, 18 km to the west of the barrier reef, monitored the radial component of ocean surface currents. Data from the two shaded areas of radar coverage were used to determine currents through the channels and behind the reef during a flood tide. Flow around Yonge Reef (*center*) shows high-speed flow between adjacent reefs with recirculation counter-flow behind the reef (*dashed lines*). The *solid lines* are interpreted via a model. The two channels on the diagram are 6.8 km apart

Surface current maps Routine observations of surface currents are carried out in the southern section of the Great Barrier Reef under the Australian Integrated Marine Observing System (IMOS) using a phased-array HF radar system. Figure 13.2 shows an unsmoothed 10-minute sample of the surface currents. The spatial structure captured in this map shows convergence zones, eddies, and jets on a scale suitable for understanding the dynamics of the ocean between and around reefs.

Lagrangian tracking and connectivity Data like those shown in Fig. 13.2 can be improved through a quality control evaluation of the data points themselves. Mantovanelli et al. (2010) accomplished this by first removing the tidal variation at each grid point, and then applying outlier removal techniques to the residuals on the 10 min data. They then reinstated the tidal signal to calculate good quality hourly averages of currents that can be used for Lagrangian tracking. This is a mathematical procedure that uses spatial maps of surface current data taken at successive time steps to produce tracks of buoyant particles. Lagrangian tracking from HF ocean radar in the southern Great Barrier Reef is now sufficiently accurate and robust to be able to investigate details of connectivity between coral reefs. These are useful in conjunction with species behavior models for tracking larvae, and in the movement of nutrients. Lagrangian tracking is also a useful technique for search-and-rescue (Ullman et al. 2006) and for pollution management. The example shown in Fig. 13.5 is the result of a simultaneous release of an instrumented drifter and the commencement of a Lagrangian track on the HF radar.

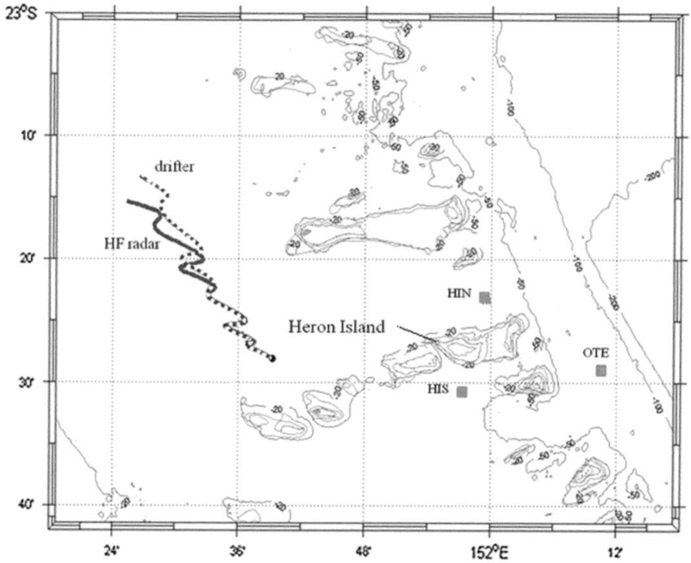

Fig. 13.5 Lagrangian tracking using an instrumented surface drifter (*dashed line*) and HF ocean radar data (*solid line*) for a release point on the continental shelf to the west of Heron Island in the southern Great Barrier Reef. The kinks are responses to semi-diurnal tides. The tracks are very close for the first 36 h and then steadily diverge over the next 30 h to a separation of about 5 km after 76 h (courtesy of A. Mantovanelli)

Turbulent kinetic energy and coral bleaching One direct application of HF ocean radar data to management challenges in coral reefs is the evaluation of turbulent kinetic energy, and its relationship to coral bleaching. Most coral bleaching occurs in conditions of low wind, current and waves, when the water column becomes stratified and the upper layer temperature becomes critical due to insolation. In a classic paper, Simpson and Hunter (1974) demonstrated that water speed and water depth define a critical condition where the water column becomes mixed by turbulent kinetic energy rather than remaining stratified. This has been utilized by DiMassa et al. (2010) to define zones within an HF radar footprint that may be classified as being mixed by turbulent kinetic energy during a 24 h period. Zones classified in this way retain a mixing characteristic irrespective of wind or waves, and are unlikely to suffer coral bleaching by a warm stratified layer forming in the upper part of the water column. The relative susceptibility to coral bleaching is a useful input to planning and management of coral reefs. The concept is illustrated in Fig. 13.6 by a map of the maximum surface current speed during a 24 h period. DiMassa et al. (2011) extend this analysis by applying the Simpson-Hunter criterion to also map the vertically mixed zones, providing further information on localized mixing regimes.

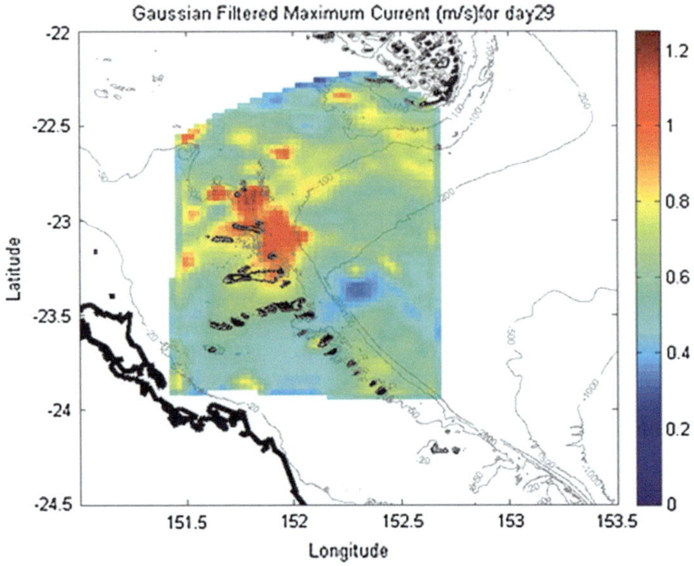

Fig. 13.6 Maximum surface current speeds within the HF radar footprint for a 24 h period in the southern Great Barrier Reef. Higher currents indicate locations with reduced susceptibility to coral bleaching (courtesy of D. DiMassa)

13.3 VHF High Resolution Radar

13.3.1 System Overview

VHF radars can have wider bandwidths than HF radars because of their higher operating frequencies. At these higher frequencies the range is normally reduced, so a typical VHF system operating at 150 MHz, with 1.5 MHz bandwidth and power output of 100 mW, has a maximum range of 4 km and spatial resolution in the range direction of typically 100 m. With azimuthal resolution of about 5°, the pixels are thus 100 × 100 m at a range of 1 km. These spatial parameters are ideal for mapping currents in smaller areas, such as reef lagoons and channels.

As examples, the VHF WERA and SeaSonde systems operate at 40–50 MHz, using phase and amplitude, respectively, to achieve the direction-finding requirement. The PortMap system operates at 60–180 MHz and uses a hybrid analysis of beam forming and phase direction-finding to determine the azimuth of incoming echo energy. These systems are all relatively compact, and increasingly so at higher frequencies, making them suitable for deployment where space is limited. For applications to observe flow in and around offshore coral reefs, the radar stations need to be mounted on an atoll or manmade platform.

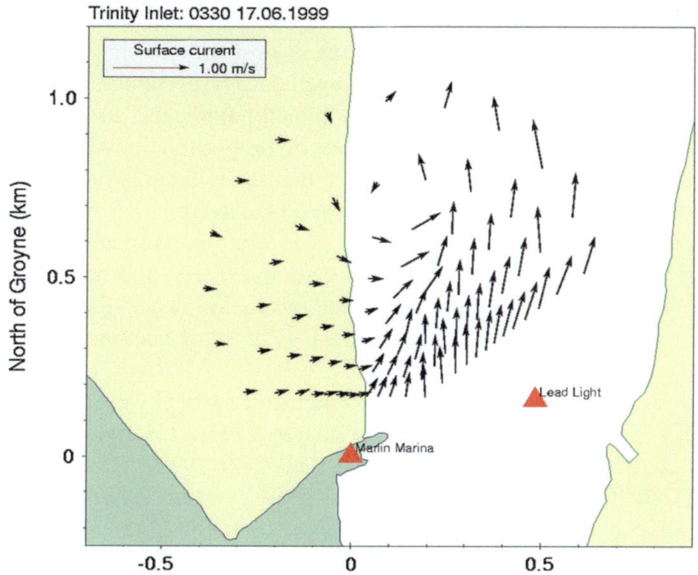

Fig. 13.7 VHF ocean radar surface current map in a tidal channel, Trinity Inlet, Cairns, Australia. The *yellow area* is a tidal flat and the *green area* is land. The two VHF radars are located on the roof of the Marina, and on a navigation post (*red triangles*) (PortMap VHF Radar, courtesy A. Prytz)

13.3.2 Example Application

Figure 13.7 shows a map of surface currents made using a PortMap VHF radar system operating at 152.2 MHz, bandwidth of 1.5 MHz, and transmitted power of 100 mW. The site is in Trinity Inlet in the Port of Cairns, Australia. The figure reveals a complex circulation feature on the ebb tide when the fast flow in the channel interacts with runoff from the tidal flats, which then produces a recirculation from the inlet onto the tidal flat further down the channel. This example illustrates the capability of high-resolution VHF systems for resolving detailed flow characteristics in a relatively small study area.

13.4 Synthetic Aperture Radar

13.4.1 Analysis and Classification Techniques

Many automated and interactive techniques have been developed to exploit the unique and/or useful properties of SAR imagery (i.e., all-weather, day/night, high-resolution, and sensitivity to small changes in ocean surface roughness) in order to measure, observe, and map quantitative and qualitative ocean surface information. A few of the more mature techniques for analyzing and classifying SAR imagery applicable to coral reef research and management are presented here.

High resolution winds SAR imagery is directly applicable to the measurement of wind because the backscatter increases in proportion to wind speed. SAR-derived winds have some unique properties that make them ideally suited for use in coral reef management: (1) they have the advantage of high resolution (on the order of a few hundred meters to 1 km); (2) accurate measurements can be made right up to the coast, and indeed even in lagoons, straits, and bays (unlike scatterometers that are affected by proximity to land); and (3) accuracy is comparable to that of scatterometers in the open ocean with better accuracy near the coast (Yang et al. 2011).

SAR winds are most commonly estimated using a Geophysical Model Function (GMF) derived originally for use with scatterometers (Monaldo et al. 2004a). Examples include CMOD-4 (Stofflen and Anderson 1997), CMOD-5 (Hersbach et al. 2007), and CMOD-IFR2 (Quilfen et al. 1998). Among other parameter inputs (Christiansen et al. 2008), these GMFs require estimation of the angle between the wind direction and the radar look direction. In order to determine this angle, it is necessary to know the wind direction, a parameter that is obtainable with scatterometers, which have more than one look angle, or from meteorological models or buoys, but which cannot be measured directly with SAR instruments. Alternatively, in the absence of other data sources, wind direction can be obtained (with 180° ambiguity) from wind-aligned features (e.g., roll vortices, island wind shadows) in the SAR image itself.

Since the existing C-band (4–8 GHz) scatterometer GMFs have all been derived for vertical transmit, vertical receive (VV) polarization, it is necessary to use a polarization ratio when applying the GMFs to horizontal transmit, horizontal receive (HH) polarization SAR imagery. Polarization ratios currently employed are a function of angle of incidence only (Thompson et al. 1998) or both angle of incidence and wind direction (Mouche et al. 2007). Wind accuracy using these GMFs and polarization ratios with C-band SAR data is in the range of ± 1.5 to ± 2.5 m/s for wind speeds less than 15 m/s when compared against buoy winds or scatterometer winds (Monaldo et al. 2001, 2004b; Xu et al. 2010). Pichel et al. (2008) similarly compared SAR CMOD5 winds derived from RADARSAT-1 imagery with buoy winds in the Bering Sea during the period November 2006 to April 2007. For winds between 0 and 15 m/s (136 matches) the mean difference was –0.02 m/s and the standard deviation was 2.23 m/s. For winds between 15 and 25 m/s (22 matches) the mean difference was 2.33 m/s and the standard deviation was 2.37 m/s. Above 25 m/s, the currently available single-polarization GMFs were determined not to have sufficient sensitivity. Cross-polarization algorithms currently under development show promise of achieving accurate winds for these higher wind speeds (Vachon and Wolfe 2010).

Oils spills Marine oil spills can damage coral reefs by killing coral outright or making them more susceptible to bleaching (Johannes et al. 1972; Haapkylä et al. 2007), as well as harming reef fish and marine mammals, birds, and turtles. Wave suppression (damping) by surfactants results in visible changes in surface roughness by producing a slick that is smoother than the surrounding wind-roughened ocean (Hu et al. 2009). SAR instruments are very sensitive to this roughness change, which can occur in the presence of even very small quantities of oil or other surfactants. The radar signal is forward scattered off the smooth ocean with decreased backscatter to the SAR instrument, resulting in a dark region in the image. The oil-mapping ability of SAR instruments is well known, and amply demonstrated during responses to both major and minor oil spills (Clemente-Colón and Pichel 2006; Clemente-Colón et al. 1997; Gade and Alpers 1999). Oil spills can be the result of accidents, intentional dumping (e.g., bilge pumping), pipeline leaks, oil platform releases, releases from sunken vessels, or coastal effluents (Clemente-Colón and Pichel 2006). Unfortunately, a dark region in a SAR image is not always indicative of an oil spill. There are many other low-backscatter features that are not oil spills (i.e., false positives) such as: areas of low wind speed; upwelling; natural oil seeps; biogenic slicks such as algal blooms and coral spawn; fish oils released during fishing and fish processing activities; and organic surfactant runoff directly from land and/or carried by rivers. Nonetheless, despite these potential confusion areas, criteria have been developed for successful remote sensing of oil slicks.

Oils can be detected, in general, when wind speeds are within 3–15 m/s (Espedal et al. 1998; Johannessen et al. 1994; Wahl et al. 1996; Wismann et al. 1998), while biogenic surfactants can usually be detected when the wind speed is between 2 and 8 m/s (DiGiacomo and Holt 2001). When winds are strong (i.e., exceeding 10 m/s for extensive periods during a period of 24 h), the oil becomes mixed into the upper ocean and slicks do not form (Simecek-Beatty and Pichel 2006; Simecek-Beatty and Clemente-Colón 2004). Differentiating between oil spills and other naturally occurring slicks requires a trained analyst and extensive knowledge of local ocean and atmospheric conditions. Fully automated detection algorithms using neural networks, fuzzy logic, wavelet transforms, and other trained analysis techniques have also achieved some success (Garcia-Pineda et al. 2009; Liu et al. 2000, 2010); however, no single-channel, single-polarization algorithm can detect oil without some false positives. Multi-frequency and multi-polarization analysis techniques, which exploit changes in polarization of the backscattered signal as a result of different scattering mechanisms for oil-dampened ocean and normal wind-roughened ocean, may help to reduce these false alarms (Gade et al. 1998; Trivero et al. 1998; Mugliaccio et al. 2009; Zhang et al. 2011).

Ocean features Since SAR instruments are very sensitive to any ocean phenomenon that alters capillary and small gravity waves at the scale of the SAR wavelength (i.e., about 5 cm for C-band), many ocean features appear on SAR imagery. These include:

- *Currents*. Since wave propagation and height are affected by horizontal shears in ocean currents, phenomena like strong currents, eddies, and river plumes show up clearly in SAR imagery, often as narrow bright features marking their boundaries (Zheng et al. 2004). Natural surfactants can also be swept along by the current, and by damping the capillary waves they can help mark circulation features, showing up as dark lines. Turbulent mixing related to currents, current shear, and eddies can modify the thermal environment of the ocean near-surface, which has relevance for coral bleaching. Dramatic increases in coastal river outflow, especially as a result of flooding, may negatively affect reefs through salinity reduction and/or increased nutrient and sediment loads (e.g., Devlin and Brodie 2005; McCulloch et al. 2003).

- *Upwelling*. Upwelling of cold water decreases the wind stress, resulting in a dark region in SAR imagery (often accompanied by slick lines after the nutrient-rich cold water leads to blooms of algae) (Clemente-Colón 2004; Li et al. 2009b).
- *Ocean fronts*. Frontal regions may show up as dark or bright lines, depending on the orientation of the front. When the front is perpendicular to the radar look direction, bright radar returns are obtained in convergent fronts and dark returns in divergent fronts (Johannessen et al. 1996; Li et al. 2005).
- *Bathymetric features*. Shallow-water bathymetric features may be imaged by SAR instruments in situations where the current interacts with the bottom topography (Li et al. 2009a).
- *Internal waves*. Internal wave packets are ubiquitous in SAR imagery of regions where there is a generation mechanism for these waves (e.g., tidal current interaction with irregular bathymetry such as a shallow sub-surface sill or islands; Li et al. 2008b; Jackson 2004; Wolanski and Deleersnijder 1998). These internal waves, which propagate along the ocean pycnocline, may have amplitudes of 30–40 m and in some cases as large as 100 m (Apel 2004). In coral reef regions, internal waves can cause both positive and negative impacts on coral health and growth. On the one hand they can bring cool, nutrient-rich subsurface water to the coral, mitigating surface warming that can lead to bleaching. On the other hand, internal waves can produce significant thermal stress, with short-term fluctuations in temperature of 10 °C, to as much as 20 °C (Wolanski et al. 2004), leading to mortality for species that cannot tolerate these fluctuations.

For the most part, ocean features are identified in SAR data using visual examination of imagery; however, considerable experience in SAR image interpretation and knowledge of the local oceanographic environment and processes operating therein are essential for correct interpretation. Often a complex mix of processes is at work in the ocean modifying the surface roughness. In addition, atmospheric boundary layer phenomena can also mask or confuse the correct interpretation of ocean features. For example, atmospheric gravity waves can be similar in appearance to internal ocean wave signatures and atmospheric fronts can mimic ocean fronts (Shuming et al. 2010). These same considerations provide serious challenges for automated ocean feature analysis algorithms, such as those that utilize wavelet transforms (Wu and Liu 2003).

Wave measurements SAR instruments are able to image the two-dimensional ocean surface wave field via three imaging processes: (1) wind waves and swell waves modulate the smaller Bragg waves, leading to changes in backscatter;

(2) changes in wave slope vary the amount of radar energy that is directly reflected; and (3) wave orbital velocities lead to Doppler shifts that influence the SAR image formation process, producing variations in image intensity called "velocity bunching." Random movement of small waves, however, also produce Doppler shifts that limit the azimuth (i.e., along-track) resolution of the image spectrum (i.e., the "azimuth cut-off"). Practically, this means that for polar-orbiting satellites, wave information for short-wavelength wind waves is limited, and accurate wave information can generally be obtained only for longer swell waves. For these longer waves, one can measure dominant swell wavelength, wave direction, wave spectra (Li et al. 2002), and significant wave height. Monitoring of swell during bleaching events can provide information on upper-ocean mixing, which may mitigate the severity of bleaching (Skirving et al. 2006).

From the two-dimensional SAR image spectrum, an inversion process utilizing an appropriate transform function can be used to measure the corresponding two-dimensional ocean surface wave spectrum. The wave spectrum can be obtained directly from the SAR image spectrum with a linear transform in the case of longer swell waves that are not very high and are travelling in the range direction. However, swell waves travelling strictly in the azimuth direction, unless they have wavelengths longer than about 300 m, are not imaged by typical space-based SAR instruments. In general, when making wave measurements for shorter and steeper waves using SAR instruments operating at altitudes typical for sun-synchronous polar-orbiting satellites, non-linear transforms must be employed, although there are still azimuth limitations (Vachon et al. 2004). In a comparison of swell wave significant wave height derived from RADARSAT-1 Standard Mode imagery with buoy wave measurements from NOAA buoy 46029 off the coast of Oregon (46.144°N, 124.51°W) the root-mean-square difference in 168 significant wave height comparisons made between 2001 and 2005 was 0.6 m (after eliminating matches differing by more than 2 standard deviations). Wave heights for these comparisons ranged from 0–4 m.

Vessel monitoring Vessel traffic in protected and use-restricted marine sanctuary areas is of interest to coral reef managers, and fishing vessel locations and concentrations are of interest to fishery managers and fishing enforcement personnel. International vessels greater than 300 tons and all passenger ships are required to carry Automatic Identification System (AIS) transmitters; however, smaller vessels and non-compliant vessels cannot be monitored via AIS. SAR instruments are the sensor-of-choice for monitoring such ships since SAR operates day and night, cloudy or clear, and under a wide range of ocean conditions.

Vessels provide three different SAR signatures under the right ocean and wind conditions: (1) direct returns; (2) wake patterns (Lyden et al. 1988); and (3) slicks (Clemente-Colón et al. 1998). Since most ocean-going vessels have steel hulls, they reflect the incoming SAR radar signal to provide a strong direct return (i.e., they are "hard targets"), typically much more intense than the surrounding ocean surface. Radar returns from vessels result from: (a) direct reflection of the radar signal from portions of the ship oriented perpendicular to the incoming signal (single bounce); (b) a double bounce from the hull and ocean surface; or (c) reflection (triple bounce) by corner reflectors in the superstructure (Pichel et al. 2004). In general, direct returns from steel vessels with a length greater than or equivalent to the resolution of a pixel in a SAR image can be detected most of the time, and vessels with lengths of about half the SAR resolution can be detected half of the time. Factors affecting the detectability of direct returns from vessels include: (i) ship characteristics—structural configuration and materials used, orientation of the ship in relation to the radar signal, and ship size; (ii) environmental conditions—sea state, wind speed, proximity to land and small islands; (iii) radar characteristics—angle of incidence, polarization, resolution, sensitivity; (iv) SAR image quality—image processing errors, beam seams, speckle noise, and nadir ambiguities; and (v) image resolution. In particular, one will see a variation in vessel detectability in each SAR image from near-range to far-range since the backscatter from the ocean surface is very large at small angles of incidence ($\sim 20°$), and significantly smaller at high angles of incidence ($\sim 50°$). The vessel backscatter, however, does not change as dramatically with incidence angle; thus, there is an increase in contrast between vessel and ocean as incidence angle increases. Automated vessel detection algorithms therefore have to account for this change in contrast within the SAR image. Many such algorithms determine localized background ocean statistics throughout the image and thereby set a dynamic threshold for detection of a vessel in different parts of the image. This allows the algorithm to maintain a constant false alarm rate (CFAR) and achieve a fairly constant detection sensitivity throughout the image (Vachon et al. 2000; Wackerman et al. 2001).

13.4.2 Example Applications

Wind energy SAR-derived wind imagery is useful for monitoring coastal wind spatial variations due to: (i) island and coastal landform wind shadowing, island gaps, and island wake formation (Beal et al. 2005); (ii) katabatic winds (Li et al. 2007); (iii) barrier jets (Winstead et al. 2006); (iv) vortex streets (Li et al. 2008a); (v) atmospheric fronts (Young et al. 2005); (vi) hurricanes and other storms (Horstmann et al. 2006; Sikora et al. 2000); (vii) convection cells, thunderstorms, and roll vortices (Sikora and Ufermann 2004); (viii) coastal and mountain lee waves (Li et al. 2011); and (ix) other atmospheric marine boundary layer phenomena.

Figure 13.8 shows local wind variations, specifically ocean surface patterns of atmospheric lower-boundary-layer roll vortices and island wakes, over the Great Barrier Reef in the vicinity of Rockhampton in eastern Australia. The region stretches from Curtis Island at the center of the southern edge of the image to Cape Manifold near the northern edge of the image. The island in the left center of the image is Great Keppel Island. The wind patterns were calculated from a RADARSAT-1 Standard Mode image covering an area 100 km on a side (original SAR image resolution of 30 m). Wind directions used in the CMOD5 algorithm were derived from the Navy Operational Global Atmospheric Prediction System

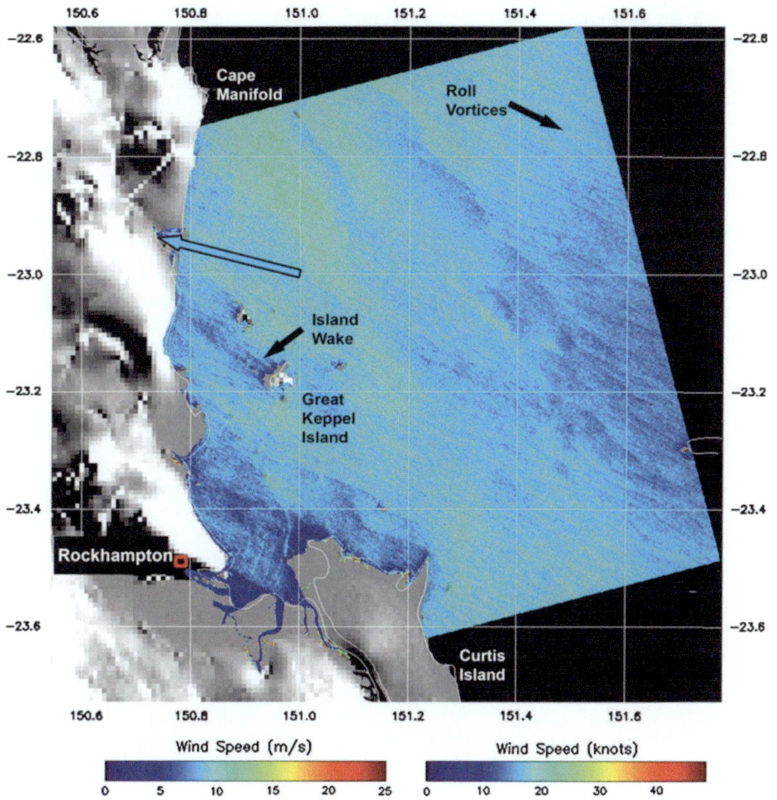

Fig. 13.8 SAR-derived wind image for 08:28 UT February 26, 2008, calculated using the CMOD5 scatterometer algorithm (Monaldo et al. 2004a) from a RADARSAT-1 Standard Mode image (image resolution 30 m, wind resolution approximately 600 m). This image was generated with software and algorithms developed by The Johns Hopkins University Applied Physics Laboratory. The outlined *blue arrow* indicates the NOGAPS wind direction used in the CMOD5 algorithm

(NOGAPS). Winds are from the southeast as denoted by the NOGAPS vector (shown by the outlined *blue arrow* originating from 23°S, 151°E) and as indicated by island wakes and atmospheric roll vortices, both of which are aligned with the wind.

Figure 13.9 depicts regional wind patterns in the Philippines using Wideswath ENVISAT Advanced SAR (ASAR) data. The region shown is the central portion of the Philippines with Luzon to the north, Panay in the center of the image, and the Sulu Sea to the south and west of Panay. Wind directions, derived from the NOAA Global Forecast System (GFS) weather model, are shown as wind barbs in the figure. Higher winds can be seen in a gap flow south of Luzon and in point wakes stretching downwind to the southwest of some of the larger islands, such as Panay (Beal et al. 2005). These higher wind regions will generate more mixing of the upper ocean layer and more turbidity in shallower waters.

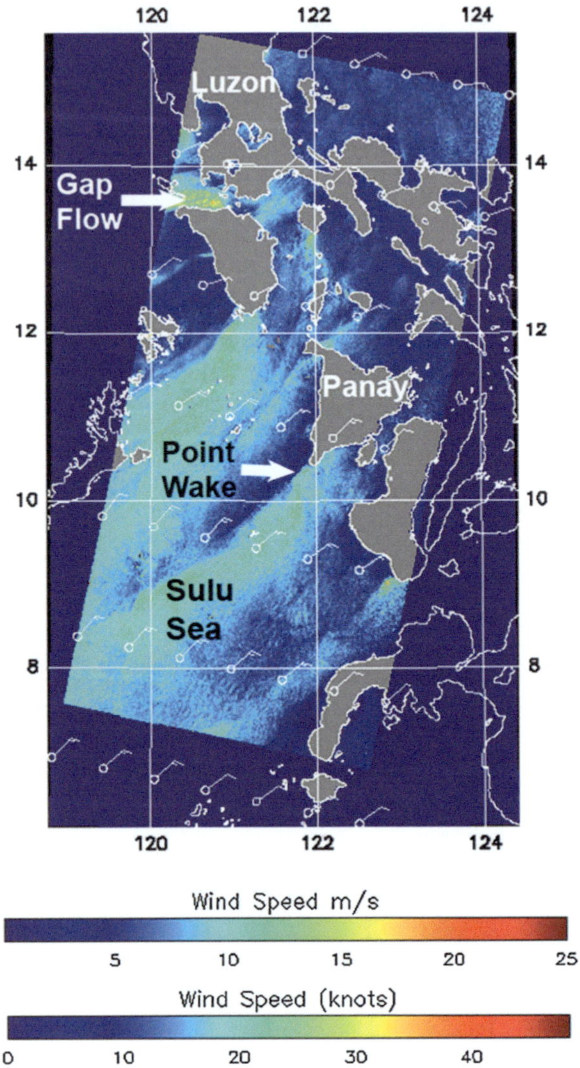

Fig. 13.9 SAR-derived wind image for 01:45 UT on January 31, 2008 with wind resolution about 500 m calculated with the CMOD5 scatterometer wind algorithm (Monaldo et al. 2004a). This image was generated with software and algorithms developed by The Johns Hopkins University Applied Physics Laboratory

Oil spill mapping Using SAR data to monitor oil releases in the vicinity of coral reefs is a practical use of SAR imagery for coral reef monitoring and management. For example, ship traffic can be monitored for intentional oil releases while underway (i.e., bilge pumping or tank cleaning). This is a very common practice in many parts of the world, particularly the South China Sea (Lu et al. 1999). Large accidental spills can also be tracked to determine if booming is required to protect reefs, wetlands, or other sensitive regions. Figure 13.10 shows oil leaking from a ship, the Solar I, which sank in the Philippines, south of the island of Guimaras, on August 11, 2006, carrying about 2 million liters of oil.

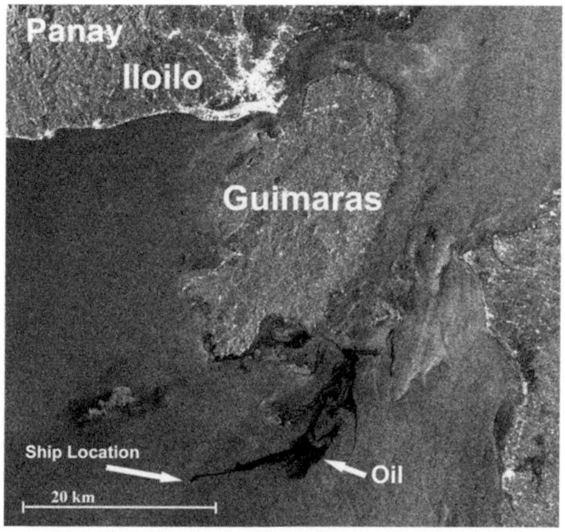

Fig. 13.10 ENVISAT Advanced SAR image from August 24, 2006 of the Solar I oil spill south of Guimaras island, east of Panay island and near the city of Iloilo in the Philippines (image © European Space Agency, 2006)

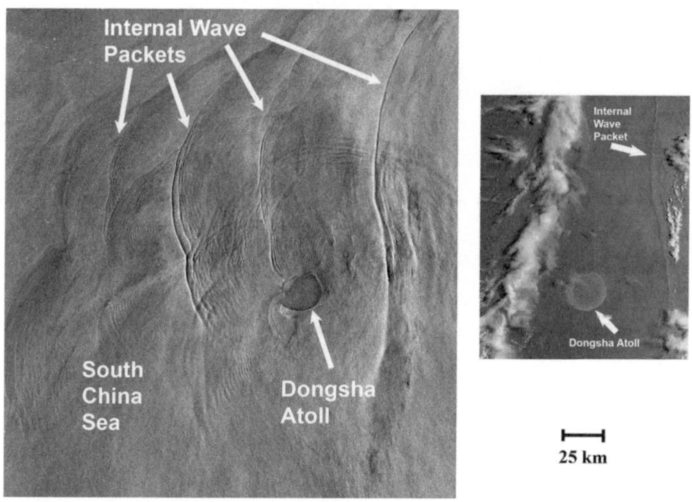

Fig. 13.11 (*left*) ENVISAT Advanced SAR (ASAR) Wideswath image (75 m resolution) of the South China Sea taken June 18, 2008 at 02:13 UT showing internal wave packets (at least 4) near Dongsha atoll (20.6°N, 116.8°E). These waves, formed by tidal interaction with the bathymetry of the Luzon Strait (located hundreds of kilometers to the east of this image), are approaching from the east (*right side* of image). One of the packets is interacting with the atoll. (SAR image © European Space Agency, 2008) (*right*) NASA Aqua MODIS true-color, 250 m resolution image of Dongsha atoll taken June 18, 2008 at 02:55 UT. Sunglint in this image allows one to see an internal wave packet to the east of the atoll. This packet corresponds to the easternmost packet in the ASAR image

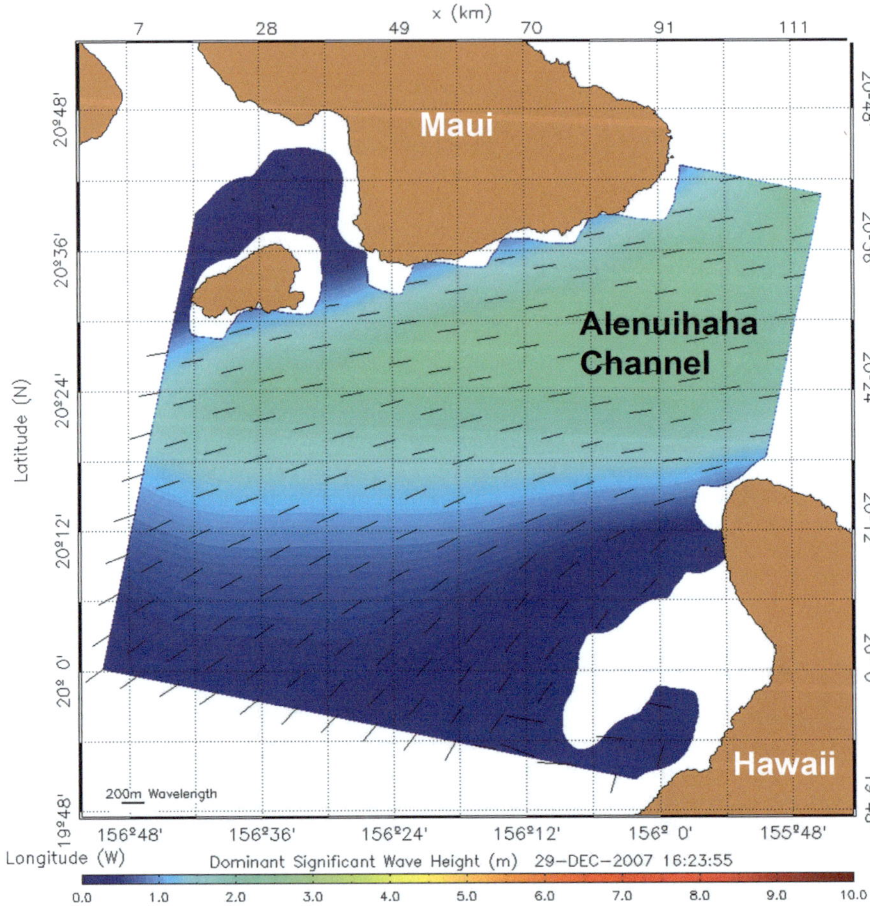

Fig. 13.12 Significant swell wave height product derived from a RADARSAT-1 Standard Mode Single Look Complex (SLC) image from 16:23 UT on 29 Dec 2007. The area shown is the Alenuihaha Channel between the islands of Maui and Hawaii in the main Hawaiian Islands. Shown in color is the dominant swell significant wave height in meters, with the swell direction and wavelength indicated by the field of lines. A scale for the lines is given in the lower *left* corner of the image. Latitudes are degrees north and longitudes are degrees west. The scale along the *top* of the image is in kilometers. This image was generated using software developed by BOOST Technologies, Brest, France (Boost is now part of Collecte Localisation Satellites). A description of the wave algorithm can be found in Collard et al. (2005)

Ocean features Ocean features analyzed from SAR imagery may provide clues to physical oceanographic processes that can impact coral reefs (i.e., currents, eddies, turbulent mixing, and convergent processes). Information can be obtained on whether upwelling and mixing of colder water will ameliorate potential or actual coral bleaching, provide extra nutrients to the coral reef ecosystem, or lead to thermal stress. Figure 13.11 illustrates an ocean feature, in this case internal waves, impacting a coral reef. Internal wave packets, propagating toward the west, refract around the Dongsha coral atoll, a national park located in the South China Sea (Zhao et al. 2008). These waves have been measured and determined to have amplitudes exceeding 50 m. Cold water intrusions resulting from these waves can lead to temperature decreases of 8 °C, variations in dissolved oxygen, and increases in chlorophyll concentration, all of which can significantly impact the coral reef ecosystem (Wang et al. 2007).

Waves Coastal wave information is useful for safety of recreational and commercial boating and fishing, coastal barge and ferry transportation, and activities such as surfing, diving and ecotourism. For coral reef management, information on the distribution and intensity of waves is important for monitoring mechanical wave action during bleaching events or destructive storms. Figure 13.12 depicts an experimental swell wave product generated from a RADARSAT-1 Single Look Complex (SLC) image showing swell wave variations in the Alenuihaha Channel between the Hawaiian islands of Maui and Hawaii. SLC images contain both amplitude and phase information needed by the wave algorithm. This two-dimensional depiction of the wave field provides valuable information regarding coastal changes in wave height, wavelength, and wave direction. In this case, the maximum swell significant wave height is about 3 m in the channel. While there are no ocean wave buoys in the region shown by the image to corroborate this estimate, an open ocean buoy, 51202, located to the west of this image at 21.417°N and 157.608°W reported a significant wave height of 3.8 m at 16:00 UT on this day (buoy data are available from the NOAA National Data Buoy Center: www.ndbc.noaa.gov).

Vessel monitoring Vessel impacts to coral reefs include destruction of coral from anchoring or running aground, introduction of invasive species, and marine pollution from oil or other chemical releases (Franklin 2008). Although at the present time obtaining SAR imagery with sufficient frequency to continuously monitor ship traffic in coral reef management areas is prohibitively expensive, it can be used to spot check AIS non-compliance, as well as to periodically monitor sensitive and restricted regions. For example, Fig. 13.13 indicates the presence of a ship within the Papahānaumokuākea Marine National Monument, a 1,530 km long chain of coral atolls and islands stretching northwest of the main Hawaiian Islands. Monitoring ship traffic within the Monument is important to enforce areas closed to fishing and boating, as well as to monitor ship traffic and possible oil spills from vessels sailing along the island chain or transiting the shipping corridor that cuts through the Monument between Pearl and Hermes Atoll and Lisianski Island (Franklin 2008). Linking of slicks and vessel targets provides insight as to whether a slick might be an intentional release of oil or a natural surfactant slick.

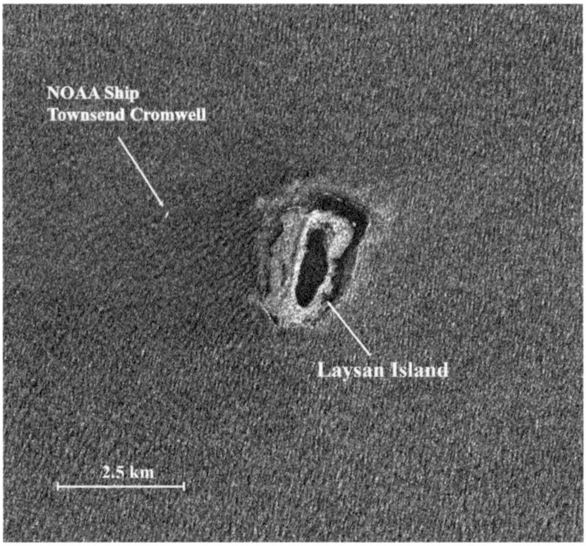

Fig. 13.13 RADARSAT-1 Standard Mode SAR Image form 17:27 UT on 9 Nov 2001 showing the NOAA Ship Townsend Cromwell operating just west of Laysan Island, an island within the Northwestern Hawaiian Islands and now part of the Papahānaumokuākea Marine National Monument. The Townsend Cromwell had a length of 49.7 m, breadth of 10.1 m, and displacement of 652 t (image © Canadian Space Agency, 2001)

13.5 Scatterometers

13.5.1 Analysis Techniques

Scatterometers are airborne or space-based microwave radars that can be used to observe surface winds. A pencil beam emitter is scanned forward and aft from the platform and the backscattered energy is received and recorded (Nadel et al. 1991). Scatterometer analysis does not use phase or Doppler shift information, and is thus a simpler technology than SAR with less processing required. There are two physical principles involved in deriving wind vectors from the intensity of microwave backscatter. The first is that Bragg scatter dominates such that microwave radar senses capillary waves on the sea surface. The second principle is an elegant, albeit empirical, linear relationship between root-mean-square (rms) wave heights (capillaries) and wind speed, identified by Cox and Munk (1954). The sea surface roughness at the centimeter-scale is produced by wind stress, which in turn is produced by the surface wind. Also, there is a directional dependence of wave heights on wind direction, which can be used to define a directional spreading function (Elfouhaily et al. 1997; Heron et al. 2006).

Figure 13.14 is a stylized illustration of the directional dependence of scattered energy. The figure shows how the scatterometer makes forward, mid, and aft scans to sample the directional pattern, so that within a few minutes the same spot on the ocean surface is observed from three different directions. An empirical Geophysical Model Function (GMF) is then used to relate the observations of backscattered energy to the direction and speed of the wind (as discussed in Sect. 13.4.1).

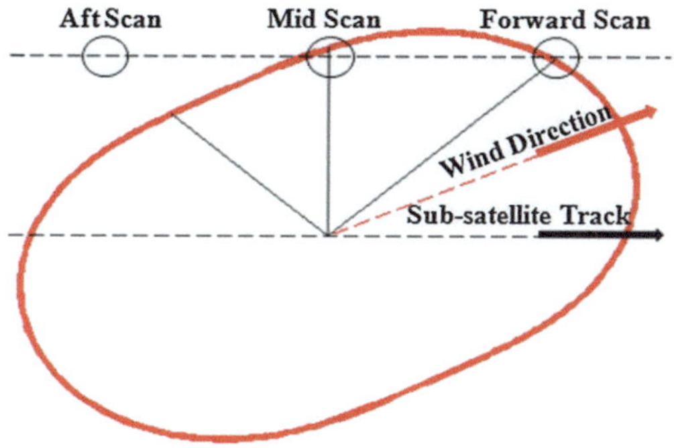

Fig. 13.14 The *oval shape* is the directional model for backscatter intensity, and the sub-satellite track is the *lower dashed line*. As the satellite moves, a single point on the sea surface is sampled from three different directions (*circles*) and the magnitude and orientation of the fitted model *oval* is transformed to a wind vector via a Geophysical Model Function

The early Seasat scatterometer operated in the Ku-band, which is susceptible to attenuation by rain, whereas recent instruments in the lower frequency C-band are less susceptible to rain attenuation. The influence of attenuation may cause problems in heavy rain with large drop sizes, such as occurs in tropical cyclones and thunderstorms. Analysis can also be adversely affected by the presence of land and/or sea ice in the target zone.

13.5.2 Example Application

The spatial resolution, or pixel size, of scatterometers is typically 25 km, though some enhanced processing can produce 12.5 km products. For coral reef management, scatterometer winds are useful to describe the surrounding environment at the reef scale, which has relevance to interconnectivity and pollution management, and in assessing the surface currents and waves. Vertical mixing of the water column by wind, currents and waves distributes solar energy from the surface and can therefore mitigate the potential for coral bleaching. Furthermore, low-wind conditions reduce evaporative cooling and sensible heat transfer (Mumby et al. 2004) and can enhance photo-degradation of colored dissolved organic matter that results in reduced shading (Manzello et al. 2006).

Recognizing the connections between low-wind periods and coral bleaching, NOAA Coral Reef Watch uses the Blended Sea Winds product of NOAA's National Climatic Data Center to identify locations experiencing periods of persistent low-wind conditions in the experimental Doldrums product (Fig. 13.15). The Blended Sea Winds product is a composite of up to six satellite observations and is produced in near real-time at 6 h, 25 km resolution. The Doldrums product provides the number of days for which the daily mean wind speed has remained below 3 m/s. Longer periods of persistent low-wind indicate higher risk of coral bleaching.

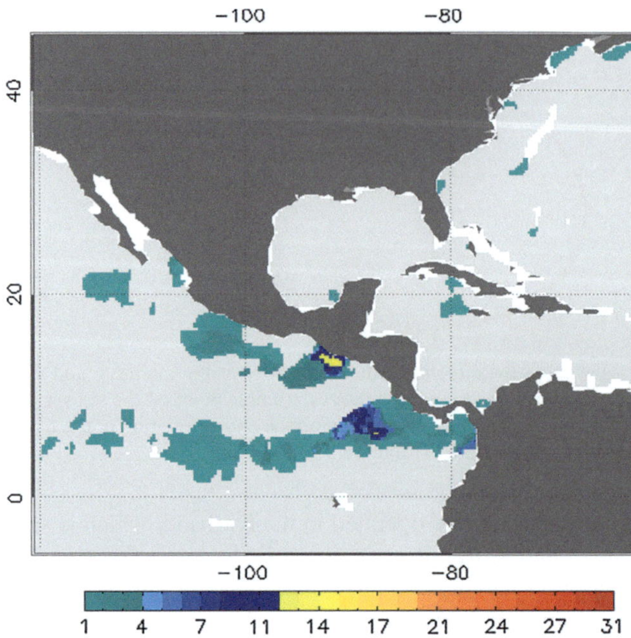

Fig. 13.15 NOAA Coral Reef Watch Experimental Doldrums Product for the period ending 30 Apr 2011 showing duration of persistent low-wind conditions for the central Americas. The color-bar represents the number of days for which the daily mean wind speed has remained below 3 m/s. The *yellow* region south of the Mexico/Guatemala border has consistently experienced low winds for the prior two weeks. *White pixels* indicate missing or insufficient data

13.6 X-band Wave Radars

X-band (8–12 GHz) wave radars represent a niche technology in ocean remote sensing. A normal nautical X-band rotating antenna is used to record the 'clutter' from ranges on the order of 1 km. The rotating antenna scans a scene every 3–4 s and the images of the wave crests are displaced between scans. The two-dimensional pattern of echo strength can be transformed to produce the two-dimensional sea surface elevation pattern, as illustrated in Fig. 13.16. X-band wave radars are well suited for observing the wave field outside a reef structure and the changes that occur as waves break on the reef front and the remnant energy appears (sometimes at a different wave frequency) in the lagoon. The applications here are for flushing dynamics and mixing at the reef front.

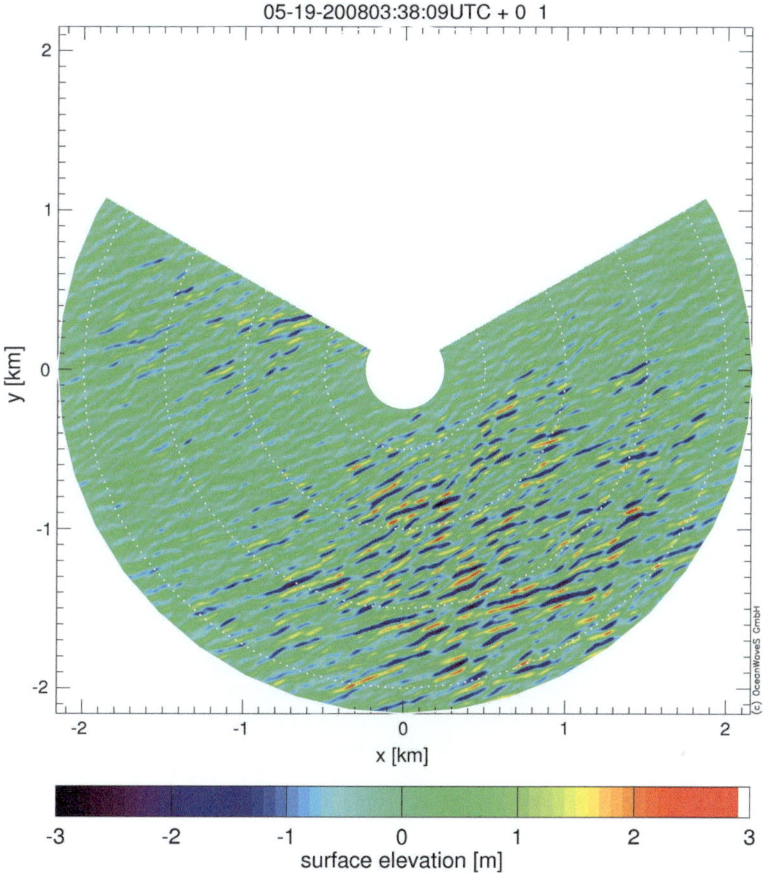

Fig. 13.16 A typical image of the 2D sea surface elevation showing wave crests and troughs over an area ~8 km^2. From these data, full directional wave spectra may be produced (© Wamos GmbH)

13.7 Conclusions and Future Directions

Radar analysis techniques and applications for the oceans and coral reefs are undergoing rapid development and implementation into operational monitoring products. This has resulted from the increasing number of personnel in government, industry, and academia who are engaged in research, commercial exploitation, and operational utilization of a rapidly increasing array of ocean radar platforms. Applications have matured rapidly during the past decade and a half, and now new radar systems are being installed or launched to directly support operational monitoring.

The deployment of HF ocean radars is in an expansion phase, with networks now established in USA, Australia and Europe. This will bring more applications to light as researchers focus on Lagrangian tracking and fine scale measurements, which are needed around coral reefs. It is expected that the application of high-resolution VHF ocean radars and microwave wave radars will also make a significant contribution to monitoring flow in and around coral reefs.

The next generation of SAR satellites, in particular the Sentinel-1 and RADARSAT Constellation Mission (RCM), will be operated and utilized as fully operational satellites rather than research/commercial satellites, which is the status of all current SAR satellites. It is anticipated that future operational, low-cost SAR imagery, with routine repeat coverage, will enable the applications discussed above to be fully utilized within the coral reef community. Emerging SAR applications will certainly include more sophisticated use of multiple polarization channels (e.g., for vessel detection and high-wind measurement) since newer satellites increasingly have dual-polarization, cross-polarization, and quad-polarization modes. There should also be increased availability of along-track interferometry for ocean current measurement, beginning with TerraSAR-X, and it is possible that coordination of SAR orbits and acquisitions may allow some multi-frequency applications (such as discrimination between oil spills and natural biogenic slicks).

Acknowledgments Data were drawn from the Australian Integrated Marine Observing System (IMOS) archive for some of the HF radar images. RADARSAT-1 SAR imagery was processed by the Alaska Satellite Facility of the University of Alaska, Fairbanks. ENVISAT imagery was processed by the European Space Agency. The authors wish to thank Frank Monaldo of The Johns Hopkins University Applied Physics Laboratory, Xiaofeng Li of GST at NOAA/NESDIS, and Christopher Jackson of Global Ocean Associates for SAR product development and processing of the figures shown and products discussed. The views, opinions, and findings contained in this paper are those of the authors and should not be construed as an official NOAA or U.S. Government position, policy, or decision.

Suggested Reading

Garello R, Romeiser R, Crout RL (2005) Special issue on synthetic aperture radar imaging of the ocean surface. IEEE J Ocean Eng 30(3):470–569

Harlan J, Terrill E, Keen C, Barrick D, Whelan C, Howden S, Kohut J (2010) The integrated observing system High-Frequency radar network: Status and local, regional and national applications. Mar Technol Soc J 44:122–132

Wyatt LR, Heron ML, Garello R (eds) (2006) Special issue on HF/VHF ocean surface radars. J Ocean Eng 31(4)

Jackson CR, Apel JR (eds) (2004) Synthetic aperture radar marine user's manual. US National Oceanic and Atmospheric Administration, Washington

Graber HC, Paduan J (1997) Special issue on high frequency radars for coastal oceanography. Oceanography 10(2)

References

Apel J (2004) Oceanic internal waves and solitons. In: Jackson CR, Apel JR (eds) Synthetic aperture radar marine user's manual. US National Oceanic and Atmospheric Administration, Washington

Barrick DE (1977) The ocean wave height non-directional spectrum from inversion of the HF sea-echo Doppler spectrum. Remote Sens Environ 6:201–227

Beal RC, Young G, Monaldo FM, Scott C, Thompson DR, Winstead NS (2005) High-resolution wind monitoring with wide swath SAR: a users guide. NOAA, Washington

Chapman RD, Graber HC (1997) Validation of HF radar measurements. Oceanography 10:76–79

Christiansen M, Hasager C, Thompson D, Monaldo F (2008) Ocean winds from synthetic aperture radar. In: Niclos R (ed) Ocean remote sensing: recent techniques and applications research Signpost, Kerala. ISBN: 978-81-308-0268-8

Clemente-Colón P (2004) Upwelling. In: Jackson CR, Apel JR (eds) Synthetic aperture radar marine user's manual. US National Oceanic and Atmospheric Administration, Washington

Clemente-Colón P, Montgomery D, Pichel W, Friedman K (1998) The use of synthetic aperture radar observations as indicators of fishing activity in the Bering Sea. J Adv Mar Sci Technol Soc 4:249–258

Clemente-Colón P, Pichel W (2006) Remote sensing of marine pollution. In: Gower JFR, Rencz AN (eds) Manual of remote sensing, 3 edn. Wiley, New York

Clemente-Colon P, Yan X-H, Pichel W (1997) Evolution of oil slick patterns as observed by SAR off the coast of Wales. In: Proceedings 3rd ERS symposium on space at the service of our environment, pp 565–568, 17–21 Mar 1997

Collard F, Ardhuin F, Chapron B (2005) Extraction of coastal ocean wave fields from SAR images. IEEE J Ocean Eng 30(3):526–533

Cook TM, De Paolo T, Terrill EJ (2007) Estimates of radial current errors from high frequency radar using MUSIC for bearing determination. IEEE OCEANS 2007, IEEE Xplore. doi: 10.1109/OCEANS.2007.4449265

Cox C, Munk WH (1954) Statistics of the sea surface derived from sun glitter. J Mar Res 13:198–227

Crombie DD (1955) Doppler spectrum of sea echo at 13.56 Mc/s. Nature 175:681–682

Devlin MJ, Brodie J (2005) Terrestrial discharge into the Great Barrier Reef Lagoon: Nutrient behavior in coastal waters. Mar Pollut Bull 51:9–22

DiGiacomo PM, Holt B (2001) Satellite observations of small coastal ocean eddies in the Southern California Bight. J Geophys Res 106(C10):22521–22544

DiMassa D, Heron ML, Mantovanelli A, Heron SF, Steinberg C (2010) Can vertical mixing from turbulent kinetic energy mitigate coral bleaching? An application of HF Ocean radar. IEEEOCEANS Sydney, IEEE Xplore

DiMassa D, Heron ML, Mantovanelli A, Heron SF (2011) HF radar: a tool for coral reef planning and management. MTS/IEEE OCEANS Kona, IEEE Xplore

Elfouhaily T, Chapron B, Katsaros K, Vandemark D (1997) A unified directional spectrum for long and short wind-driven waves. J Geophys Res 102:15781–15796

Espedal HA, Johannessen OM, Johannessen JA, Dano E, Lyzenga DR, Knulst JC (1998) COASTWATCH'95: ERS 1/2 SAR detection of natural film on the ocean surface. J Geophys Res 103:24969–24982

Franklin EC (2008) An assessment of vessel traffic patterns in the Northwestern Hawaiian Islands between 1994 and 2004. Mar Pollut Bull 56:136–162

Gade M, Alpers W (1999) Using ERS-2 SAR for routine observation of marine pollution in European coastal waters. Sci Total Environ 237–238:441–448

Gade M, Alpers W, Huhnerfuss H, Masuko H, Kobayashi T (1998) Imaging of biogenic and anthropogenic ocean surface films by the multifrequency/multipolarization SIR-C/X-SAR. J Geophys Res 103(C9):18851–18866

Garcia-Pineda O, Zimmer OB, Howard M, Pichel W, Li X, MacDonald IR (2009) Using SAR images to delineate ocean oil slicks with a texture-classifying neural network algorithm (TCNNA). Can J Remote Sens 35(5):411–421

Haapkylä J, Ramade F, Salvat B (2007) Oil pollution on coral reefs: a review of the state of knowledge and management needs. Vie et Milieu—Life and Environment 57:91–107

Heron SF, Heron ML (1998) A comparison of algorithms for extracting significant wave height from HF radar ocean backscatter spectra. J Atmos Ocean Technol 15:1157–1163

Heron ML, Marrone P (2010) Wind direction manifestation on HF Ocean radar echoes. IEEE OCEANS Sydney, IEEE Xplore

Heron ML, Prytz A (2002) Wave height and wind direction from the HF coastal ocean surface radar. Can J Remote Sens 28:385–393

Heron ML, Prytz A (2011) The data archive for the phased array HF radars in the Australian Coastal Ocean radar network. IEEE OCEANS Santander, IEEE Xplore

Heron ML, Skirving WJ, Michael KJ (2006) Ocean wave slope models for short wave remote sensing data analysis. IEEE Trans Geosci Remote Sens 44(7):1962–1973

Hersbach H, Stoffelen A, de Haan S (2007) An improved C-band scatterometer ocean geophysical model function: CMOD5. J Geophys Res 112(C3):C03006

Horstmann J, Koch W, Thompson DR, Graber HC (2006) Hurricane winds measured with synthetic aperture radar. Proceedings 2006 IEEE international geoscience remote sensing symposium, IEEE Xplore, Denver

Hu C, Li X, Pichel WG, Muller-Karger FE (2009) Detection of natural oil slicks in the NW Gulf of Mexico using MODIS imagery. Geophys Res Lett 36:L01604. doi:10.1029/2008GL036119

Jackson CR (2004) An atlas of internal solitary-like waves and their properties, 2nd edn. Global Ocean Associates, Alexandria. Available online at: http://www.internalwaveatlas.com. Accessed 02 Nov 2011

Johannes RE, Maragos JE, Coles SL (1972) Oil damages corals exposed to air. Mar Pollut Bull 3:29–30

Johannessen JA, Shuchman RA, Wackerman C, Digranes G, Lyzenga D, Johannessen OM (1994) Detection of surface current features with ERS-1 SAR. Proceedings second ERS-1 symposium—space at the service of our environment, ESA SP-361, Hamburg, Germany, pp 565–569, 11–14 Oct 1993

Johannessen JA, Shuchman RA, Digranes G, Lyzenga DR, Wackerman C, Johannessen OM, Vachon PW (1996) Coastal ocean fronts and eddies imaged with ERS 1 synthetic aperture radar. J Geophys Res 101(C3):6651–6667

Kohut JT, Roarty HJ, Glenn SM (2006) Characterizing observed environmental variability with HF Doppler radar surface currents mappers and acoustic Doppler current profilers. IEEE J Ocean Eng 31:876–884

Li X, Pichel W, He M, Wu S, Friedman K, Clemente-Colon P, Zhao C (2002) Observation of hurricane-generated Ocean swell refraction at the gulf stream North Wall with the RADARSAT-1 synthetic aperture radar. IEEE Trans Geosci Remote Sens 40(10):2131–2142. doi:10.1109/TGRS.2002.802474

Li X, Zheng W, Pichel WG, Zou C-Z, Clemente-Colon P (2007) Coastal katabatic winds imaged by SAR. Geophys Res Lett 34:L03804. doi:10.1029/2006GL028055

Li X, Zheng W, Zou C-Z, Pichel WG (2008a) A SAR observation and numerical study on ocean surface imprints of atmospheric vortex streets. Sensors 8:3321–3334. doi:10.3390/s80533212008

Li X, Zheng W, Yang X, Li Z, Pichel WG (2011) Sea surface imprints of coastal mountain lee waves imaged by SAR. J Geophys Res. doi:10.1029/2010JC006643

Li X, Li C, Xu Q, Pichel W (2009a) Sea surface manifestations of signatures of along-tidal-channel underwater ridges imaged by SAR. Trans Geosci Remote Sens 47(8):2467–2477

Li X, Zhao Z, Pichel WG (2008b) Internal solitary waves in the northwestern South China Sea inferred from satellite images. Geophys Res Lett 35(L13605). doi:10.1029/2008GL034272

Li X, Li C, He M (2009b) Coastal upwelling observed by multi-satellite sensors. Science in China Series D: Earth Sci 52(7):1030–1038. doi:10.1007/s11430-009-0088-x

Li X, Li C, Pichel WG, Clemente-Colón P, Friedman K (2005) Synthetic aperture radar imaging of axial convergence fronts in Cook Inlet, Alaska. IEEE J Oceanic Eng. doi:10.1109/JOE.2005.857510

Liu AK, Wu SY, Tseng WY, Pichel WG (2000) Wavelet analysis of SAR images for coastal monitoring. Can J Remote Sens 26(6):494–500

Liu P, Zhao C, Li C, He M, Pichel WG (2010) Identification of ocean oil spills in SAR imagery based on fuzzy logic algorithm. Int J Remote Sens 31(17):4819–4833

Lu J, Lim H, Liew SC, Bao M, Kwoh LK (1999) Oil pollution statistics in Southeast Asian waters compiled from ERS SAR imagery. Earth Obs Quart ESA Publ EOQ 61:13–17

Lyden JD, Hammond RR, Lyzenga DR, Shuchman RA (1988) Synthetic aperture radar imaging of surface ship wakes. J Geophys Res 93(C10):12293–12303

Mantovanelli A, Heron ML, Prytz A (2010) The use of HF radar surface currents for computing Lagrangian trajectories: benefits and issues. In: Proceedings IEEE Oceans 2010, 24–27 May 2010, Sydney, Australia

Manzello D, Hendee JC, Ward D, Hillis-Starr Z (2006) An evaluation of environmental parameters coincident with the partial bleaching event in St. Croix, US Virgin Islands 2003. Proceedings 10th International Coral Reef Symposium, Okinawa, Japan, pp 709–717

McCulloch M, Fallon S, Wyndham T, Hendy E, Lough J, Barnes D (2003) Coral record of increased sediment flux to the inner Great Barrier Reef since European settlement. Nature 421:727–730

Monaldo F, Kerbaol V, Clemente-Colón P, Furevik B, Horstmann J, Johannessen J, Li X, Pichel W, Sikora T, Thompson D, Wackerman C (2004a) The SAR measurement of ocean surface winds: an overview. Proceedings second workshop on coastal and marine applications of synthetic aperture radar, Svalbard, Norway, SP-565, European Space Agency, pp 15–32, 8–12 September, 2003

Monaldo F, Thompson D, Beal R, Pichel WG, Clemente-Colón P (2001) Comparisons of SAR-derived wind speed with model predictions and ocean buoy measurements. IEEE Trans Geosci Remote Sens 3(12):2587–2600

Monaldo FM, Thompson DR, Pichel WG, Clemente-Colón P (2004b) A systematic comparison of QuikSCAT and SAR ocean surface wind speeds. IEEE Trans Geosci Remote Sens 42(2):283–291

Mouche AA, Chapron B, Reul N (2007) A simplified asymptotic theory for ocean surface electromagnetic wave scattering. Waves Random Complex Media 17(3):321–341

Mugliaccio M, Nunziata F, Gambardella A (2009) On the copolarised phase difference for oil spill observation. Int J Remote Sens 30(6):1587–1602

Mumby PJ, Skirving W, Strong AE, Hardy JT, LeDrew EF, Hochberg EJ, Stumpf RP, David LT (2004) Remote sensing of coral reefs and their physical environment. Mar Pollut Bull 48:219–228

Nadel F, Freilich MH, Long DE (1991) Spaceborne radar measurement of wind velocity over the ocean—an overview of the NSCAT scattereometer system. Proc IEEE 79:850–866

Pichel WG, Clemente-Colón P, Wackerman C, Friedman K (2004) Ship and wake detection. In: Jackson CR, Apel JR (eds) Synthetic aperture radar marine user's manual. US National Oceanic and Atmospheric Administration, Washington

Pichel WG, Li X, Monaldo F, Sikora T, Jackson C (2008) High-velocity wind measurements using synthetic aperture radar, 2008. Proceedings international geoscience and remote sensing symposium (IGARSS 2008), Boston, 7–11 July 2008

Quilfen YB, Chapron B, Elfouhaily T, Katsaros K, Tournadre J (1998) Observation of tropical cyclones by high-resolution scatterometry. J Geophys Res 103:7767–7786

Sikora T, Friedman KS, Pichel WG, Clemente-Colón P (2000) Synthetic aperture radar as a tool for investigating polar mesoscale cyclones. Weather Forecast 15(6):745–758

Sikora TD, Ufermann S (2004) Marine atmospheric boundary layer cellular convection and longitudinal roll vortices. In: Jackson CR, Apel JR (eds) Synthetic aperture radar marine user's manual. US National Oceanic and Atmospheric Administration, Washington

Simecek-Beatty D, Clemente-Colón P (2004) Locating a sunken vessel using SAR imagery: detection of oil spilled from the SS Jacob Luckenbach. Int J Remote Sens 25(11):2233–2241
Simecek-Beatty D, Pichel WG (2006) RADARSAT-1 synthetic aperture radar analysis for M/V Selendang Ayu oil spill. Proceedings twenty-ninth arctic and marine oilspill program (AMOP) technical seminar vol 2. Vancouver, British Columbia, Canada, pp 931–949, 6–8 June 2006
Simpson JH, Hunter JR (1974) Fronts in the Irish Sea. Nature 250:404–406
Shuming L, Li Z, Yang X, Pichel WG, Yu Y, Zheng Q, Li X (2010) Atmospheric frontal gravity waves observed in satellite SAR images of the Bohai Sea and Huanghai Sea. Acta Oceanol Sin 29(5):35–43. doi:10.1007/s13131-010-0061-8
Skirving WJ, Heron ML, Heron SF (2006) The hydrodynamics of a bleaching event: Implications for management and monitoring. In: Phinney JT et al (eds) Coral reefs and climate change: science and management. Am Geophys Union, Washington
Stofflen A, Anderson D (1997) Scatterometer data interpretation: measurement and inversion. J Atmos Ocean Technol 14:1298–1313
Thompson DR, Elfouhaily TM, Chapron B (1998) Polarization ratio for microwave backscattering from the ocean surface at low to moderate incidence angles. Proceedings 1998 international geoscience remote sensing symposium seattle WA, IEEE
Trivero P, Fiscella B, Gomez F, Pavese P (1998) SAR detection and characterization of sea surface slicks. Int J Remote Sens 19(3):543–548
Ullman DS, O'Donnell J, Kohut J, Fake T, Allen A (2006) Trajectory prediction using HF radar surface currents: Monte Carlo simulations of prediction uncertainties. J Geophys Res 111:C12005. doi:10.1029/2006JC003715
Vachon P, Thomas SJ, Cranton J, Edel HR, Henschel MD (2000) Validation of ship detection by the RADARSAT synthetic aperture radar and the ocean monitoring workstation. Can J Remote Sens 26:200–212
Vachon P, Monaldo M, Holt B, Lehner S (2004) Ocean surface waves and spectra. In: Jackson CR, Apel JR (eds) Synthetic aperture radar marine user's manual. US National Oceanic and Atmospheric Administration, Washington
Vachon P, Wolfe J (2010) C-Band cross-polarization wind speed retrieval. IEEE Geosci Remote Sens Lett 99:456–459. doi:10.1109/LGRS.2010.2085417
Wackerman C, Friedman K, Pichel WG, Clemente-Colón P, Li X (2001) Automatic detection of ships in RADARSAT-1 SAR imagery. Can J Remote Sens 27:568–577
Wahl T, Skøelv Å, Anderssen T, Pedersen JP, Andersen JH, Follum OA, Strøm GD, Bern TI, Hamnes H, Solberg R (1996) Radar satellites: A new tool for pollution monitoring in coastal waters. Coastal Manag 24:61–71
Wang Y-H, Dai C-F, Chen Y–Y (2007) Physical and ecological processes of internal waves on an isolated reef ecosystem in the South China Sea. Geophys Res Lett 34:L18609
Winstead NS, Colle B, Bond N, Young G, Olson J, Loescher K, Monaldo F, Thompson D, Pichel WG (2006) Using SAR remote sensing, field observations, and models to better understand coastal flows in the Gulf of Alaska. Bull Am Meteorol Soc 87:787–800
Wismann V, Gade M, Alpers W, Huhnerfuss H (1998) Radar signatures of marine mineral oil spills measured by an airborne multi-frequency radar. Int J Remote Sens 19:3607–3623
Wolanski E, Deleersnijder E (1998) Island-generated internal waves at Scott Reef, Western Australia. Cont Shelf Res 18:1649–1666
Wolanski E. Colin P, Naithani J, Deleersnijder E, Golbuu Y (2004) Large amplitude, leaky, island-generated, internal waves around Palau, Micronesia. Estuar Coastal Shelf Sci 60:705–716
Wu SY, Liu AK (2003) Towards an automated ocean feature detection, extraction, and classification scheme for SAR imagery. Int J Remote Sens 24(5):935–951
Wyatt LR, Holden GJ (1994) HF radar measurement of multi-modal directional wave spectra. Glob Atmos Ocean Syst 2:265–290
Xu Q, Lin H, Li X, Zuo J, Zheng Q, Pichel W, Yuguang L (2010) Assessment of an analytical model for sea surface wind speed retrieval from spaceborne SAR. Int J Remote Sens 3(4):993–1008

Yang X, Li X, Zheng Q, Gu X, Pichel WG, Li Z (2011) Comparison of ocean surface winds retrieved from QuikSCAT scatterometer and Radarsat-1 SAR in offshore waters of the US West Coast. IEEE Geosci Remote Sens Lett 8(1):163–167. doi:10.1109/LGRS.2010.2053345
Young GS, Sikora TD, Winstead NS (2005) Use of synthetic aperture radar in fine-scale surface analysis of synoptic-scale fronts at sea. Weather Forecast 20:311–327
Young IR, Black KP, Heron ML (1994) Circulation in the ribbon reef region of the Great Barrier Reef. Cont Shelf Res 14:117–142
Zhao Y, Liu AK, Hsu M-K (2008) Internal wave refraction observed from sequential satellite images. Int J Remote Sens 29(21):6381–6390
Zhang B, Perrie W, Li X, Pichel WG (2011) Mapping sea surface oil slicks using RADARSAT-2 quad-polarization SAR image. Geophys Res Lett. doi:10.1029/2011GL047013
Zheng Q, Clemente-Colón P, Yan X-H, Liu WT, Huang NE (2004) Satellite SAR detection of Delaware Bay plumes: Jet-like feature analysis. J Geophys Res 109(C3):C03031. doi:10.1029/2003JC002100

Section V
Effective Use of Remote Sensing in Science and Management

Chapter 14
Validation

Chris M. Roelfsema and Stuart R. Phinn

Abstract Adequate understanding of the validity of information contained in coral reef remote sensing products is required to support research and management decisions. This chapter introduces accuracy measures commonly applied to two types of coral related maps: discrete (e.g., benthic cover type) and continuous (e.g., percent coral cover). A critical review of 80 coral reef remote sensing mapping publications presents the approaches and metrics commonly used to measure accuracy. The literature review shows that few studies report accuracy information at all, and when obtained, 'overall accuracy' was the most commonly used accuracy measure. Variations in accuracy levels were not only a result of actual differences in map accuracy, but are likely also due to: spatial complexity of benthic features present in the study area; distribution of the calibration and validation samples relative to each other; and the level of detail measured for each sample. As a result, accuracy measures from different studies should be compared with caution and with due attention to how the measures were derived. This chapter enables scientist and managers to understand, design and interpret validation procedures for image-based maps of coral reef environments.

C. M. Roelfsema (✉) · S. R. Phinn
Biophysical Remote Sensing Group, School of Geography, Planning and Environmental Management,
The University of Queensland, Queensland 4072, Australia
e-mail: c.roelfsema@uq.edu.au

S. R. Phinn
e-mail: s.phinn@uq.edu.au

J. A. Goodman et al. (eds.), *Coral Reef Remote Sensing*,
DOI: 10.1007/978-90-481-9292-2_14,

14.1 Introduction

Spatial information and spatial modeling are increasingly used by scientists and managers to evaluate options for making decisions in research and conservation activities. Previous chapters discussed how various passive and active remote sensing sensors and their related applications are used to derive environmental data for coral reef environments, specifically information on benthic or substrate composition (Phinn et al. 2012), biophysical processes (Weeks et al. 2006; Pittman et al. 2009; Scopélitis et al. 2010), characteristics of coral reef assemblages (Knudby et al. 2011), and/or efficiency of marine park management (Newman et al. 2007). Measuring and understanding the sources of associated errors contained within each of these image-based map or spatial products is essential to determine the error levels and reliability of the output products. As a result, scientists, technicians, and managers need to understand the commonly used validation processes, resulting accuracy measures and their relative comparability for remote sensing products (Congalton and Green 1999; Foody 2002; Andréfouët 2008; Foody 2011).

This chapter focuses on the process of validating remote sensing outputs, where we define validation as measuring the difference between a mapped feature type or value from a remotely sensed image and an accepted reference feature or value at the same location and time. The reference data can be in the form of an existing map or data collected from the field. We focus on three types of biotic and abiotic environmental information: (1) composition of coral reefs at varying levels of detail (e.g., substrate or benthic cover type, geomorphic zones, biotopes, or reef community composition); (2) the form or shape of the coral reef (e.g., bathymetry and topographic complexity); and (3) the biophysical properties of the environment around the coral reef (e.g., sea surface temperature, sea surface winds, concentrations of organic material, underwater light fields).

Each of these environmental variables exhibit variations over specific spatial scales, in terms of minimum size and maximum extent, and in terms of structure or concentration over a range of temporal scales (Hatcher 1997). As a result, mapping each feature and measuring appropriate "reference" values for validation requires recognition and consideration of the spatial and temporal scales associated with the variable being mapped (Fig. 14.1). For example, coral reef composition and biophysical properties can change at different time and spatial scales, from minutes over small areas (e.g., chlorophyll concentration in water column) to months over larger areas (e.g., macro algae cover). Additionally, the available extent and scale of different environmental variables do not always coincide. For instance, the extent to which water properties are mapped (e.g., SST, chlorophyll concentrations) are often significantly larger than the extent to which reef properties are mapped (e.g., benthic composition, depth).

The discrete versus continuous nature, and the temporal and spatial variability, of the coral reef environmental variables being mapped and monitored influence several aspects of the validation process. For example, different validation

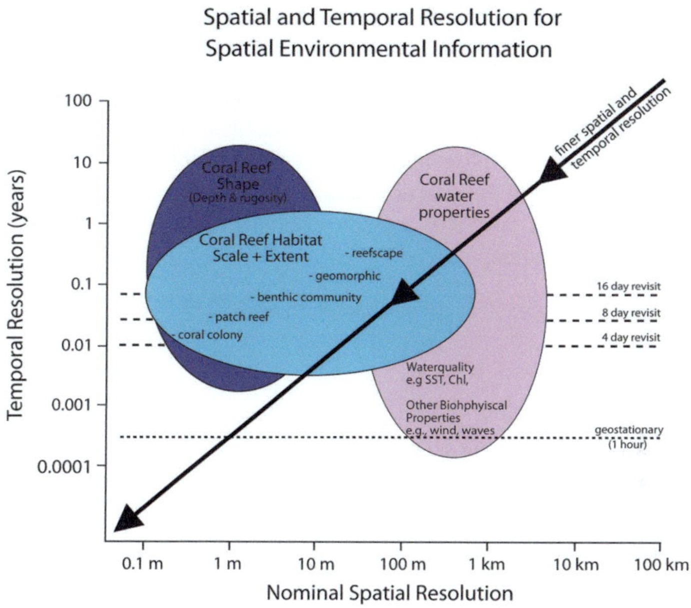

Fig. 14.1 Spatial feature size and temporal scale coral reef feature composition and biophysical properties (modified from Phinn et al. 2010)

approaches are required for a discrete map with relatively stable feature types (e.g., coral reef geomorphic zones) as compared with a continuous map depicting a dynamic environmental variable (e.g., sea surface temperature). The differences in these two primary types of maps are defined as follows:

- *Discrete Maps*: In these maps individual pixels are assigned a unique category to identify reef composition (e.g., substrate or benthos type). These maps are used to describe reef/non-reef areas, geomorphic zones, communities and habitats of specific organisms.
- *Continuous Maps*: These are quantitative maps, where each pixel is a quantitative measure of a biophysical property (e.g., SST, depth, water column constituent concentration, optical properties, or percent live coral cover).

The aim of this chapter is to provide coral reef scientists and managers with an overview and comparison of the accuracy measures commonly reported for remote sensing derived map products. We first provide a brief description of sampling design (Sect. 14.2.1) and then present the main accuracy assessment methods for discrete (Sect. 14.2.2) and continuous (Sect. 14.2.3) remote sensing image-based maps. This is followed by a literature review assessing the most commonly applied accuracy assessment approaches and metrics used to assess maps of coral reef systems (Sect. 14.3).

14.2 Sampling Design and Accuracy Measures

To determine an unbiased accuracy measure of a map, independent reference data is collected and compared with the mapped data. Before collecting the reference data, however, a sampling design needs to be defined, which answers the questions of how, what and where the reference data is to be collected. The comparison of map data with this reference data is then used for calculating discrete and continuous mapping accuracies.

14.2.1 Sampling Design

A carefully developed sampling design, in terms of the number, location and distribution of samples among mapped thematic classes or continuous values, is needed for collecting reference data, as it will determine the validity, statistical power and overall cost of the accuracy assessment (Stehman and Czaplewski 1998; Congalton and Green 1999). The sampling design is described by sample unit size, number of samples, spatial distribution of samples, and data collection process at each sample unit (Stehman and Czaplewski 1998; Congalton and Green 1999). An additional consideration is the number of validation samples per mapped class in relation to the area covered by each class (Green et al. 2000).

Sampling units, which can be points or areas (e.g., pixels or cluster of pixels), define the spatial extent of the reference data used to calibrate and validate a map product and its map classes (Stehman and Czaplewski 1998). In terms of the number of samples required for each map class or continuous map value to produce a statistically valid analysis, a balance is required between what is statistically sound in terms of probability sampling, and what is practically achievable (Congalton and Green 1999). Ideally, the number of samples can be determined based on the expected accuracy, the allowable error, and the expected confidence level (Fitzpatrick-Lins 1981). Though the most rigorous in design, this idealized approach of probability sampling is not commonly used, as it requires large sample numbers and high spatial sample distribution, which typically cannot be sampled with available resources, particularly given the logistical difficulties of fieldwork in coral reef environments (Stehman 2001). These logistical challenges include: limited resources (e.g., boats, skilled people and equipment); access to the sample areas (e.g., water depth, water clarity, currents, tides, dangerous marine animals); and/or extent and remoteness of sampling area. As a compromise, Congalton (1991) suggests a 'rule of thumb' of a minimum of 50 validation samples per discrete mapping category. This study further advised the minimum sample number should increase when the study area is larger than 4,000 km^2 or when more than 12 categories are mapped. This approach has been adopted as the default sampling design in the majority of satellite image-based land-cover mapping applications published to date (Foody 2002; Jensen 2005; Lillesand et al. 2008) and is

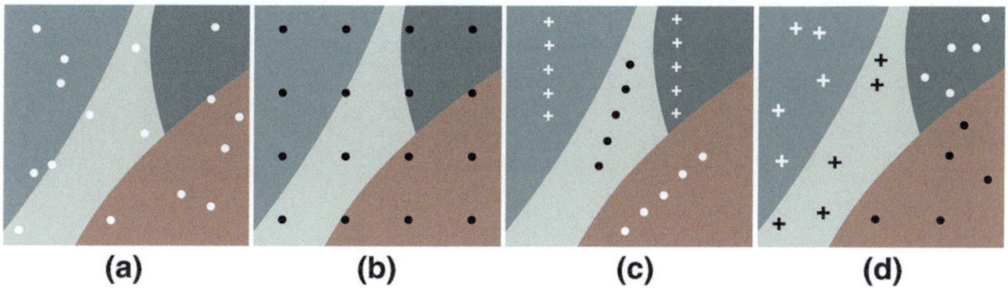

Fig. 14.2 Example of four commonly used probabilistic sampling schemes: **a** simple random, **b** systematic, **c** stratified systematic unaligned and **d** stratified random

commonly followed in coral reef remote sensing applications, as indicated by the literature review results presented later in this chapter.

With respect to the spatial distribution of sample units, in order to have a random selection of independent samples, a procedure needs to be applied to assure that the different mapping categories in a given study area have equal probabilities of being sampled. Common probability sampling schemes are (Fig. 14.2): simple random, systematic, stratified random, and stratified systematic unaligned sampling (Congalton and Green 1999). Random sampling is the most statistically robust (Congalton and Green 1999; Stehman 1999); however, it requires large numbers of sample units and is often not applicable due to logistical challenges associated with implementing this approach in a marine environment. For example, access to points by boats may be restricted by water depth, surface roughness or observation capability for the observer. Stratified random sampling schemes are more frequently implemented in marine environments, as field sampling can be designed within the limits of these practical logistics (Stehman and Czaplewski 1998).

14.2.2 Accuracy of Discrete Maps

Accuracy measures for discrete maps are commonly derived using an error matrix (Table 14.1), which tabulates the level of agreement between the thematic class at a location in the image-based map and the same location in the reference data (Congalton and Mead 1983; Story and Congalton 1986; Ma and Redmond 1995).

The accuracy of each mapping category is described by the individual class accuracies, or according to the user's and producer's accuracies, which are derived from the error matrix (Congalton 1991). The producer's accuracy is the probability of a reference data point being classified correctly, whereas the user's accuracy is the probability of a classified pixel being correct. The user's and producer's accuracies are then used to assess misclassification characteristics, such as where pixels have been erroneously excluded from a class (omission) versus pixels erroneously included in a class (commission). Overall accuracy is calculated by

Table 14.1 Example of overall, producer's and user's accuracies displayed in an error matrix, which tabulates the level of agreement between the thematic class at a location in the image (rows) and the same location in the reference data (columns): overall accuracy = sum of the diagonal values divided by the sum of all values; producer accuracy= column total divided by diagonal value for a specific category; and user accuracy = row total divided by diagonal value for a specific category

	Reference data							Row total	User accuracy
	Seagrass cover				Sand	Deep	Mangrove		
	1–10 %	10–40 %	40–70 %	70–100 %					
Image data									
Seagrass cover									
1–10 %	179	177	61	1	103	0	0	521	34 %
10–40 %	161	333	343	69	87	0	0	993	34 %
40–70 %	47	297	542	217	22	0	0	1,125	48 %
70–100 %	11	27	168	403	2	0	0	611	66 %
Sand	41	198	68	3	760	2	0	1,073	71 %
Deep	0	0	2	31	0	914	0	947	97 %
Mangrove	1	5	1	12	2	0	276	297	93 %
Column total	440	1,037	1,186	736	976	916	276		
Producer accuracy	41 %	32 %	46 %	55 %	78 %	100 %	100 %	Overall accuracy	61 %

dividing the total number of pixels that were correctly classified (i.e., assigned to the correct class; principal axis in Table 14.1), by the total number of validation pixels or reference points used to construct the error matrix (Congalton 1991). Understanding the overall, user's and producer's accuracy values is essential for interpreting thematic maps, determining if they are useable for a specific application, and understanding which image map classes are mapped more accurately than others.

Although the overall accuracy is the most commonly used accuracy measure, it does not take into account both errors of commission and omission for all individual classes. As a result, overall accuracy tends to overestimate accuracy (Ma and Redmond 1995). As a result, adjustments to the overall accuracy are often made to account for the chance agreement or accuracy resulting from a completely random assignment of pixels to mapping categories (Lillesand et al. 2008). This is referred to as 'adjusted' overall accuracy and is represented by Kappa (Congalton and Mead 1983) or Tau (Ma and Redmond 1995) values. Although adjusted accuracy measures were commonly used in past remote sensing studies, recent research advises that they are not as useful as thought due to significant flaws in their assumptions and methods and it is now advised to report only overall accuracy (Foody 2011; Pontius Jr and Millones 2011).

14.2.3 Accuracy of Continuous Maps

Continuous maps, in comparison to discrete maps, require alternative accuracy measures as they do not consist of a set of fixed thematic classes, but rather represent a biophysical property (e.g., depth) that falls within a given range for every pixel (e.g., 0–30 m water depth). In general, continuous mapping methods are applied for mapping biophysical properties that vary across an image, examples of which include percentage cover, depth, and water column properties such as chlorophyll biomass and light attenuation. The most commonly used accuracy measures for continuous maps are the root-mean square error (RMSE) and regression coefficient (i.e., coefficient of determination), which are both used to quantify the co-variance between reference data and mapped data.

Because RMSE and linear regression have been commonly applied to various coral reef datasets, map users can easily compare results with the published literature to evaluate the relative quality of the map product. As with thematic accuracy assessment data, the reference sample set for continuous maps should also be a statistically sound representation of the continuous parameter.

Effective validation of continuous variables requires collection of field measurements coinciding with acquisition of the remote sensing data, which is often challenging in dynamic environments, or impractical in remote or very large coral reefs. As a result, point-to-point comparison of field measures and image data can be limited. Therefore, alternative statistical parameters (e.g., distribution functions) need to be estimated to assess the accuracy of dynamic continuous variables.

Practically, this means that a series of field measurements from known locations are used to model a distribution function of the measured parameter over time and space. This function can then be used to estimate a value at the time of the remote sensing data acquisition (e.g., water depth as a function of tidal variations), which then can be compared with the measured value derived from the remote sensing data (Gregg and Casey 2004).

14.3 Validation Literature Review

The methods used to collect reference data and validate image products described above were mostly derived from terrestrial remote sensing, but they are also used in coral reef remote sensing studies (Ma and Redmond 1995; Stehman and Czaplewski 1998; Foody 2002). To provide context as to how these methods are routinely applied in coral reef environments, the following section presents a review of 80 papers (Table 14.2; Roelfsema 2009). The 80 remote sensing papers focused on coral reef and seagrass mapping in coastal environments, including 66 published in the period 2002–2009, 9 in 1992–2002, and 5 in 1976–1992. Of the 80 publications, 41 were published in remote sensing journals and 39 in other journals related to reef or coastal environments (e.g., Coral Reefs, Journal of Coastal Research, Estuaries). The information from the review covers: (1) mapping approaches used (Sect. 14.3.1; Table 14.3); (2) the associated sampling designs used (Sect. 14.3.2; Table 14.4); and (3) accuracy measures used for specific mapping purposes (Sect. 14.3.3; Table 14.5).

14.3.1 Mapping Approaches

To characterize the types of mapping approaches applied in coral reef environments, we divided 80 peer-reviewed studies on coastal and marine remote sensing into two main habitats, coral reefs (54 papers) and seagrass (26 papers). We then subdivided papers on coral reefs based on the reef zones studied: barrier, platform and fringing reefs (7 papers); barrier and platform reefs (14 papers); fringing reefs (26 papers); and lagoons (8 papers). The areal extent covered in each of the papers varied from: 0–10 km^2 (13 papers); 10–100 km^2 (37 papers); 100–360 km^2 (15 papers) and >360 km^2 (10 papers). Studies included in the 0–10 km^2 group were mainly proof-of-concept studies for developing specific new applications for coral reef habitat mapping (e.g., Rowlands et al. 2008).

The various mapping approaches were characterized according to: remote sensing data type, calibration and validation data type(s), and mapping purpose (Table 14.3). Studies were also grouped by type of image data, method for collecting calibration and validation data, and mapping technique used to transform image data to a map product. It should be noted that some of the 80 papers used

Table 14.2 Compendium of publications included in remote sensing methods and validation review (Roelfsema 2009): numbers represent citations used in Tables 14.3–14.5

[1] Capolsini et al. (2003)	[41] Maeder and Narumalani (2002)
[2] Cuevas-Jimenez and Ardisson (2002)	[42] Mishra et al. (2006)
[3] Fornes et al. (2006)	[43] Purkis and Riegl (2005)
[4] Newman et al. (2007)	[44] Riegl et al. (2005)
[5] Young et al. (2008)	[45] Purkis et al. (2008)
[6] Isoun et al. (2003)	[46] Houk and van Woesik (2008)
[7] Andréfouët and Dirberg (2006)	[47] Benfield et al. (2007)
[8] Andréfouët et al. (2003)	[48] Phinn et al. (2008)
[9] Dekker et al. (2005)	[49] Joyce et al. (2004)
[10] Call et al. (2003)	[50] Roelfsema et al. (2009)
[11] Mumby and Edwards (2002)	[51] Schweizer et al. (2005)
[12] Pasqualini et al. (2005)	[52] Klonowski et al. (2007)
[13] Peneva et al. (2008)	[53] Palandro et al. (2008)
[14] Riegl and Purkis (2005)	[54] Prada et al. (2008)
[15] Mumby et al. (1998)	[55] Lesser and Mobley (2007)
[16] Roelfsema et al. (2002)	[56] Louchard et al. (2003)
[17] Roelfsema et al. (2006)	[57] Cassata and Collins (2008)
[18] Sheppard et al. (1995)	[58] Matarrese et al. (2004)
[19] Zharikov et al. (2005)	[59] Kvernevik et al. (2002)
[20] Wabnitz et al. (2008)	[60] Pergent et al. (2002)
[21] Dierssen et al. (2003)	[61] Dahdouh-Guebas et al. (1999)
[22] Habeeb et al. (2007)	[62] Kutser et al. (2006)
[23] Palandro et al. (2003)	[63] Vanderstraete et al. (2006)
[24] Rowlands et al. (2008)	[64] Jordan et al. (2005)
[25] Bouvet et al. (2003)	[65] Kendrick et al. (2002)
[26] Chauvaud et al. (1998)	[66] Moore et al. (2002)
[27] Hochberg and Atkinson (2003)	[67] Andréfouët and Guzman (2005)
[28] Goodman and Ustin (2007)	[68] Lauer and Aswani (2008)
[29] Chauvaud et al. (2001)	[69] Ahmad and Neil (1994)
[30] Armstrong (1993)	[70] Ackleson and Klemas (1987)
[31] Meehan et al. (2005)	[71] Aswani and Lauer (2006)
[32] Gullstrom et al. (2006)	[72] Alexander (2008)
[33] Franklin et al. (2003)	[73] Benton and Newman (1976)
[34] Naseer and Hatcher (2004)	[74] Jupp et al. (1985)
[35] Sagawa et al. (2008)	[75] Andréfouët et al. (2005)
[36] Garza-Perez et al. (2004)	[76] Lathrop et al. (2006)
[37] Andréfouët et al. (2004)	[77] Lennon and Luck (1989)
[38] Bainbridge and Reichelt (1998)	[78] Murdoch et al. (2007)
[39] Bertels et al. (2008)	[79] Orth et al. (2006)
[40] Holmes et al. (2007)	[80] Pasqualini et al. (2000)

more than one combination of data types and/or processing/mapping technique as part of a comparison study or as additional information.

Table 14.3 demonstrates that, of the 80 studies reviewed, the main remote sensing data type used was medium spatial resolution multispectral (e.g., Landsat Thematic Mapper) (27 papers), followed by aerial photography (19 papers), high spatial resolution multispectral (e.g., Ikonos) (17 papers), hyperspectral (e.g., CASI, Hyperion) (11 papers), and acoustic data (e.g., Roxan) (6 papers). From the 54 coral reef habitat mapping papers, 20 papers used moderate spatial resolution

Table 14.3 Summary of mapping approach components (habitat, study area size, remote sensing imagery, calibration/validation method, and mapping/processing method) used in 80 scientific publications on coral reef and seagrass mapping using remote sensing

Area type		Number of papers	Remote sensing data					Calibration validation data				Processing focus					Reference
Enviroment type	Area size(km^2)		MRMS	HRMS	Hyper spectral	Aerial	Acoustic	Spot check	Transect	Video	Other	Supervised	Unsupervised	Spectral + radiative	delineation	Other	
Coral reef environments																	
Barrier, platform, fringing	–10																
	10–100	2		1		1		2				1				1	[7, 8]
	100–360	1				1		1								0	[18]
	360–	4	4					1	1		2	2			1	1	[34, 67, 74, 75]
Barrier, platform	–10	4	2	2				2	2			3	1				[1, 23, 37, 38]
	10–100	8	5	1	2			2	5		1	4		1	1	2	[16, 25, 38, 45, 49, 53, 62, 69]
	100–360	2	2					1	1			1				1	[30, 63]
	360–																
Fringing reef	–10	9		3	3	3		5		3	1	5	1	3			[2, 4, 6, 22, 24, 36, 55, 56, 72]
	10–100	14	2	5	3	1	3	5	3	6		2	3	3	2	4	[11, 14, 15, 27, 28, 32, 41, 42, 43, 54, 57, 58, 59, 61]
	100–360	2		1			1		1	1					1	1	[47, 64]
	360–																
Lagoon	–10	1			1			1						1			[21]
	10–100	3	2	1				1		1	1	1	1		1		[10, 46, 68]
	100–360	3				3		1			2				2	1	[33, 71, 78]
	360–	1	1						1			1					[51]
Segrass environment																	
	–10	4		2	1	1		2		2		2		1		1	[3, 5, 35, 52]
	10–100	10	5		1	3	1	5	1	1	3	4	1			5	[9, 12, 13, 26, 40, 44, 60, 70, 76, 77]
	100–360	7	2	1		3	1	2	1	1	3	3			4		[17, 29, 48, 65, 73, 79, 80]
	360–	5	2			3		3	2			2			1	2	[19, 20, 31, 50, 66]
Total		80	27	17	11	19	6	34	18	15	13	31	7	9	14	19	

Remote sensing data types include medium spatial resolution multispectral (MRMS), high spatial resolution multispectral (HRMS), hyperspectral, aerial and acoustic. Calibration and validation data types include spot check, transect, video and other (local knowledge, aerial photographs and undefined). Processing types include supervised classification, unsupervised classification, spectral analysis and radiative transfer modeling (spectral+radiative), manual delineation, and other imagery, 14 papers used high spatial resolution imagery, and 9 papers used aerial photographs. For seagrass mapping (26 papers), 10 papers used aerial photographs, 7 papers used moderate spatial resolution imagery, and 3 papers used high spatial resolution imagery.

The choice of sensor type was often based on what type of map product was required. Some of the trade-offs, and relevant examples, related to this decision include:

- *Spectral resolution*: Multispectral imagery such as Landsat has three bands in visible range versus hyperspectral imagery such as CASI, which can have a predetermined number of bands up to 512.
- *Temporal resolution*: Landsat imagery is acquired every 16 days over the last 3 decades, whereas high spatial resolution imagery such as Ikonos needs to be specifically scheduled and has only been available since 2002.
- *Spatial resolution*: MODIS imagery is a resolution of 250 × 250 m per pixel, whereas Ikonos is 4 × 4 m.
- *Areal extent*: Landsat imagery covers an area of 185 × 185 km, whereas QuickBird only covers around 12 × 12 km.

The main calibration and validation data sources used in these studies were primarily spot check or point data (34 papers), followed by transect data without video (18 papers), video transects (15 papers), and other data sources such as local knowledge (13 papers). From the coral reef habitat mapping papers (54 papers), 22 papers used spot check methods and 25 papers used video transects. For seagrass mapping (26 papers), 12 papers used spot check and 8 used video transects. The tradeoff for selecting the different types of calibration or validation data collection techniques was most often based on the available means (e.g., people, equipment, boats, remoteness) and the information required (e.g., live coral presence versus species composition). In some instances, existing data that was collected for other purposes was deemed suitable for use in a different calibration and validation application (Roelfsema et al. 2006; Roelfsema and Phinn 2010).

The most common method for classifying habitats was supervised classification (31 papers), followed by manual delineation (14 papers), spectral analysis and radiative transfer modeling (9 papers), and unsupervised classification (7 papers). The remaining 19 papers represent those that used various combinations of processing techniques, or techniques not frequently applied, such as regression analysis (2 papers) and object based classification (3 papers). Methods chosen for classification depended on the available software, equipment and knowledge and skills of the producers. Unsupervised classification, for example, can be readily achieved with GIS and remote sensing software packages (e.g., ARCGIS, ENVI, IMAGINE); however, object based image analysis is mostly limited to costly additional software that requires specialized training of the operator.

The Millennium Global Coral Reef Mapping Project (Andréfouët et al. 2005) covered the largest extent of coral reefs, and the largest for seagrass was conducted by Wabnitz et al. (2008), which mapped seagrass extent throughout the Caribbean.

Other notable mapping examples covering a large spatial extent include 22,000 km^2 of seagrass mapping in Shark Bay, Western Australia (Bruce 1997) and a 345,400 km^2 coral reef habitat map based on Landsat TM data for the Great Barrier Reef, Australia by the Great Barrier Reef Marine Park Authority (Lewis et al. 2003). However, despite their scale, these last two studies were not included in the literature review as they are not published in readily accessible scientific journals or reports.

This review of mapping approaches was followed by an experiment where eight different mapping approaches were applied to create habitat maps for three different reef environments. The mapping approaches incorporated different combinations of remote sensing image type, calibration/validation method, and the processing method applied. The approaches were assessed according to cost, time, and the knowledge and skills needed to produce each map. The final map products were analyzed with respect to perceived relevance by map users (Roelfsema et al. 2008). Results demonstrated how variations in coral reef environments, and differences in intended map purpose, affected the user's selection of a suitable mapping approach.

14.3.2 Sampling Design

Of the 80 papers reviewed, 19 used a sampling scheme based on stratified random sampling, 16 used a stratified and non-random sampling, 5 used simple random sampling, and the remaining 32 papers did not provide sufficient information to determine the type of sampling scheme used (Table 14.4). Of the 40 papers that reported sampling schemes, maps presenting the location of the calibration and validation samples were included in only 36. Ideally, the number and distribution of samples would be determined based on the statistical rules previously suggested (Sect. 14.2.1); however, that is not always practical due to logistical challenges. It was therefore not surprising that the sampling schemes were mostly stratified, with fewer sample locations used than was statistically required, which is the inherent trade off when faced with having to overcome the logistical challenges.

A total of 15 papers reported that they used more than 50 recommended validation samples per mapping category, 28 reported using less than 50, and the remaining 37 did not provide the number of samples. The statistically optimum number of required sample points per mapping category cannot always be achieved, particularly for mapping categories that cover small areas and are thus harder to actually find for sampling purposes. This was the case for Chauvaud et al. (1998) who created a map with 32 categories over an area extent of 240 km^2 with 95.7 % overall accuracy. Statistically at least 1,600 sampling points should have been visited (50 for each category), however it was not possible to identify sufficient sample areas for every category and just 111 samples were assessed.

Table 14.4 Characteristics of the types of coral reef and seagrass environments assessed in 80 scientific publications on coral reef and seagrass mapping using remote sensing, listed with mapping area size, classification focus, sampling design for collecting validation data, sampling unit, and relative number of samples collected (>50 or <50)

Area type					Total number of papers	Sampling design reported							Reference
Environment type	Area size (km^2)	Classification focus				Sampling scheme			Sampling Unit		Number of samples		
		Coral	Seagrass	Geomorphic		Stratified random	Stratified not random	random	Point	area	>50	<50	
Coral reef environments													
Barrier, platform, fringing	−10												
	10–100	2			2		1		1	1	2		[7, 8]
	100–360	1			1		1			1	1		[18]
	360–	1		3	4		1			1	1		[34, 67, 74, 75]
Barrier, platform	−10	3		1	4	1	2		1	2	2		[1, 23, 37, 38]
	10–100	7		1	8	4	1		2	3	1	4	[16, 25, 38, 45, 49, 53, 62, 69]
	100–360	1	1		2								[30, 63]
	360–												
Fringing reef	-10	8	1		9	3	1	1	5	1		4	[2, 4, 6, 22, 24, 36, 55, 56, 72]
	10–100	12	1	1	14	2	4	2	9	1		7	[11, 14, 15, 27, 28, 32, 41, 42, 43, 54, 57, 58, 59, 61]
	100–360	2			2		1		2			1	[47, 64]
	360–												
Lagoon	-10		1		1	1			1			1	[21]
	10–100	3			3	1			2	1		3	[10, 46, 68]
	100–360	2	1		3			1	1				[33, 71, 78]
	360-			1	1			1		1		1	[51]
Segrass environment													
	−10		4		4	2	1		3	1	3		[3, 5, 35, 52]
	10–100	1	9		10	2	2		5	3	2	4	[9, 12, 13, 26, 40, 44, 60, 70, 76, 77]
	100–360	1	6		7	2			1	1	2	1	[17, 29, 48, 65, 73, 79, 80]
	360–		5		5	1	1		3		1	2	[19, 20, 31, 50, 66]
Total		44	29	7	80	19	16	5	36	17	15	28	

The time and cost of collecting calibration and validation samples is often the limiting factor in developing comprehensive accuracy assessments (Green et al. 2000; Stehman 2001; Andréfouët et al. 2008). These costs can be reduced by:

- combining field efforts through collaboration of different monitoring agencies (Roelfsema et al. 2009);
- applying a field sampling approach that can easily be split into independent data sets for calibration and validation (Roelfsema and Phinn 2010; Andréfouët 2008);
- utilizing cross validation methods, where a majority subset of samples are used for calibration and the remainder for validation, applied in an iterative fashion such that calibration/validation is repeated using different subsets of samples (Andréfouët et al. 2004).

Studies with insufficient resources to collect independent calibration and validation set can alternatively use all of the samples for both calibration and validation, resulting in a 'pseudo' accuracy assessment (Schweizer et al. 2005; Roelfsema et al. 2006). The pseudo accuracy is higher than the normal accuracy, since it describes how well the training samples are classified using the same sample set. Although this approach is generally not recommended, it can provide some measure of accuracy when no independent validation data is available (Congalton and Green 1999).

14.3.3 Accuracy Measures

The 80 studies reviewed were categorized according to whether their main application purpose was related to: (1) management or (2) research/science (Table 14.5). In terms of management applications, habitat maps were created either for reporting the extent and composition of resources (e.g., Lauer and Aswani 2008) or changes in resource extent and composition (e.g., Palandro et al. 2008). The most common mapping applications were for assessment of resource extent and composition (27 papers) and for assessing images for coral reef extent and composition (16 papers). The science applications tended to focus on whether different types of remotely sensed images and image processing techniques could be used to effectively map a specific property of coral reefs (e.g., algal biomass; Andréfouët et al. 2004). Science applications also focused on improving remote sensing approaches by studying changes in mapping accuracy produced by variations in image type (Phinn et al. 2008), processing method (Purkis 2005; Benfield et al. 2007), and inclusion of additional field data and acoustic based image data sets (Pasqualini et al. 2000).

Of the 80 publications only 38 reported mapped area size, number of categories mapped, map accuracy and sensor type. Table 14.5 shows that the most common map accuracy measures reported were overall accuracy (39 papers), Kappa (15 papers) and Tau (3 papers). The user and producer accuracies were provided in 25

Table 14.5 Summary of the types of reported accuracy measures for mapping applications provided in 80 scientific publications on coral reef and seagrass mapping using remote sensing. The studies are grouped by target audience and mapping application, as well as the level of detail given regarding validation methods. References in bold provided rich explanation of the validation method applied

Mapping Applications	Number of papers	Accuracy measure provided					Sample number reported	Validation method explanation			References
		Overall	Kappa	Tau	Error matrix	producer and user		Rich	Pore	None	
Manager											
Resource extent & composition	27	9	5		7	7	11	2	11	14	[4, 19, 20, 26, 29, 31, 32, 34, 49, 50, 51, 57, 60, 61, 63, 64, 66, 67, 68, 71, 74, 75, 76, 77, 78, 79]
Change in resource extent & composition	5	2	1			1			3	2	[9, 23, 46, 53, 65]
Scientist-technecian											
Develop mapping technique for coral reef extent & composition	11	8			2	3	9	1	7	3	[5, 7, 8, 16, 17, 18, 22, 24, 33, 37, 45]
Assess different image types for coral reef extent & composition	16	11	4	2	2	9	10	5	6	5	[1, 2, 3, 11, 15, 27, 30, 35, 41, 42, 44, 48, 69, 70, 72, 73]
Assess processing techniques for coral reef extent & composition	9	5	3	1	4	3	7	3	3	3	[10, 12, 13, 25, 36, 40, 43, 47, 59]
Assess field/acoustic data for coral reef extent & composition	4	1	1		2	1	2		2	2	[14, 38, 54, 80]
Assess radiative transfer based processing techniques for coral reef extent & composition	8	3	1		4	2	3	2	3	3	[6, 21, 28, 39, 52, 55, 56, 62]
Total	80	39	15	3	21	26	42	13	35	32	

papers, of which 21 also reported the error matrix. In general, the papers that provided an error matrix focused on just a few maps and only one site (Lewis et al. 2003).

In terms of study repeatability, 13 papers contained a high degree of detail in their explanation of methods, to the extent that the methods could be repeated. A total of 35 papers contained limited or non-repeatable methods, and 32 provided no description of methods. Some of the differences may be due to the intended audience of the paper, who may be either managers, scientists or technicians. Of the 32 papers focused on resource management applications, 11 provided an overall accuracy and only 2 provided repeatable methods. Of the 48 papers focused on a scientific/technical perspective, 28 studies reported an overall accuracy, and 11 provided repeatable accuracy assessment methods.

Applications to define resource extent and composition of coral reefs were the subject of 27 publications, 25 of which provided little or no information on validation. Of these 25 publications, 19 were published in biology, conservation and management journals, suggesting that the importance was placed on the remote sensing product and not on the process of how the map was created. This implies that the process is undervalued. A thorough understanding of the methods and validation process of the mapped products will result in more reliable and better informed management decisions. Therefore, when requesting remote sensing products and studies, the users of mapping products should require a report that includes both an explanation of the mapping approach used and a thorough description of the validation process, including sample design, sample unit, sample number, sample distribution, error matrix and accuracy measures.

Amongst the various applications, very limited descriptions of validation data collection and accuracy assessment were provided in papers that focused on use of benthic field data and acoustic image data, as well as those using radiative transfer models with analytic or semi-analytic solutions. These papers focused mainly on retrieval of bathymetric information or substrate reflectance. As a result, detailed accuracy information was provided mostly for the validation of bathymetry, but was limited for the validation of benthic habitat maps. Additionally, error propagation should not be neglected and needs attention when developing systems that merge different sensor types (Aitken et al. 2010). The fusion algorithms that merge different data types to create new information products can improve overall product accuracy, but will also integrate systematic errors introduced by the individual sensors.

A synthesis of all 80 papers was conducted to reveal how the reported validation characteristics varied by sensor type (Fig. 14.3). Results show that medium spatial resolution multispectral sensors (e.g., Landsat Thematic Mapper) were mostly used to map large areas (ranging from 16 to 21,377 km^2; average of 2,362 km^2) with a limited number of categories (typically around 6) and an average accuracy of 65 %. Smaller areas were mapped using high spatial resolution hyperspectral (average area of 46 km^2 with 80 % accuracy) and multispectral (average area of 44 km^2 with 72 % accuracy). However, the average number of categories mapped was higher for the medium spatial resolution multispectral sensor (13) than for the hyperspectral sensors (8). Aerial photography was used for a variety of area sizes, averaging an area of 71 km^2, 73 % accuracy and 14 categories.

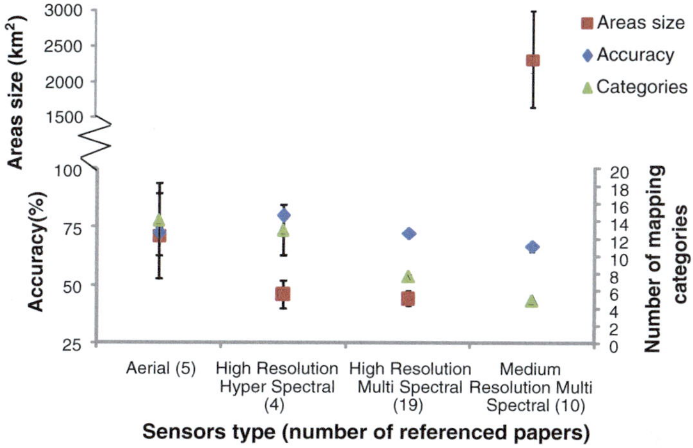

Fig. 14.3 Average size of study area mapped, mean number of categories mapped, and mean accuracy of the output map, per remote sensing sensor type used for coral reef or sea grass habitat mapping, derived from 38 out of 80 peer-reviewed scientific publications. The sensor types include: aerial photography; high spatial resolution hyperspectral; high spatial resolution multispectral; and medium spatial resolution multispectral sensors. Error bars represent standard error for each sensor type

14.3.4 Validation Limitations

The review of the 80 papers demonstrated that in many cases limited information is being reported in the literature on the processes used to collect and analyze data for accuracy assessment of coral reef maps. Our review of a wide variety of coral reef mapping papers (Tables 14.3, 14.4, 14.5) showed that only 13 out of 80 papers provided sufficient information to repeat the validation process (e.g., sample unit, number, location, scheme), and to adequately assess and compare the accuracy values (i.e., include the error matrix, and the accuracy measure for the overall map and its map categories). Similar findings have also been derived from assessments of calibration and validation approaches used in terrestrial vegetation mapping (Trodd 1995; Foody 2002).

The reviewed literature was dominated by areas smaller than 100 km^2, and did not focus on larger areas, which are more common for management needs and which typically contain a variety of reefs that need to be mapped at the same time. In fact, a majority of the papers were based on a single small reef area, potentially limiting the effectiveness of the application to larger areas. It is therefore suggested that reef remote sensing studies should include testing and validation on a collection of reefs and reef types and/or on large areas of reefs.

Spot check and transect surveys were the most commonly used calibration and validation field methods as both showed to be applicable in various environments and circumstances (Roelfsema and Phinn 2010). Supervised classification procedures in combination with medium spatial resolution multispectral imagery was the most commonly applied mapping approach. The implication is that the majority of classification methods cannot be applied completely unsupervised, indicating that field knowledge is required for calibration as much as for validation.

As a caveat, the review itself has limitations, which need to be taken into account when assessing the results. Firstly, the 80 papers represent just the published scientific literature and not the gray literature. The gray literature could reveal additional findings since these are often reporting operational mapping applications in comparison to scientific research applications. Secondly, papers were assigned to just a single category in each of the tables assessed. However, some of these papers could be assigned to more than one category since they compared habitat mapping for a variety of environments, image data types, calibration and validation field methods, mapping approaches, or accuracy measures. Thirdly, the review focused only on the most common accuracy measures and sampling designs. It did not discuss less commonly used, alternative approaches, such as regression analysis (Mumby et al. 1999) and the hold-out method (Andréfouët et al. 2004). Although the reported accuracy information provided in the reviewed literature is limited, it is surmised that in most instances the authors have more information than what was reported, and did not include the information due for example to length restrictions of scientific papers.

14.4 Conclusions and Recommendations

This chapter highlights the need for all coral reef mapping applications to contain detailed information on accuracy assessment methods. Missing information on the sampling design and accuracy assessment measures can lead to misinterpretation of the results, inability to compare with other mapping projects, and potentially incorrect conclusions by scientists and/or managers. To achieve better communication and better results, the accuracy reporting should include at a minimum:

- *Description of the study area*: extent, depth, and range of water clarity conditions.
- *Sampling design*: sample unit size, number of samples, spatial distribution of samples (map of study site with sample locations), and data collection process at each sample unit.
- *Accuracy measure used for continuous data*: RMSE, correlation coefficient or other.
- *Accuracy measure used for thematic data*: error matrix, individual thematic class accuracies (user/producer accuracy) and the map accuracy.
- *Methodology*: An explanation of the chosen sampling design, accuracy measure and references samples.

The chapter indicated that the costs of collecting independent validation data were relatively high to the total cost of habitat mapping project and a major factor in reducing the number of samples collected (Green et al. 2000; Stehman 2001;

Andréfouët 2008). The validation sampling in coral reef environment can take a long time and result in high costs due to a combination of boat travel time, boating, diving and/or snorkeling requirements, variable sea state and weather conditions. To reduce the cost and increase the number of validation samples, alternative approaches for acquiring validation information can be utilized, such as community-based validation of land cover maps using the GeoWikki portal (Fritz et al. 2009), or validation of ocean color products through the SeaWiFS Bio-optical Archive and Storage System (SEABASS), which is populated with both voluntary and funded data (Fargion et al. 2004; Bailey and Werdell 2006).

This chapter also discussed validation processes, where mapped data sets were compared with reference data sets for a specific location and time. However, it did not discuss alternative approaches to assess the accuracy of a map in case no or insufficient reference data is available. Congalton (1991) suggested that in place of a quantitative accuracy assessment, accuracy could be tested through visual inspection, non site-specific analysis, difference image creation, or error budgeting. For example, validation of ocean color products is not only based on matching image data with field observations (Bailey and Werdell 2006), but also on algorithm assessment using long time-series, single location reference data sets (Werdell et al. 2007), assessing temporal trends (Campbell et al. 1995), and comparison of trends derived with different processing algorithms (Campbell et al. 1995).

Although the chapter focused predominantly on validation of coral reef and seagrass habitat mapping, the results and conclusions are also applicable to other image-derived maps, such as water depth, rugosity and water properties. Obviously, there are differences in regards to field data collection for the different environmental information types. Some of these differences include:

- Depth and rugosity require specialized equipment and processing techniques, such as single or multi-beam SONAR, which are boat based.
- Water properties require sampling strategies that take into account the large extent of the water surrounding the coral reef and the continuous three dimensional movements and mixing of water molecules and particles in the water column through tides, currents and wind.
- Maps presenting continuous properties (e.g., cholorphyll concentration) require sampling of the properties such that measurements represent a normalized distribution of the continuous property.
- Maps presenting thematic properties (e.g., habitat classis, geomorphic zones) require sampling in the field for each thematic class to be mapped at the full extent of the coral reef assessed.

To better understand the accuracy of coral reef maps and to improve approaches in the future, additional research should create models to help predict and interpret these accuracies for specific coral reef environments and particular image types. For habitat information, this model should combine prediction of habitat map accuracy in relation to the sensor type (Andréfouët et al. 2003), benthic spatial complexity (Gustafson 1998), detail of mapping categories required (Mumby et al.

1997), and sampling design (Stehman and Czaplewski 1998). A "draft" of such a model is conceptualized in Fig. 14.4 based on the findings of previous research by the authors. Although the "draft" model will require further research to validate the trend lines, it does express the findings presented in this chapter. For example: (1) predicted habitat map accuracy declines when the number of habitat mapping categories increases for a specific sensor (Phinn et al. 2010); (2) predicted habitat map accuracy increases when spatial resolution is increased relative to a set number of mapping categories (Roelfsema et al. 2008); (3) and predicted habitat map accuracy decreases when the benthic spatial complexity increases relative to a set area size and specific sensor type (Roelfsema and Phinn 2010).

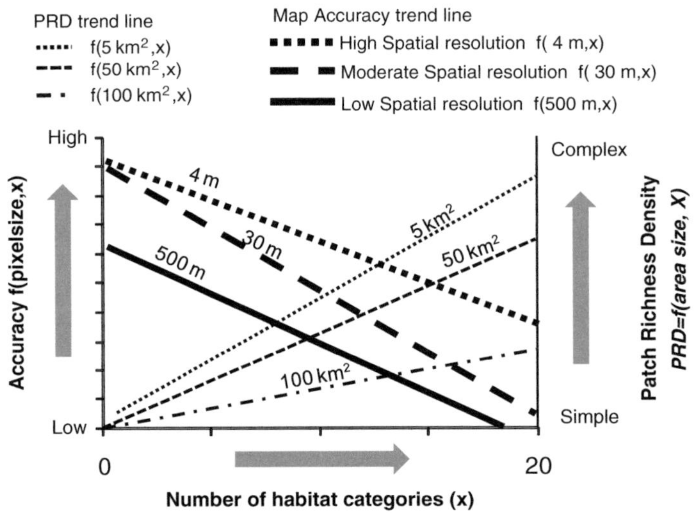

Fig. 14.4 A "draft" model conceptualizing the relationships between: accuracy of benthic habitat map; the number of habitat mapping categories; the size of the mapped study area; the pixel size of the sensor type used to create the habitat map; and the benthic spatial complexity expressed in patch richness density (PRD)

The conceptualization of accuracy and influencing factors, as presented for habitat mapping in Fig. 14.4, should be created for the other environmental information types studied, such as bathymetry and water column properties. Models like these can help managers of coral reef environments better assess what type of mapping approach is suited for a specific application, and also assist in interpreting the accuracies associated with a map.

Acknowledgments This work was funded through ARC Discovery Project—Innovative Coral Reef Mapping, University of Queensland, University of South Pacific, World Bank GEF Coral Reef Target Research—Remote Sensing working group, South Pacific Applied Geoscience Committee, Coral Cay Conservation, and Coral Reef Initiative for the Pacific. Bill Aalbersberg, James Comley and Leon Zann for their support and assistance. The fourteen participants of the map user assessment. Assitance from D. Kleine and the people Navakavu and Dravuni Qoliqoli.

Suggested Reading

Andréfouët S (2008a) Coral reef habitat mapping using remote sensing: a user versus producer perspective. Implications for research, management and capacity building. J Spatial Sci 53(1):113–129

Congalton RG, Green K (1999) Assessing the accuracy of remotely sensed data: principles and practices. Boca Rotan FL, Lewis Publishers, p 137

Foody GM (2011) Classification accuracy assessment. IEEE Geosci Remote Sens Soc Newsl, 8–13

Roelfsema CM, Phinn SR (2010a) Integrating field data with high spatial resolution multispectral satellite imagery for calibration and validation of coral reef benthic community maps. J Appl Remote Sens 4:043527

Roelfsema CM, Phinn SR (2008) Evaluating eight field and remote sensing approaches for mapping the benthos of three different coral reef environments in Fiji. In: proceedings of SPIE Asia Pacific remote sensing conference—remote sensing of inlands, coastal and oceanic water, Noumea, New Caledonia, 7150:17–21 Nov 2008

Stehman SV (2001a) Statistical rigor and practical utility in thematic map accuracy assessment. Photogrammetric Eng Remote Sens 67(6):727–734

References

Ackleson SG, Klemas V (1987) Remote-sensing of submerged aquatic vegetation in lower Chesapeake Bay—a comparison of landsat MSS to TM imagery. Remote Sens Environ 22(2):235–248

Ahmad W, Neil DT (1994) An evaluation of landsat-thematic-mapper (Tm) digital data for discriminating coral-reef zonation—heron-reef (GBR). Int J Remote Sens 15(13):2583–2597

Aitken J, Ramnath V, Feygels V, Mathur A, Kim M, Park JY, Tuell G (2010) Prelude to CZMIL: seafloor imaging and classification results achieved with charts and the rapid environmental assessment (REA) processor. Algorithms and technologies for multispectral, hyperspectral, and ultraspectral imagery Xvi. SS Shen and PE Lewis. Bellingham, Spie-Int Soc Opt Eng, 7695

Alexander D (2008) Remote sensing and the coast: development of advanced techniques to map nuisance macro-algae in estuaries. New Zealand Geographer, 64(2):157–161

Andréfouët S, Dirberg G (2006) Cartographie et inventaire du système récifalde Wallis, Futuna et Alofi par imagerie satellitaire Landsat 7 ETM+ et orthophotographies aériennes à haute résolution. In Report Conventions: Sci Mer Biol, Noumea. p 53

Andréfouët S, Guzman HM (2005) Coral reef distribution, status and geomorphology-biodiversity relationship in Kuna Yala (San Blas) archipelago, Caribbean Panama. Coral Reefs 24(1):31–42

Andréfouët S, Kramer P, Torres-Pulliza D, Joyce KE, Hochberg EJ, Garza-Perez R, Mumby PJ, Riegl B, Yamano H, White WH, Zubia M, Brock JC, Phinn SR, Naseer A, Hatcher BG, Muller-Karger FE (2003) Multi-site evaluation of IKONOS data for classification of tropical coral reef environments. Remote Sens Environ 88(1–2):128–143

Andréfouët S, Muller-Karger FE, Chevillon C, Brock JC, Hu C (2005) Global assessment of modern coral reef extent and diversity for regional science and management applications: a view from space. In 10th International coral reef symposium, Okinawa, Japan: Internat Coral Reef Soc, pp 1732–1745

Andréfouët S, Zubia M, Payri C (2004) Mapping and biomass estimation of the invasive brown algae *Turbinaria ornata* (Turner). Agardh and *Sargassum mangarevense* (Grunow) setchell on heterogeneous tahitian coral reefs using 4 m resolution IKONOS satellite data. Coral Reefs 23(1):26–38

Andréfouët S, Hochberg EJ, Chevillon C, Muller-Karger FE, Brock JC, Hu C (2005) Multi-scale remote sensing of coral reefs. Remote sensing of coastal aquatic environments: technologies, techniques and applications. In: RL Miller, CE Del Castillo, BA McKee, Springer, pp 299–317

Andréfouët S (2008b) Coral reef habitat mapping using remote sensing: a user versus producer perspective. Implications for research, management and capacity building. J Spatial Sci 53(1):113–129

Armstrong RA (1993) Remote sensing of submerged vegetation canopies for biomass estimation. Int J Remote Sens 14(3):621–627

Aswani S, Lauer M (2006) Benthic mapping using local aerial photo interpretation and resident taxa inventories for designing marine protected areas. Environ Conserv 33(3):263–273

Bailey SW, Werdell PJ (2006) A multi-sensor approach for the on-orbit validation of ocean color satellite data products. Remote Sens Environ 102(1–2):12–23

Bainbridge SJ, Reichelt RE (1998) An assessment of ground truth methods for coral reef remote sensing data. In: Proceedings of the 6th international coral reef symposium, Townsville, p 439–444

Benfield SL, Guzman HM, Mair JM, Young JAT (2007) Mapping the distribution of coral reefs and associated sublittoral habitats in Pacific Panama: a comparison of optical satellite sensors and classification methodologies. Int J Remote Sens 28(22):5047–5070

Benton AR, Newman JRM (1976) Color aerial photography for aquatic plant monitoring. J Aquat Plant Manage 14:14–16

Bertels L, Vanderstraete T, Coillie V, Knaeps E, Sterckx S, Goossens R, Deronde B (2008) Mapping of coral reefs using hyperspectral CASI data; a case study: Fordata, Tanimbar. Indonesial Internat J Remote Sens 29(8):2359–2391

Bouvet G, Ferraris J, Andréfouët S (2003) Evaluation of large-scale unsupervised classification of New Caledonia reef ecosystems using landsat 7 ETM+ imagery. Oceanolog Acta 26(3):281–290

Bruce EM (1997) Application of spatial analysis to coastal and marine management in the shark bay world heritage area. Ph.D, University of Western Australia, Western Australia Perth

Call KA, Hardy JT, Wallin DO (2003) Coral reef habitat discrimination using multivariate spectral analysis and satellite remote sensing. Int J Remote Sens 24(13):2627–2639

Campbell JW, Blaisdell JM, Darzi M (1995) Level-3 seaWiFS data products: spatial and temporal binning algorithms. NASA Tech Memo 104566, Hooker SB, Firestone ER, JG Acker. Greenbelt Maryland, NASA Goddard Space Flight Center, p 32

Capolsini P, Andréfouët S, Rion C, Payri C (2003) A comparison of landsat ETM+, SPOT HRV, ikonos, ASTER, and airborne MASTER data for coral reef habitat mapping in South Pacific Islands. Can J Remote Sens 29(2):187–200

Cassata L, Collins LB (2008) Coral reef communities, habitats, and substrates in and near sanctuary zones of Ningaloo Marine Park. J Coastal Res 24(1):139–151

Chauvaud S, Bouchon C, Maniere R (2001) Thematic mapping of tropical marine communities (coral reefs, seagrass beds and mangroves) using SPOT data in Guadeloupe Island. Oceanolog Acta 24:S3–S16

Chauvaud S, Bouchon C, Maniere R (1998) Remote sensing techniques adapted to high resolution mapping of tropical coastal marine ecosystems (coral reefs, seagrass beds and mangroves). Int J Remote Sens 19(18):3625–3639

Congalton RG, Mead RA (1983) A quantitative method to test for consistency and correctness in photo interpretation. Photogrammetric Eng Remote Sens 49(1):69–74

Congalton RG (1991) A review of assessing the accuracy of classifications of remotely sensed data. Remote Sens Environ 37:35–46

Congalton RG, Green K (1999) Assessing the accuracy of remotely sensed data: principles and practices. Lewis Publishers, Boca Rotan, p 137

Cuevas-Jimenez A, Ardisson PL (2002) Mapping shallow coral reefs by colour aerial photography. Int J Remote Sens 23(18):369–371

Dahdouh-Guebas F, Coppejans E, Van Speybroeck D (1999) Remote sensing and zonation of seagrasses and algae along the Kenyan coast. Hydrobiologia 400:63–73
Dekker AG, Brando VE, Anstee JM (2005) Retrospective seagrass change detection in a shallow coastal tidal Australian lake. Remote Sens Environ 97(4):415–433
Dierssen HM, Zimmerman RC, Leathers RA, Downes TV, Davis CO (2003) Ocean color remote sensing of seagrass and bathymetry in the Bahamas banks by high-resolution airborne imagery. Limnol Oceanogr 48(1):444–455
Fargion GS, BA Franz, Kwiatkowska E, Pietras C (2004) SIMBIOS program in support of ocean color missions: 1997–2003. Ocean remote sensing and imaging: II. In: Proceedings SPIE, pp 49–60
Fitzpatrick-Lins K (1981) Comparison of sampling procedures and data-analysis for a land-use and land-cover map. Photogrammetric Eng Remote Sens 47(3):343–351
Foody GM (2002) Status of land cover classification accuracy assessment. Remote Sens Environ 80:185–201
Foody GM (2011) Classification accuracy assessment. IEEE Geosci Remote Sens Soc Newsl, 8–13
Fornes A, Basterretxea G, Orfila A, Jordi A, Alvarez A, Tintore J (2006) Mapping Posidonia oceanica from IKONOS. J Photogrammetry Remote Sens 60(5):315–322
Franklin EC, Ault JS, Smith SG, Luo J, Meester GA, Diaz GA, Chiappone M, Swanson DW, Miller SL, Bohnsack JA (2003) Benthic habitat mapping in the Tortugas region, Florida. Mar Geodesy 26:19–34
Fritz S, McCallum E, Schill C, Perger C, Grillmayer R, Achard F, Kraxner F, Obersteiner M (2009) Geo-Wiki.Org: the use of crowdsourcing to improve global land cover. Remote Sens Environ 1:345–354
Garza-Perez JR, Lehmann A, Arias-Gonzalez JE (2004) Spatial prediction of coral reef habitats: integrating ecology with spatial modeling and remote sensing. Mar Ecol-Prog Ser 269:141–152
Goodman J, Ustin SL (2007) Classification of benthic composition in a coral reef environment using spectral unmixing. J Appl Remote Sens, p 17
Green EP, Mumby PJ, Edwards AJ, Clark CD (2000) Remote sensing handbook for tropical coastal management. Paris, UNESCO, 316
Gregg WW, Casey NW (2004) Global and regional evaluation of the seawifs chlorophyll data set. Remote Sens Environ 93(4):463–479
Gullstrom M, Lunden B, Bodin M, Kangwe J, Ohman MC, Mtolera MSP, Bjork M (2006) Assessment of changes in the seagrass-dominated submerged vegetation of tropical Chwaka Bay (Zanzibar) using satellite remote sensing. Estuar Coast Shelf Sci 67(3):399–408
Gustafson EJ (1998) Quantifying landscape spatial pattern: what is the state of the art? Ecosyst 1(2):143–156
Habeeb RL, Johnson CR, Wotherspoon S, Mumby PJ (2007) Optimal scales to observe habitat dynamics: a coral reef example. Ecol Appl 17(3):641–647
Hatcher BG (1997) Coral reef ecosystems: how much greater is the whole than the sum of the parts? Coral Reefs 16(5):S77–S91
Hochberg EJ, Atkinson MJ (2003) Capabilities of remote sensors to classify coral, algae, and sand as pure and mixed spectra. Remote Sens Environ 85(2):174–189
Holmes KW, Van Niel KP, Kendrick GA, Radford B (2007) Probabilistic large-area mapping of seagrass species distributions. Aquat Conserv-Mar Freshwater Ecosyst 17(4):385–407
Houk P, van Woesik R (2008) Dynamics of shallow-water assemblages in the Saipan Lagoon. Mar Ecol-Prog Ser 356:39–50
Isoun E, Fletcher C, Frazer N, Gradie J (2003) Multi-spectral mapping of reef bathymetry and coral cover Kailua Bay. Hawaii Coral Reefs 22(1):68–82
Jensen JR (2005) Introductory digital image processing: a remote sensing perspective, 3rd edn. Prentice Hall, p 316

Jordan A, Lawler M, Halley V, Barrett N (2005) Seabed habitat mapping in the Kent Group of islands and its role in marine protected area planning. Aquat Conservat-Mar Freshwater Ecosyst 15(1):51–70
Joyce KE, Phinn SR, Roelfsema CM, Neil DT, Dennison WC (2004) Combining landsat ETM plus and reef check classifications for mapping coral reefs: A critical assessment from the southern Great Barrier Reef. Aust Coral Reefs 23(1):21–25
Jupp DLB, Mayo KK, Kuchler DA, Claasen DVR, Kenchington RA, Cuerin PR (1985) Remote sensing for planning and managing the Great Barrier Reef of Australia. Photogrammetria 40:21–42
Kendrick GA, Aylward MJ, Hegge BJ, Cambridge ML, Hillman K, Wyllie A, Lord DA (2002) Changes in seagrass coverage in cockburn sound, Western Australia between 1967 and 1999. Aquat Bot 73(1):75–87
Klonowski WM, Fearns PR, Lynch MJ (2007) Retrieving key benthic cover types and bathymetry from hyper spectral imagery. J Appl Remote Sens p 1
Knudby A, Roelfsema C, Jupiter S, Lyons M, Phinn S (2011) Mapping fish community variables by integrating field and satellite data, object-based image analysis and modeling in a traditional Fijian fisheries management area. Remote Sens 3(3):460–483
Kutser T, Miller I, Jupp DLB (2006) Mapping coral reef benthic substrates using hyperspectral space-borne images and spectral libraries. Estuar Coast Shelf Sci 70(3):449–460
Kvernevik TI, Akhir MZM, Studholme J (2002) A low-cost procedure for automatic seafloor mapping, with particular reference to coral reef conservation in developing nations. Hydrobiologia 474(1–3):67–79
Lathrop RG, Montesano P, Haag S (2006) A multi-scale segmentation approach to mapping seagrass habitats using airborne digital camera imagery. Photogrammetric Eng Remote Sens 72(6):665–675
Lauer M, Aswani S (2008) Integrating indigenous ecological knowledge and multi-spectral image classification for marine habitat mapping in Oceania. Ocean Coast Manag 51(6):495–504
Lennon P, Luck P (1989) Seagrass mapping using landsat tm data. In: Asian conference on remote sensing. Asian association on remote sensing, Kuala Lumpur
Lesser MP, Mobley CD (2007) Bathymetry, water optical properties, and benthic classification of coral reefs using hyperspectral remote sensing imagery. Coral Reefs 26(4):819–829
Lewis A, Lowe D, Otto J (2003) Remapping the Great Barrier Reef position magazine. South Pacific Science Press International, 46–49
Lillesand TM, Kiefer RW, Chipman JW (2008) Remote sensing and image interpretation. Wiley, Danvers, p 756
Louchard EM, Reid RP, Stephens FC, Davis CO, Leathers RA, Downes T (2003) Optical remote sensing of benthic habitats and bathymetry in coastal environments at Lee Stocking Island, Bahamas: a comparative spectral classification approach. Limnol Oceanogr 48(1):511–521
Ma ZK, Redmond RL (1995) Tau-coefficients for accuracy assessment of classification of remote-sensing data. Photogrammetric Eng Remote Sens 61(4):435–439
Maeder J, Narumalani S, Rundquist DC, Perk RL, Schalles J, Hutchins K, Keck J (2002) Classifying and mapping general coral-reef structure using Ikonos data. Photogrammetric Eng Remote Sens 68(12):1297–1305
Matarrese A, Mastrototaro G, D'onghia G, Maiorano P, Tursi A (2004) Mapping of the benthic communities in the Taranto Seas using side scan sonar and an underwater video camera. Chem Ecol, 20(5):377–386
Meehan AJ, Williams RJ, Watford FA (2005) Detecting trends in seagrass abundance using aerial photograph interpretation: problems arising with the evolution of mapping methods. Estuaries 28(3):462–472
Mishra D, Narumalani S, Rundquist D, Lawson M (2006) Benthic habitat mapping in tropical marine environments using QuickBird multispectral data. Photogrammetric Eng Remote Sens 72(9):1037–1048
Moore KA, Wilcox DJ, Orth RJ (2002) Analysis of the abundance of submersed aquatic vegetation communities in the Chesapeake Bay. Estuaries 23(1):115–127

Mumby PJ, Edwards AJ (2002) Mapping marine environments with IKONOS imagery: enhanced spatial resolution can deliver greater thematic accuracy. Remote Sens Environ 82(2–3):248–257
Mumby PJ, Green EP, Clark CD, Edwards AJ (1997) Reef habitat assessment using (CASI) airborne remote sensing. In: 8th International coral reef symposium, Panama, pp 1499–1502
Mumby PJ, Green EP, Edwards AJ, Clark CD (1999) The cost-effectiveness of remote sensing for tropical coastal resources assessment and management. J Environ Manage 55(3):157–166
Mumby PJ, Green EP, Clark CD, Edwards AJ (1998) Digital analysis of multispectral airborne imagery of coral reefs. Coral Reefs 17(1):59–69
Murdoch TJT, Glasspool AF, Outerbridge M, Ward J, Manuel S, Gray J, Nash A, Coates KA, Pitt J, Fourqurean JW, Barnes PA, Vierros M, Holzer K, Smith SR (2007) Large-scale decline in offshore seagrass meadows in Bermuda. Mar Ecol-Prog Ser 339:123–130
Naseer A, Hatcher BG (2004) Inventory of the Maldives' coral reefs using morphometrics generated from landsat ETM+ imagery. Coral Reefs 23(1):161–168
Newman CM, Knudby AJ, LeDrew EF (2007) Assessing the effect of management zonation on live coral cover using multi-date IKONOS satellite imagery. J Appl Remote Sens, p 1
Orth RJ, Luckenbach ML, Marion SR, Moore KA, Wilcox DJ (2006) Seagrass recovery in the Delmarva Coastal Bays. USA Aquat Bot 84(1):26–36
Palandro DA, Andréfouët S, Hu C, Hallock P, Muller-Karger FE, Dustan P, Callahan MK, Kranenburg C, Beaver CR (2008) Quantification of two decades of shallow-water coral reef habitat decline in the Florida Keys National Marine Sanctuary using landsat data (1984–2002). Remote Sens Environ 112(8):3388–3399
Palandro D, Andréfouët S, Dustan P, Muller-Karger FE (2003) Change detection in coral reef communities using Ikonos satellite sensor imagery and historic aerial photographs. Int J Remote Sens 24(4):873–878
Pasqualini V, Clabaut P, Pergent G, Benyoussef L, Pergent-Martini C (2000) Contributions of side scan sonar to the management of Mediterranean littoral ecosystems. Int J Remote Sens 21(2):367–378
Pasqualini V, Pergent-Martini C, Pergent G, Agreil M, Skoufas G, Sourbes L, Tsirika A (2005) Use of SPOT 5 for mapping seagrasses: an application to Posidonia oceanica. Remote Sens Environ 94(1):39–45
Peneva E, Griffith JA, Carter GA (2008) Seagrass mapping in the northern Gulf of Mexico using airborne hyperspectral imagery: a comparison of classification methods. J Coastal Res 24(4):850–856
Pergent G, Djellouli A, Hamza A, Ettayeb K, El Mansouri A (2002) Characterization of the benthic vegetation in the Farwà Lagoon (Libya). J Coastal Conserv 8(2):119–126
Phinn SR, Roelfsema CM, Dekker A, Brando V, Anstee J (2008) Mapping seagrass species, cover and biomass in shallow waters: an assessment of satellite multi-spectral and airborne hyper-spectral imaging systems in Moreton Bay (Australia). Remote Sens Environ 112:3413–3425
Phinn SR, Roelfsema CM, Stumpf R (2010) Remote sensing: discerning the promise from the reality. In: Longstaff BJ, Carruthers TJB, Dennison WC, Lookingbill TR, Hawkey JM, Thomas JE, Wicks EC, Woerner J (eds) Integrating and applying science: a handbook for effective coastal ecosystem assessment. IAN Press, Cambridge
Phinn SR, Roelfsema CR, Mumby P (2012) Multi-scale object based image analysis for mapping coral reef geomorphic and ecological zones. Internat J Remote Sens
Pittman SJ, Costa BM, Battista TA (2009) Using lidar bathymetry and boosted regressiontrees to predict the diversity and abundance of fish and corals. J of coast Res 25(6): 27–38
Pontius RG Jr, Millones M (2011) Death to kappa: birth of quantity disagreement and allocation disagreement for accuracy assessment. Int J Remote Sens 32(15):4407–4429
Prada MC, Appeldoorn RS, Rivera JA (2008) Improving coral reef habitat mapping of the Puerto Rico insular shelf using side scan sonar. Mar Geodesy 31(1):49–73
Purkis SJ (2005) A 'reef-up' approach to classifying coral habitats from IKONOS imagery. IEEE Trans Geosci Remote Sens 43(6):1375–1390
Purkis SJ, Graham NAJ, Riegl BM (2008) Predictability of reef fish diversity and abundance using remote sensing data in Diego Garcia (Chagos Archipelago). Coral Reefs 27(1):167–178

Purkis SJ, Riegl B (2005) Spatial and temporal dynamics of Arabian Gulf coral assemblages quantified from remote-sensing and in situ monitoring data. Mar Ecol-Prog Ser 287:99–113

Riegl BM, Moyer RP, Morris L, Virnstein R, Dodge RE (2005) Determination of the distribution of shallow-water seagrass and drift algae communities with acoustic seafloor discrimination. Rev Biol Trop 53:165–174

Riegl BM, Purkis SJ (2005) Detection of shallow subtidal corals from IKONOS satellite and QTC View (50, 200 kHz) single-beam sonar data (Arabian Gulf; Dubai, UAE). Remote Sens Environ 95(1):96–114

Roelfsema CM, Dennison WC, Phinn SR (2002) Spatial distribution of benthic microalgae on coral reefs determined by remote sensing. Coral Reefs 21(3):264–274

Roelfsema CM, Phinn SR (2010b) Integrating field data with high spatial resolution multispectral satellite imagery for calibration and validation of coral reef benthic community maps. J Appl Remote Sens 4:043527

Roelfsema CM, Phinn SR, Udy N, Maxwell P (2009) An integrated field and remote sensing approach for mapping seagrass cover, moreton bay. Aust J Spatial Sci 54(1):45–62

Roelfsema CM, Phinn SR (2008) Evaluating eight field and remote sensing approaches for mapping the benthos of three different coral reef environments in Fiji. In: proceedings of SPIE Asia Pacific remote sensing conference—remote sensing of Inlands, coastal and oceanic water, Noumea, New Caledonia, vol 7150:G6800 17–21 Nov 2008

Roelfsema CM, Joyce KE, Phinn SR (2006) Evaluation of benthic survey techniques for validating remotely sensed images of coral reefs. In: Proceedings 10th international coral reef symposium Okinawa

Roelfsema CM, Phinn SR, Dennison WC, Dekker A, Brando V (2006b) Monitoring toxic cyanobacteria L. majuscula in moreton bay, Australia by integrating satellite image data and field mapping. Harmful Algae 5:45–56

Rowlands GP, Purkis SJ, Riegl BM (2008) The 2005 coral-bleaching event Roatan (Honduras): use of pseudo-invariant features (PIFs) in satellite assessments. J Spatial Sci 53(1):99–112

Sagawa T, Mikam A, Komatsu T, Kosaka N, Kosako A, Miyazaki S, Takahashi M (2008) Mapping seagrass beds using IKONOS satellite image and side scan sonar measurements: a Japanese case study. Int J Remote Sens 29(1):281–291

Schweizer D, Armstrong RA, Posada J (2005) Remote sensing characterization of benthic habitats and submerged vegetation biomass in Los Roques Archipelago National Park, Venezuela. Internat J Remote Sens 26(12):2657–2667

Sheppard CRC, Matheson K, Bythell JC, Murphy P, Myers CB, Blake B (1995) Habitat mapping in the Caribbean for management and conservation: use and assessment of aerial photography. Aquat Conserv-Mar Freshwater Ecosyst 5(4):277–298

Scopélitis J, Andréfouët S, Phinn S, Arroyo L, Dalleau M, Cros A, Chabanet P (2010) The nextstep in shallow coral reef monitoring: combining remote sensing and in situ approaches. MarPollut Bull 60(11): 1956–1968

Stehman SV, Czaplewski RL (1998) Design and analysis for thematic map accuracy assessment: fundamental principles. Remote Sens Environ 64:331–344

Stehman SV (1999) Basic probability sampling designs for thematic map accuracy assessment. Int J Remote Sens 20(12):2423–2441

Stehman SV (2001b) Statistical rigor and practical utility in thematic map accuracy assessment. Photogrammetric Eng Remote Sens 67(6):727–734

Story M, Congalton RG (1986) Accuracy assessment: a users perspective. Photogrammetric Eng Remote Sens 52(3):397–399

Trodd NM (1995) Uncertainty in land cover mapping for modelling land cover change. RSS95 remote sensing in action, Nottingham, 1138–1145

Vanderstraete T, Goossens R, Ghabour TK (2006) The use of multitemporal landsat images for the change detection of the coastal zone near Hurghada. Egypt Internat J Remote Sens 27:3645–3655

Wabnitz CC, Andréfouët S, Torres-Pulliza D, Muller-Karger FE, Kramer PA (2008) Regional-scale seagrass habitat mapping in the wider Caribbean region using landsat sensors: applications to conservation and ecology. Remote Sens Environ 112(2008):3455–3467

Weeks SJ, Barlow R, Roy C, Shillington FA (2006) Remotely sensed variability of temperatureand chlorophyll in the southern benguela: upwelling frequency and phytoplankton response. Afr J of Mar Sci 28(34): 493–509

Werdell PJ, Franz BA, Bailey SW, Harding LW, Feldman GC (2007) Approach for the long-term spatial and temporal evaluation of ocean color satellite data products in a coastal environment, Art. no. 66800G. Coastal Ocean Remote Sensing. Frouin RJ, Lee Z. 6680:G6800–G6800

Young DR, Clinton PJ, Specht DT, DeWitt TH, Lee H (2008) Monitoring the expanding distribution of nonindigenous dwarf eelgrass Zostera japonica in a Pacific Northwest USA estuary using high resolution digital aerial orthophotography. J Spatial Sci 53(1):87–97

Zharikov Y, Skilleter GA, Loneragan NR, Taranto T, Cameron BE (2005) Mapping and characterising subtropical estuarine landscapes using aerial photography and GIS for potential application in wildlife conservation and management. Biol Conserv 125:87–100

Chapter 15
Science and Management

Stacy Jupiter, Chris M. Roelfsema and Stuart R. Phinn

Abstract Coral reef scientists and managers are increasingly relying on remote sensing data to provide information on biophysical processes of reefs and to help identify optimum management strategies for reef resources. For these users, we provide some guidelines to identify which remote sensing tools and data should be used to address coral reef research and management questions. We additionally discuss: opportunities to reconcile the sometimes conflicting needs of producers and users of coral reef information; data requirements and limitations for specific coral reef management applications; and trade-offs between production costs and accuracy of coral reef remote sensing data products. Finally, we provide several in-depth examples of current uses of remote sensing data to: provide resources inventories for prioritizing areas for management; develop spatially explicit models of reef fish assemblage characteristics; and monitor and respond to threats (e.g., from terrestrial runoff, crown-of-thorns outbreaks, oil spills and ship groundings). Throughout, we emphasize ways that remote sensing can be cost-effectively integrated within coral reef management programs to improve the quality of information on which management decisions are based.

S. Jupiter (✉)
Wildlife Conservation Society, Fiji Country Program, 11 Ma'afu Street, Suva, Fiji
e-mail: sjupiter@wcs.org

C. M. Roelfsema · S. R. Phinn
School of Geography, Planning and Environmental Management,
University of Queensland, Brisbane, QLD 4072, Australia
e-mail: c.roelfsema@uq.edu.au

S. R. Phinn
e-mail: s.phinn@uq.edu.au

J. A. Goodman et al. (eds.), *Coral Reef Remote Sensing*,
DOI: 10.1007/978-90-481-9292-2_15,

15.1 Introduction

The science of coral reef remote sensing has emerged, in many respects, in response to management needs. Recent, rapid advances in processing systems, data storage capacity, and ease of data dissemination via the internet have made remote sensing data and data products integral parts of many early warning systems (e.g., NOAA Coral Reef Watch Bleaching Outlook; Chap. 12), threat assessments (e.g., from oil spills; Chap. 13) and monitoring campaigns (e.g., environmental change detection; this chapter) in both the developed and developing world (Mumby et al. 2004a). The confidence in coral reef remote sensing science has similarly grown such that managers and decision makers now depend on remote sensing products to develop, implement, assess and adapt management strategies and government policies.

Frequently there is still a broad gap between managers' expectations and the ability of remote sensing science to address their needs (Andréfouët 2008). These gaps may exist due to cost, lack of technical capacity to use data, unrealistic expectations and/or over-engineered products being used for more routine applications. Thus, in order to optimize the application of remote sensing science for coral reef management, there needs to be ongoing communication between the engineers who produce sensors, researchers who produce data products, and managers and decision makers who use those products. Such communication is instrumental for identifying which remote sensing tools are most appropriate to address given coral reef research and management questions, as well as specifying the best possible acquisition, operational and cost requirements necessary for effective application.

In this chapter, we discuss how to fill some of these communication gaps by presenting guidelines for selecting appropriate coral reef remote sensing data and products to address research and management questions. In addition, we describe some of the current limitations of remote sensing data in order to help bring the loftier expectations for coral reef management back down to Earth. Lastly, we provide clear examples of how remote sensing can be used from local to oceanic basin scales to: (1) provide baseline inventories; (2) predict threats to marine habitats; (3) assess ecosystem response to disturbance; and (4) understand ecological and biophysical dynamics at a range of spatial and temporal scales. For each of the examples, we describe how coral reef remote sensing has been used to build scientific understanding and to develop and implement management strategies.

15.2 Research and Management Needs

Most coral reef remote sensing research products have direct applications for management, and the past few decades have seen significant advances in integration of remote sensing products with other data to inform management decisions. Coral reef managers often rely on remotely sensed data to provide information on:

baseline ecosystem condition; habitat resource inventories; ecosystem vulnerability, threats and damage following disturbance; priority locations for management; and the effectiveness of management interventions. Meanwhile, applied research focuses on the development and improvement of existing tools and methods to support the information needs of coral reef managers, and pure research continues to build an understanding of how the environment functions from organism to landscape/seascape scales.

The specific research and management needs of coral reef scientists and managers will ultimately determine the choice of imagery and processing tools required. To assist with making these decisions, the sections below:

- Present guidelines for framing research and management questions;
- Highlight the sometimes conflicting needs of producers and users of coral reef information derived from remotely sensed data, and suggest ways to reconcile these needs to improve coral reef management;
- Discuss data requirements and limitations for specific coral reef management applications; and
- Identify trade-offs between production costs and accuracy of reef information derived from remote sensing image data.

15.2.1 Framing the Question

The choice of imagery and analysis techniques largely depends on the focus of the research or management efforts. For research, scientists first need to identify a specific question. For example, Andréfouët and Payri (2000) were interested in evaluating whether coral reef habitat maps and field data from reefs around Moorea Island, French Polynesia, could be integrated to assess organic and inorganic metabolism of an entire reef system. They needed to first find imagery which covered the scale of their research site (35 km^2) and from which they could reliably identify distinct habitats with differing metabolic rates (requiring multispectral, relatively high spatial resolution data).

Research questions can also be utilized to refine processing techniques that can improve information for management. For example, Mumby et al. (1998) asked whether the combination of water column correction and contextual editing improved the accuracy of shallow water coral reef and seagrass habitat classifications in the Turks and Caicos Islands of the Caribbean using a variety of different multispectral sensors. Their comparisons showed that, indeed, both processing steps together substantially improved map accuracy from airborne and satellite imagery and could be applied with little extra cost but much improved quality for management decisions (Mumby et al. 1998).

For management questions, managers need to first identify the management objectives and targets and only then ask whether coral reef remote sensing data are appropriate to inform a management strategy. For example, threat analysis techniques (e.g., Salafsky and Margoluis 1999) were used in Fiji by NGOs (Wildlife Conservation Society, WWF, Wetlands International-Oceania) to identify the simplified conceptual model of threats to coral reefs shown in Fig. 15.1. Once the conceptual model was delineated, the NGOs could easily identify where remote sensing tools and products could be of assistance to help local communities protect coral reef biodiversity and reduce declines of reef resources. Remote sensing data were accordingly used by these groups in Fiji to: help identify critical watersheds for management to maintain ecosystem linkages between catchment and reef (Jenkins et al. 2010); identify alternate configurations for marine protected area networks that minimize costs to fishers (Adams et al. 2011); and predict the spatial distribution of reef fish biomass and diversity for management of fisheries resources (Knudby et al. 2011).

Once the research or management question is clearly identified, the choice of whether to use remote sensing data, and which remote sensing platform and/or products are most appropriate, will depend on attributes of the environment to be mapped and the capacity of the mapping organization (Phinn 1998). This includes items such as:

- The scale of the research or management area;
- The environmental conditions in the area to be mapped;
- The minimum size of features of interest;
- The minimum measurement precision and accuracy required;
- Temporal requirements for assessing change; and
- The organizational capacity to process new data or use existing data products (Phinn et al. 2010).

Table 15.1 and Fig. 15.2 modify the approaches of Phinn (1998) and Phinn et al. (2000) for identifying considerations to be taken into account when selecting image data and processing approaches to deliver maps of key biophysical properties of coral reefs.

15.2.2 User Versus Producer Needs

While many coral reef remote sensing scientists direct their research towards improving products for management applications, there is still some sentiment that coral reef remote sensing work has been too experimental and small in scale to be applied to broad (100–1,000 s of km) management units covering areas across a high diversity of geomorphic zones and benthic communities (Hopley et al. 2007). Andréfouët (2008) notes that this problem of scale has largely arisen due to divergence between the needs of reef map producers versus reef map users. Scientific map producers who focus on publication in scientific journals are obliged to come up with new and original mapping tools whose robustness are often tested over small areas with high volumes of field data acquired at great cost (Andréfouët 2008). While there is certainly a place for scientific innovation to push the field of

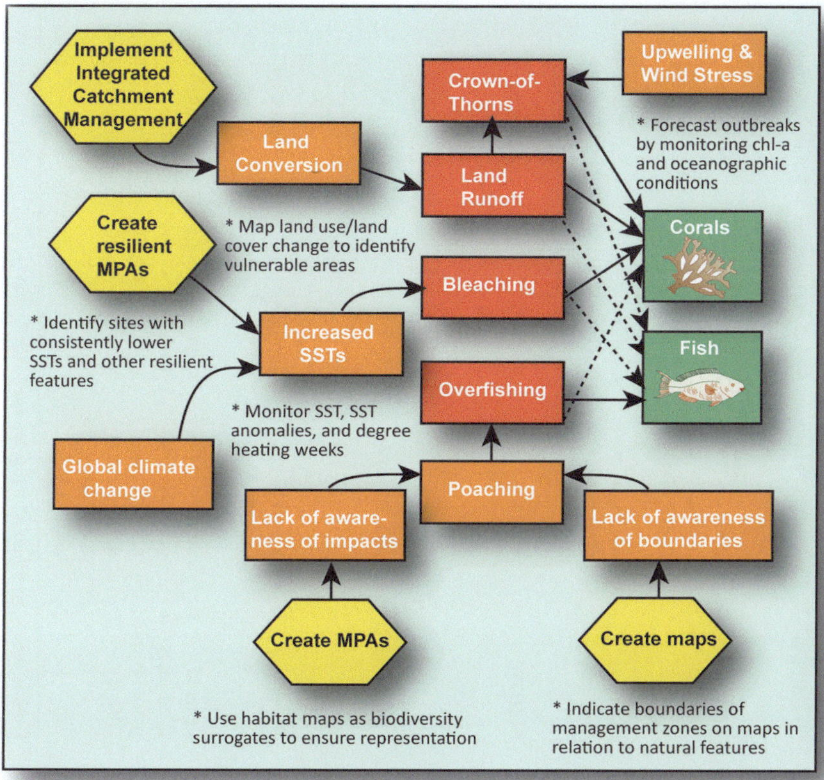

Fig. 15.1 Basic conceptual model depicting some of the major direct (*red rectangles*) and indirect (*orange rectangles*) threats to coral reef management targets (*green boxes*). *Black* text indicates points where remote sensing tools could be used to monitor threats and implement management strategies (*yellow hexagons*). *Solid arrows* indicate direct links and *dashed arrows* indicate indirect links between threat and coral reef management target

coral reef remote sensing in new directions, Andréfouët (2008) challenges reef map producers to additionally create products that address the main needs of reef map users: (1) comprehensiveness; (2) accuracy; (3) repeatability; and (4) cost effectiveness.

With respect to comprehensiveness, map classification systems need to be broad enough to cover all representative habitats in a given region if they are to be used effectively for spatially explicit research and management applications, such as community structure analysis, productivity assessments and protected area design. As a model for this approach, a hierarchical classification scheme was developed for the global Millennium Coral Reef Mapping Project (imars.usf.edu/MC/), including more than 800 different geomorphological units that can be extracted from Landsat 7 Enhanced Thematic Mapper imagery (Andréfouët et al. 2006).

Table 15.1 Consideration criteria to guide the selection of remote sensing data needs for research and management

Important considerations to guide selection of data	Mapping or monitoring requirements				
Scale of the research or management area	0–50 km^2	50–250 km^2	250–500 km^2	500–5,000 km^2	5,000–50,000 km^2
Access to study site	Easy	Only on demand	Difficult (too remote or too many sites)	Dangerous	No access
Minimum size of features of interest to be mapped	Extremely fine (<5 m)	Fine (5–20 m)	Medium (20–250 m)	Coarse (250–100 m)	Extremely coarse (>1,000 m)
Minimum measurement precision (i.e., smallest measurable change in % coral cover)	0–5 %	5–10 %	10–25 %	25–50 %	Anything
Minimum measurement accuracy (e.g., required acceptable agreement between estimated and actual value of live coral cover)	Anything	Low (10–40 %)	Medium (40–70 %)	High (70–90 %)	Very high (90–100 %)
Temporal requirements for assessing change	≤1 day	Weekly	Monthly	Annually	>Annually
Organizational capacity to process and use data	None (Can use maps only)	Limited (Can integrate products into GIS)	Some (Can use but not process data products for multiple applications)	High (Can process data and integrate into other applications)	Very high (Can serve as data provider for other organizations)

Table adapted from University of Queensland's online Remote Sensing Toolkit, www.gpem.uq.edu.au/cser-rstoolkit

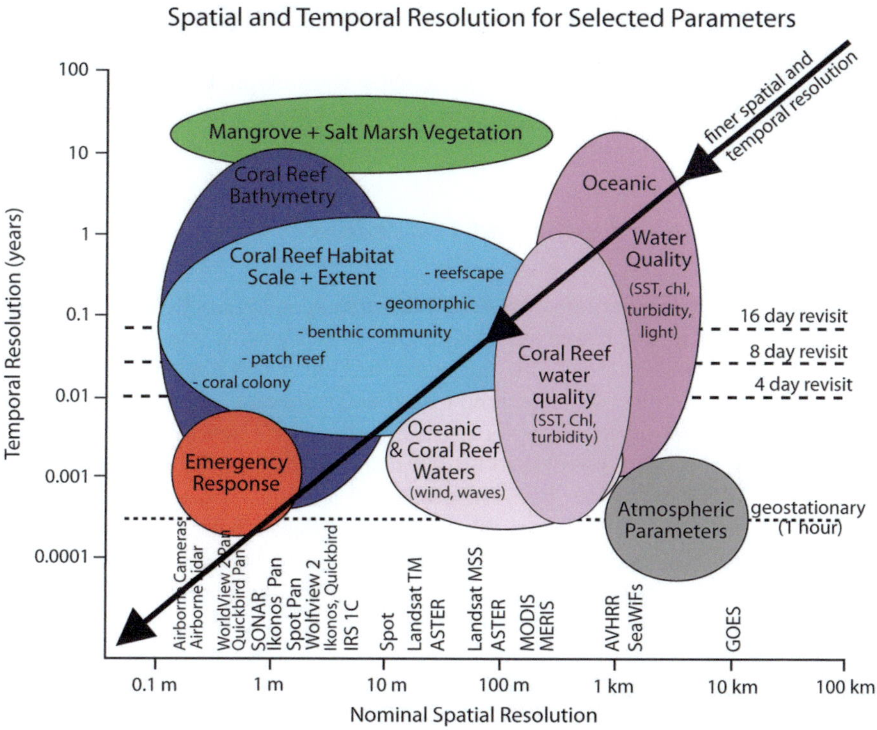

Fig. 15.2 Spatial and temporal scales of coral reef mapping and monitoring applications in relation to the pixel size and temporal frequency of commercially available airborne and satellite image data (modified from Phinn et al. 2010)

The model has already been applied for reef resource inventories in the Maldives (Naseer and Hatcher 2004) and to assist in definition of regional-scale baseline seagrass cover for the broader Caribbean (Wabnitz et al. 2008).

With regards to accuracy, both the scientists and managers who use remote sensing products need to have confidence that map classifications represent their true habitats (or benthic cover types and communities) on the ground and that image-derived environmental variables (e.g., sea surface temperature, chlorophyll-a) are both accurate and precise compared with values measured from the field. For example, in order to be able to confidently scale-up reef metabolic rates, Andréfouët and Payri (2000) needed to trust that the satellite derived habitats represented a reasonable approximation of their true coverage and that the classification system was able to discriminate effectively between habitats with highly varied metabolic rates. User confidence in remote sensing methods and products can be built by showing consistent strong correlation between field measurements and image maps of the same measurements from multiple sites around the world. For instance, operational products such as the coral bleaching HotSpot and coral bleaching Degree Heating Week produced by the U.S. National Oceanographic and Atmospheric Administration (NOAA) Coral Reef Watch Bleaching Outlook program were only released after successful early warning predictions of bleaching events were first established using experimental products (Liu et al. 2003).

15.2.3 Data Requirements and Limitations

A summary of the most common research and management applications for coral reef remote sensing are displayed in Table 15.2 along with some of the corresponding sensor options for producing the required data products. The gaps in the table also indicate some of the current spatial and temporal limitations of remote sensing to meet user needs. Some of these limitations are described below.

In most cases, it is not possible to discriminate benthic, sessile organisms to the species level. For example, because corals all contain similar pigments and because colonies typically occur at the scale of less than 0.5 m^2, it is infeasible to provide comprehensive coral species diversity maps. For the same reasons, it is not possible to consistently distinguish spectrally similar, but functionally distinct categories, such as hard and soft corals. This may have implications if researchers or managers are trying to scale-up properties or processes (e.g., carbonate production) derived from reef building hard corals that are spectrally indistinguishable for non-calcifying organisms. These problems can partially be addressed by integrating data from multiple sensor systems, incorporating a combination of spatial and spectral analysis techniques, and linking field data describing dominant components of community structure to habitat classes (Mumby 2001).

Mapping and monitoring the extent of coral bleaching has become a strong priority given the increasing frequency and severity of events in the past few decades. As corals expel their pigmented zooxanthellae during a bleaching event, the resulting strong change in color would suggest that remote sensing at visible wavelengths would be useful for quantifying extent of coral mortality (Clarke et al. 2000; Andréfouët et al. 2002). However, its application has been limited given the patchiness of events, the varying intensities and susceptibilities of coral to bleaching, the inability to detect changes deeper than just shallow depths, and the need for imagery to be acquired within specific timeframes (Andréfouët et al. 2002). Moreover, it is often difficult to acquire imagery over bleached reefs prior to algal colonization of dead coral surfaces, making differentiation of dead versus live coral difficult (Clarke et al. 2000). These temporal limitations are particularly challenging when relying on data from satellite sensors with low revisit frequencies, which makes the likelihood of acquiring cloud-free images from the tropics directly following a disturbance event fairly low.

Table 15.2 Capabilities of current types of remote sensing platforms to be used for various coral reef remote sensing applications

Platform	Boat		Aircraft			Satellite			
Sensor type	*Acoustic*	*Laser*	*Laser*	*Hyperspectral*	*Aerial photograph*	*Multispectral (high res)*	*Multispectral (med res)*	*Multispectral (low res)*	*Radiometer (low res)*
Example of sensor	Roxanne		Lidar	CASI, HyMap		Ikonos, QuickBird	Landsat, SPOT	SeaWiFs, MODIS	AVHRR
Coral species									
Coral & algal cover				√	√	√	√		
Reef community	√	?	?	√	√	√			
Occurrence of bleaching		?	?	√?	√	√?			
Structural complexity	√	√	√	√		√			
Reef geomorphology	√		√	√	√	√	√	√	
Habitat diversity				√		√			
Community structure change detection				√	√	√	√?		
Location of reefal habitat	√	√	√	√	√	√	√	√	
Bathymetry	√		√	√	√?	√	√	√?	
Water quality (e.g., chl a)[a]				√			√	√	
Sea surface temperature								√	√

√ indicates routine use; √? indicates few examples to date; ? indicates possible but untested (adapted from Mumby et al. 2004a)

[a] Note that multispectral sensors cannot be used to map chlorophyll a concentrations in turbid, coastal environments

15.2.4 Balancing Costs and Product Quality

Managers often accept maps at face value without any recognition of the underlying methods used and associated product quality. Additionally, many maps are, in fact, far more complex than required by the management needs of the users, which can be as basic as simply guiding selection of monitoring sites. A recent study by Roelfsema and Phinn (2008) found that three main items affected user selection of a suitable mapping approach: (1) variation in coral reef environments; (2) purpose of the habitat map; and (3) resource management requirements. The study evaluated the accuracy, cost, and relevance of eight commonly used benthic cover mapping approaches applied in three different coral reef environments in Fiji. The mapping approaches differed from each other by: image type; level of image correction; detail of calibration and validation field data; and/or classification method (e.g., manual digitizing, supervised classification). Each approach and resulting output map was assessed quantitatively, by calculating map accuracy, production time and cost, and qualitatively, by subjective evaluation by users from local marine monitoring agencies in Fiji. In this case study, analysis of results for a 14 km^2 fringing reef showed that users: (1) perceived that maps showing the highest detail best represented the reef study area, even though they were sometimes more detailed than required for their mapping purposes; (2) rated most maps evenly when provided with the associated cost and time to create maps, without receiving information on map accuracy; and (3) favored detail and higher accuracy above cost and production when all information was provided (Fig. 15.3).

It is important to note, however, that there is no unique answer to which mapping methods and techniques will be most appropriate and cost-effective for each study area and application. The time and cost involved in mapping and monitoring coral reefs can vary between mapping approaches, study questions, and the environmental characteristics of the study area, such as extent, heterogeneity, water depth variations, water clarity range, and remoteness (Mumby et al. 1999; Roelfsema and Phinn 2008). Time and cost, therefore, need to be considered across three main components: (1) project planning; (2) field work; and (3) processing, analysis and reporting. In addition, cost calculations must include personnel, image and equipment requirements, as well as skill and experience levels needed. Given that coral reef management agencies often operate on a limited budget, in such cases consideration should also be given to collaborations with research centers or other government agencies within and external to their countries, which may have greater technical capacity and resources to produce higher quality map products.

15.3 Example Applications

The sections below describe several examples of how remote sensing data are actively being used for coral reef management and monitoring applications. In each case, we describe the context of the application, the types of data that are

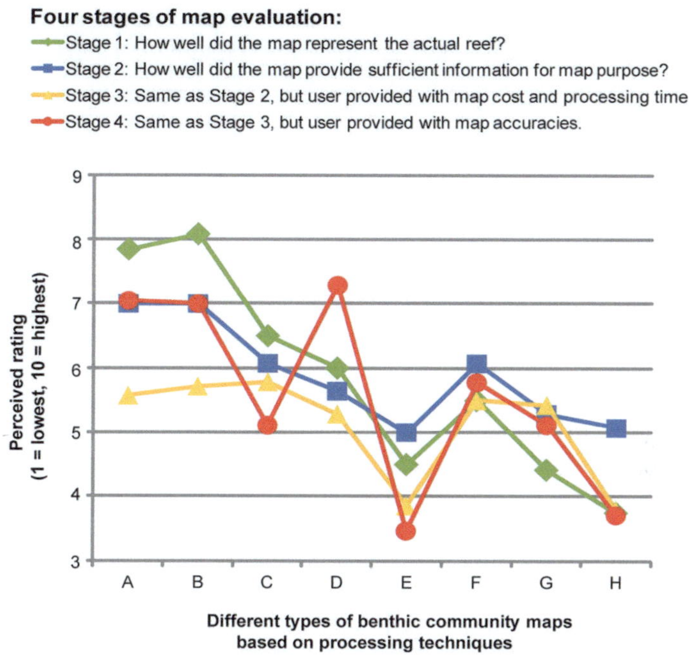

Map	A	B	C	D	E	F	G	H
Cost (US$)	4214	4179	2363	3000	2604	2447	578	823
Time (Days)	20	18.5	6.5	9.5	18	19	7	5.5
Accuracy (%)	57	56	40	61	25	37	32	22

Fig. 15.3 Example of user assessments for benthic coral reef community maps from Navakavu, Fiji, with their associated cost, time and accuracy. The mapping approaches were: (*A*) supervised classification of fully atmospherically corrected QuickBird image data integrated with detailed field data; (*B*) supervised classification of basic corrected QuickBird image data integrated with detailed field data; (*C*) supervised classification of basic corrected QuickBird image data integrated with local research knowledge; (*D*) supervised classification of basic corrected QuickBird image data integrated with basic field data; (*E*) supervised classification of basic corrected Landsat TM image data integrated with detailed field data; (*F*) manual delineation of QuickBird RGB imagery and detailed field data; (*G*) manual delineation of QuickBird RGB imagery and local expert knowledge; and (*H*) manual delineation of Landsat TM image data and local community knowledge (Roelfsema and Phinn 2008)

required, and how the products are used to inform management decisions. Section 15.3.1 discusses the use of baseline data and data products for compiling reef inventories and prioritizing areas for protection on local to global scales, using both scientific and local knowledge to inform mapping. Section 15.3.2 details recent techniques developed to spatially predict reef fish assemblage characteristics across entire reefs and the uses of these tools for fisheries management.

Section 15.3.3 provides four specific examples of how remote sensing data and products are used to monitor and respond to reef threats. Lastly, Sect. 15.3.4 discusses challenges and opportunities for monitoring temporal change on coral reefs.

15.3.1 Resource Management

In order to make informed decisions about how management strategies, development proposals, and policies for use of the marine and coastal zone will affect coral reef resources and biodiversity, decision-makers need to be aware of what resources exist at local, national and regional levels. Often these resource inventories are required by law through an environmental impact assessment (EIA) prior to approval of development projects in the coastal zone. For example, the Environment Management (EIA Process) Regulations 2007 under Fiji's Environment Management Act 2005 require "a description of the environmental setting of the site of the proposal, including a statement of environmental resources in the area before the implementation of the activity or undertaking, and a projection or estimation of changed environmental circumstances that may occur as a result of the activity or undertaking". Coral reef remote sensing can provide these baselines and, when field data are available, the image classifications can potentially be used to scale up site-based abundance and biomass measures of species or species groups (Edwards 2000; Andréfouët et al. 2005a).

On a global scale, maps of coral reef habitat are available through the United Nations Environment Program—World Conservation Monitoring Centre (UNEP-WCMC) publication of the World Atlas of Coral Reefs (Spalding et al. 2001). Note that the original publication had a few limitations, namely that the described habitat largely represented only emergent reef crest and just 30 % of the reefs within the atlas were mapped from source data with resolution better than 1:250,000 scale. However, the UNEP-WCMC data have been updated through partnership with the U.S. National Aeronautic and Space Agency (NASA) to correct positional errors and the extent of reef areas covered. These reef maps, in combination with data from the Millennium Coral Reef Mapping Project (MCRMP), are available online through ReefBase (www.reefbase.org).

On a national scale, resource inventories are essential to expand national marine protected area networks in support of commitments as signatories to the Convention on Biological Diversity (CBD). Through the CBD's Programme of Work on Protected Areas, parties to the convention are required to assess gaps in current protected area networks and identify candidate sites to fill those gaps (Dudley and Parish 2006). For these national and regional planning exercises, accurate metrics of coral reef habitats are needed in order to properly account for management progress against national targets (Naseer and Hatcher 2004). For example, in 2005 the Fiji Government declared a commitment to effectively manage 30 % of its inshore and offshore marine areas by 2020. As the MCRMP data were not yet available for Fiji, reef habitats digitized from aerial photographs by the Department of Lands were used in combination with current management area boundaries to determine that, by 2010, Fiji had already achieved effective management of marine biodiversity on 10–22 % its coral reefs (Mills et al. 2011). Maps have also been created for each province across Fiji showing the amount of additional protection required to meet objectives for each of the major inshore marine habitats (i.e., mangroves, intertidal, fringing reefs, other reefs, and other benthic substrate). These maps have been used to help provincial administrators identify sites to add to the national MPA network (Fig. 15.4).

The above example represents a coarse spatial-scale approach to marine spatial planning and makes some assumptions that the broad habitats serve as acceptable surrogates for the range of marine biodiversity actually present (Margules and Pressey 2000). When higher resolution habitat and field data are available, other approaches can be used to incorporate finer scale ecological processes and biological diversity metrics into marine protected area network design. For example, Mumby (2001) presents a technique to measure β-dissimilarity of habitats within a given window of image pixels: the resulting map layer can be used to ensure representation of biologically dissimilar habitats as well as hotspots of habitat diversity within MPA networks. In another case, Edwards et al. (2010) combine information on the relative contribution of habitats to fish biomass, scaled up using

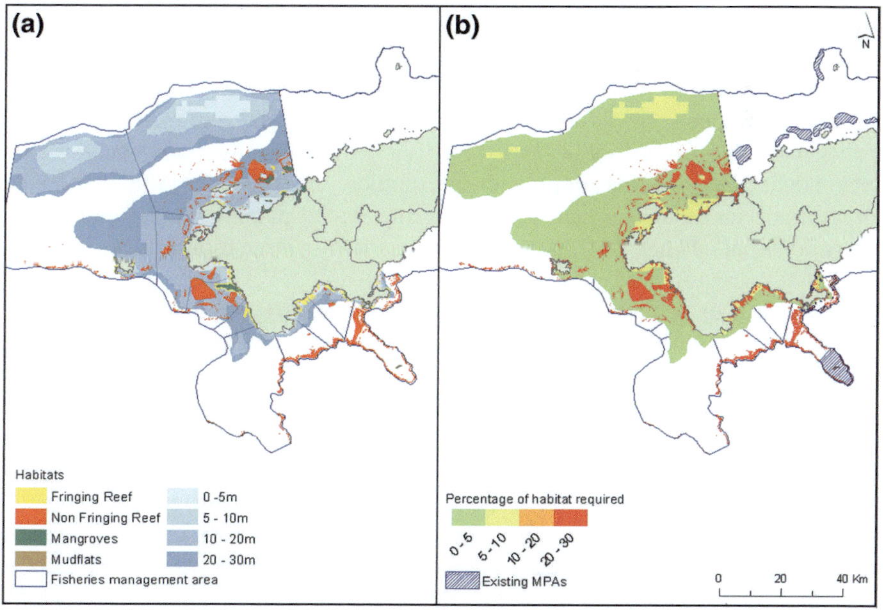

Fig. 15.4 Output maps from Fiji's national marine gap analysis for Bua Province, Vanua Levu. **a** Distribution of coastal and marine habitats within traditional fisheries management areas. **b** Amount of each habitat that remains to be protected to meet national biodiversity targets, shown with the location of existing protected areas in the region as of September 2010

field data and models from Mumby et al. (2004b) and the mapped habitat proximities to mangroves, to evaluate options for MPA design in Belize.

In much of the developing world, field data are often limited, acquisition of high spatial resolution satellite data may be cost-prohibitive, and technical capacity of agency staff for processing images may be low. However, governments often have archives of aerial photographs available for free or low cost which can be locally interpreted. For instance, Aswani and Lauer (2006) used local fisherfolk to trace representative habitats on print-outs of aerial photographs covering the Roviana and Vonavona regions of the western Solomon Islands. The researchers found that local knowledge of fish distributions within each habitat was robust compared with field data and therefore could be used to build MPAs that are sensitive to local needs (Aswani and Lauer 2006). In fact, in cases where there is strong local tenure over marine resources, community remote sensing approaches to resource inventories that match local perceptions may be more readily used for management applications than maps produced by external agencies.

15.3.2 Predictive Mapping of Fish Assemblages

A relatively new application of coral reef remote sensing data has emerged using remote sensing-derived information about habitat type and structure to predict spatial characteristics of reef community fish assemblages. Given the high costs associated with collecting field data and the difficulty of accessing certain remote reefs (Mumby et al. 1999), the ability to confidently predict relative differences in abundance, biomass, species richness and diversity of coral reef fish across broad spatial scales has obvious advantages for fisheries management and conservation. Such advantages include improved information for prioritizing locations for conservation and management through marine spatial planning and the ability to assess potential changes to fish communities following large disturbances that alter reef habitat structure.

New advances in passive and active sensor technology with very high spatial resolution allow quantification of reef habitat and structural variables at a scale to which fish may respond. Factors that contribute to site-specific composition of coral reef fish assemblages include: biogeographic patterns (Thresher 1991); habitat area (Bellwood and Hughes 2001); ecological processes such as recruitment, competition and predation (Carr et al. 2002); environmental stochasticity (Connolly et al. 2005); and interactions between fish and habitat across a range of scales (Friedlander and Parrish 1998; Lara and Gonzalez 1998). Coral reef remote sensing scientists have developed techniques (discussed in previous chapters) to quantify at least five habitat variables from remote sensing data which are shown to be related to fish community structure, including: (1) depth; (2) structural complexity; (3) substrate type; (4) habitat diversity; and (5) live coral cover.

Multiple studies have used bathymetry and topographic complexity derived from LiDAR to develop predictive relationships of coral reef rugosity with fish assemblage characteristics (Chap. 6; Wedding et al. 2008; Pittman et al. 2009). Studies have also used multispectral data (e.g., QuickBird, IKONOS) to derive depth and structural complexity (Purkis et al. 2008; Knudby et al. 2010a, 2011), which results in less precise measurements than LiDAR but is considerably cheaper and requires less pre-processing time. In addition, the multispectral data can also be used to create maps of substrate type, habitat diversity and live coral cover (Harbourne et al. 2006; Knudby et al. 2010a, 2011).

In order to create spatial predictions, models must be developed relating field data to remote sensing derived parameters. General linear models (GLM) have been used by Wedding et al. (2008) with reasonable correlations; however, fish may exhibit non-linear relationships with habitat variables. For example, using data from Chumbe and Bawe reefs in Tanzania, Knudby et al. (2010b) found non-linear relationships between fish species richness and live coral cover, coarse rugosity, and depth range. Amongst the most commonly used models (including GLM), general additive models (GAM), support vector machine, and regression tree models have performed the best for predicting fish species richness, abundance and biomass. Bootstrap aggregating (bagging) tree-models were consistently the best predictors of Tanzanian fish assemblages from IKONOS-derived habitat

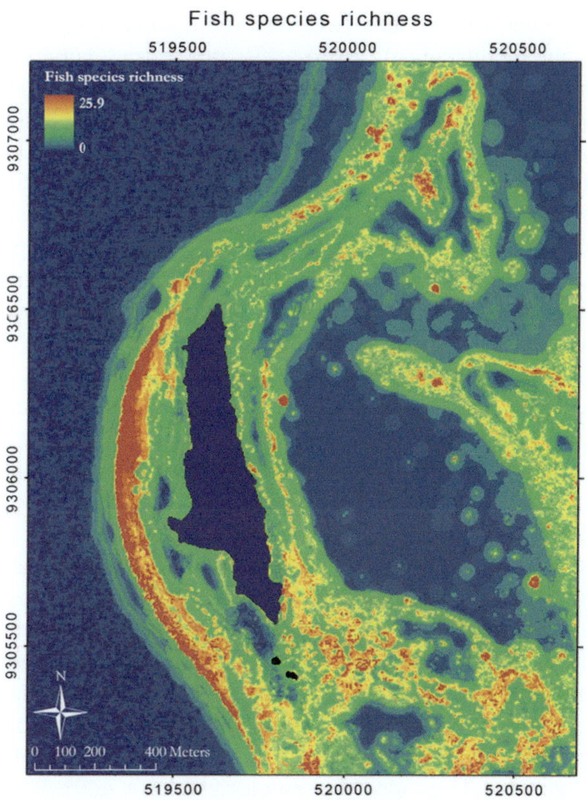

Fig. 15.5 Map showing predicted fish species richness around Chumbe Island, Tanzania, based on IKONOS-derived habitat variables and 'bagging' spatial prediction model (Knudby 2009)

variables (Fig. 15.5; Knudby et al. 2010a), while random forest tree-models were the best predictors of Fijian fish assemblages from both QuickBird and IKONOS-derived habitat measures (Knudby et al. 2011). Most studies to date have shown the strongest relationships between fish species richness and remote sensing-derived structural complexity exists at relatively small spatial scales (2.5–42.5 m; Kuffner et al. 2007; Purkis et al. 2008; Wedding et al. 2008; Pittman et al. 2009; Knudby et al. 2010a); however, there is some evidence to suggest that coarser scale (200–225 m) rugosity may also contribute to total fish biomass (Pittman et al. 2009; Knudby et al. 2010a).

At present, the Wildlife Conservation Society in Fiji is using outputs of predictive models of fish species richness and food fish biomass, in concert with data on reef resilience and opportunity costs to fishers, to optimize MPA network design. The results are being used to provide recommendations to community managers on options for reconfiguring the existing MPA network to reduce conflict and improve food security for the future.

15.3.3 Threat and Damage Assessments

Remote sensing data have become rapidly integrated into many local and global scale monitoring efforts to assess threats and damage to coral reefs. In this section, we present four brief case studies highlighting different applications for coral reef monitoring and management: (1) water quality assessments in the Great Barrier Reef lagoon; (2) predictions of crown-of-thorns (*Acanthaster planci*) outbreaks in the Pacific; (3) oil spill threat mapping in the Gulf of Mexico; and (4) damage assessments following a ship grounding in Australia.

Water quality. The Queensland Government in Australia has set an ambitious goal "to ensure that by 2020 the quality of water entering the Reef from adjacent catchments has no detrimental impact on the health and resilience of the Great Barrier Reef" (State of Queensland 2009). In order to achieve this goal, the management partners need to: assess changes to sediment and nutrient loads entering the Great Barrier Reef lagoon; identify reefs at risk from flood plumes; and assess reef ecosystem condition over time. Catchment models such as SedNet and ANNEX, which incorporate digital elevation models and land use and land cover maps derived from remote sensing data, can be used for estimating river end loads of sediments and nutrients (McKergow et al. 2005a, b). Such maps can be periodically updated to monitor temporal change based on land cover change in response to management initiatives. New techniques have been recently used in the Wet Tropics region to identify reefs at risk based on frequency of exposure to flood plumes mapped and classified from remote sensing data (Devlin and Schaffelke 2009). The boundaries of primary, secondary and tertiary plumes were defined for each flood level based on threshold concentrations of chlorophyll-a and colored dissolved organic matter (CDOM) derived from freely available MODIS Aqua and Terra data (oceancolor.gsfc.nasa.gov). Results were combined in a GIS to create an exposure map indicating the relative threats to various marine habitats, including coral reefs (Devlin and Schaffelke 2009). Change detection methods can be employed to assess shallow reef community response along a distance gradient from river mouths. Integrated results can then be used to evaluate the effectiveness of catchment management strategies identified within the Reef Water Quality Protection Plan (Queensland 2009).

Crown-of-thorns starfish. Periodic outbreaks of the coral-eating crown-of-thorns (COTS) starfish have caused extensive damage to reefs across the Pacific and Indian Oceans. While some authors have suggested that nutrient enriched river runoff contributes to enhanced *A. planci* survival (Birkeland 1982; Brodie et al. 2005), frequent outbreaks occurring nearly simultaneously on small, distant Pacific islands suggest that alternative mechanisms may also contribute. Recent research has shown successful hindcasting of COTS outbreaks across the Pacific using a combination of SeaWiFs and MODIS surface chlorophyll-a data and QuickSCAT Ekman transport (upwelling) and wind stress data (Houk et al. 2007; Houk and Raubani 2010). All of the data are freely available through the NOAA Oceanwatch Central Pacific program (oceanwatch.pifsc.noaa.gov) and can potentially be incorporated into the NOAA Coral Reef Watch program to provide early warning outbreak forecasts, which could be of aid to local managers.

Oil spills. The catastrophic leak in 2010 from the Deepwater Horizon oil rig in the U.S. waters of the Gulf of Mexico highlighted the urgency for real-time monitoring of the spreading oil plume as a threat to marine habitats and species. To regularly update managers and the general public of the plume's path, NOAA and the University of New Hampshire's Coastal Response Research Center established the Emergency Response Management Application (ERMA) web-based platform (gomex.erma.noaa.gov/erma.html), containing interactive maps incorporating data and products from NOAA's National Environmental Satellite, Data and Information Service (NESDIS; www.nesdis.noaa.gov). These data layers included radar, MODIS, and aerial images that defined the extent of the spill, plus forecasted environmental conditions for winds, waves, currents, and precipitation. Together, the data gave managers an opportunity to assess in advance whether their habitats were going to be affected by the spill and prepare for any necessary management action.

Ship groundings. Ship groundings typically impact reefs on a relatively smaller scale than large oils spills, as described above; nevertheless, they can damage the reef significantly. The damage can be direct, through the ship contact with the bottom, or indirect, through leaks of fuel or other contaminants. For example, in April 2010, the 230 m long bulk coal carrier *Shen Neng* 1 ran aground in a minimum of 10 m water depth on Douglas Shoal, which is located in the Capricorn Bunker Group of the southern Great Barrier Reef, Australia. Weighted with approximately 68,000 tonnes of coal and 950 m^3 of fuel, the ship caused extensive damage while traversing the reef, resulting in a 3 km $\times$ 250 m scar. Fifty percent of this area was directly damaged, which included the deposition of antifouling paint. The Australian Maritime Safety Authority monitored the site remotely using thermal and infrared sensors on planes to check for spills, while the Great Barrier Reef Marine Park Authority conducted acoustic surveys and in-water photo and video transects surveys to monitor the damage. Passive and active remote sensing, in various combination with direct in-water damage assessment using photo or video, have now become part of a common portfolio of tools to successfully assess damage from ship groundings and monitor recovery over space and time.

15.3.4 Monitoring Temporal Changes

The previous case studies have shown how remote sensing can be used to provide maps of coral reef properties at specific points in time with defined levels of detail and areal extent. The logical extension of these applications is to reproduce the maps of reef extent, geomorphic zonation, benthic communities, fish habitat, etc., at different dates and then map and quantify the nature of changes observed within the coral reef over time. This feeds into one of the key requirements for science and management: defining what is there and how it is changing over time.

Prior chapters have clearly demonstrated the ability of remote sensing to be used at an operational level by government agencies, private companies and NGOs for mapping the extent of reefs, biophysical composition and processes within reefs, and the physical-biological oceanographic and meteorological parameters affecting reef extent and growth. The majority of the satellite and airborne image datasets collected to produce these maps can just as readily obtain repeat coverage of the same area. This enables reef scientists to map changes in reef composition, physical features and controlling processes to assess: the type and nature of changes occurring; the level of natural variations and disturbances; and the direct or indirect impact of human actions and management decisions (Jupp et al. 1985; Dustan et al. 2001; Andréfouët and Riegel 2004; Knudby et al. 2007). While mapping changes in reef extent, composition or habitat has predominantly been completed using two successive dates of images, research applications are emerging for mapping these features from long time series of satellite images, aerial photography and field survey data. Below are two general examples of image change and trend detection studies used for understanding and managing reefs. The first example focuses on changes in characteristics mapped between two successive dates, and the second application covers a longer time series of multiple (100–1,000 s dates) images.

There are relatively few publications in the peer-reviewed literature covering operational mapping of changes in coral reefs between two successive dates. Most of the published studies report results from the map producer's point of view, focusing on the techniques used, levels of accuracy obtained, and inference of the natural or human processes that produced the observed changes. The most commonly reported examples have been for geomorphic zones and benthic communities, mainly using Landsat Thematic Mapper and now higher spatial resolution multispectral satellites (e.g., Dustan et al. 2001; Andréfouët et al. 2001, 2005b; Palandro et al. 2003, 2008; Andréfouët 2008; Knudby et al. 2009). These studies have primarily focused on the context of assessing impacts of disturbances such as cyclones and bleaching (Yamano and Tamura 2004; Collier and Humber 2007). More specific studies are emerging using a combination of multispectral datasets, higher spatial resolution images, airborne LiDAR bathymetry, and field data to map changes in percentage-cover of coral versus algae and to define specific habitat zones for predicting fish, shell-fish or shell population changes associated with environmental shifts or management actions (e.g., Hardy 1999; Collier and Humber 2007; Scopélitis et al. 2009, 2011). Satellite images and aerial photography have also been used to map changes in mean sea-level on coral reefs and their atolls (Yamano et al. 2000, 2007; Yamano 2007).

Most recent research advances in the area of "change detection analysis" are now using time series analysis of satellite image and aerial photograph archives with 10 and 100 s of images to map changes in the composition and biophysical properties of a coral reef (Purkis and Riegl 2005; Knudby et al. 2009; Scopélitis et al. 2009, 2011). This allows assessment of natural dynamics of a reef, including disturbance impact and recovery, in terms of its benthic communities, habitat zones, and the amount of coral cover. With access to aerial photographic archives often extending to the 1930s and satellite image archives (Landsat MSS and TM) to the early 1970s on a global basis, there is substantial potential to extend applications in this area. Most importantly, these types of datasets and analyses enable natural variability and processes to be identified and teased apart from the potential positive or negative effects of human activities.

Remote sensing change detection has been used most frequently for reef monitoring and management over large areas and time scales in relation to assessing the biological and physical controls of coral and algal growth, and for tracking the condition of physical or biological disturbances that will damage reefs. As shown in Chaps. 11, 12, satellite SST time series data and knowledge of thermal conditions required for bleaching in certain regions is used to provide daily updates of the likelihood of bleaching in an area. This near-real-time approach is based on a long-term global archive of SST data. The long-term archive enables mean, seasonal, or monthly SST to be defined, while the near-real-time can be compared to the mean or standard conditions to quantify anomalies, and then adjusted to match the SST anomaly required for bleaching. Other operational applications include prediction and tracking of events that can lead to significant disturbances on the reef, such as physical impacts (e.g., from tropical cyclone induced waves) or river flood plumes. Each of these requires a long time series of images at appropriate temporal scales to track the disturbance. These applications are, however, usually confined to satellite imagery that provides daily global coverage, hence pixel sizes range from 250 m to 1 km. Once a disturbance event is detected, the typical approach has been to obtain higher spatial resolution data and implement techniques similar to those listed above. Research developments in this area now focus on multiple biological and physical oceanographic parameters, along with new parameters such as acidity, salinity and aragonite saturation levels, to allow more detailed assessment of reef growth (Kayanne et al. 2005; Moses et al. 2009).

Table 15.3 Examples of freely available remote sensing data and data products applicable for coral reef remote sensing

Application/product	Source
NOAA Coral Reef Watch	http://coralreefwatch.noaa.gov/satellite/bleachingoutlook/index.html
Disease outbreak risk	Includes links to free online data from NASA, NOAA/NESDIS, and others
Light stress damage	
Short-term SST trends	
Ocean acidification	
Bleaching outlook	
NOAA National Environmental Satellite Data and Information Service (NESDIS)	http://www.nesdis.noaa.gov/
Sea surface temperature	http://www.osdpd.noaa.gov/ml/ocean/sst.html
Ocean surface current analyses—real time	http://www.oscar.noaa.gov/index.html
Bathymetry	http://www.ngdc.noaa.gov/mgg/bathymetry/relief.html
Ocean color	http://www.osdpd.noaa.gov/ml/ocean/color.html
Sea surface height	http://www.osdpd.noaa.gov/ml/ocean/ssheight.html
Winds	http://www.osdpd.noaa.gov/ml/air/wind.html
NASA Physical Oceanography DAAC	http://podaac.jpl.nasa.gov/
Gravity	http://podaac.jpl.nasa.gov/gravity
Sea surface temperature	http://podaac.jpl.nasa.gov/SeaSurfaceTemperature
Sea surface salinity	http://podaac.jpl.nasa.gov/SeaSurfaceSalinity
Ocean winds	http://podaac.jpl.nasa.gov/OceanWind%20
Ocean currents & circulation	http://podaac.jpl.nasa.gov/OceanCurrentsCirculation
Ocean surface topography	http://podaac.jpl.nasa.gov/OceanSurfaceTopography
NASA MODIS Level 1 Data	http://ladsweb.nascom.nasa.gov/index.html
USGS Earth Resources Observation and Science Center	http://glovis.usgs.gov/
Landsat and MODIS data	
Millennium Coral Reef Mapping Project	http://www.imars.usf.edu/MC/index.html
Output products	
Millennium Coral Reef Landsat Archive	http://oceancolor.gsfc.nasa.gov/cgi/Landsat.pl
Reefbase	http://www.reefbase.org/gis_maps/download.aspx
GIS datasets	
Google Earth	http://earth.google.com/
University of Queensland Marine Remote Sensing Toolkit	http://www.gpem.uq.edu.au/cser-rstoolkit

15.4 Conclusions and Recommendations

Remote sensing data and data products can be invaluable tools for addressing coral reef research and management questions. However, research and management needs will best be met by choosing imagery and data products that best suit the scale of the question and capacity of the users. Given that cost is often a limiting factor for monitoring and assessment programs, managers should make use of the increasing amounts of freely available data products and tools (Table 15.3). In addition, development of collaborative agreements between research and management organizations can often be extremely useful for tailoring research to the needs of coral reef managers, and for training staff and managers in how to use and interpret the data. Finally, new techniques and tools developed to produce environmental information should be clearly documented and communicated within the management organization to ensure that the knowledge will continue to be applied and improved by new staff and staff staying in longer term roles.

Acknowledgments The authors would like to thank A. Knudby and R. Weeks for providing figures and the contributors to the University of Queensland's marine spatial remote sensing toolkit, which is referenced throughout the text. C. Roelfsema acknowledges the support of staff at the University of the South Pacific and Navakavu community members for participation in the Fiji study on end-user map evaluation.

Suggested Reading

Andréfouët S (2008) Coral reef habitat mapping using remote sensing: a user vs producer perspective. Implications for research, management and capacity building. J Spat Sci 53:113–129

Phinn SR (1998) A framework for selecting appropriate remotely sensed data dimensions for environmental monitoring and management. Int J Remote Sens 19:3457–3463

Phinn S, Roelfsema C, Stumpf RP (2010) Remote sensing: discerning the promise from the reality. In: Dennison WC (ed) Integrating and applying science: a handbook for effective coastal ecosystem assessment. IAN Press, Cambridge

References

Adams VM, Mills M, Jupiter SD, Pressey RL (2011) Improving social acceptability of marine protected area networks: a method for estimating opportunity costs to multiple gear types in both fished and currently unfinished areas. Biol Conserv 144:350–361

Andréfouët S (2008) Coral reef habitat mapping using remote sensing: a user vs producer perspective. Implications for research, management and capacity building. J Spat Sci 53:113–129

Andréfouët S, Payri C (2000) Scaling-up carbon and carbonate metabolism of coral reefs using in situ data and remote sensing. Coral Reefs 19:259–269

Andréfouët S, Riegel S (2004) Remote sensing: a key tool for interdisciplinary assessment of coral reef processes. Coral Reefs 23:1–4

Andréfouët S, Müller-Karger FE, Hochberg EJ, Hu C, Carder KL (2001) Change detection in shallow coral reef environments using Landsat 7 ETM+ data. Remote Sens Environ 78:150–162

Andréfouët S, Berkelmans R, Odriozola L, Done T, Oliver J, Müller-Karger F (2002) Choosing the appropriate spatial resolution for monitoring coral bleaching events using remote sensing. Coral Reefs 21:147–154

Andréfouët S, Gilbert A, Yan L, Remoissenet G, Payri C, Chancerelle Y (2005a) The remarkable population size of the endangered clam *Tridacna maxima* assessed in Fangatau Atoll (Eastern Tuamotu, French Polynesia) using in situ and remote sensing data. ICES J Mar Sci 62:1037–1048

Andréfouët S, Hochberg EJ, Chevillon C, Müller-Karger FE, Brock JC, Hu C (2005b) Multi-scale remote sensing of coral reefs. In: Miller RL, Del Castillo CA, McKee BA (eds) Remote sensing of coastal aquatic environments: technologies, techniques and applications. Springer, Dordrecht

Andréfouët S, Müller-Karger FE, Robinson JA, Kranenburg CJ, Torres-Pulliza D, Spraggins SA, Murch B (2006) Global assessment of modern coral reef extent and diversity for regional science and management applications: a view from space. In: Proceedings of the 10th International Coral Reef Symposium, pp 1732–1745

Aswani S, Lauer M (2006) Benthic mapping using local aerial photointerpretation and resident taxa inventories for designing marine protected areas. Environ Conserv 33:263–273

Bellwood DR, Hughes TP (2001) Regional-scale assembly rules and biodiversity of coral reefs. Science 292:1532–1534

Birkeland C (1982) Terrestrial runoff as a cause of outbreaks of *Acanthaster planci* (Echinodermata: Asteroidea). Mar Biol 69:175–185

Brodie J, Fabricius K, De'ath G, Okaji K (2005) Are increased nutrient inputs responsible for more outbreaks of crown-of-thorns starfish? An appraisal of the evidence. Mar Pollut Bull 51:266–278

Carr MH, Anderson TW, Hixon MA (2002) Biodiversity, population regulation, and the stability of coral-reef fish communities. Proc Natl Acad Sci U S A 99:11241–11245

Clarke CD, Mumby PJ, Chisholm JRM, Jaubert J, Andréfouët S (2000) Spectral discrimination of coral mortality states following a severe bleaching event. Int J Remote Sens 21:2321–2327

Collier JS, Humber SR (2007) Time-lapse side-scan sonar imaging of bleached coral reefs: a case study from the Seychelles. Remote Sens Environ 108:339–356

Connolly SR, Hughes TP, Bellwood DR, Karlson RH (2005) Community structure of corals and reef fishes at multiple scales. Science 309:1363–1365

Devlin M, Schaffelke B (2009) Spatial extent of riverine flood plumes and exposure of marine ecosystems in the Tully coastal region, Great Barrier Reef. Mar Freshw Res 60:1109–1122

Dudley N, Parish J (2006) Closing the gap. Creating ecologically representative protected area systems: a guide to conducting the gap assessments of protected area systems for the convention on biological diversity. Technical series no. 24. Secretariat of the convention on biological diversity, Montreal, Canada. http://www.cbd.int/doc/publications/cbd-ts-24.pdf. Accessed 24 May 2012

Dustan P, Dobson E, Nelson G (2001) Landsat Thematic Mapper: detection of shifts in community composition of coral reefs. Conserv Biol 15:892–902

Edwards AJ (2000) Assessment of coastal marine resources: a review. In: Edwards AJ (ed) Remote sensing handbook for tropical coastal management. UNESCO, Paris

Edwards HJ, Elliott IA, Pressey RL, Mumby PJ (2010) Incorporating ontogenetic dispersal, ecological processes and conservation zoning into reserve design. Biol Conserv 143:457–470

Friedlander AM, Parrish JD (1998) Habitat characteristics affecting fish assemblages on a Hawaiian coral reef. J Exp Mar Biol Ecol 224:1–30

Harbourne AR, Mumby PJ, Zychaluk K, Hedley JD, Blackwell PG (2006) Modeling the beta diversity of coral reefs. Ecology 87:2871–2881

Hardy JT (1999) Coral reef monitoring with airborne LIDAR. In: Proceeding of the International Workshop on the Use of Remote Sensing Tools for Mapping and Monitoring Coral Reefs. NOAA, CSC/NEDIS/ICLARM, Honolulu, Hawaii

Hopley D, Smithers S, Parnell K (2007) The geomorphology of the Great Barrier Reef: development, diversity, and change. Cambridge University Press, Cambridge

Houk P, Raubani J (2010) *Acanthaster planci* outbreaks in Vanuatu coincide with ocean productivity, furthering trends throughout the Pacific Ocean. J Oceanogr 66:435–438

Houk P, Bograd S, van Woesik R (2007) The transition zone chlorophyll front can trigger *Acanthaster planci* outbreaks in the Pacific Ocean: historical confirmation. J Oceanogr 63:149–154

Jenkins AP, Jupiter SD, Qauqau I, Atherton J (2010) The importance of ecosystem-based management for conserving migratory pathways on tropical high islands: a case study from Fiji. Aquat Conserv 20:224–238

Jupp DLB, Mayo KK, Kuchler DA, Claasen DVR, Kenchington RA, Guerin PR (1985) Remote sensing for planning and managing the Great Barrier Reef of Australia. Photogrammetria 40:21–42

Kayanne H, Hata H, Kudo S, Yamano H, Watanabe A, Ikeda Y, Nozaki K, Kato K, Negishi A, Saito H (2005) Seasonal and bleaching-induced changes in coral reef metabolism and CO_2 flux. Global Biogeochem Cycles 19:GB3015

Knudby AJ (2009) Remote sensing of reef fish communities. PhD thesis, University of Waterloo, Waterloo, Canada

Knudby A, LeDrew E, Newman C (2007) Progress in the use of remote sensing for coral reef biodiversity studies. Prog Phys Geogr 31:421–434

Knudby A, Newman C, Shaghude Y, Muhando C (2009) Simple and effective monitoring of historic changes in nearshore environments using the free archive of Landsat imagery. Int J Appl Earth Obs Geoinf 12:S116–S122

Knudby A, LeDrew E, Brenning A (2010a) Predictive mapping of reef fish species richness, diversity and biomass in Zanzibar using IKONOS imagery and machine-learning techniques. Remote Sens Environ 114:1230–1241

Knudby A, Brenning A, LeDrew E (2010b) New approaches to modelling fish-habitat relationships. Ecol Model 221:503–511

Knudby A, Roelfsema C, Lyons M, Phinn S, Jupiter S (2011) Mapping fish community variables by integrating field and satellite data, object-based image analysis and modeling in a traditional Fijian fisheries management area. Remote Sens 3:460–483

Kuffner IB, Brock JC, Grober-Dunsmore, Bonito VE, Hickey TD, Wright CW (2007) Relationships between reef fish communities and remotely sensed rugosity measurements in Biscayne National Park, Florida, USA. Environ Biol Fish 78:71–82

Lara EN, Gonzalez EA (1998) The relationship between reef fish community structure and environmental variables in the southern Mexican Caribbean. J Fish Biol 53:209–221

Liu G, Strong AE, Skirving W (2003) Remote sensing of sea surface temperature during 2002 Great Barrier Reef bleaching event. EOS 84:49–56

Margules CR, Pressey RL (2000) Systematic conservation planning. Nature 405:243–253

McKergow LA, Prosser IP, Hughes AO, Brodie J (2005a) Sources of sediment to the Great Barrier Reef World Heritage Area. Mar Pollut Bull 51:200–211

McKergow LA, Prosser IP, Hughes AO, Brodie J (2005b) Regional scale nutrient modelling: exports to the Great Barrier Reef World Heritage Area. Mar Pollut Bull 51:186–199

Mills M, Jupiter S, Pressey RL, Ban NC, Comley J (2011) Incorporating effectiveness of community-based management in a national marine gap analysis for Fiji. Conserv Biol 25:1155–1164

Moses CS, Andréfouët S, Kranenburg CJ, Müller-Karger FE (2009) Regional estimates of reef carbonate dynamics and productivity using Landsat 7 ETM+, and potential impacts from ocean acidification. Mar Ecol Prog Ser 380:103–115

Mumby PJ (2001) Beta and habitat diversity in marine systems: a new approach to measurement, scaling and interpretation. Oecologia 128:274–280

Mumby PJ, Clark CD, Green EP, Edwards AJ (1998) Benefits of water column correction and contextual editing for mapping coral reefs. Int J Remote Sens 19:203–210

Mumby PJ, Green EP, Edwards AJ, Clark CD (1999) The cost-effectiveness of remote sensing for tropical coastal resources assessment and management. J Environ Manag 55:157–166

Mumby PJ, Skirving W, Strong AE, Hardy JT, LeDrew EF, Hochberg EJ, Stumpf RP, David LT (2004a) Remote sensing of coral reefs and their physical environment. Mar Pollut Bull 48:219–228

Mumby PJ, Edwards AJ, Arias-Gonzalez JE, Lindeman KC, Blackwell PG, Gall A, Gorczynska MI, Harborne AR, Pescod CL, Renken H, Wabnitz CCC, Llewellyn G (2004b) Mangroves enhance the biomass of coral reef fish communities in the Caribbean. Nature 427:533–536

Naseer A, Hatcher B (2004) Inventory of the Maldives' coral reefs using morphometrics generated from Landsat ETM+ imagery. Coral Reefs 23:161–168

Palandro D, Andréfouët S, Dustan P, Müller-Karger FE (2003a) Change detection in coral reef communities using Ikonos satellite sensor imagery and historic aerial photographs. Int J Remote Sens 24:873–878

Palandro D, Andréfouët S, Müller-Karger FE, Dustan P, Hu C, Hallock P (2003b) Detection of changes in coral reef communities using Landsat 5/TM and Landsat 7/ETM+ Data. Can J Remote Sens 29:207–209

Palandro DA, Andréfouët S, Hu C, Hallock P, Müller-Karger FE, Dustan P, Callahan MK, Kranenburg C, Beaver CR (2008) Quantification of two decades of shallow-water coral reef habitat decline in the Florida Keys National Marine Sanctuary using Landsat data (1984–2002). Remote Sens Environ 112:3388–3399

Phinn SR (1998) A framework for selecting appropriate remotely sensed data dimensions for environmental monitoring and management. Int J Remote Sens 19:3457–3463

Phinn SR, Menges C, Hill GJE, Stanford M (2000) Optimising remotely sensed solutions for monitoring, modelling and managing coastal environments. Remote Sens Environ 73:117–132

Phinn S, Roelfsema C, Stumpf RP (2010) Remote sensing: discerning the promise from the reality. In: Dennison WC (ed) Integrating and applying science: a handbook for effective coastal ecosystem assessment. IAN Press, Cambridge

Pittman SJ, Costa BM, Battista TA (2009) Using lidar bathymetry and boosted regression trees to predict the diversity and abundance of fish and corals. J Coast Res 53:27–38

Purkis SJ, Riegl B (2005) Spatial and temporal dynamics of Arabian Gulf coral assemblages quantified from remote-sensing and in situ monitoring data. Mar Ecol Prog Ser 287:99–113

Purkis SJ, Graham NAJ, Riegl BM (2008) Predictability of reef fish diversity and abundance using remote sensing data in Diego Garcia (Chagos Archipelago). Coral Reefs 27:167–178

Roelfsema CM, Phinn SR (2008) Evaluating eight field and remote sensing approaches for mapping the benthos of three different coral reef environments in Fiji. In: Proceedings of SPIE Asia Pacific Remote Sensing Conference—Remote Sensing of Inland, Coastal and Oceanic Water. Noumea, New Caledonia

Salafsky N, Margoluis R (1999) Threat reduction assessment: a practical and cost-effective approach to evaluating conservation and development projects. Conserv Biol 13:830–841

Scopélitis J, Andréfouët S, Phinn S, Chabanet P, Naim O, Tourrand C, Done T (2009) Changes of coral communities over 35 years: integrating in situ and remote-sensing data on Saint-Leu Reef (la Réunion, Indian Ocean). Estuar Coast Shelf Sci 84:342–352

Scopélitis J, Andréfouët S, Phinn SR, Done T, Chabanet P (2011) Coral colonisation of a shallow reef flat in response to rising sea-level: quantification from 35 years of remote sensing data at Heron Island, Australia. Coral Reefs 30:951–965

Spalding MD, Ravilious C, Green EP (2001) World atlas of coral reefs. UNEP-WCMC in association with the University of California Press, Berkeley

State of Queensland (2009) Reef water quality protection plan 2009 for the Great Barrier Reef World Heritage Area and adjacent catchments. Reef Water Quality Protected Plan Secretariat, Brisbane, Australia. http://www.reefplan.qld.gov.au/about/assets/reefplan-2009.pdf. Accessed 24 May 2012

Thresher RE (1991) Geographic variability in the ecology of coral reef fishes: evidence, evolution, and possible implications. In: Sale PF (ed) The ecology of fishes on coral reefs. Academic, San Diego

Wabnitz CCC, Andréfouët S, Torres-Pulliza D, Müller-Karger FE, Kramer PA (2008) Regional-scale seagrass habitat mapping in the Wider Caribbean region using Landsat sensors: applications to conservation and ecology. Remote Sens Environ 112:3455–3467

Wedding LM, Friedlander AM, McGranaghan M, Yost RS, Monaco ME (2008) Using bathymetric lidar to define nearshore benthic habitat complexity: implications for management of reef fish assemblages in Hawaii. Remote Sens Environ 112:4159–4165

Yamano H (2007) The use of multi-temporal satellite images to estimate intertidal reef-flat topography. J Spat Sci 52:73–79

Yamano H, Tamura M (2004) Detection limits of coral reef bleaching by satellite remote sensing: simulation and data analysis. Remote Sens Environ 90:86–103

Yamano H, Kayanne H, Yamaguchi T, Kuwahara Y, Yokoki H, Shimazaki H, Chikamori M (2007) Atoll island vulnerability to flooding and inundation revealed by historical reconstruction: Fongafale Islet, Funafuti Atoll, Tuvalu. Global Planet Change 57:407–416

Yamano H, Kayanne H, Yonekura N, Kudo K (2000) 21 year changes of back reef coral distribution: causes and significance. J Coast Res 16:99–110

Index

J. A. Goodman et al. (eds.), *Coral Reef Remote Sensing*,
DOI: 10.1007/978-90-481-9292-2,

Q

R

MIX
Papier aus verantwortungsvollen Quellen
Paper from responsible sources
FSC® C105338

If you have any concerns about our products,
you can contact us on
ProductSafety@springernature.com

In case Publisher is established outside the EU,
the EU authorized representative is:
Springer Nature Customer Service Center GmbH
Europaplatz 3, 69115 Heidelberg, Germany

Printed by Libri Plureos GmbH
in Hamburg, Germany